Polymer Materials for Energy and Electronic Applications

Polymer Materials for Energy and Electronic Applications

Huisheng Peng
Xuemei Sun
Wei Weng
Xin Fang
Department of Macromolecular Science and Laboratory of Advanced Materials at Fudan University
Shanghai, China

ELSEVIER

AMSTERDAM • BOSTON • HEIDELBERG • LONDON
NEW YORK • OXFORD • PARIS • SAN DIEGO
SAN FRANCISCO • SINGAPORE • SYDNEY • TOKYO
Academic Press is an imprint of Elsevier

Academic Press is an imprint of Elsevier
125 London Wall, London EC2Y 5AS, United Kingdom
525 B Street, Suite 1800, San Diego, CA 92101-4495, United States
50 Hampshire Street, 5th Floor, Cambridge, MA 02139, United States
The Boulevard, Langford Lane, Kidlington, Oxford OX5 1GB, United Kingdom

Library of Congress Cataloging-in-Publication Data
A catalog record for this book is available from the Library of Congress

British Library Cataloguing-in-Publication Data
A catalogue record for this book is available from the British Library

ISBN: 978-0-12-811091-1

For information on all Academic Press publications
visit our website at https://www.elsevier.com/

Publisher: Jonathan Simpson
Acquisition Editor: Simon Tian
Editorial Project Manager: Vivi Li
Production Project Manager: Lisa Jones
Designer: Mark Rogers

Typeset by Thomson Digital

Contents

Preface

The advancement of material is a main sign for the history of human society, and the use of synthetic polymers has brought enormous progress since the early 1900s, particularly, in the field of electronics. Polymers are originally used as electrical insulating and protective materials in electronics due to a good dielectric property, easy fabrication, low cost, and broad source. Along with the discovery and vigorous development of conducting polymers, they have inspired increasing interests for the use as electrically active components in electronics. They can be now found in diverse devices for photoelectric and thermoelectric conversion, electrochemical energy storage, electroluminescence, and electrical sensing.

With the advancement of modern electronics that needs small size, light weight, high flexibility, and high efficiency, polymers that can simultaneously satisfy the above requirements become more and more important and have been widely investigated in recent decades. Out of questions, the performances of electronic devices are significantly influenced by the used polymers as well as their preparation processes. For instance, a lot of efforts have been recently made to develop flexible polymer solar cells, and the conducting polymers and their structures and interfaces that depend on the fabrication processes are found to be critical to the power conversion efficiency. Increasing interests are then attracted to synthesize new polymer materials and design novel structures to improve the electronic performances. Numerous attempts are also made in producing conducting composites through the addition of conducting polymers and inorganic nanomaterials to general polymers to achieve higher performances.

This book intends to carefully highlight the main advancements in polymer materials for electronic applications. It first overviews the features of polymers and their applications in energy and electronics (Chapter 1), and introduces the synthesis, structures and properties of polymers (Chapters 2 and 3) as well as their composites (Chapter 4). Then, four families of electronic devices based on polymers will be described, that is, energy harvesting devices (Chapter 5), energy storage devices (Chapter 6), light-emitting devices (Chapter 7), and electrically driving sensors (Chapter 8). Particularly, the flexible electronic devices will be highlighted to demonstrate the importance of polymers in electronics (Chapter 9). Finally, the future development and challenge of polymers for energy and electronics will be provided (Chapter 10). Each chapter focuses on the relationship between polymer structure/property and performance of electronic device in the respective fields.

This book is organized to be accessible to the readers in the field of polymer science, chemistry, physics, materials science, electrical engineering, microelectronics, and energy.

It is useful to senior scientists such as faculty members and senior engineers at industry, and it will be particularly useful to junior scientists and newcomers such as postdoctoral fellows, graduate students, and undergraduate students to learn the basic knowledge on polymers and electronic materials and devices. We sincerely hope that these young scientists can find new ideas and opportunities from the introduced areas in this book. Limited by our experience and ability, there may be many imperfections and omissions in this book. We will greatly appreciate critical and suggestive comments from the readers.

Finally, we strongly hope to express our gratitude to many postdoctoral fellows and graduate students who joined in the writing of this book. They are Hao Sun, Guangxi Huang, Qiang Gao, Longbin Qiu, Xunliang Cheng, Zhitao Zhang, Peining Chen, Bingjie Wang, Jing Zhang, Yiming Li, Ye Zhang, Sisi He, Yifan Xu, Songlin Xie, Lie Wang, Yang Zhao, and Jingyu Cao. We sincerely thank Editor Dr. Simon Tian for the warm support in publishing this book.

1

Introduction

1.1 History of Polymers

The earliest record of human being using polymer materials can be traced back before the Christian era. Ancient people used straw, a natural polymer material to blend with slurry to enhance the mechanical strength for the construction of stable houses. The preparation and weaving technologies of natural silks were also well developed to produce delicate costumes in ancient China. However, it is not until 1920 that the concept of polymer is presented, and the rapid development of polymer materials had started from then on (Staudinger, 1920). Insulating polymers including but not limited to polyethylene, polypropylene, polyvinyl alcohol, and poly(methyl methacrylate) are widely used in our daily life attributed to their promising advantages, such as high machinability, light weight and low cost compared to the metal and ceramic materials. In particular, it is available to accurately control the properties of polymer materials by tuning the moieties and structures of backbones and side chains to satisfy a variety of applications.

Before the discovery of conducting polymers, polymers had long been convinced to be insulating, so they were typically used as separating and sealing materials in electronic devices. It was rare to consider them as electrode or active materials or both that should be electrically conducting. As a result, the emergence of conducting polymers (also called conjugated polymers) has rapidly attracted vigorous interests from both academy and industry since the discovery of polyacetylene in 1977 (Shirakawa et al., 1977). Due to high and tunable electrical conductivities, conducting polymers, such as poly(3,4-ethylenedioxythiophene), polypyrrole, and polyaniline have exhibited high potentials in serving as electrodes and active materials for various energy and electronic devices. For instance, semiconducting polymers with tunable band gaps are ideal candidates as the semiconducting layers of optoelectronic devices, such as organic solar cells. Due to the broad applications of polymer materials and the limited space of this book, we will mainly shed light on the use of polymer materials for energy harvesting, energy storage, light emitting, and sensing. Particularly, flexible energy and electronic devices based on polymer materials will be highlighted as an important trend in modern electronics.

To well understand the characteristics of polymers for electronic applications, the general synthetic methods are described in Chapter 2, and the relationships between structure and property are then presented in Chapter 3. The necessity to introduce a second phase into polymers is also carefully discussed on the basis of solving the remaining challenges of bare polymers in Chapter 4. The representative electronic applications of polymers are highlighted in the following chapters.

1.2 Energy Harvesting Based on Polymers

Energy harvesting devices typically convert the other forms of energy in environment to electricity. Photovoltaic, thermoelectric, piezoelectric, and triboelectric devices have been mostly explored to collect light, thermal, and mechanical energies due to the easy operation and low cost in use. Conducting polymers are widely explored as counter electrodes in dye-sensitized solar cell and p-type conducting components in thermoelectric generators. In particular, conducting polymers can serve as almost all the components including light absorber, electron donor, acceptor, and/or hole transporting material in polymer solar cells. With the advancement of molecular engineering to tune the microscopic morphologies and electrical properties of the polymer materials, the device performances have witnessed a continuous and rapid enhancement in both the energy conversion efficiency and stability. Polymer solar cells are discussed here as a demonstration.

At the early development of polymer solar cells, a planar p–n junction structure represented the mainstream in mimicking conventional silicon-based solar cells. However, the obtained devices demonstrated poor photovoltaic performances due to the long distance between the exciton and junction interface and insufficient light absorption due to the thin light absorber. It was not until 1995 that the dilemma was overcome with the discovery of a novel bulk heterojunction in which donor and acceptor form interpenetrated phases. Poly[2-methoxy-5-(2′-ethylhexyloxy)-*p*-phenylene vinylene] was blended with C_{60} or its derivatives to form the bulk heterojunction. A much improved power conversion efficiency of 2.9% was thus achieved under the illumination of 20 mW/cm^2. (Yu et al., 1995). The emergence of the donor/acceptor bulk-heterojunction structure had boosted the photovoltaic performances of polymer solar cells. Currently, a maximal power conversion efficiency of 10.6% had been reported on the basis of synthesizing appropriate polymer materials and designing a tandem structure (You et al., 2013). The detailed discussions are provided in Chapter 5.

Although it is rare to develop insulating polymers as electrode or active materials in solar cells, they may be effectively used as the charge-generating layers in building high-performance piezoelectric and triboelectric devices. For instance, Teflon, polyimide, and polydimethylsiloxane that demonstrate the most negative polarity are the basic components in triboelectric devices, and the surface modification is widely explored to enhance the contact area. For instance, with the use of a template-based method, polymer substrates with nanostructured surface containing pyramid-, square-, or hemisphere-based patterns can be easily produced. A triboelectric layer with a smooth surface can only generate an open circuit voltage of ~3 V; after the formation of nanostructured surface as a comparison, the open-circuit voltage can be efficiently increased to ~20 V, and the short-circuit current was meanwhile increased (Fan et al., 2012a,b).

1.3 Energy Storage Based on Polymers

The evergrowing energy needs and unsustainable fossil-fuel resources demand more reasonable and efficient strategies for energy management, including the utilization of renewable energy sources and development of novel storage techniques. To make the best use of energy generated from energy harvesting devices, it is crucial to design efficient electrical facilities for advanced energy storage. Supercapacitor (also called electrochemical capacitor) and lithium ion battery represent two mostly explored systems that well meet the requirements of next generation energy storage devices. They generally share a sandwiched structure comprised of two electrodes, an electrolyte, and a separator to avoid possible short circuit. Polymer materials, as the basic elements of these components, play irreplaceable roles in both structural and functional aspects.

A supercapacitor demonstrates the combined advantages of high power density and satisfactory energy density, which have bridged the gap between traditional dielectric capacitor and battery. It is proposed as a promising alternative and competitor for the advanced energy storage in a foreseeable future (Patrice et al., 2014). Conducting polymers, such as poly(3,4-ethylenedioxythiophene), polypyrrole, and polyaniline, have been widely explored to provide pseudocapacitances that can largely improve the energy storage capacitance via conducting redox reactions. However, their limited electrical conductivities require the incorporation of other conducting materials, such as carbon nanomaterials to guarantee an acceptable rate performance. For instance, only ~20% of the initial capacitance can be maintained for bare polyaniline-based supercapacitors when the applied current density was increased from 0.3 to 1.0 A g^{-1}, attributed to a limited electrical conductivity of 0.05 S cm^{-1} for bare PANI. When the conductivity was enhanced to 570 S cm^{-1} by the incorporation of highly conducting aligned carbon nanotubes, a much improved rate performance was demonstrated with over 90% of the initial capacitance remained at the same current density range (Lin et al., 2013). Apart from conducting polymers, insulating polymer materials including poly(vinyl acetate), polyvinylidene fluoride, and polyethylene oxide are extensively investigated as the scaffolds of gel electrolytes for the production of all-solid-state supercapacitors, which contribute to a higher stability during operation and deformation.

As another important energy storage device, lithium ion battery demonstrates a higher energy density (10–100 times) compared with the supercapacitor, and it has been already widely used at both industry and our life. Polymer materials are used as active materials and assistant agents in electrodes, thin separators and effective electrolytes for the construction of lithium ion batteries. Among numerous conducting polymers, there are five main types of polymers including polyacetylene (Nigrey et al., 1981) polyphenylene (Shacklette et al., 1985), polyaniline (MacDiarmid et al., 1987), polypyrrole (Mermilliod et al., 1986), and polythiophene (Liu et al., 2012) that have been intensely investigated. Except for polyacetylene, the other four can be synthesized by either chemical or electrochemical oxidation polymerization to participate lithium ion storage processes. Polymer materials can be also incorporated with other materials to form efficient composites to further improve the energy storage performances. The more information is carefully discussed in Chapter 6.

1.4 Light Emitting and Sensing Devices Based on Polymers

Polymer light-emitting devices, such as polymer light-emitting diodes and polymer light-emitting electrochemical cells have attracted broad interests due to their large potentials in serving as next generation displays and illuminants. Compared with the small molecule counterparts, conducting polymers demonstrate high film-forming property that enables a uniform deposition via solution-based methods, such as spin-coating and printing, which are favored in industrial applications. Poly(*p*-phenylene vinylene), polyfluorene, poly(*p*-phenylene), and polycarbazole have been mainly studied, and the morphology control tuning the solubility and stability is intensely conducted to achieve better device performances. For instance, poly(*p*-phenylene) films were generally prepared via precursor routes due to the insolubility in common organic solvents. To enhance the solubility, one solution lies in the synthesis of ladder-type conducting polymers, which leads to a higher solubility ascribed to a better coplanarity (Tasch et al., 1996). In addition, a full color display can be realized by regulating the molecular structure to control the HOMO-LUMO energy gap, and small molecule doping is also well adopted in this field to achieve desired luminous properties. The related details are covered in Chapter 7.

Apart from light emitting, strain and color changes can be also driven by electric energy, with which the facilities are generally called electromechanical actuator and electrochromic devices, respectively. Electromechanical actuators directly convert electrical energy to mechanical energy, so they can be used for robotics, artificial muscles, and biomimetic devices. An electrochromic device generates reversible color changes in response to electricity, which enables it useful in various displays, smart windows, and electronic skins. Polymer materials, especially conducting polymers, are promising for high-performance electromechanical actuators and electrochromic devices on the basis of tunable redox states under electricity besides the discussed light weight and flexibility previously. To further enhance the actuating or electrochromic properties, it is generally necessary to incorporate polymer materials with other highly conducting materials, such as carbon nanomaterials to accelerate the response speeds. For instance, polyaniline had been incorporated with aligned multiwalled carbon nanotubes via an electrochemical deposition method to serve as composite electrodes with a high conductivity ranged from 10^2 to 10^3 S cm^{-1}, which enabled rapid and reversible electrochromic processes within several seconds (Chen et al., 2014). Moreover, limited by the actuating mechanisms, for example, electrothermally and electrochemically driven actuators suffer from slow responsiveness and weak cyclic life, and electrochemically driving actuators are limited by the electrolyte media and electrostatically driving actuators require ultrahigh voltages, it is important to develop electromechanical actuators on the basis of new mechanisms to realize programmable and sophisticated motions. The latest achievements along this direction are provided in Chapter 8.

1.5 Flexible Energy and Electronic Devices Based on Polymers

Flexibility and integration have been regarded as the mainstreamed trends for the development in the next generation electronics (Pan et al., 2016; Sun et al., 2016). Current energy and electronic devices that are generally rigid, heavy, and bulky can hardly meet the requirements of flexible electronics. Furthermore, in some specific applications, such as wearable devices, the energy and electronic devices are required to be highly stable during bending, twisting, and even stretching. To this end, to design and fabricate flexible energy and electronic devices is highly important for the promotion of modern electronics. There are several key factors that determine the flexibility of the devices. First, the thickness of each component layer is required to be as low as possible, in order to fabricate a thin device that can be easily deformed. Second, compact contacts at the interfaces among different components are critical for the stability, otherwise crack and separation may happen during deformation. Last but not the least, the constitutive component should be stable enough in property, for example, the electrical conductivity that directly determines the final performance and stability of the devices need to be maintained during deformation. Fortunately, many polymer materials well meet the aforementioned requirements, and they have been widely explored as promising candidates in building flexible electronic devices with high performances. Particularly, a novel family of energy and electronic devices with a one-dimensional configuration has been developed in the last decade. They rapidly attracts broad interests on the basis of their high flexibility, weavability, and stability, and may potentially solve the current challenges in flexible devices (Peng, 2015; Chen et al., 2013, Yu et al., 2015; Sun et al., 2014a,b). To this end, polymer materials have been confirmed to serve as good building blocks for fiber-shaped energy and electronic devices on the basis of their high processability, which contributes to a uniform deposition and controllable thickness on highly curved fiber surfaces. Besides, with the assistance of some functional polymers, we can design fiber-shaped energy and electronic devices with stretchability (Yang et al., 2013), self-healing (Sun et al., 2014b), and shape-memory (Deng et al., 2015) capabilities, which have largely enriched their function and integration. A careful description is made in Chapter 9.

1.6 Challenges and Perspectives

Despite of many inspiring achievements using polymer materials to fabricate energy and electronic devices, there exist some formidable challenges clouded before we can acclaim their success in this emerging field.

1.6.1 Stability

Considering the practical applications of energy and electronic devices, the stability of each component plays an important role in determining the lifetime of the device. For instance, the lifetimes of polymer-based photovoltaics fall generally in the range of minutes to a few days under operation in air (Neugebauer et al., 2000; Krebs et al., 2005), which is much shorter compared with silicon-based solar cells in excess of 25 years. The degradation is inevitable when organic materials are exposed to strong illumination, elevated temperatures, electric currents, reactive metal electrodes, oxygen and water, and the main reactions concerning performance degradation are photoreduction of the conjugated polymer by reacting with the aluminum electrode, diffusion of the aluminum into the active layer, and reaction with molecular oxygen throughout the cell. By chemical modification and structure optimization, for example, inserting a C_{60} layer as the barrier layer between the active layer and aluminum electrode to hinder their reaction, the lifetime can be increased to 20,000 h under the exclusion of water and oxygen (Krebs and Spanggaard, 2005). On the other hand, some conducting polymers suffer from different levels of capacitance loses due to irreversible redox reactions that occur during cyclic charge and discharge processes. Enhancing the chemical and electrochemical stability of conducting polymers can efficiently simplify the fabrication process and further extend the lifetime of the device to meet application requirements.

1.6.2 Electrical Conductivity

The relatively low electrical conductivities of the conducting polymers have limited the applications and device performances to some extent. For instance, planar supercapacitors derived from bare polyaniline films with a low electrical conductivity of 0.05 S cm^{-1} demonstrated a low specific capacitance of ~26 F g^{-1} at 0.5 A g^{-1}. Once the electrical conductivity was increased to 570 S cm^{-1} by incorporating aligned carbon nanotube sheets, a high specific capacitance of ~110 F g^{-1} had been achieved under the same condition, and the rate performance was also improved (Lin et al., 2013). Increasing the electrical conductivities of conducting polymers can reduce the internal resistance of the resulting energy devices and accelerate the response speed and performance of the obtained electronic devices. Therefore, a lot of efforts are made to synthesize novel polymers and enhance their crystallizations for good morphologies.

1.6.3 Temperature Tolerance

For polymer materials, the relatively poor temperature tolerance is a severe problem that hinders their applications. Common polymer substrates, such as polyethylene and polyethylene terephthalate can only withstand a maximal temperature of 80 and 150°C, respectively. They are obviously too low for high-temperature fabrication methods, such as chemical vapor deposition and thermal annealing. This dilemma is currently compromised by conducting a transfer process where the devices are first fabricated at high temperatures and then transferred onto polymer substrates at low temperatures. Obviously, the transfer consequences are highly dependent on the fabrication process, and may largely increase the individual variation of the resulting devices. Furthermore, it has largely increased the fabrication complexity and cost, which can hardly meet the requirements of industrial fabrication. The solution of the aforementioned problem lies in the advancement of low-temperature processing methods, and it is also critical to enhance the temperature tolerance of polymer materials, which will meanwhile extend the operating temperature range of the resulting devices.

1.6.4 Scale-up Production

Although a variety of polymer materials has realized successfully industrial production, the scale-up production of polymer materials for energy and electronic device applications is still challenging. Distinguished from commercially available polymer materials, the characteristics and morphologies of the polymers used in building energy and electronic devices are required to be highly customized and stable aiming at high device performances. The conventional synthetic methods may be unavailable, particularly for a scale-up production of flexible devices with highly curved or soft substrates, which calls for more efforts to explore efficient and continuous production methods. As a result, it will become even more important for scale-up production concerning the future development of polymer-based energy and electronic devices.

References

Chen, T., Qiu, L., Yang, Z., Peng, H., 2013. Novel solar cells in a wire format. Chem. Soc. Rev. 42, 5031–5041.

Chen, X., Lin, H., Chen, P., Guan, G., Deng, J., Peng, H., 2014. Smart, stretchable supercapacitors. Adv. Mater. 26, 4444–4449.

Deng, J., Zhang, Y., Zhao, Y., Chen, P., Cheng, X., Peng, H., 2015. A shape-memory supercapacitor fiber. Angew. Chem. Int. Ed. 54, 15419–15423.

Fan, F.R., Tian, Z.Q., Wang, Z.L., 2012a. Flexible triboelectric generator. Nano Energ. 1, 328–334.

Fan, F.R., Lin, L., Zhu, G., Wu, W., Zhang, R., Wang, Z.L., 2012b. Transparent triboelectric nanogenerators and self-powered pressure sensors based on micropatterned plastic films. Nano Lett. 12, 3109–3114.

Krebs, F.C., Spanggaard, H., 2005. Significant improvement of polymer solar cell stability. Chem. Mater. 17, 5235–5237.

Krebs, F.C., Carl, J.E., Cruys-Bagger, N., Andersen, M., Lilliedal, M.R., Hammond, M.A., Hvidt, S., 2005. Lifetimes of organic photovoltaics: photochemistry, atmosphere effects and barrier layers in ITO-MEHPPV:PCBM-aluminium devices. Sol. Energ. Mat Sol. Cells 86, 499–516.

Lin, H., Li, L., Ren, J., Cai, Z., Qiu, L., Yang, Z., Peng, H., 2013. Conducting polymer composite film incorporated with aligned carbon nanotubes for transparent, flexible and efficient supercapacitor. Sci. Rep. 3, 1353.

Liu, L., Tian, F., Wang, X., Yang, Z., Zhou, M., Wang, X., 2012. Porous polythiophene as a cathode material for lithium batteries with high capacity and good cycling stability. React. Funct. Polym. 72, 45–49.

MacDiarmid, A.G., Yang, L.S., Huang, W.S., Humphrey, B.D., 1987. Polyaniline: electrochemistry and application to rechargeable batteries. Synth. Met. 18, 393–398.

Mermilliod, N., Tanguy, J., Petiot, F., 1986. A study of chemically synthesized polypyrrole as electrode material for battery applications. J. Electrochem. Soc. 133, 1073–1079.

Neugebauer, H., Brabec, C., Hummelen, J.C., Sariciftci, N.S., 2000. Stability and photodegradation mechanisms of conjugated polymer/fullerene plastic solar cells. Sol. Energ. Mat Sol. Cells 61, 35–42.

Nigrey, P.J., MacInnes, Jr., D., Nairns, D.P., Macdiarmid, A.G., Heeger, A.J., 1981. Lightweight rechargeable storage batteries using polyacetylene, (CH)*x* as the cathode-active material. J. Electrochem. Soc. 128, 1651–1654.

Pan, S., Ren, J., Fang, X., Peng, H., 2016. Integration: an effective strategy to develop multifunctional energy storage devices. Adv. Energ. Mater. 6, 1501867.

Patrice, S., Yury, G., Bruce, D., 2014. Materials science. Where do batteries end and supercapacitors begin? Science 343, 1210–1211.

Peng, H., 2015. Fiber-Shaped Energy Harvesting and Storage Devices. Springer, Berlin, Heidelberg.

Shacklette, L.W., Toth, J.E., Murthy, N.S., Baughman, R.H., 1985. Polyacetylene and polyphentlene as anode materials for nonaqueous secondary batteries. J. Electrochem. Soc. 132, 1529–1535.

Shirakawa, H., Louis, E.J., Macdiarmid, A.G., Chiang, C.K., Heeger, A.J., 1977. Synthesis of electrically conducting organic polymers: halogen derivatives of polyacetylene,(CH)*x*. J. Chem. Soc., Chem. Commun. 474, 578–580.

Staudinger, H. 1920. Über polymerisation. Berichte der deutschen chemischen Gesellschaft (A and B Series), 53. pp. 1073–1085.

Sun, H., You, X., Deng, J., Chen, X., Yang, Z., Ren, J., Peng, H., 2014a. Novel graphene/carbon nanotube composite fibers for efficient wire-shaped miniature energy devices. Adv. Mater. 26, 2868–2873.

Sun, H., You, X., Jiang, Y., Guan, G., Fang, X., Deng, J., Chen, P., Luo, Y., Peng, H., 2014b. Self-healable electrically conducting wires for wearable microelectronics. Angew. Chem. Int. Ed. 53, 9526–9531.

Sun, H., Fu, X., Xie, S., Jiang, Y., Peng, H., 2016. Electrochemical capacitors with high output voltages that mimic electric eels. Adv. Mater. 28, 2070–2076.

Tasch, S., Niko, A., Leising, G., Scherf, U., 1996. Highly efficient electroluminescence of new wide band gap ladder-type poly(*para*-phenylenes). Appl. Phys. Lett. 68, 1090–1092.

Yang, Z., Deng, J., Chen, X., Ren, J., Peng, H., 2013. A highly stretchable, fiber-shaped supercapacitor. Angew. Chem. Int. Ed. 52, 13695–13699.

You, J., Dou, L., Yoshimura, K., Kato, T., Ohya, K., Moriarty, T., Emery, K., Chen, C.-C., Gao, J., Li, G., 2013. A polymer tandem solar cell with 10.6% power conversion efficiency. Nat. Commun. 4, 1446.

Yu, G., Gao, J., Hummelen, J.C., Wudl, F., Heeger, A.J., 1995. Polymer photovoltaic cells enhanced efficiencies via a network of internal donor acceptor heterojunctions. Science 270, 1789–1791.

Yu, D., Qian, Q., Wei, L., Jiang, W., Goh, K., Wei, J., Zhang, J., Chen, Y., 2015. Emergence of fiber supercapacitors. Chem. Soc. Rev. 44, 647–662.

2

Synthesis and Design of Conjugated Polymers for Organic Electronics

2.1 Synthesis of Conjugated Polymers

2.1.1 Introduction

Before the great discovery of the first highly conducting polymer, it is always believed always that polymers are insulated. Insulating polymers, such as polyethylene (PE), polypropylene, polyvinyl alcohol, or polymethyl methacrylate, have very important influence on human life in the past 100 years. However, due to the insulating property, they can only be used in several fields of electronics, for example, electrolyte, separator or piezoelectric materials. By contrast, the emergence of conducting polymer, which also called conjugated polymers or organic metals, has aroused broad and sustaining interest for various optoelectronic devices in the last few decades. Highly conducting polymers are good candidates for electrode materials and thermoelectric materials. And semiconducting polymer materials have brought about the reform of electronics and motivated the development of polymer light-emitting diodes (PLEDs), polymer solar cells (PSCs), organic/polymer field-effect transistors, and polymer electrochromic displays (PECDs).

Most traditional insulating polymers have already achieved industrial production for many years, hence their synthesis methods are very mature, and not included in this chapter. In comparison, the research about the synthesis of conjugated polymers is very active up to now (Reynolds, 2014). Chemists have made a lot of efforts to improve traditional polymerization methods and develop new synthetic routes by virtue of the broad prospect of application and commercial value of conjugated polymer.

2.1.2 Common Methods for Constructing Carbon–Carbon Double Bond

2.1.2.1 Synthesis of Polyacetylenes

The first conducting polymer is polyacetylene **1**. Its high electrical conductivity upon doping with controlled amounts of Cl_2, Br_2, I_2, and AsF_5 was discovered by Hideki Shirakawa, Alan Heeger, and Alan MacDiarmid, which helped to create the field of organic conducting polymers (Chiang et al., 1977). One of the most common methods to obtain polyacetylene is using titanium and aluminum catalysts, which are known as Ziegler–Natta catalysts, to catalyze the polymerization of gaseous acetylene (Scheme 2.1A) (Ito et al., 1974). However, unsubstituted polyacetylene is unprocessable due to its inherent insolubility and infusibility. Hence, considerable research efforts have been directed toward obtaining **1** from workable, easily processable polymers by means of precursor polymer route.

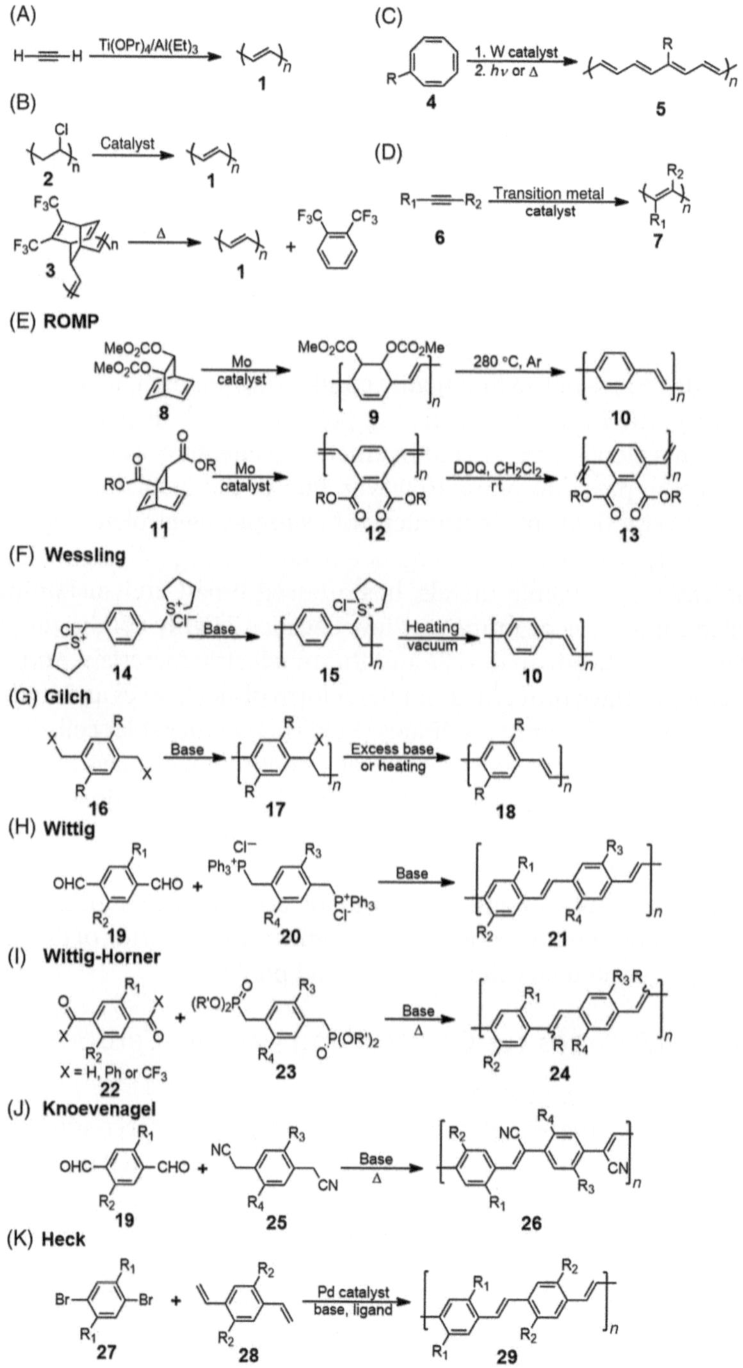

SCHEME 2.1 (A–D) Synthetic methods of polyacetylenes. (E–K) Synthetic methods of poly(phenylene-vinylene)s.

Dehydrohalogenation of poly(vinyl chloride) **2** was proposed first, but the polyacetylene prepared by this route generally possesses relatively short and irregular segments (Bowley et al., 1985). Thermal conversion of precursor polymers is a more effective method for synthesizing long polyacetylene chains. In the Durham precursor route, precursor polymer **3** (Scheme 2.1B) which is soluble in common organic solvents is converted to polyacetylene through a heat-induced reverse Diels–Alder reaction (Edwards and Feast, 1980). By controlling the conversion reaction conditions (temperature, time, pressure, mechanical stress, and solvent), the morphology of the final product can be regulated. Another alternative approach is introducing solubilizing groups to the main chain of the polymer partially, which can not only increase the solubility but also maintain the conjugated degree of the polymer backbone. Grubbs and coworkers developed a very facile method called ring opening metathesis polymerization (ROMP) to prepare a variety of polyacetylene derivatives with linear and branched alkyl chains in 1993 (Gorman et al., 1993). At the presence of well-defined tungsten-based olefin metathesis catalyst, a family of partially substituted polyacetylenes **5** is synthesized via the ROMP of monosubstituted 1,3,5,7-cyclooctatraenes (COT, **4**), which is illustrated in Scheme 2.1C. Finally, most members of the obtained polyacetylenes are highly conjugated and soluble. Unlike partially substituted polyacetylenes, mono- or disubstituted polyacetylenes **7** are always prepared from the corresponding acetylenic monomers **6** by Ziegler–Natta type polymerization with $MoCl_6$, WCl_6 or $TaCl_5$ based catalysts or by using organorhodium (I) complexes (Scheme 2.1D) (Liu et al., 2009). Among these transition-metal catalysts, Rh catalysts are found to polymerize monosubstituted acetylenes to give *cis*-stereoregular polymers in high molecular weight and low polydispersity, with some also effecting the living polymerization of substituted acetylenes (Kishimoto et al., 1999).

2.1.2.2 Synthesis of Poly(phenylenevinylene)s and Poly(arylene vinylene)s

ROMP can also be used to prepare the precursors of poly(phenylenevinylene)s (PPVs, **10**). Reported in the first time by Grubbs and coworkers, a bicyclo[2.2.2]octadiene monomer **8** was treated with a ruthenium olefin metathesis catalyst to give a precursor polymer **9** (Scheme 2.1E) (Conticello et al., 1992). The number average molecular weight of the obtained polymer **9** was up to 63,000 and its Polydispersity index (PDI) is 1.3. The ^{1}H NMR spectrum indicates that the *cis*/*trans* double bond ratio in the precursor polymer **9** is about 1:1. Conversion of **9** to **10** can be achieved by heating at 280°C. Interestingly, the mixture of *cis* and *trans* double bonds in the precursor polymer **9** was mostly isomerized to *trans* during the thermal elimination treatment. Subsequently, the same group reported the synthesis of ester-substituted PPVs (Wagaman and Grubbs, 1997). As shown in Scheme 2.1E the polymerization of 2,3-dicarboxybarrelene **11** via ROMP yields the precursor polymer **12**, which is converted into diester-substituted poly(1,4-phenylenevinylenes) **13** by aromatizing the cyclohexadiene rings using 2,3-dichloro-5,6-dicyano-1,4-benzoquinone (DDQ) at room temperature. ^{1}H NMR and IR analyzes indicate that the double bonds in **13** are all *trans* after the oxidation of DDQ. Besides, siloxy paracyclophene (Miao and Bazan, 1994) or paracyclophane-diene (Yu and Turner, 2006) monomer has also been used to prepare precursor polymers of PPVs via ROMP. Although ROMP synthesis of PPV derivatives typically provides the best control, the preparation of the starting monomers is difficult with moderate to low yields, which is especially a problem for the strained cyclophane monomers.

It is well known that the first PLED was made in Cambridge using conjugated polymer PPV as electroluminescence (EL) material in 1989 (Burroughes et al., 1990). From then on, PPVs and other poly(arylene vinylene)s (PAVs) or related copolymers become the most representative electroluminescence conjugated polymers. Including the aforementioned ROMP method, PPVs can be synthesized by a variety of methods. Among them, sulfinyl precursor route is the most popular one, which was introduced by Wessling and Zimmerman (Wessling and Zimmerman, 1968) and later modified and optimized by several groups (Garay et al., 1993; Garay and Lenz, 1992; Lenz et al., 1988; Burn et al., 1992; Garay and Lenz, 1989; Tokito et al., 1990). Scheme 2.1F shows the standard preparation of PPVs (Burn et al., 1992). Treating the bis-sulfonium salt monomer **14** with base (slightly less than 1 equiv) in a protic solvent at around 0°C induces the formation of the sulfonium precursor polymer **15.** If 1 equiv of base is used, some of the sulfonium groups on the precursor polymer would be eliminated. Solutions of **15** (with a number average molecular weight, M_n, of about 100,000) are extremely viscous even at concentrations of 1% w/w, which facilitates to prepare a highly uniform film as needed by spin-casting or doctor blade. Consequently the precursor polymer **15** is converted in situ into **10** by heating to high-temperature (300°C) under vacuum condition (10^{-6} mbar) for a period of several hours. And the elimination conditions have a substantial influence on the properties of the thin film material particularly in device applications. In some cases, sulfonium group of **15** is liable to take place nucleophilic substitution reaction with water to form insoluble gel, which is not good for structural control. To solve this problem, a modified Wessling synthesis is treating **15** with refluxing methanol to afford an uncharged precursor polymer (Halliday et al., 1993). Subsequent processing can obtain more highly ordered PPVs than the original route. Besides, other modified Wessling procedures using sulfur-based leaving groups, such as sulfinyl (Louwet et al., 1992), sulfonyl (Louwet et al., 1995) dithiocarbamate (Henckens et al., 2006), and xanthate (Son et al., 1995) have all been reported for the synthesis of a range of PPV derivatives. Nevertheless Wessling and other modified routes have some obvious drawbacks including the generation of toxic side products during the elimination process and uncontrollable structural defects or broad molecular weight distribution arising from thermal conversion or oxidation.

Up until now, the most convenient method for the preparation of PPV derivatives is the Gilch route, which was first used by Gilch and optimized by Swatos in 1990 (Gilch and Wheelwright, 1966). A one-pot straightforward condensation of 1,4-bis-halomethylated benzene derivatives **16** at the presence of a large excess of base, usually KO^{t}Bu, can directly produce PPV or its derivatives **18** and hydrogen chloride as byproduct (Scheme 2.1G). It is also possible to treat the monomer **16** with slightly less than 1 equiv. of alkoxide base to obtain the precursor polymer **17**, which can then be converted to **18** by the addition of more base or by heating. Due to smaller size of the halide relative to sulfur-based leaving groups, this route is particularly suitable for introducing more bulky groups into the phenylene ring of PPV skeleton without affecting the reaction. And the overall process shortens the preparation of PPVs by two steps compared to the Wessling route so consequently increases the yield. The molecular weights of the polymers obtained by Gilch route are generally high, which can lead to the formation of insoluble gels or precipitation during polymerization. One solution is to control the molecular weight by carrying out anionic polymerizations, which uses 4-methoxyphenol as the initiator to initiate the polymerization of monomer in the base solution (Neef and Ferraris, 2000). In addition, structural defects still inevitably exist in Gilch route due to the partially random couplings (Becker et al., 1999). These couplings will decrease the conjugation of the PPV backbone and ultimately attenuate the desired electronic and optical properties of the resulting polymers. Despite these disadvantages, the Gilch route is widely used for PPV synthesis in industry and has been demonstrated in continuous processing (Seyler et al., 2012).

Precursor approaches require postpolymerization modifications, which often require multistep synthesis and result in undesirable side reactions. Thus direct routes including Wittig/Wittig–Horner condensation, Knoevenagel condensation or Heck coupling reaction are developed to prepare PPVs with less structural defects. These routes are highly compatible with functional groups and can provide control over molecular design, allowing the fine tuning of the final optoelectronic properties.

The Wittig method was first reported in the 1960s by McDonald and Campbell for the preparation of PPVs (McDonald and Campbell, 1960). The reactions are always carried out between the bisphosphonium salt **20** and bisaldehyde **19** in the presence of a strong base to yield a mixture **21** of *cis*- and *trans*-isomers, which can potentially influence the fluorescence quantum efficiency of the polymer (Scheme 2.1H). The Wittig–Horner condensation replaces the phosphonium salts **20** with phosphonates **23** (Scheme 2.1I) (Hörhold and Opfermann, 1970). The higher reactivity of the ylide intermediates and facile removal of water-soluble phosphonate byproduct are the two advantages of this method. The resulting PPVs exhibit higher degrees of polymerization and predominately *trans* configuration of double bond. This approach has been successfully applied to the synthesis of regioregular PPVs with different lateral functionalizations, such as poly[2-methoxy-5-(2-ethylhexyl)-1,4-phenylene vinylene] (MEH-PPV). When the aromatic dialdehyde **22** is replaced by a diketone, PPVs with functionalized vinylene linkages can be prepared (Hörhold and Helbig, 1987).

Knoevenagel condensation is always employed to synthesize CN-PPVs **27** (Scheme 2.1J) (Lenz and Handlovits, 1960; Funke and Schütze, 1964). The polymerization takes place typically between benzylic nitriles **25** and terephthaldehydes **19** in the presence of an excess of a strong base (tetrabutylammonium hydroxide or potassium *tert*-butoxide) in a mixture of aprotic and protic solvents (typically poor solvents for the polymers as reaction medium). This synthetic strategy is not only particularly tolerant to functional groups, but the monomers are also readily available or easily accessible from the corresponding dihalomethylene derivatives. Molecular weights obtained by these methods are generally not very high due to the side reaction of nucleophiles to the vinylene bond.

Heck coupling reaction is one of the most important routes for the synthesis of PPVs via the palladium-catalyzed arylation of olefins by aryl halides (Heitz et al., 1988; Greiner and Heitz, 1988). The typical reaction is showed in Scheme 2.1K. The copolymers **29** are obtained by the reaction of 1,4-divinyl-2,5-dialkyloxybenzene **28** with dibromide monomers **27** in *N,N*-dimethylformamide (DMF), in the presence of triethylamine, tri(*o*-tolyl) phosphine and a catalytic amount of palladium acetate. Under these conditions, PPVs with M_n of approximately 10,000–20,000 can be obtained. This reaction offers advantages over other methods due to its mild reaction conditions and high compatibility to functional groups, producing a low level of defects and favoring high control over stereoselectivity (*cis*/*trans* ratio). When starting from suitable vinylic and aryl halide monomers, copolymers with defined alternating structures can be obtained. Hence, a large number of examples of Heck (or Heck-type) reactions for the synthesis of various copolymers containing the *p*-phenylenevinylene unit are reported in the last decade (Grimsdale et al., 2009).

2.1.3 Common Methods for Constructing Carbon–Carbon Single Bond

The synthesis of conjugated polymers is highly dependent on the effective carbon–carbon single bond generation between unsaturated carbons in aromatic molecules. Aromatic units in conjugated polymers can be benzene, aniline, pyrrole, thiophene, carbazole, naphthalenediimide, perylenediimide (PDI), or their derivatives, etc. Although monomers are various, their synthetic methods can be mainly classified into chemical and electrochemical polymerizations. Chemical polymerization includes chemical oxidative polymerization and metal-catalyzed coupling condensation.

2.1.3.1 Chemical Oxidative Polymerization and Electrochemical Polymerization

Intrinsically conducting polymers, such as polyaniline (PANI), polypyrrole (PPy), or polythiophene (PT) as well as their doped forms are always prepared from the corresponding monomers by chemically or electrochemically oxidative polymerization (Waltman and Bargon, 1986; Toshima and Hara, 1995). These methods are widely used in the preparation of complex electrodes with carbon materials. Other unsubstituted or β-substituted monomers with structural similarity also use these two approaches to synthesize the corresponding conjugated polymers. Electrochemical polymerization is carried out in a monomer solution using a three electrode system. When applying a high anodic potential, the monomer is oxidized at the electrode surface, leading to the polymerization thereafter. As the deposited polymer is in an oxidized conducting state, polymerization continues at the surface of the electrodeposited polymer until the potential is stopped. Hence the thickness of polymer film can be controlled by adjusting the polymerization time and the film always has a relatively uniform morphology. But the following disadvantage is the limited surface area of the working electrode, which is an obstacle for large-scale production and application. In contrast, chemical polymerization does not need to consider this problem, because this method only uses oxidizing agent, such as iron (III) chloride or ammonium persulfate to initiate the polymerization of monomers in an appropriate solution. Nevertheless, the conjugated polymer obtained by chemical polymerization is powder, and the poor solubility makes it difficult to spin-cast into a thin film.

2.1.3.2 Metal-Catalyzed Coupling

2.1.3.2.1 KUMADA–CORRIU AND NEGISHI COUPLING

Actually, chemical or electrochemical polymerization always creates random couplings for asymmetric monomers (Leclerc et al., 1989). Here we take poly(3-alkylthiophene)s (P3ATs) as an example to clarify the issues since they are the most abundantly employed conjugated polymers for organic electronics. As shown in Fig. 2.1A, there are three regioisomeric couplings for P3ATs: head-to-tail (HT), head-to-head (HH), and tail-to-tail (TT) couplings, and the degree of regioregular HT arrangement is called regioregularity. Usually, the HT regioregularity of P3ATs obtained by chemical or electrochemical polymerization is only 50–80%. It has been demonstrated that HH and TT couplings imposed steric hindrance on the backbones of P3ATs, so the planarity and $\pi-\pi$ stacking of the polymer skeletons are reduced, resulting in a destruction of high conductivity and other desirable properties for

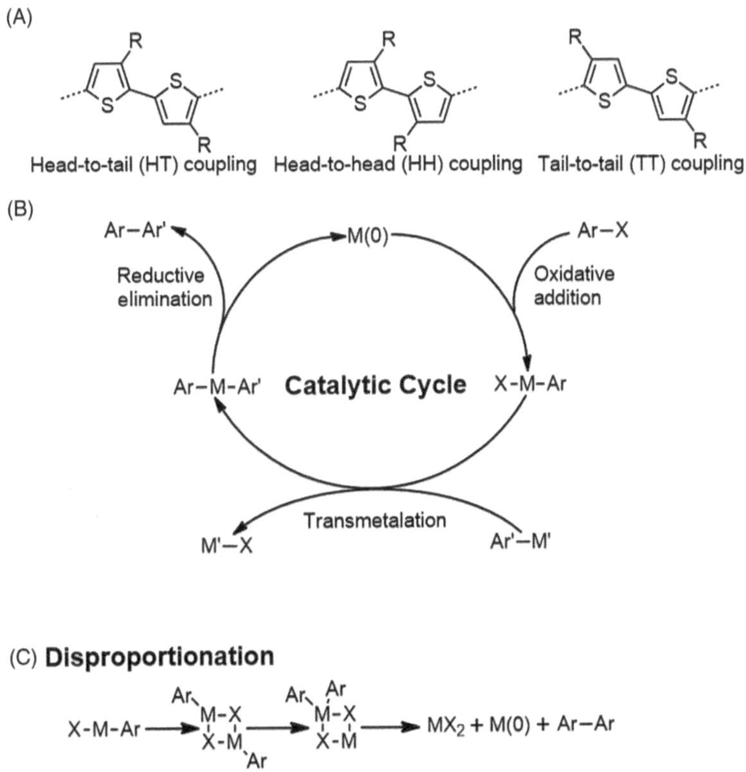

FIGURE 2.1 (A) Regioisomeric couplings of 3-alkylthiophenes. (B) Catalytic cycle of transition-metal-catalyzed reactions. (C) Disproportionation reaction in the catalytic cycle.

P3ATs. On the contrary, less steric hindrance enables HT regioregular P3ATs to form well-defined, 3-D polycrystalline structures by solid-state self-assembly. These kinds of structures are able to create effective charge carrier pathways of both interchain and intrachain, leading to high electrical properties. For instance, the mobility of regioirregular poly(3-hexylthiophene) (P3HT) is only $\sim 10^{-5}$ $cm^2 \cdot (V \cdot s)^{-1}$, but that of regioregular P3HT rises up to 0.2 $cm^2 \cdot (V \cdot s)^{-1}$. To conclude, only HT regioregular P3ATs are highly desirable. Therefore, new approaches should be developed to produce conjugated polymers with high regioregularity from asymmetric monomers.

The first route to synthesize HT regioregular P3ATs is known as the McCullough method (Scheme 2.2A) (McCullough and Lowe, 1992; McCullough et al., 1993). This method treats 2-bromo-3-alkylthiophene **30** with LDA (lithium diisopropylamide) at −78°C, followed by an addition of $MgBr_2 \cdot Et_2O$ to generate 2-bromo-5-bromomagnesio-3-alkylthiophene intermediate **31**. Quenching studies indicate that 98–99% of the desired intermediate **31** and less than 1–2% of 2,5-exchanged isomer **32** are produced. The further addition of a catalytic amount of $Ni(dppp)Cl_2$ catalyzes the reaction of intermediate to produce HT regioregular P3ATs **33** (44–66% yield). The regioregularity of the product is 98–100% revealed by NMR study, and the M_n is typically 20,000–40,000 with polydispersities (PDI) of around 1.4. This procedure was later modified by replacing $MgBr_2 \cdot Et_2O$ with $ZnCl_2$, which results in the better properties of HT regioregular P3ATs.

Soon after the report of the McCullough method, the Rieke method (Scheme 2.2B) was published (Chen and Rieke, 1992). In this method, the starting material is changed to 2,5-dibromo-3-alkylthiophenes **34**, which reacts with "Rieke zinc" at low temperature to yield a mixture of two isomeric organozinc intermediates **35** and **36** in a ratio of 90:10 directly. The intermediates are polymerized in the presence of $Ni(dppe)Cl_2$. The yield of P3ATs **33** is increased to 75%, and the polymer parameters (M_n = 24,000–34,000, PDI = 1.4) are maintained in a comparison to McCullough method.

Then in 1999, Grignard metathesis (GRIM) method (Scheme 2.2C) was reported, (Loewe et al., 1999) which is better than previous two methods. In this method, the generation of organometallic intermediates does not need the highly reactive reagent LDA and can be carried out smoothly at room temperature or reflux. Treating 1 equiv of any Grignard reagent (R'MgX) with 2,5-dibromo-3-alkylthiophenes **34** can obtain a mixture of intermediates **37** and **38** in a ratio of 88:12 to 80:20 (Loewe et al., 2001). The following polymerization of the intermediates can finish in less than 1 h after the addition of $Ni(dppp)Cl_2$. Although the ratio of the undesirable to desirable isomers is higher in the GRIM method, compared with the McCullough and Rieke methods, the resulting P3ATs **33** still has a very high regioregularity of >99% HT couplings. The typical M_n is 20,000–35,000 with a very low PDI of 1.2–1.4. These advantages make GRIM method become the most convenient and economical way to produce high quality regioregular P3ATs at large scale. Follow-up studies of the mechanism reveal that GRIM method is not only a chain-growth process (Yokoyama et al., 2004) but also a *quasi*-living process (Iovu et al., 2005; Sheina et al., 2004). In this *quasi*-living polymerization, $Ni(dppp)Cl_2$ is believed to act as an initiator rather than a catalyst. That is to say, only one end of the polymer chain can realize chain growth, which

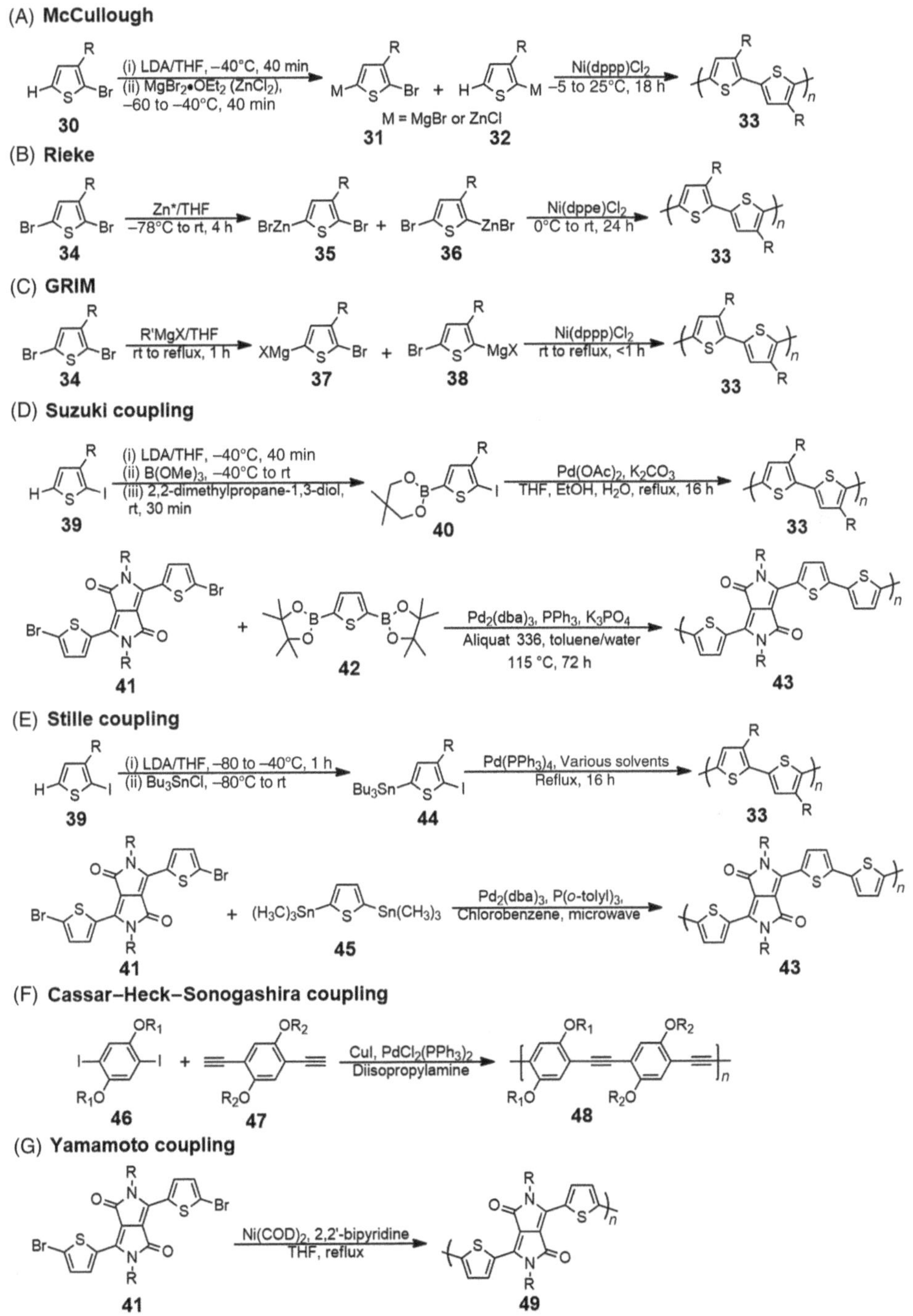

SCHEME 2.2 (A–G) Metal-catalyzed coupling for constructing carbon–carbon single bond.

facilitates the regulation of the molecular weight of the polymer through controlling the reaction time and the amount of Ni catalyst. Furthermore, sequential addition of different 3-alkylthiophene monomers could yield various block copolymers. Detailed mechanistic studies promote the fast development of GRIM method, whose application has extended to other conjugated polymers of different chain architectures.

In the previous preparations of P3ATs, all the polymerization of the intermediates undergoes nickel-catalyzed cross-coupling polycondensation. Essentially, both McCullough and GRIM methods belong to Kumada reaction, while Rieke method is Negishi reaction. Actually, transition-metal-catalyzed cross-coupling reactions have become a particularly powerful tool in the field of organic synthesis at the present time, especially for the synthesis of conjugated polymers with tailored structures (Babudri et al., 2004; Marzano et al., 2014; Pron et al., 2010; Osaka and McCullough, 2008; Carsten et al., 2011; Sakamoto et al., 2009). In general, the mechanism of transition-metal-catalyzed involves a catalytic cycle of three consecutive steps: oxidative addition reaction across the C—X bond of an electrophile, transmetalation with a main group organometallic nucleophile, and reductive elimination step leading to the coupled product, and the regenerated active catalyst (Fig. 2.1B). The most commonly employed transition-metal catalysts are nickel- or palladium-based complexes, although other metals have also been used. And the organometallic nucleophiles can be Grignard reagents (Kumada–Corriu) (Tamao et al., 1972; Corriu and Masse, 1972), Zinc reagents (Negishi) (King et al., 1977), boron reagents (Suzuki–Miyaura) (Miyaura and Suzuki, 1979; Miyaura et al., 1979), stannyl reagents (Stille) (Stille, 1986), or copper reagents (Cassar–Heck–Sonogashira) (Sonogashira, 1975). As a result, successive transformations during catalytic cycles is capable to increase the conjugation length. Regioregularity of polymers is realizable as long as the monomeric molecules possess both electrophilic and nucleophilic centers.

Cross-coupling reactions of organomagnesium (the Kumada–Corriu coupling) or organozinc (the Negishi coupling) reagents with aromatic/heteroaromatic halides in the presence of transition metal complexes (palladium or nickel) are exploited as a synthetic route to obtain PT and polyphenylenes for photoelectric applications (Osaka and McCullough, 2008). Regioregular P3ATs can be efficiently obtained via these two methods as mentioned previously. Several polyarylene and PAV type polymers have been prepared by Kumada coupling of the dibromomagnesium derivative with the appropriate dihaloarenes. The polymers were obtained in 40–73% yield and the weight-average molecular weights (M_w) were 6350–11,300 (Babudri et al., 2002). In these reaction, Pd(dppf)Cl_2 is chosen as the best catalyst to obtain the polymers with high yields and molecular weights. Actually, different metal complex has different influence on the homocoupling of the monomers, which is the side reaction in the polymerization step. It is observed that Ni complexes [such as Ni(dppp)Cl_2] with bidentate ligand gave exclusively cross-coupled products, affording highly regioregular P3ATs, whereas with Pd complexes bearing monodentate ligand a random mixture of cross-coupled and homocoupled products is produced. In addition, although the two coupling reactions are similar to each other, the lower reactivity of the organozinc reagents with respect to the Grignard reagents results in a moderate tolerance of some functional groups. Thus the Negishi coupling (Rieke method) has been used to prepare ester- or cyano-functionalized P3AT copolymers.

2.1.3.2.2 SUZUKI–MIYAURA AND STILLE COUPLING

Among metal-catalyzed cross-coupling reactions, Suzuki–Miyaura and Stille reactions are the most widely used methods. Mild experimental conditions, high tolerance toward several functional groups, good stability of orgaometallic nucleophiles (compared with Grignard reagents, organolithium reagents, and others) to oxygen and water are some of the mutual superiorities of Suzuki–Miyaura and Stille reaction. Suzuki–Miyaura reaction is the Pd-catalyzed cross-coupling between aromatic boron derivatives and vinyl or aryl halides in the presence of a base (Scheme 2.2D). Typically, the boron reagents can be introduced by reacting an aryl-lithium or Grignard species with borate ester, or via Suzuki coupling between halogenated aromatics and diboronyl ester/diboronic acid. The boron reagents can be boronic acids or boronic esters, especially pinacol boronic ester. Boronic acids tend to dehydrate and form cyclic trimeric anhydride (boroxines), which makes it difficult to purify the diboronic acid mononers. Boronic esters show less reactivity than boronic acids, although they have better solubility and stability than boronic acids. Suzuki polymerization always requires a two-phase system with toluene or xylene as the organic phase and K_2CO_3 or K_3PO_4 solution as the basic aqueous phase. But the solubility of the polymer may rapidly decrease as molecular weight increases by virtue of two-phase system, which reduces the yield and molecular weight and enhances the polydispersity of the product. An alternative solution to improve the outcome of Suzuki–Miyaura cross-coupling in heterogeneous media is the use of a phase-transfer agent, such as Aliquat 336. Moreover, catalysts have great impact on the molecular weight, yield, and end groups of the products. $Pd(PPh_3)_4$ is commonly employed in Suzuki reaction, but PPh_3 can be easily oxidized into Ph_3PO by a trace amount of oxygen. Another common catalyst, $Pd(PPh_3)_2Cl_2$, with a reduced amount of PPh_3 is also used, but it is even more prone to oxygen attack. Hence, in situ generation of the active catalytic species Pd(0) from Pd(II) complex with a labile ligand, such as $Pd(dba)_2$ (dba, dibenzylideneacetone), and a suitable amount of phosphine ligand is preferred as a useful alternative to the use of air-sensitive Pd(0) phosphine complexes. Phosphine ligands have many options, such as PPh_3, $^tBu_3P$, and $P(o\text{-tolyl})_3$. It is demonstrated that stronger donating ligands are more easily oxidized, leading to formation of palladium black (and the oxide of the ligand). Thus, the catalyst stability is greater by using "weak" ligands, for example, $P(o\text{-tolyl})_3$. Besides, sometimes the phosphine-free catalyst $Pd(OAc)_2$ is also employed in Suzuki–Miyaura aryl coupling, which has a marked positive effect on the molecular weight of the copolymer with respect to the phosphine palladium complex.

Instead of using boron reagents to activate the sp^2 carbon, the Stille reaction employs stannyl reagents to serve the same function in the absence of a base (Scheme 2.2E). The organotin compound can be conveniently attached to the aromatic halides by butyllithium and trimethyltin chloride. The catalytic cycles in Stille coupling are similar to Suzuki–Miyaura coupling, so Pd-based catalysts $Pd(PPh_3)_4$, $Pd(PPh_3)_2Cl_2$, and $Pd(dba)_2$ with a suitable amount of PPh_3 or P(*o*-tolyl)$_3$ are also commonly employed. The major drawback of Stille coupling is the toxicity of tin-containing monomers and the byproducts, especially in comparison with the boron reagents and byproducts from Suzuki coupling. On the other hand, the tin-containing monomers can usually be prepared ultrapure via recrystallization, while it is often difficult to purify boron reagents. Meanwhile, Stille coupling possesses more tolerance for a wide range of functional groups than Suzuki coupling. As Suzuki reaction needs to be performed in basic conditions, which requires more complex protection-deprotection strategies for monomers that may be labile to basic environment. Finally yields and molecular weights of the polymers prepared via Stille reaction are always higher than Suzuki reaction due to its two-phase system. It is noteworthy that stannyl groups functionalized benzene exhibits low reactivity toward aryl halides under Stille coupling conditions. Therefore, Stille coupling is more suitable for thiophene-based aromatic monomers substituted with stannyl groups, whereas Suzuki coupling is a better option for polycondensation of benzene-based compounds functionalized with boronic groups.

Stille and Suzuki reactions have been reported for preparing HT regioregular P3ATs. Iraqi et al. have ever investigated the synthesis of HT regioregular P3ATs through the Stille reaction, using 2-iodo-3-hexyl-5-tri-*n*-butylstannylthiophene **44** with a variety of solvents (Scheme 2.2E) (Iraqi and Barker, 1998). In all cases HT regioregular P3ATs with greater than 96% HT couplings were obtained. Number averaged molecular weights of the products **33** were 10,000–16,000 with PDI of 1.2–1.4 after purification, but the yields were rather low (<50%). However, the Suzuki coupling of 3-octyl-2-iodo-5-boronatothiophene **40** in the presence of $Pd(OAc)_2$ led to highly regioregular poly(3-octylthiophene) **33** with a 51% yield (Scheme 2.2D) (Guillerez and Bidan, 1998). After removal of short length oligomers, the resulting polymer contained 96–97% HT couplings and the M_w was 27,000. It can be seen that P3ATs produced by both Stille and Suzuki coupling have lower molecular weights than Rieke or GRIM method. However, it does not reduce the important role of these two methods. As they are specifically appropriate for preparing alternating copolymers of donor (D)–acceptor (A) type by using two distinct monomers with various structures. Recently, an interesting electron acceptor building block, diketopyrrolopyrrole (DPP) has attracted considerable attentions for the construction of a large number of high-performance copolymers for OFETs and PSCs (Li et al., 2013). Suzuki or Stille coupling is always employed to synthesize these DPP-based copolymers. For example, copolymer **43**, first reported by Janssen et al., was prepared by Suzuki coupling with a yield of 84% and M_n of 54,000 (PDI = 3.15) (Scheme 2.2D) (Bijleveld et al., 2009). The synthesis of this copolymer **43** was performed using the 2,5-thiophenebis(boronic ester) **42**, which was coupled with the dibromo-DPP monomer **41** using a Pd(0) complex generated in situ from $Pd_2(dba)_3/PPh_3$, K_3PO_4, and Aliquat 336 in toluene/water. A much lower M_n of 10,000 (PDI = 2.4) and yield of 69% had been obtained when $Pd(PPh_3)_4$ was used as the catalyst, which clearly reveals the influence of catalyst system. Then DeLongchamp et al. reported another synthetic route of **43**, where **41** was copolymerized with 2,5-bis(trimethylstannyl) thiophene **45** by microwave-assisted Stille cross-coupling in chlorobenzene (Scheme 2.2E)

(Zhang et al., 2011). The palladium catalyst was also generated in situ from $Pd_2(dba)_3$ in the presence of tri(*o*-tolyl)phosphine ($P(o\text{-tolyl})_3$) as a ligand. The use of a bulky phosphine ligand afforded **43** with very high molecular weights (M_n = 104,000, PDI = 3.0) in 68% yield. The authors attributed the significantly improved mobility values to the higher molecular weight of their polymer compared with the polymer reported by Janssen et al. (Bijleveld et al., 2009).

Similarly, a large number of π-conjugated moieties with ditin or diboronic acid (or ester) functionalities can be used as building blocks to couple with diborominated aromatic monomers (not merely DPP-based monomers) via Stille or Suzuki coupling. In other words, Suzuki (Sakamoto et al., 2009) or Stille (Carsten et al., 2011) coupling has become a versatile synthetic methodology and has been widely applied to the synthesis of numerous conjugated polymers.

2.1.3.2.3 CASSAR–HECK–SONOGASHIRA AND YAMAMOTO COUPLING

As shown in Scheme 2.2F, Pd-catalyzed Cassar–Heck–Sonogashira coupling of terminal alkynes and aromatic halides (bromides or iodides) is a rather special reaction, which is the most widely used method for C—C bond formation between a sp and a sp^2 carbon. The standard experimental conditions require the use of Pd(II) ($Pd(PPh_3)Cl_2$) or Pd(0) ($Pd(PPh_3)_4$) complexes as catalyst, an amine as base, triethylamine, piperidine, morpholine, or pyrrolidine being the most used, while CuI as the cocatalyst, the function of which is probable to form copper σ- or π-acetylides and activate the alkyne to the step of transmetalation (Bunz, 2000). The main drawback in the polymers prepared by this method is the formation of butadiyne defects. To reduce the defects, Pd(0) catalysts are preferred to Pd(II) complexes. In conclusion, this process represents the preferred methodology for incorporating conjugated structures containing triple bonds into the polymer backbone, which are known as poly(phenyleneethynylene)s (PPE) and, more generally, poly(aryleneethynylene)s (PAE). PAEs can be viewed as the dehydro analogs of PAVs, in which the ethyne groups impart an additional rigidity to the polymer structure. The rigid structure of the polymer backbone may facilitate the intermolecular charge transport and render the polymer better photostability. The other route to synthesize PAEs with high yields and polymerization degrees is the transition-metal-catalyzed metathesis of diynes. The representative catalytic conditions are molybdenum hexacarbonyl and 4-chlorophenol in 1,2-dichlorobenzene (Kloppenburg et al., 1998).

Besides the reactions mentioned previously, the nickel-mediated Yamamoto dehalogenation coupling reaction also provides a feasible route to prepare conjugated polymers from dihaloaromatic monomers, although only random copolymers can be obtained (Yamamoto et al., 1992). Generally the homocoupling of the dihaloaromatic monomers is promoted by Ni(0) catalyst [$Ni(cod)_2$ or generated in situ by reduction of $NiCl_2$ with zinc in the presence of triphenylphosphine]. In addition, 2,2′-bipyridine is normally used as a promoter of the polymerization process. For example, the simplest DPP-based D–A polymer **49** has been directly synthesized through this coupling reaction of dibromo DPP-based monomer **41** (Scheme 2.2G) (Li et al., 2012).

2.1.3.2.4 A NEW TOOL-DIRECT (HETERO)ARYLATION POLYCONDESATION

Although transition-metal-catalyzed cross-coupling reactions represented by Suzuki and Stille polymerizations are vital for the synthesis of π-conjugated polymers in the last decades, they still have several inevitable shortcomings. Conventionally, transition metal catalyzed polymerization needs to prepare two types of arenes: one arene is substituted with a leaving group (I, Br, OTf, etc.), while the other contains an organometallic moiety, such as —$B(OR)_3$, —SnR_3, —ZnR, or —MgX. Therefore, the introduction of these functional groups undoubtedly increases the reaction steps and the isolation and purification of the functional monomers consumes a great deal of time. Moreover, some compounds, such as Grignard reagents and stannyl reagents are unstable and very toxic, respectively. For these reasons, an economically attractive and ecologically benign method as an alternative to the traditional cross-coupling methodologies will be in high demand. In the very recent years, this new method for the synthesis of conjugated polymers, which is named direct arylation polycondensation (DArP) has emerged and becomes more and more popular (Mercier and Leclerc, 2013; Rudenko and Thompson, 2015; Facchetti et al., 2012; Okamoto et al., 2013b). DArP method, just as its name implies, is the direct coupling of a prefunctionalized arene bearing a leaving group with an arene C—H bond to form new C—C bonds. In the reaction system, transition metal catalyst is used for C—H bond activation and base is added to neutralize the stoichiometric amount of acid formed. In fact, C—H activation methods have been realized for over 50 years (Alberico et al., 2007), but until recently their function to synthesize conjugated polymer has been corroborated.

Lemaire et al firstly tried to use DArP to synthesize conjugated polymer in 1999 (Sévignon et al., 1999). They used four 2-iodothiophene derivatives **39** to produce the corresponding P3ATs (Scheme 2.3A). The general procedure consists of reacting the 2-iodothiophene derivative with catalytic amounts of palladium acetate and equimolar amounts of tetrabutylammonium bromide and K_2CO_3 as a base in DMF at 80°C. In all cases, the yields of the polymers were satisfactory but the molecular weights were relatively low, only oligomeric products with M_w about 6000 were obtained. The disappearance of ^1H NMR resonance signal at 6.98 ppm indicates the formation of HT regioregular P3ATs with about 90% regioregularity.

Over 10 years later, Ozawa et al. successfully polymerized 2-bromo-3-hexylthiophene **30** to obtain P3HT with high molecular weight and high yield by using DArP (Scheme 2.3B) (Wang et al., 2010). The polymerization conditions include palladium-based Herrmann's catalyst, *tris*(2-dimethylaminophenyl)phosphine as a ligand, Cs_2CO_3 as a base and tetrahydrofuran (THF) as a solvent. The reaction temperature is 125°C, which is above the boiling points of the solvent, so the reaction should be conducted in a pressurized vessel. After characterization, the obtained polymer was proved to be HT regioregular P3HT with high molecular weight (M_n = 30,600, PDI = 1.60) and high regioregularity (98%) in quantitative yields (99%). These results have brought about a strong interest in DArP from many researchers to this day. Now Ozawa conditions have been adopted with or without modifications by many research groups for the synthesis of a variety of conjugated polymers.

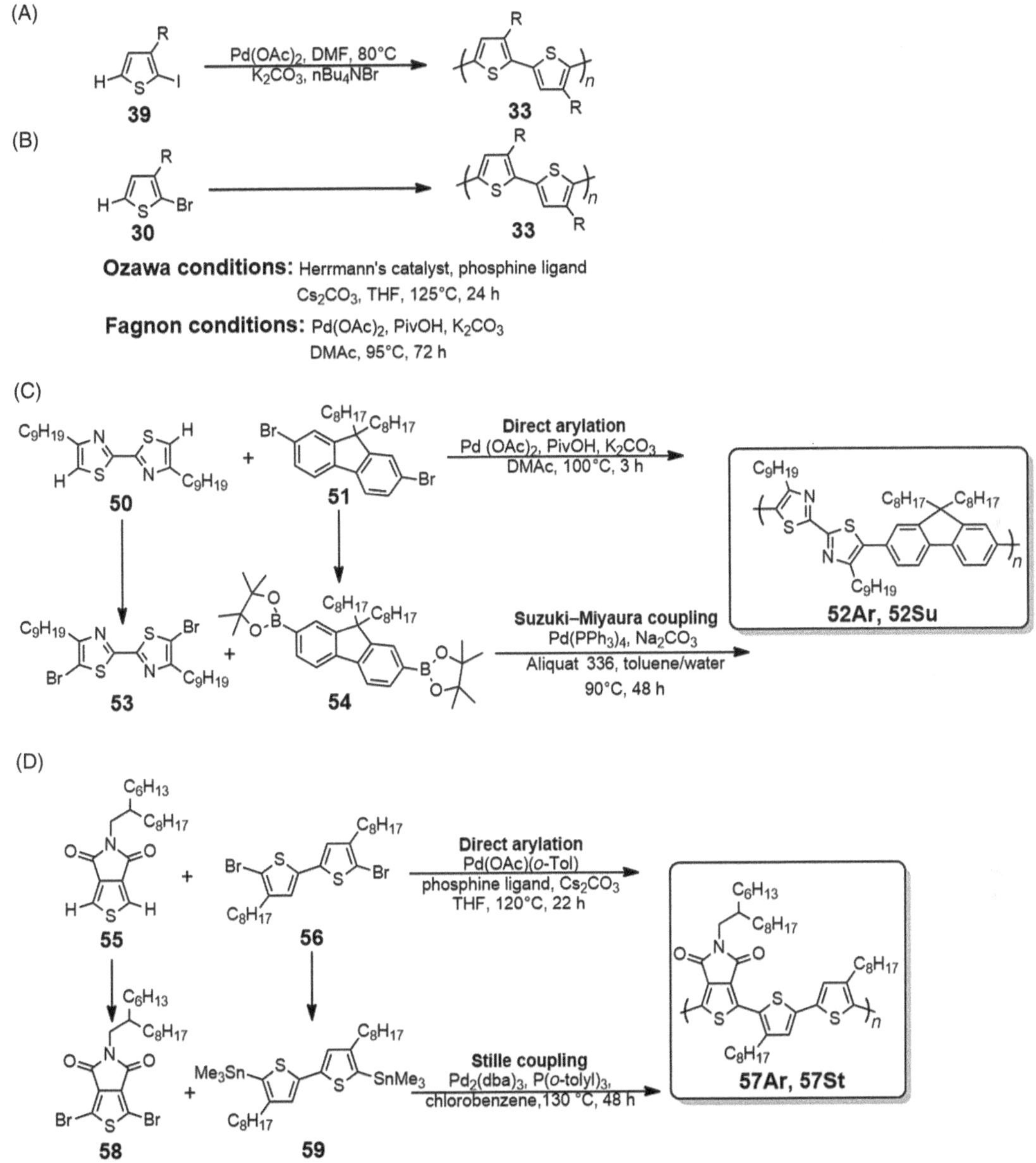

SCHEME 2.3 (A) The first attempt to utilize DArP for preparation of conjugated polymers. (B) Reaction conditions of DArP approach. (C) Comparison of the Suzuki versus DArP approach. (D) Comparison of the Stille versus DArP approach.

What is interesting is that the synthesis of conjugated polymers via DArP has two distinct sets of reaction conditions: one is Ozawa conditions and another is called Fagnou conditions. The Fagnou conditions was originally developed by Fagnou et al. to synthesize small molecules via direct arylation in 2006 (Lafrance and Fagnou, 2006) and then later extended to conjugated polymers. In this method, the catalyst is palladium acetate and phosphine ligand is not compulsory, solvent can be *N,N′*-dimethylacetamide (DMAc) or toluene, base is K_2CO_3 (Scheme 2.3B). And the most important thing lies that a carboxylic acid (usually, pivalic acid) should be employed as a carboxylate ligand/proton shuttle. Compared with the Ozawa conditions, the reaction temperature of Fagnou conditions is lowered to 70–110°C, which allows the polymerization to be performed at ambient pressure. We can easily find that Fagnou conditions are similar to the condition used by Lemaire et al. in 1999 except the presence of a carboxylate additive. Now it has been demonstrated that without the help of carboxylate additive the C—H activation reaction will not be realized because carboxylate additive is cardinal to induce an obvious decrease of transition state energy (Gorelsky et al., 2012; Gorelsky et al., 2008).

DArP is also a convenient way of preparing alternating conjugated copolymers. Copolymerization of 4,4′-dinonyl-2,2′-bithiazole **50** with 2,7-dibromo-9,9-dioctylfluorene **51** to give poly[(4,4′-dinonyl-2,2′-bithiazole-5,5′-diyl)-(9,9-dioctylfluorene-2,7-diyl)] **52Ar** via DArP was reported by Kanbara group (Scheme 2.3C) (Lu et al., 2012). They carried out the polycondensation under various conditions. By adjusting the catalyst amount, ligand and polycondensation time, they obtained the optimized conditions for polycondensation after comparing the yield and molecular weight of **52Ar**. The reaction was conducted in the presence of $Pd(OAc)_2$ (2 mol.%), pivalic acid (30 mol.%), and K_2CO_3 (2.5 equiv.) in DMAc for 12 h at 100°C, giving **52Ar** with a molecular weight of 45,900 (PDI = 2.32) in 90% yield. In this reaction, the addition of phosphine ligand gave a lower yield. **52Su** has also been prepared by other group, which was synthesized for 2 days, with a molecular weight of 18,400 (PDI = 2.5) in 69% yield via the Suzuki–Miyaura coupling reaction (Scheme 2.3C) (Lee et al., 2005). These results reveal DArP is more efficient than traditional cross-coupling reactions in terms of the molecular weight of the polymer and reaction time.

In 2012, Leclerc group reported the synthesis of thieno[3,4-*c*]pyrrole-4,6-dione (TPD)-based polymers **57Ar** by both direct heteroarylation polycondensation method and Stille coupling (Scheme 2.3D) (Berrouard et al., 2012). The Stille reaction was carried out in chlorobenzene for 2 days at 130°C, with $Pd_2(dba)_3$ as a catalyst and P(*o*-tolyl)$_3$ as a ligand. The yield of the desired copolymer **57Ar** was 71% but the M_n was only approximately 9000 (PDI = 1.6). This relatively low molecular weight may be related to the loss of some functional groups during the Stille polymerization. In parallel, for the preparation of **57Ar** using DArP, they tested several reaction conditions. The optimal results were obtained with ligand *tris*(2-methoxyphenyl)phosphine (8 mol.%) and catalyst Pd(OAc)(*o*-Tol) (4 mol.%). A high M_n of 56,000 (PDI = 2.6) and a yield of 96% were obtained. To further characterize and compare these copolymers prepared by two methods, optical absorption, ^{1}H NMR spectroscopy, thermal analysis, and thin-film X-ray diffraction were investigated. The ^{1}H NMR spectra are similar to those reported previously while X-ray diffraction patterns show the same features for the polymers prepared by two methods, which mean that **57Ar** and **57St** are similarly organized either in the solution or solid state. Their UV/Vis absorption spectra exhibit similar features with an absorption maximum at 464 nm (**57St**) and 474 nm (**57Ar**) in $CHCl_3$, and the thermal analysis indicates **57Ar** has greater crystallinity and higher melting point. These results are related to a higher molecular weight of **57Ar**, indicating that DArP method can lead to an increase in polymer molecular weight. Hence, the aforementioned examples clearly demonstrate the potential of the DArP method to synthesize various π-conjugated homo- and copolymers.

Although DArP for the synthesis of regioregular polymers as well as alternating copolymers has been realized for only several years, it shows a few promising advantages compared with the traditional methods. (1) The high costs and difficulties associated with the synthesis and purification of the organometallic reagents are reduced, therefore it is more economical. (2) Instable or high toxic organometallic reactants are no longer necessary, which makes the reaction very convenient to manipulate. (3) Unlike typical polymerization reactions that give rise to stoichiometric amounts of waste, the only byproduct of DArP is acid, so it is more ecological. (4) The molecular weights of the polymers prepared by DArP are generally higher than those obtained by classical transition-metal-catalyzed reactions.

However, together with other polymerization approaches, DArP does not yield polymers with perfectly ideal structures. The resulting polymers are known to contain a certain amount of homocoupling regiodefects or branched and cross-linked polymers, which originate from side reactions during the cross-coupling process. In the transition-metal-catalyzed reaction, the cross-coupling between an aryl nucleophile and an aryl electrophile is what we need, while the homocoupling of aryl nucleophiles and the homocoupling of aryl electrophiles produce the homocoupling regiodefects. For DArP, it is commonly accepted that the arylpalladium (and arylnickel) species which are generated after the oxidative addition step (Fig. 2.1B) have a tendency to undergo aryl-group exchange, which is called disproportionation. In other words, the arylpalladium species undergo transmetallation with the same arylpalladium species and finally produce the electrophile homocoupling byproducts, as illustrated in Fig. 2.1C (Yagyu et al., 2001; Remy et al., 2010; Suzaki and Osakada, 2003). In addition, the presence of reducing impurities or oxidative impurities (e.g., residual oxygen) in the reaction system can also induce the homocoupling of aryl electrophiles or nucleophiles, respectively. Looking from the previous literature, decreasing the reaction temperature or catalyst load and increasing the bulk of phosphine ligand could be a practicable way to control the homocoupling regiodefects.

DArP method has another kind of defects, branching defects (Okamoto et al., 2013a), which are absence in other classic methods of polymerization, such as Stille, Suzuki, Negishi, Kumada, etc. Five-membered ring monomers, such as thiophene, pyrrole, or furan have two categories of protons, that is, 2- or 5-position is α-proton and 3- or 4-position is β-proton. The monomers in transition-metal-catalyzed reactions have only one reactive organometallic nucleophilic group, which reacts with the leaving group, leading to linear chains of the polymers. Hence, the reaction is not related to the unsubstituted α- or β-proton. However, the situation is different for DArP method. DArP method is capable to activate more than one kind C—H bond if it is present in the monomers. As for the thiophene monomers, although α-proton is more reactive than β-proton, they all can be activated by the catalytic system. Activation of α-proton leads to linear polymer chains, which is deemed desirable for optoelectronic application. However, activation of β-proton leads to branching defects and further leads to cross-linking of polymer chains, which will result in lower polymer yields due to the low solubility. Branching defects always give rise to the disruption of polymer chain packing and their decreased ability to self-assemble into ordered structures. These changes impose greatly influence on the electronic and optoelectronic properties of target polymers, which are negative effects in most cases (Rudenko et al., 2014). In order to restrain branching defects, adjusting the reaction temperature, catalyst load, and carboxylate additive are conventional methods. Other strategies include introducing modified groups to block β-positions, such as, 3,4-dimethylthiophene and 3,4-ethoxylenedioxythiophene, or using directing groups on the monomer to ensure selective α-proton activation. Besides five-membered ring monomers, formation of branching points has also been observed in other monomers containing different C—H bonds which exhibit comparable dissociation energies. However, from another perspective, unselective C—H activation of different C—H bonds can be utilized to fulfill the purposes of preparing hyperbranched polymers.

After several years of evolution, direct (hetero)arylation polymerization has demonstrated its full potential to produce a huge variety of conjugated polymers in an environmentally benign, inexpensive, and straightforward way. Some recent reports have successful incorporated known or novel materials prepared by this method into organic electronic devices, such as OFETs, PLEDs, and PSCs. But it is noteworthy that DArP method is influenced by multiple factors, such as type of base and solvent. Moreover, the most efficient catalytic condition may vary for different monomers. Therefore, in order to obtain conjugated polymers with high yields and molecular weights and low PDIs, experimental parameters (base, solvent, reaction temperature, additive, catalyst loading, and ligand) have to be carefully optimized for each process. To summarize, application of direct (hetero)arylation polycondensation to polymerization is still in its infancy. However, it is believed that DArP may become a typical polymerization strategy for conjugated polymers with the rapidly growing efforts on this methodology in recent years.

2.1.4 Conclusion

The synthetic methods have great influences on the purity, molecular weight and regioregularity of the conjugated polymers, which are directly relevant to their performances in electronic devices. In this section, we briefly review the most commonly used synthetic methods of conjugated polymers. At the same time, the advantages and limitations of these methodologies are also discussed.

2.2 Design of Conjugated Polymers for Organic Electronics

2.2.1 Introduction

Conjugated polymers have attracted broad academic and industrial interest for various optoelectronic applications. Conjugated polymers possess obvious advantages toward inorganic materials and small-molecule organic materials, such as light weight, low cost, and flexibility. Furthermore, most polymer materials are soluble in common solvents thus polymer films can be prepared through solution process and printing, such as spin-coating, ink-jet and roll-to-roll, which is a critical issue for the realization of large-scale manufacture of printed electronics. In order to obtain the high-performance of polymer electronic devices, regulation of the processing technology parameters and manufacturing methods is a feasible approach, but the whole optimization process is complicated and the reproducibility is unsatisfying, which is not favorable for the application to large-scale production. In contrast, rational design and optimization of the molecular structure of the conjugated polymers will be more convenient and effective.

2.2.2 Rational Design of High-Performance Conjugated Polymers for PSCs

Bulk heterojunction (BHJ) PSCs are commonly composed of a blend layer of a conjugated polymer donor and a fullerene derivative acceptor sandwiched between a poly(3,4-ethylenedioxythiophene) (PEDOT)/poly(styrenesulfonate) (PSS) modified indium tin oxide (ITO) positive electrode and a low work function metal negative electrode. The most ubiquitously acceptor for current BHJ solar cell research is [6,6]-phenyl-C_{61}-butyric acid methyl ester (PCBM) or its higher fullerene analog C_{70} PCBM ($PC_{71}BM$) (Hummelen et al., 1995; Wienk et al., 2003). Therefore, the present studies of the PSCs are mainly focused on the design and synthesis of high-efficiency conjugated polymer donor photovoltaic materials.

The properties desired for ideal p-type conjugated polymers are: (1) proper highest occupied molecular orbital (HOMO)/lowest unoccupied molecular orbital (LUMO) energy levels and high hole mobility to promise efficient charge transfer, (2) a narrow bandgap to guarantee sensitivity to broadband spectrum, (3) compatibility with fullerene acceptor for the generation of nanoscaled morphology, (4) good solubility in common solvent for easy fabrication of film, (5) good stability to withstand degradation. Actually these mentioned requirements are not isolated with each other. For instance, the change of LUMO and HOMO energy levels will alter the energy bandgap at the same time. As a result, it is a formidable task to improve all the factors together and much work remains to be done.

From the view of polymer structure, conjugated polymer photovoltaic materials are composed of π-conjugated backbone, flexible side chains, and the substituents. The physical properties of the conjugated polymer, such as energy levels, bandgap, and inter-/intramolecular interactions depend a lot on the polymer backbone. Thus the conjugated backbone of the polymer must be properly designed for the further development of PSCs. On the other hand, side chains have been primarily utilized as solubilizing groups in organic photovoltaic conjugated polymers. But then it is found that polymer absorption, emission, energy level, carrier mobility, nanoscale morphology, and interfacial contact also have relationship with the side chains of conjugated polymers. We can say that sorting out suitable side chains has the same importance as sorting out conjugated backbones. Last but not least, substituents are also a potential method to adjust the properties of conjugated polymers, and then enhance the performance of the PSCs.

Due to the immense amount of work performed in this area, this section could not cover all the relevant research. Thus readers can refer to some reviews for a more detailed introduction about the design of conjugated polymers for PSCs (Cheng et al., 2009; Zhou et al., 2012; Mei and Bao, 2014; Li, 2012).

2.2.2.1 Backbones

To carefully rectify the bandgap of conjugated polymers used in solar cell, a lot of efforts have focused on their backbones (Roncali, 2007; Roncali, 1997). Homopolymer, such as polyphenylene **60**, PPV **10**, and PT **61** (Fig. 2.2A) consists of a single repeating aromatic unit, which has two possible resonance structures for the ground state with nondegenerate energy. One is called the aromatic form, which can transfer into the quinoid form via π-electron delocalization along the conjugated chain. Since the quinoid structure destructs the aromaticity and decreases the stabilization energy, this kind of structure possesses narrower bandgap and higher energy in contrast to aromatic form. **60** has a high bandgap of c.

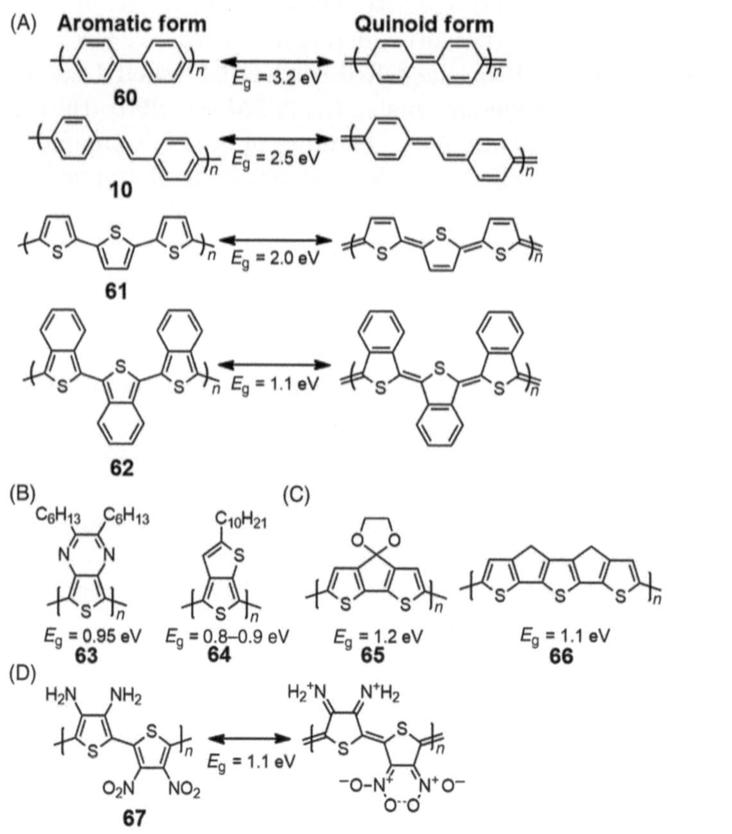

FIGURE 2.2 (A) Aromatic and quinoid resonance forms of poly(phenylene) (**60**), poly(*p*-phenylenevinylene) (**10**), polythiophene (**61**), and polyisothianaphthene (**62**). (B) Chemical structures of **63** and **64**. (C) Chemical structures of **65** and **66**. (D) Chemical structure of **67** with its resonance structure.

3.2 eV since the benzene rings have a high degree of aromaticity. In contrast, the bandgap of 10 is decreased to c. 2.4 eV since a double bond is inserted into the backbone to offset the influence brought by the benzene rings and as a result the aromaticity is reduced. In comparison to benzene, thiophene has lower aromaticity and narrower bandgap of 2.0 eV since it can adopt a quinoid form easier. The most creative way to effectively increase the quinoid character of **61** is to fuse another aromatic ring at the β-position of the thiophene ring. This concept was first illustrated by polyisothianaphthene **62** (Fig. 2.2B), which has a narrow bandgap as low as 1.1 eV (Wudl et al., 1984). In this polymer, benzene has a larger aromatic resonance energy than thiophene, thereby the thiophene ring in **62** is more likely to keep the quinoid form while maintaining the aromaticity of benzene to some extend. In a similar approach, polymers **63** (Pomerantz et al., 1993) and **64** (Pomerantz and Gu, 1997) (Fig. 2.2B) also reveal a low bandgap of 0.95 eV and 0.8–0.9 eV, respectively.

Planarization between adjacent aromatic units so as to extend conjugation and facilitate delocalization is also a useful strategy to reduce the bandgap. A straightforward method is to fix the adjacent aromatic units with the help of covalent chemical bonding. For example, bithiophene unit is rigidified by grafting a dioxolane moiety at the bridging sp^3 carbon to obtain polymer **65** (Fig. 2.2C) with a low bandgap of 1.2 eV (Brisset et al., 1994). Another polymer **66** (Fig. 2.2C), where thiophene units are extended to terthiophene through bridged methylene groups, exhibits a remarkably low bandgap of 1.1 eV (Roncali and Thobie-Gautier, 1994). Generally speaking, the increase of conjugation length will lead to the decrease of energy difference between LUMO and HOMO. But this effect will become unconspicuous when the number of monomer units gets to a saturated value. Therefore, unlimited extension of the conjugation length is not necessary by virtue of a limited reduction of the bandgap.

A more powerful strategy to lower the bandgap and tune the electronic energy levels of the conjugated polymers is introducing an alternative structure of conjugated electron-rich donor (D) and electron-deficient acceptor (A) into one polymer backbone (Havinga et al., 1993). Usually, the strong push-pull interaction between donor and acceptor leads to the internal charge transfer, which is favored for the delocalization of π-electrons and therefore makes it easier for the generation of quinoid mesomeric structure ($D–A \rightarrow D^+ = A^-$) at the polymer backbone. These effects lead to a narrowing of the optical bandgap of the D–A polymer according to the rules of perturbation theory. The representative example of this strategy is copolymer **67** (Fig. 2.2D) of 3,4-aminothiophene and 3,4-nitrothiophene reported by Tour et al. (Zhang and Tour, 1998). The bandgap of the polymer is reduced to 1.1 eV with the help of the high extend of quinoid and zwitterionic character, which is caused by the coexistence of electron-withdrawing and electron-donating groups. In addition, the D–A copolymers possess an intramolecular charge transfer absorption band at longer wavelength direction, so the absorption of the copolymers is broadened, which results in the enhancement of the utilization efficiency of sunlight. Furthermore, D–A polymers have highly localized HOMO and LUMO energy levels on donor and acceptor moiety, respectively. This character brings beneficial opportunities to adjust the energy level and bandgap through appropriate choosing of donor and acceptor units. Due to aforementioned advantages, D–A polymers have become the most popular research field of PSCs, so we will focus on the design in this section.

A low HOMO energy level and a narrow bandgap of D–A polymer is required in the application of BHJ solar cells to guarantee high open-circuit voltage (V_{oc}) and short-circuit current (J_{sc}). Therefore, a "weak donor" should be employed to maintain a low HOMO energy level, which is almost independently decided by the donor of the D–A polymers. In contrast, a "strong acceptor" is needed to reduce the bandgap of the polymers via intramolecular charge transfer. So a "weak donor–strong acceptor" strategy is proposed to design ideal D–A polymers. As a matter of convenience, some commonly used donors and acceptors are summarized in Figs. 2.3A and 2.3B, respectively.

Fluorene has good physical and chemical stability, high charge carrier mobility, and high absorption coefficients. At the same time, its rigid, planar backbone leads to the good intermolecular interaction and stacking of fluorene-related conjugated polymers. Furthermore, the fluorene unit is a very weak electron donor (HOMO energy level around −5.5 eV) because of the relatively electron-deficient benzene units, so V_{oc} of polyfluorene (PF)-based BHJ solar cells can generally reach around 1 V. With these notable features, fluorene becomes one of the most popular donor units used in D–A polymers for PSC. For example, PFDTBT (**68** in Fig. 2.4A) based BHJ cells showed a V_{oc} of 1.04 V and a power conversion efficiency (PCE) value of 2.2% (Svensson et al., 2003). However, these polymers usually have relatively large bandgaps, which are not ideal for efficient light harvesting.

Structurally analogous to fluorine, cyclopentadithiophene (CPDT) derivatives, where two thiophene units are tied and rigidified by a covalent carbon, also have attracted considerable interests. Due to the two thiophene rings, CPDT unit is much more electron-rich than fluorene, leading to a much stronger intramolecular charge transfer between the CPDT donor unit and the acceptor unit. Thus, the bandgap of the corresponding D–A polymers display a significant decrease. However, the relatively high HOMO energy level of the CPDT units inevitably results in a low V_{oc} (0.5–0.6 V). When the carbon atom at the 9-position of the CPDT unit is replaced with a silicon atom, a new type of building block, namely, dithienosilole evolves. Since carbon and silicon have similar electron negativity, dithienosilole has a similar electron-donating ability with CPDT. This explains the fact that HOMO levels and bandgaps of Si-PCPDTBT (**71** in Fig. 2.4A) and PCPDTBT (**70** in Fig. 2.4A) are nearly identical (Coffin et al., 2009). However, the seemingly small change from carbon to silicon leads to a better stacking of the polymer backbone demonstrated by the absorption spectra and grazing incidence X-ray study of polymer thin films. Subsequently, this enhanced aggregation helps to reduce the formation of charge-transfer complexes and improve the hole and electron mobility. Thus, the solar cell made from **71** had a pronounced increase in both V_{oc} and J_{sc} with the same processing condition, leading to a high PCE of 5.9% (vs. 2.7% for **70**) (Coffin et al., 2009).

Similar to the beneficial molecular design on going from CPDT to dithienosilole, fluorene is also further modified by substituting the carbon atom with a silicon bridge to generate silafluorene. Compared to **68**, PSiF–DTBT (**69** in Fig. 2.4A) showed both higher J_{sc} and a PCE (5.4%), though the V_{oc} was slightly decreased (Wang et al., 2008). However,

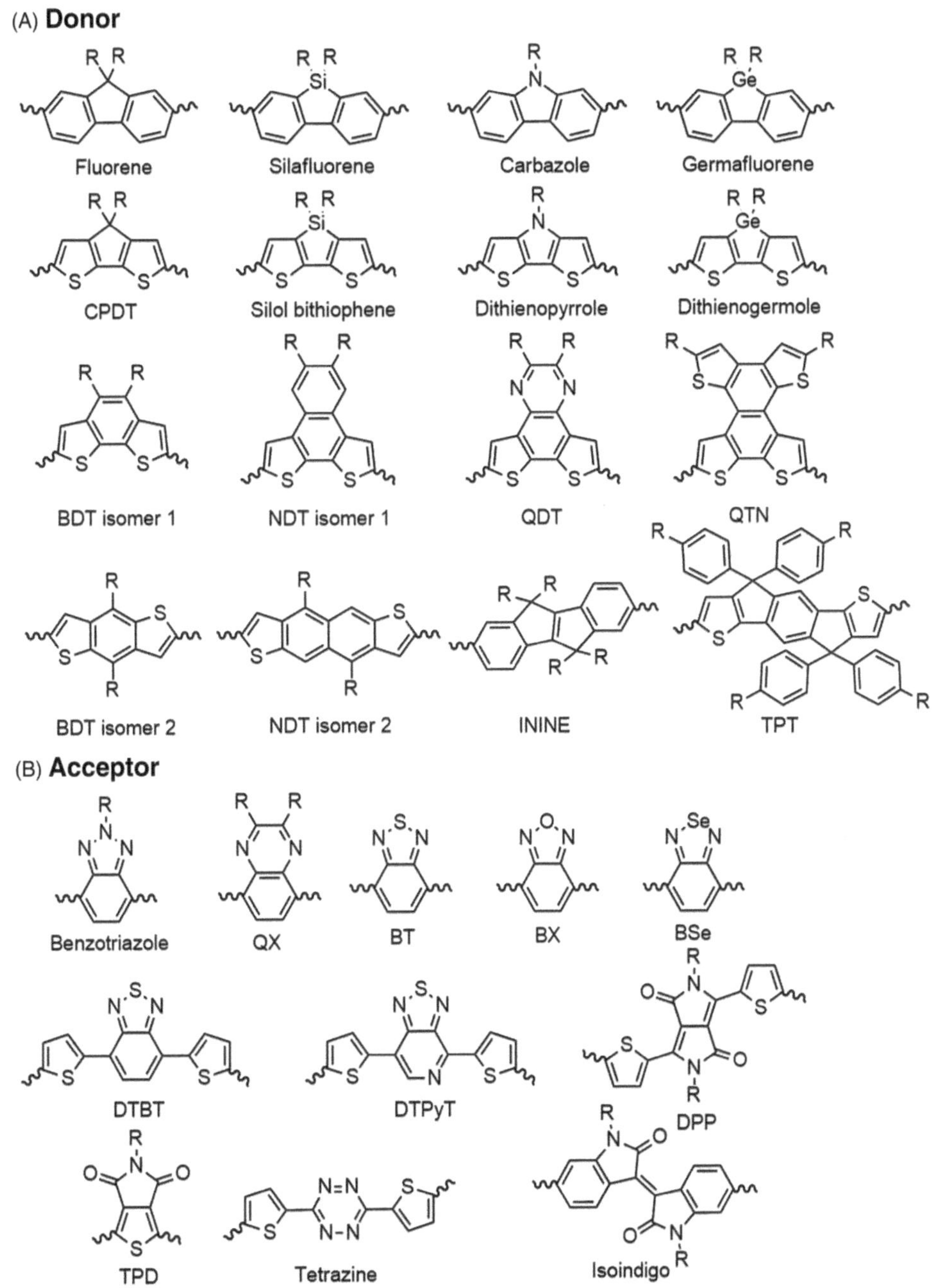

FIGURE 2.3 Chemical structures of some donor (A) and acceptor (B) units discussed in Section 2.2.2.1.

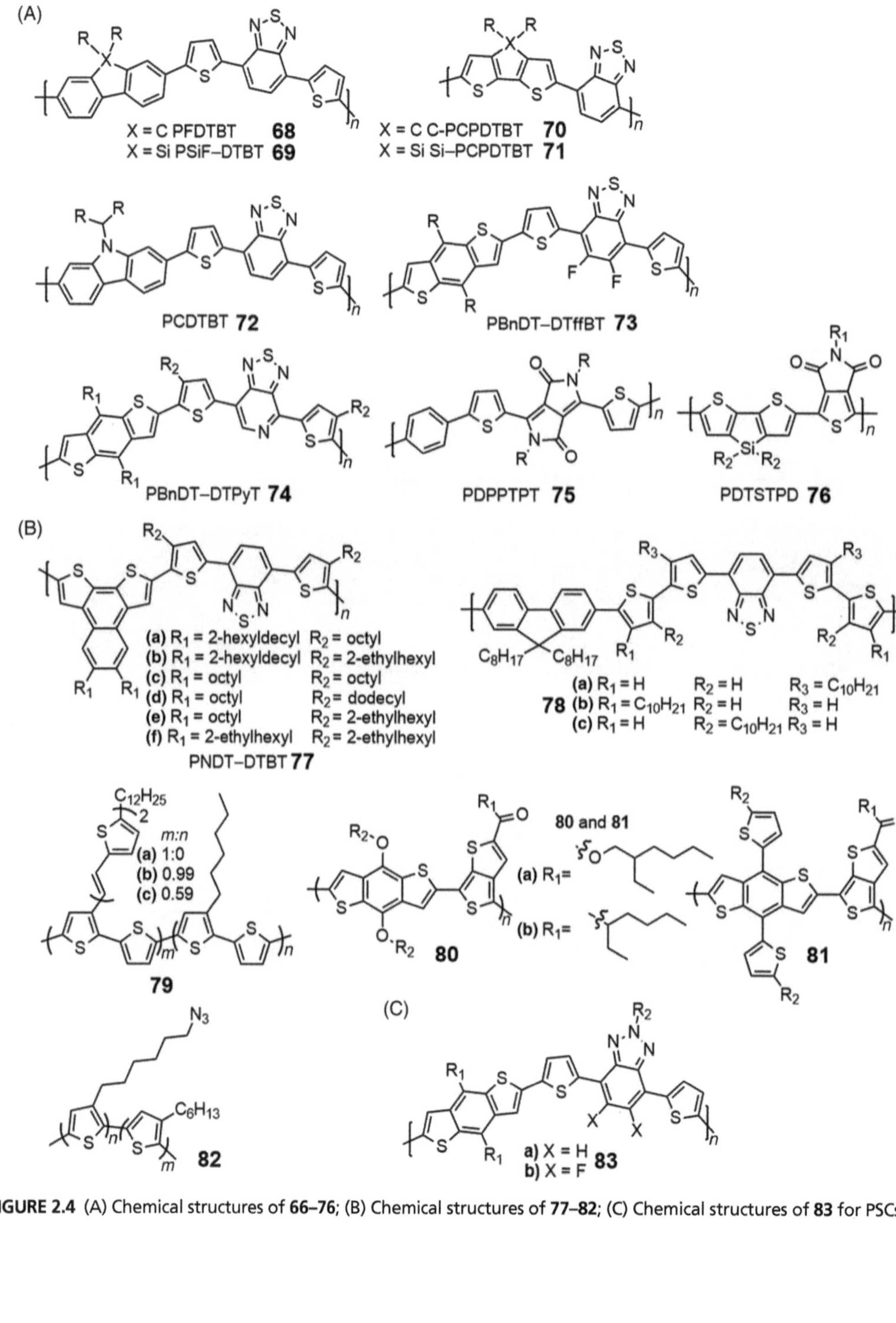

FIGURE 2.4 (A) Chemical structures of **66–76**; (B) Chemical structures of **77–82**; (C) Chemical structures of **83** for PSCs.

substituting the center carbon in the fluorene skeleton with a nitrogen atom (i.e., converting the fluorene into the carbazole) leads to undesired electronic and optical properties. Electronically, the carbazole unit is more electron-rich than fluorene so the HOMO energy levels of carbazole-based D–A polymers would be closer to the ideal HOMO level. Moreover, carbazole derivatives have been widely used as p-type semiconductors due to their excellent thermal and photochemical stabilities, relatively high charge mobility, and easy availability. As a result, a carbazole-based polymer PCDTBT (**72** in Fig. 2.4A) had a V_{oc} of 0.88 V, a J_{sc} of 10.6 mA cm^{-2}, and high fill factor (*FF*) of about 66%, resulting in a PCE of over 6% in the BHJ devices (Park et al., 2009). Besides, by changing the bridging atoms in fluorene or CPDT, dithienopyrrole, dithienogermole, or germafluorene have also been explored as donor unit to prepare D–A polymers (Fig. 2.3A).

Actually, fluorene or CPDT unit discussed previously can be considered as fused benzene or thiophene rings by a bridging atom, respectively. This is an effective method to extend conjugation and facilitate delocalization, resulting in a narrow bandgap and a broad light absorption. Meanwhile, the resulting coplanar geometries and rigid structures can suppress the rotational disorder around interannular single bonds and lower the reorganization energy, which help to enhance the charge mobility. When benzene unit and two thiophene units are fused together, a new donor benzodithiophene (BDT) (and its isomers) can be obtained. In comparison with fluorene and CPDT, BDT units offer two advantages: the center benzene ring can be alkylated easily to improve the solubility of resulting polymers, and the thiophene units besides is able to induce less steric hindrance toward acceptor units, increasing the planarity of the backbone. Moreover, both of the isomers of BDT exhibit weaker electron-donating ability than that of the CPDT. As a result, some BDT-based polymers possess a HOMO level that is closer to the ideal value and a small amount of them have indicated PCE over 60% in BHJ devices. Specially, BHJ devices constructed by PBnDT–DTffBT (**73** in Fig. 2.4A) have demonstrated a V_{oc} about 0.9 V and a J_{sc} over 12 mA·cm^{-2}, resulting in a PCE that is more than 7% (Zhou et al., 2011). By utilizing extended π-conjugation method, more donor units with suitable HOMO energy level can be obtained. For example, indenoindene (ININE), thiophene/phenylene/thiophene (TPT), naphthalenedithiophene (NDT), dithienoquinoxaline (QDT), and quadrathienonaphthalene (QTN) have been reported in the design of D–A polymers (Fig. 2.3A).

Compared to the donor units, the acceptor units are of equal importance in controlling the energy levels and bandgaps of conjugated polymers. The most commonly employed acceptor unit is the 2,1,3-benzothiadiazole (BT). The two N atoms in BT could possibly form intermolecular hydrogen bonding, leading to a more planar backbone. Along with its strong electron-accepting ability, BT acceptor unit has been successfully used to synthesize D–A polymers with low bandgaps and good photovoltaic properties. The representative example comes from **71**, which has been shown ahead (Coffin et al., 2009).

Adding one thienyl group on both sides of the BT converts BT into DTBT, which has a few more advantages when compared with the original BT unit. The two flanking thienyl units eliminate the otherwise severe steric hindrance between the BT unit and donor aromatic units, which can increase the planarity and conjugation of the polymer backbone. It is well known that a more planar conjugated backbone is good for the chain-chain interactions among polymers in the solid state, thereby improving the charge carrier (usually hole) mobility. On the other hand, compared to BT unit, the flanking thienyl units in BDT can provide additional possible positions for further modification of electronic properties and solubility of conjugated polymers, with minimal steric hindrance involved if properly introduced.

Similar to the design of donor units, the electronic properties of acceptors can also be fine-tuned by replacing specific atoms. There are several units which can be treated as structurally related to the BT, for example, benzotriazole, quinoxaline (Qx), benzoxadiazole (BX), and 2,1,3-benzoselenadiazole (BSe) (Fig. 2.3B). However, these acceptors usually do not perform as well as BT-based analogs, mainly due to their relatively weak electron acceptability. In contrast, by replacing the benzene in the BT unit with pyridine, the new acceptor, thiadiazolo[3,4-*c*]-pyridine (PyT) becomes a relatively strong acceptor. Therefore, polymers incorporating PyT unit and "weak donors" show noticeably reduced LUMO levels, slightly reduced HOMO levels, and thus lower bandgaps than those of their BT counterparts. Polymer PBnDT–DTPyT (**74** in Fig. 2.4A) achieved a high PCE of 6.3% in its BHJ device, demonstrating the great utility of PyT acceptor moiety in designing high-performance photovoltaic materials (Zhou et al., 2010).

DPP and TPD have also been used as acceptors in a number of D–A conjugated polymers in the past several years. The electron-withdrawing amide groups render DPP and TPD strong electron acceptability. Their planar structure can improve the extended delocalization of π-electrons along the backbone and enhance the polymer chain–chain interactions, which helps to increase the hole mobility. Therefore, DPP- or TPD-based D–A polymers always have noticeable photovoltaic properties, for example, Janssen and coworkers reported a DPP-based polymer PDPPTPT (**75** in Fig. 2.4A) with a PCE as high as 5.5% (Bijleveld et al., 2010); and more recently, a PCE up to 7.3% was produced from TPD-based polymer PDTSTPD (**76** in Fig. 2.4A) (Chu et al., 2011). In addition, the other strong acceptors, such as tetrazine and isoindigo are also employed to construct low-bandgap polymers.

2.2.2.2 Side Chains

Recently, more and more studies have started to focus on side chain engineering, and various flexible chains have appeared in conjugated polymers. These side chains can be categorized based on their compositions, namely, alkyl, hybrid, ionic, oligoether, fluoroalkyl, and latently reactive side chains. In this section, we will introduce some representative side chains used in conjugated polymers, and some recent examples in which side chain engineering has been successfully employed to improve device performance will be highlighted.

Alkyl chains are the most commonly used side chains in conjugated polymers. They can be easily introduced to the conjugated polymers, which could improve the solubility and processing of the polymer without changing the conjugation of the main chain. Linear and branched alkyl chains are the two forms of alkyl chains. Hexyl, octyl, and dodecyl chains are the most frequently used linear chains. In other words, linear alkyl chains with even numbers of carbon atoms are more widely used than those with odd-numbered carbon atoms, presumably due to their commercial availability. It has been observed that branched alkyl chains have superior solubility than linear alkyl chains when both chains have identical molecular formulas, but linear chains can promote interchain interdigitation due to less steric hindrance. The most commonly used branched alkyl chains include 2-ethylhexyl, 2-hexyldecyl, 2-octyldodecyl, and 2-decyltetradecyl. These chains with more than one methylene group through the branching point always have good solubilizing capability without disrupting the $\pi-\pi$ interactions, which are particularly important when designing conjugated polymers for charge transport.

The size and length of various alkyl chains are different. Will these differences affect the photovoltaic properties of the corresponding polymers? The first quantitative analysis of the influence from the size of side chains on photovoltaic properties of polymer-fullerene solar cells was carried out by You and coworkers (Yang et al., 2010). They synthesized a series of polymers PNDT–DTBT (**77a–f** in Fig. 2.4B) with identical conjugated backbone but different alkyl chains (i.e., size and branching). The results demonstrate that the side chains significantly impacts the photovoltaic characteristics of the corresponding BHJ devices with variations as much as 100%, depending upon the length and shape of these alkyl chains. Among the six polymers, **77b** with the long and branched side chains (C10,6–C6,2) displayed the highest V_{oc} of 0.81 V in its BHJ cells, while **77c** contained two slim linear side chains exhibited the lowest V_{oc} of 0.41 V. It was suggested that the weak $\pi-\pi$ interactions among the conjugated polymer backbones enhanced the V_{oc}. However, the charge transport is adversely affected when the $\pi-\pi$ interactions are weakened, generating a lower J_{sc}. The BHJ cells based on the C6,2–C6,2 polymer demonstrated a PCE of 3.36% with V_{oc} of 0.69 V and J_{sc} of 10.67 mA·cm^{-2} because an appropriate balance between V_{oc} and J_{sc} was generated with the help of both short and branched side chains.

The influence of side chain length has been systematically researched as well. For example, Gadisa and coworkers studied the morphological, bipolar charge-carrier transport, and photovoltaic characteristics of regioregular P3AT:PCBM blends (Gadisa et al., 2009). The P3ATs include poly(3-butylthiophene) (P3BT), poly(2-pentylthiophene) (P3PT) and P3HT. Solar cells with these blends delivered similar order of photo-current yield, while the *FF* values were 0.529, 0.624, and 0.675 for P3BT:PC$_{61}$BM (1:0.8), P3PT:PC$_{61}$BM (1:1) and P3HT:PC$_{61}$BM (1:1), respectively. But as the side chain length increased, V_{oc} increased mildly from 0.539 to 0.549 and 0.574 V. Bipolar measurements made by field-effect transistor shows a decrease in the hole mobility and an increase in the electron mobility with increasing alkyl chain length, which is positively related to the degree of phase separation revealed by morphological studies. Since balanced charge transport is only achieved in the P3HT:PCBM blend, it possesses a superior photovoltaic property.

It is also found that the position of alkyl chain has an obvious effect on the molecular weights and photovoltaic performance of the resulting polymers. Cao et al. synthesized a series of internal donor–acceptor type of copolymers **78** containing BT and four thiophenes incorporating side chains on different position as shown in Fig. 2.4B (Wang et al., 2009).

78b was obtained with a significantly lower molecular weight due to the steric hindrance effect of the alkyl chains close to the fluorene units during polymerization, which resulted in low PCEs of related BHJ cells when compared with 1.82% for **78a** and 2.63% for **78c** based BHJ solar cells.

All the mentioned examples indicate that the charge transport, as well as V_{oc}, J_{sc}, and *FF* can be modulated by the change of π–π interactions and interchain interdigitation caused by the length, size, and position of the side chains. Unfortunately, it is still difficult to predict the proper chain bulkiness and length for a given conjugated backbone due to the complex morphological behaviors of these polymers and the effects of molecular weight and purity.

Hybrid side chains are more various compared to alkyl chains since the various decorated functional groups are able to bring special characters. Actually hybrid side chains comprise of two categories: the first category possess functional groups directly tied to the backbones and this category can be further divided into electron-donating, electron-withdrawing and conjugated side chain; the second category possess terminally functionalized side chains.

Electron-donating side chains are primarily chosen to raise the HOMO energy levels by donating some electron density to the conjugated polymer backbones through connecting moieties, even though they may also affect LUMO levels. Furthermore, these side chains are able to keep the backbone to a coplanar structure with the help of intramolecular interactions. While if the moiety is bigger than a methylene unit, steric hindrance will be introduced to disrupt the π–π interactions. Alkoxyl side chains are the most frequently used electron-donating side chains.

Electron-withdrawing side chains are able to withdraw electron density from π-conjugated backbones and the energy level of the material is lowered. Similar to electron-donating moieties, electron-withdrawing side chains can also generate intramolecular interactions.

For broadening the absorption spectra of the conjugated polymers, a 2-D conjugated concept can be utilized by attaching a conjugated side chain on polymer backbones. Li and Hou et al. had synthesized a series of 2-D conjugated PT derivatives **79a–c**, as shown in Fig. 2.4B (Hou et al., 2006). The absorption spectrum of the resulting polymers can be tuned by controlling the ratio of the conjugated versus linear side chains. **79c** with bi(thienylenevinylene) conjugated side chains shows a broad absorption plateau in the visible region from 350 to 650 nm, and its absorbance is stronger than that of P3HT in the range 350–480 nm. In addition, the HOMO energy level of **79c** drops by c. 0.2 eV in comparison with P3HT, which is beneficial to higher V_{oc} of the PSCs. The PSC device based on **79c**/PCBM demonstrated a PCE of 3.18% with a higher V_{oc} of 0.72 V. In contrast, the PSC based on P3HT displayed PCE of 2.55% and V_{oc} of 0.60 V under the same experimental conditions.

The scope of this method was extended to include a donor/acceptor-type conjugated polymer by replacing the alkoxy group with an alkylthienyl group on the donor unit BDT (Huo et al., 2011). The 2-D conjugated **81a** and **81b** (Fig. 2.4B) exhibited higher thermal stabilities, red-shifted absorption spectra, lower HOMO levels, significantly higher hole mobility, and greatly improved photovoltaic properties, in comparison with the two corresponding alkoxy substituted copolymers **80a** and **80a**. For example, the PCE of the PSC based on **81b** reaches 7.59%, which is superior to that of 6.43% based on **80b**. These results reveal that the 2-D conjugated concept will be a promising method for the design of high-performance photovoltaic copolymers.

Unlike the first hybrid side chain category, terminally functionalized hybrid side chains have functional groups at the chain ends. These side chains can be used to change the film's morphology without affecting the polymers' electronic properties directly, because these functional moieties are far away from the conjugated backbones.

Ionic, oligoether, or fluoroalkyl side chains are also introduced to the conjugated polymers to improve their solubility, charge-carrier transport and photovoltaic characteristics. However, the need for polymer films with improved morphological stability (e.g., solvent resistance and thermal and photochemical stabilities) and patterning capabilities is important as well. Therefore, new polymers with various latently reactive side chains are developed. Kim et al. synthesized PT with azide groups attached to the end of the alkyl chain (**82** in Fig. 2.4B) (Kim et al., 2012). Photo-cross-linking of **82** dramatically improved the solvent resistance of the active layer without disrupting the molecular ordering and charge transport. Furthermore, the BHJ organic photovoltaics containing this polymer showed an average PCE above 3.3% after 40 h annealing at an elevated temperature of 150°C, which represents one of the most thermally stable organic solar cells reported to date. Although various latently reactive side chains have been reported, they have not been successfully implemented in high-performance conjugated polymers yet, especially for organic solar cells or organic field-emission transistors. However, these developments are crucial for future commercialization of organic electronics with required stability and reliability.

2.2.2.3 Substitutes

Substituents can be used to tune energy level, bandgap, molecular interaction, and even morphology as well. You et al. reported two polymers **83a** and **83b** (Fig. 2.4C) incorporating BnDT as the donor and either benzotriazol (HTAZ) or its fluorinated analog (FTAZ) as the acceptor (Price et al., 2011). The energy gap of both polymers (2.0 eV) is slightly higher than that of P3HT (1.9 eV). Surprisingly, in spite of a higher bandgap, the high molecular weight together with the high hole mobility of **83b** elevates the J_{sc} of the BHJ devices to be over 12 mA·cm^{-2}. **83b** based BHJ devices demonstrated higher J_{sc} and *FF* than the devices constructed by the polymers that didn't have fluorine substituents (PBnDT–HTAZ, **83a**) at comparable device thickness. The highest PCE could reach 7.1% in BHJ solar cells of **83b**:PC$_{61}$BM. Remarkably, when the thickness of active layer reached 1 μm, the PCE still maintained over 6%. Several other studies also reveal that the fluorine substitute could contribute to an enhanced PCE performance (Chen et al., 2009; Son et al., 2011). These interesting effects of substituents on physical properties of conjugated polymers are still under active investigation.

2.2.3 Rational Design of High-Performance Conjugated Polymers for PLEDs

The first light-emitting diode (LED) using a conjugated polymer as emissive material was reported by Burroughes et al. in 1989 using PPV (Burroughes et al., 1990). Later, Heeger and coworkers developed a solution-processable fabrication method for PLED (Gustafsson et al., 1992). Since then PLEDs have attracted much attention in both academic and industrial communities by virtue of their potential applications in displaying and solid state lighting. The color of light, quantum efficiency of light emission, turn-on voltage, and stability of the PLED are the main requirements for the application as commercial light emitting devices, therefore rational design of conjugated polymers is essential for good performance. For display applications, full color of light emitting need to be realized. According to optical principles, all the colors can be achieved by mixing the three principle colors (red, green, and blue), so considerable efforts have been devoted to developing red, green, and blue light-emitting polymers in the last 20 years. Nowadays, all three principle colors have been demonstrated in PLEDs. However, compared with green and red polymers, blue light-emitting polymers have poorer electroluminescence property and stability for commercial use, which becomes the obstacle of the full color display. In addition, it is well-known that blue light can be converted to green or red with proper dyes by fluorescence resonance energy transfer, which means a blue PLED alone may generate all colors while green or red cannot be converted to blue by the same method. So, we first focus on the design of blue light-emitting materials in this section.

2.2.3.1 Blue Light-Emitting Polymers

At present, widely investigated blue light-emitting polymers mainly include PFs, PTs, PPVs, and poly(*para*-phenylene)s (PPPs). Among these polymers, PFs are a class of promising blue light-emitting materials due to their attractive properties. So, we take them as examples to discuss the design strategies for blue light-emitting materials, which are also helpful for other color light-emitting materials. PFs exhibit high photoluminescence (PL) quantum yields, charge-carrier mobility, and good processability. Meanwhile, they can be easily chemically modified by introducing functional groups to the side groups at the C-9 positions with other conjugated units. Therefore, they are widely used as active materials in PLEDs and PSCs. However, the application of PFs in PLEDs has been hampered by the appearance of long-wavelength green emission, leading to the problems of color purity and stability. Hence, careful design of the structure of the fluorene-based conjugated polymers is an important issue concerning the realization of real blue-emitting PLEDs.

The first PLED for blue color emission, poly(9,9′-di-*n*-hexylfluorene) (**84a** in Fig. 2.5A), was fabricated in 1991 (Ohmori et al., 1991). The EL spectrum of **84a** exhibits the emission maximum at 470 nm with a shoulder at 420 nm and a full width at half maximum (FWHM) of c. 200 nm. The broad emission spectrum was ascribed to the interchain interaction between different chromophores. In order to reduce the interchain interaction, Huang et al. first introduced spirobifluorene units into PFs and synthesized a soluble

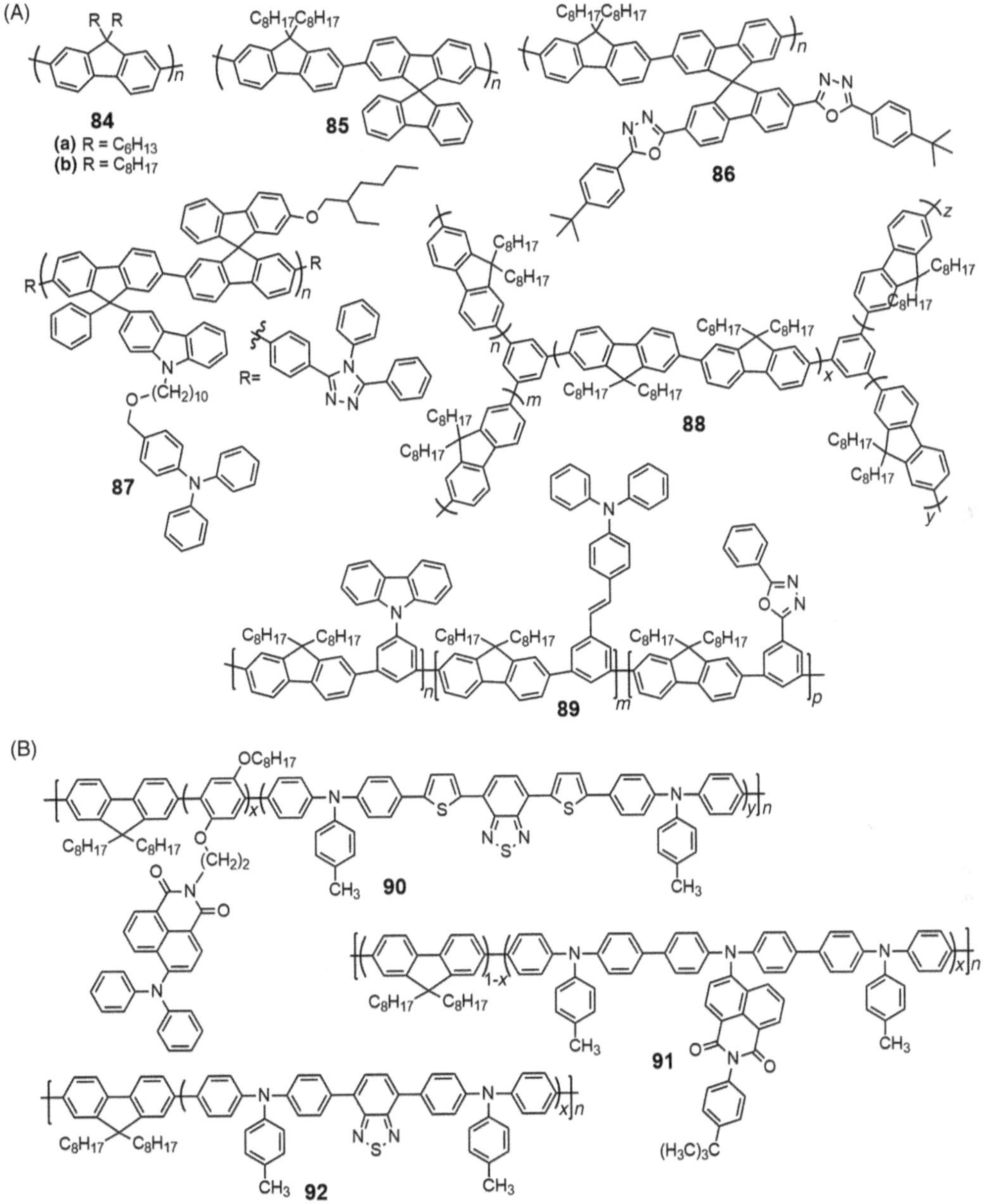

FIGURE 2.5 (A) Chemical structures of blue light-emitting polymers **84–89**. (B) Chemical structures of white light-emitting polymers **90–92**.

spiro-functionalized polymer **85** (Yu et al., 2000). The dihedral angle between two connected fluorene moieties is 90 degree due to the steric hindrance. This structural character minimizes the close packing of spiro-annulated molecules in the solid state, leading to good PLED performance. The T_g of **85** is 105°C, which is much higher than that of poly(9,9′-di-*n*-octylfluorene) (**84b** in Fig. 2.5A) (75°C) (Grell et al., 1997). The higher T_g can help to improve the spectral stability. Furthermore, FWHM of the EL spectrum of **85** is narrower than conventional PFs, so pure blue emission can be obtained. The pure blue emission can be attributed to the amorphous structure and less aggregation of the polymer in the solid state on account of the steric hindrance of the spirostructure. These results demonstrate that the introduction of spirostructure is a promising approach to realize pure blue-emitting PFs. It is well known that a balance of the injection and transport for both holes and electrons in the polymer material is necessary for highly efficient electroluminescent polymers. But most of the conjugated polymers used as light-emitting materials are p-type polymers, which possess greater hole-transporting property than electron-transporting. Hence, electron-deficient unit is inserted into the polymer backbone to improve the electron affinity of an emitting polymer. In 2005, Huang et al. attached oxadiazole branches to a PF backbone through the spirobifluorene units to obtain polymer **86** (Fig. 2.5A) (Zhu et al., 2005). The oxadiazole branches act as an electron transport channel, while the PF backbone is a channel that prefers hole-transporting. The LUMO energy level of polymer **86** is significantly tuned along with a slight change of the HOMO energy level, in comparison with **84b**, which reveals that the electron affinity has been improved. The steric and rigid structure of oxadiazole branches further reduces the interchain interactions to increase the T_g (167°C) and improve the stability.

By carefully choosing the electron transport and hole transport moieties, Chen and coworkers (Huang et al., 2012) synthesized polymer **87** (Fig. 2.5A). They integrated the triphenylamine (TPA) and carbazole in the same side chain of spiropolyfluorene with logical spatial and energetic sequence of these moieties to establish graded route for more efficient hole injection; they also incorporated the electron transport moiety with strong electron-withdrawing capability, triazole (TAZ), on both chain ends to give favorable arrangement in space and energy for electron injection. These two factors allow the corresponding single layer device to exhibit deep blue emission with commission internationale de l'eclairage (CIE) coordinates (0.16, 0.07) and external quantum efficiency (η_{ext}) up to 7.28%, which is the highest value among the deep blue polymer fluorescent diodes ever reported. These properties demonstrate that introduction of multiple charge transport moieties for efficient injection of charge carriers is a promising concept for designing highly electroluminescent polymers.

Hyperbranch also attracts a lot of interest for developing efficient electroluminescent devices. Since the hyperbranched and globular structure can depress the interchain interaction and excimer formation so as to make the materials form high-quality amorphous films, and consequently improve the stability, light-emitting efficiency, and emission color purity. Bo and coworkers reported the first example of hyperbranched polymers used as blue light-emitting materials in 2004 (Li and Bo, 2004). They adopted a novel "AB_2 + AB" approach based on Suzuki coupling reaction for the synthesis of polymer **88** (Fig. 2.5A). The obtained hyperbranched polymers are of high molecular weight and good solubility in common organic solvent, meanwhile the blue-green emission band can be completely suppressed in the annealed film. Similarly, oxadiazole, 1,3,5-triazine or other electron-deficient group has been introduced to the hyperbranched polymer as a branching unit to improve the charge transport-balance properties of PFs. Dendronized polymers are also proved to be an efficient way to suppress the formation of aggregates/excimers of PFs main chains because of the shielding effect provided by the dendritic side chains on the conjugated backbones. Peng and coworkers designed and synthesized a series of PFs with dendritic functional carbazole as well as oxazole side chains in 2009 (Peng et al., 2009). These copolymers show good thermal properties and solubility in organic solvents. The photoluminescent and electroluminescent emission color quality are improved greatly due to less aggregates of main chains of PFs with the steric hindrance of dendritic functional carbazole and oxazole units. The HOMO energy level of the polymer is raised while the LUMO level is reduced due to the bipolar nature of the carbazole and oxadiazole system. Therefore, the high electronic properties afforded by dendronized functional carbazole and oxadiazole side chains can give rise to efficient injection and transport of holes and electrons. Ultimately, the best devices with the external quantum efficiencies of 1.03% were obtained.

Dopant/host concept is another alternative idea to realize pure blue light emission, which was demonstrated by Alex et al. in 2007 (Huang et al., 2007). They synthesized a series of fluorene-based blue-light-emitting copolymers **89** (Fig. 2.5A). In these polymers, the large-bandgap UV blue-light-emitting polymer poly[2,7-(9,9′-dioctylfluorene)-*alt*-1,3-(5-carbazolphenylene)] (PFCz) was developed as the host with a small amount of blue-light-emitting 4-*N,N′*-diphenylaminostilbene (DPS) units incorporated as dopant. A small amount of electron-transporting oxadiazole (OXD) units were also incorporated into the polymer main chain to improve the electron-transporting ability. Due to charge-trapping and energy-transfer mechanisms, a small amount of narrow-bandgap units (dopants) can greatly improve the device performance of the PF derivatives. Moreover, in these systems, the dopant is covalently connected to the host to realize molecular dispersion of the dopant, which avoids the aggregation/self-quenching and the phase separation problems. Consequently, the device based on these copolymers displayed very stable pure blue light electroluminescent emission with.

Besides, replacing the carbon atom at C-9 position of fluorenes with other atoms (such as silicon, germanium, and phosphorus) may produce excellent blue emitting materials, which originates from the particular interactions between the heteroatom and the π-conjugated backbone.

2.2.3.2 White Light-Emitting Polymers

In recent years, more and more efforts are attracted to polymer white light-emitting materials and diodes (PWLED) (Wu et al., 2009; Tang et al., 2013; Ying et al., 2014). Its main applications are the back-lighting sources for liquid–crystal displays, full color flat-panel displays and next-generation solid-state lighting. Among these applications, lighting is of special interest since about 20% of the total generated electric power has been consumed by lighting. Compared with white light-emitting devices based on small organic molecules, the relatively poorer performance of WPLEDs, especially lower efficiencies, hinders their commercialization. In other words, polymer white light-emitting materials with high-performances still need to be explored.

In practice, WPLEDs can be obtained by various approaches to achieve simultaneous, properly balanced emission covering the entire visible light spectrum (400–700 nm). These methods include multiple component emissive layers containing an appropriate ratio of red, green and blue (RGB) phosphorescent or fluorescent dopants, polymer blends containing RGB emitting species and stacked RGB multilayer structures. Although several good results had been reported, these PWLEDs are generally confronted with the problem of phase separation, leading to declined device efficiency and poor reproducibility. Hence, the design and synthesis of a single white emitting polymer capable of white emission from simultaneous blue, green and red emission or complementary emission has received more and more attention. The basic strategy to obtain such a polymer is to embed the narrow-bandgap unit in the conjugated polymer main chain and then it can contain several different chromophores. In this system, the chromophores can be considered as homogeneous dispersion at the molecular level in a polymer matrix. By adjusting the ration of chromophores to a certain level, incomplete energy transfer will take place, leading to a simultaneous emission from all chromophores. As the energy transfer efficiency is high, the requirement for low-bandgap chromophore is little, ranging from one ten thousandth to one thousandth (molar ratio). Thus the emissive spectrum of each chromophore overlaps, creating a continuous spectrum close to the standard white light. As chromophore can be embedded into the polymer host chain and also can be attached to the polymer backbone as side chain, in the view of materials design and synthesis, this approach is flexible.

Wang et al. first reported a single polymer **90** for white light emitting in 2005 (Fig. 2.5B) (Liu et al., 2005). They embedded a red chromophore in the PF main chain, and attached the green chromophore to the polymer backbone as a side chain. That is to say, red, green, and blue chromophores are incorporated in a single polymer at the same time. By adjusting the ratio of the chromophore, they obtained highly efficient PWLED with the efficiency up to 1.59 cd A^{-1} and brightness up to 3786 cd m^{-2}. The spectral stability of the PWLED was also high. When the voltage was changed, the spectrum remained unchanged in color coordinates (0.31, 0.34).

Later, a binary white emission system (**91** in Fig. 2.5B) in which certain amounts of yellow emitting 1,8-naphthalimide derivatives were chemically incorporated into the PF main chain had been achieved (Tu et al., 2006). The optimized electroluminescent device emitted white light with the color coordinates (0.32, 0.36). The spectra were maintained when the voltage varied, with a luminous efficiency of 3.8 cd A^{-1} and a power efficiency of 2.0 lm W^{-1}. Although the efficiencies are not very high, the results clearly demonstrate that both the matchup of emission wavelengths between the blue and orange emitting species and their relative intensities are critical to obtain white emission with high color purity. The polymer backbone could be further optimized to enhance the luminescence efficiencies

(Liu et al., 2005). The orange chromophore (1,8-naphthalimide derivative, with a PL quantum efficiency, $\Phi_{PL} = 0.25$) was replaced by a more efficient chromophore, 4,7-bis(4-(*N*-phenyl-*N*-(4-methylphenyl)amino)phenyl)-2,1,3-benzothiadiazole unit (TPABT, $\Phi_{PL} = 0.76$) (Liu et al., 2006). The WPLED based on the obtained polymer **92** generated a pure white emission with CIE coordinates of (0.35, 0.34), an luminous efficiency (LE) of 8.99 cd A^{-1} and a PE of 5.75 lm W^{-1}, which is a dramatic improvement over the yellow chromophore (3.8 cd A^{-1} and 2.0 lm W^{-1}). On the other hand, the WPLED demonstrated much lower LE of 2.53 cd A^{-1} and PE of 0.73 lm W^{-1} when efficient blue emitting polymer host PF ($\Phi_{PL} = 0.55$) was displaced by poly-(9,9-dioctyl-2,7-fluorene-alt-2,5-bis(hexyloxy)-1,4-phenylene (PFB, $\Phi_{PL} = 0.30$), indicating that efficient blue and orange emitting polymers should exist at the same time.

Most of the single white-light-emitting polymers are achieved based on the partial energy transfer and charge trapping on dopant in the EL process. A novel strategy was recently proposed to realize single white EL polymer (**93–95** in Fig. 2.6A) based on the mechanism of electron trapping on host in 2009 (Guo et al., 2009). Phosphonate-functionalized PF had been selected as a blue light-emitting polymer host and three heterocyclic compounds 2,1,3-BT, BSe, and 2,1,3-naphthothiadiazole (NT) as the narrow-bandgap dopants. The phosphonate groups on the PF side chains display stronger electron affinity than the dopant units, and the electrons are therefore confined mostly in main chain during the EL process. As a result, the electron transferring from main chain to dopant is effectively suppressed, leading to individual emission from host and dopant in the polymer. Thus, the doping concentration of narrow-bandgap chromophore in host chain was raised to centesimal level. The high work function metal Al without an additional electron-injection layer was employed as electrode in the fabrication of the WPLED due to the introduction of phosphonate. Among the three polymers, **95** showed the best white EL with CIE coordinates of (0.34, 0.35) at the 8 V. This new strategy affords more opportunities to develop single white-light-emitting polymers.

The aforementioned single white polymers only focused on incorporating narrow-bandgap fluorescent chromophores into the polymer backbone or side chain, thus only singlet excitons were utilized during the electroluminescent processes. In fact, triplet phosphorescent chromophore can also be used for developing white emitting polymers. Cao and coworkers reported a strategy to realize white light-emitting from a single polymer **96** (Fig. 2.6A) by simultaneously introducing fluorescent and phosphorescent chromophores into the polymer chain (Jiang et al., 2006). In this polymer, blue emitting PF was employed as host, the green 2,1,3-BT chromophore was incorporated into the polymer backbone as a dopant, and iridium(III) bis(2-phenylquinolyl-*N*,$C^{2'}$) acetylacetonate (IrPhq) as a triplet iridium complex was introduced to function as a red emitter. By adjusting the ratio of copolymerizing chromophores a pure white EL could be obtained. The results reveal that the white emission based on the copolymers was stable upon change of applied voltages. The highest luminous efficiency of single-layer devices made by the series of materials was 6.1 cd A^{-1}, and the optimized color coordinates (0.32, 0.33) are very close to white equal energy point. This work opens up a new path for white emitting polymers by utilizing both singlet and triplet excitons.

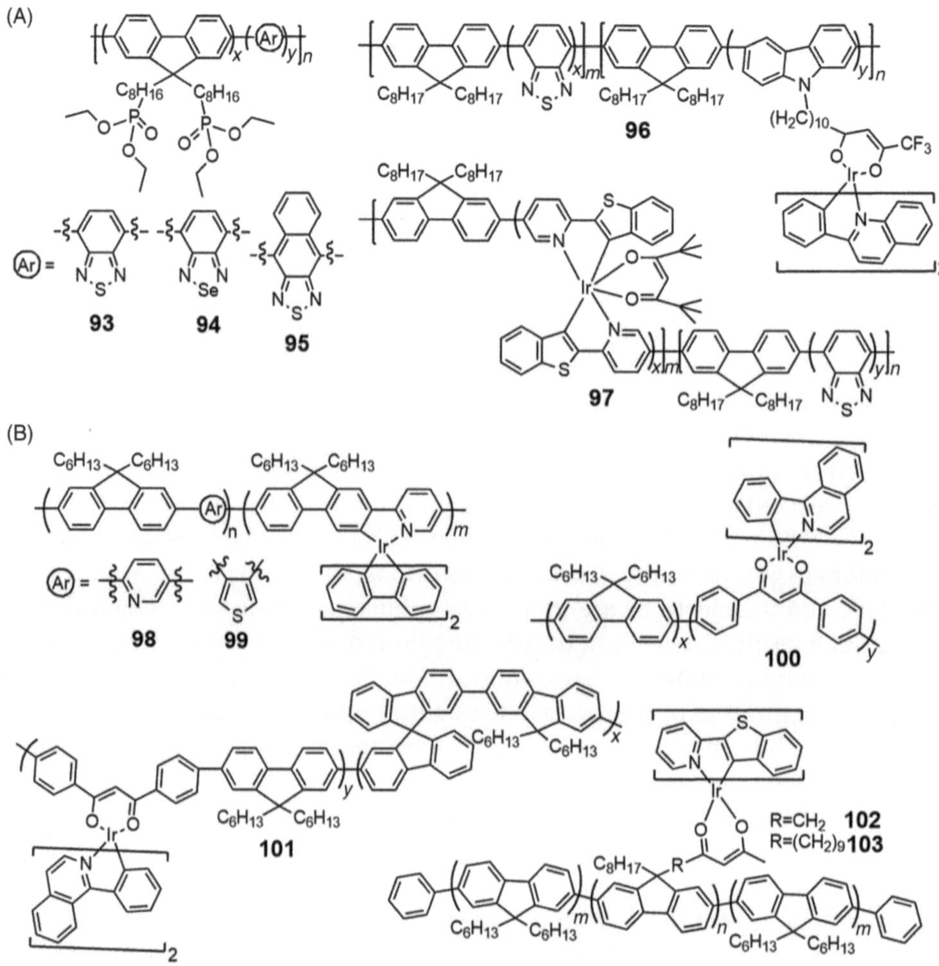

FIGURE 2.6 (A) Chemical structures of white light-emitting polymers **93–97**. (B) Chemical structures of electrophosphorescent polymers **98–103**.

Another strategy to realize single white-light-emitting polymers by mixing fluorescent and phosphorescent emission is to introduce both singlet and triplet emitters into the polymer backbone. Peng and Yang et al. still took blue emitting PF as host material, and the main chain was doped with green emitting BT and red emitting iridium complex [$(btp)_2Ir(tmd)$] to synthesize polymer **97** (Fig. 2.6A) by Suzuki polycondensation (Zhen et al., 2006). The devices showed a peak external quantum efficiency (EQE) of 3.7%, a peak LE of 3.9 cd A-1, and nearly ideal CIE coordinates of (0.33, 0.34). The white-light emissions from these copolymers were stable at all of the applied voltages, thus indicating that this approach is feasible to prepare a white emitting single polymer.

Osmium complexes as less accessible heavy metal can also be chosen to act as phosphorescent emitter for WPLED, comparatively to iridium complexes. The radiative lifetime of osmium complexes is shorter than the corresponding Ir systems, which should reduce the extent of triplet–triplet annihilation for devices that have a higher dopant concentration or under a higher current density, leading to better properties. By covalently attaching fluorescent green emitting BT units and a phosphorescent red emitting osmium complex (Os(bpftz)) to the blue emitting PF backbone, Su et al. prepared a white electrophosphorescent copolymer with CIE coordinates of (0.37,0.30) (Chien et al., 2008). The white device based on the resulting polymer showed a maximum LE of 10.7 cd A^{-1}, with corresponding external quantum efficiency of 5.4%, which is one of the highest values reported so far for a white emitting single polymer by simultaneously utilizing both singlet and triplet excitons.

2.2.3.3 Phosphorescent Polymer Light-Emitting Materials

It is known that only the singlet exciton is emissive in fluorescent OLEDs or PLEDs, and thus most energy is wasted through the nonemissive triplet exciton. As the phosphor is able to induce effective coupling of spin-orbital and then improves the intersystem crossing from singlet excited state to triplet excited state, electrophosphorescence constructed on phosphors is capable to reach nearly 100% internal quantum efficiency theoretically (Baldo et al., 1998). Therefore, phosphorescent PLEDs (PhPLEDs) are the important subject of recent investigations (Gong et al., 2012). Generally, there are two approaches to fabricate PhPLEDs, one is blending a phosphor into a polymer matrix and the other is covalently bonding a phosphor into a polymer backbone to obtain a single electrophosphorescent polymer. The latter approach is an efficient strategy to avoid phase separation and concentration quenching. However, it also suffers from the problem of triplet energy back transfer from the incorporated phosphor to the polymer backbone just as in the blending system, leading to a significant decrease in device efficiency. Thus, rational design of the molecular structure is crucial for this class of materials.

To explore the relationship between the triplet energy of the polymer main chain and triplet quenching of phosphors by the polymer backbone, Schulz et al. synthesized conjugated fluorene-*alt*-pyridine and fluorene-*alt*-thiophene polymers containing phosphorescent iridium complex (**98** and **99** in Fig. 2.6B) (Schulz et al., 2006). When the 2,5-linked pyridine group **98** was replaced with the 3,4-linked thiophene group **99**, the absorption and emission spectra of the conjugated polymers were blue shifted. Accordingly, the triplet energy increased significantly from 2.13 eV **98** to 2.88 eV **99** by incorporation of a 3,4-linked thienyl group, about 0.7 eV higher than the triplet energy of the embedded phosphor (2.22 eV), thereby decreasing triplet energy transfer from the phosphor to the polymer backbone. As a result, phosphorescent quantum yields of films increased from 0.05 **98** to 0.20 **99**, and the external quantum efficiencies of the devices were enhanced from 0.32% **98** to 0.84% **99**. These results reveal that enhancing the triplet energy of the polymer backbone can effectively reduce quenching of the phosphor by the main chain so as to improve the device efficiency.

Similar to the design of blue light-emitting polymer, introducing sterically hindered groups to the phosphor or the polymer host is also an efficient approach to reduce the triplet energy back transfer. With this in mind, Yang and coworkers introduced the sterically hindered spirobifluorene units into the PF backbone to synthesize the copolymer **100** (Fig. 2.6B) (Zhang et al., 2009). PL lifetime of **101** is 0.68 μs, which is higher than that of the homofluorene backbone **100** (0.45 μs). However, the triplet energy levels of **101** (2.18 eV) and **100** (2.10 eV) are close to each other. Therefore, the intrachain triplet energy back transfer should not be the most important factor to induce the difference of triplet emission lifetime. Namely the reason why 101 has longer lifetime can be ascribed to the steric hindrance introduced by the spirobifluorene segments which restrain the aggregation of polymer chains. Consequently, the chance for the attached phosphor to be in contact with the nearby main chain is decreased, resulting in less interchain triplet energy back transfer. The device using **101** as the emitting layer exhibited a maximal external quantum efficiency of 3.21%, which is much higher than that of **100** (0.74%). Apparently, the promoted efficiency is derived from the introduction of sterically hindered spirobifluorene units on the backbone.

Because the strong distance dependence of the triplet energy back transfer from the attached phosphor to the polymer backbone, prolonging the distance between the triplet centers of the phosphor and the polymer backbone could be an alternative approach to reduce the triplet energy back transfer. As shown in Fig. 2.6B, Holmes et al. synthesized two copolymers **102** and **103** with red-emitting phosphorescent iridium complexes attached either directly (spacerless) or through an octamethylene-tethered [—$(CH_2)_8$—] chain at the 9-position of a 9-octylfluorene host (Evans et al., 2006). In these two copolymers, the triplet energy of the attached phosphor (2.0 eV) is close to that of the PF backbone (2.1 eV), so the triplet energy back transfer can happen. The photoluminescent lifetime of the spacerless copolymer **102** (3.8 μs) is shorter than that of the corresponding octamethylene-tethered copolymer **103** (4.3 μs). These discoveries demonstrate that it is the close proximity of attached phosphor to the host PF in **102** that promotes the back transfer of triplet energy in the system. Differently, **103** has octamethylene tether that provides spatial control of phosphor related to polymer backbone and reduces triplet back transfer of triplet energy from iridium complex to PF. Phosphorescent quantum yield (22%) and electroluminescence efficiency (2.0%) of the octamethylene-tethered copolymers **103** doubled those of the spacerless copolymers **102** (12 and 1.1%, respectively), which is consistent with the suppression of the back transfer of triplets from the phosphor to the PF backbone in **103**. Thus, the incorporation of a long alkyl chain between the polymer host and phosphorescent guest is an efficient design principle for achieving higher efficiencies in PhPLEDs for which the triplet energy levels of the host and guest are similar.

2.2.4 Rational Design of Conjugated Polymers for Electrochromic Devices

Electrochromic materials is a remarkable and productive research area over the last three decades since it has potential applications in smart window products, e-papers, optical shutters, transmissive and reflective displays, self-darkening mirror devices, and optical memories. The electrochromic effect has been observed in metal oxides (e.g., WO_3), molecular organic electrochromes (e.g., viologen derivatives), and electrodeposited conjugated polymers films. Among them, conjugated polymers were proven to be fascinating materials due to their numerous advantages over the other electrochromic systems, such as different colors obtained from the same material at different redox states, ease of bandgap/color-tuning via structure modification, fast switching, superior coloration efficiencies, high mechanical flexibility, potentially low cost, and ease of processing. Through rational structure design, electrochromic conjugated polymers with new colors, higher contrasts, and improved ambient stability can be successively obtained. Actually, some of the most effective strategies described in the construction of other organic electronic-based devices which are mentioned previously have also been applied to polymeric electrochromic devices (PECD). Detailed reviews on electrochromic conjugated polymers and the different strategies to tune the optical transitions have been recently published (Gunbas and Toppare, 2012; Beaujuge and Reynolds, 2010; Amb et al., 2011; Beverina et al., 2014), and consequently we will just highlight several representative approaches in this section to clarify the molecular design of conjugated polymers.

2.2.4.1 Backbones

Similar to PSC applications, narrow-bandgap polymers are also attractive in PECDs. Such polymers are easy to oxidize and to be charge neutralized which is needed for high stabilities. Furthermore, lower bandgaps make them show colored reduced forms and possess faintly colored or colorless oxidized forms, which is a suitable characteristic for a polymer-based electrochromic device (ECD). One of the most important examples is PEDOT **104a** (Fig. 2.7A), which shows a bandgap of around 1.6 eV (Gaupp et al., 2002). Due to the relatively narrow bandgap, **104a** possesses a highly transmissive sky-blue oxidized state and a distinctive dark-blue reduced state. The contrast ratio and the coloration efficiencies (CE) of **104a** are 54% and 137 $cm^2 \cdot C^{-1}$ at full switch, respectively.

Bandgap reduction can be achieved through using many methods. Taking PTs, for example, polymerization of fused heterocycles containing thiophene and their derivatives is a practicable method to obtain low-bandgap polymers. Sotzing and coworkers have synthesized a conjugated polymer **105** (Fig. 2.7A) with a strikingly low bandgap of 0.85 eV from the monomer thieno[3,4-*b*]thiophene (Lee and Sotzing, 2001; Sotzing and Lee, 2002). **105** was sky-blue in the reduced form and was both colorless and highly transparent in the oxidized state with a coloration efficiency of 160 $cm^2 \cdot C^{-1}$ at 800 nm. Stability studies indicated that **105** was highly stable to redox switching between its neutral and oxidized states. Later, they replaced the thiophene ring in **105** with a furan ring to obtain poly(thieno[3,4-*b*]furan) **106** (Fig. 2.7A) (Kumar et al., 2008; Kumar et al., 2006). Polymer **106** had a low bandgap of 1.04 eV, which is lower than that of PEDOT but comparatively higher than that of **105**. **106** showed a pale blue neutral state and a more transparent pale blue oxidized state with transmittance of 62 and 72%, respectively, indicating a little subtle change in transparency level between redox states. These unique properties qualify this polymer as a candidate for transparent conductor or ion-storage layers of electronic devices.

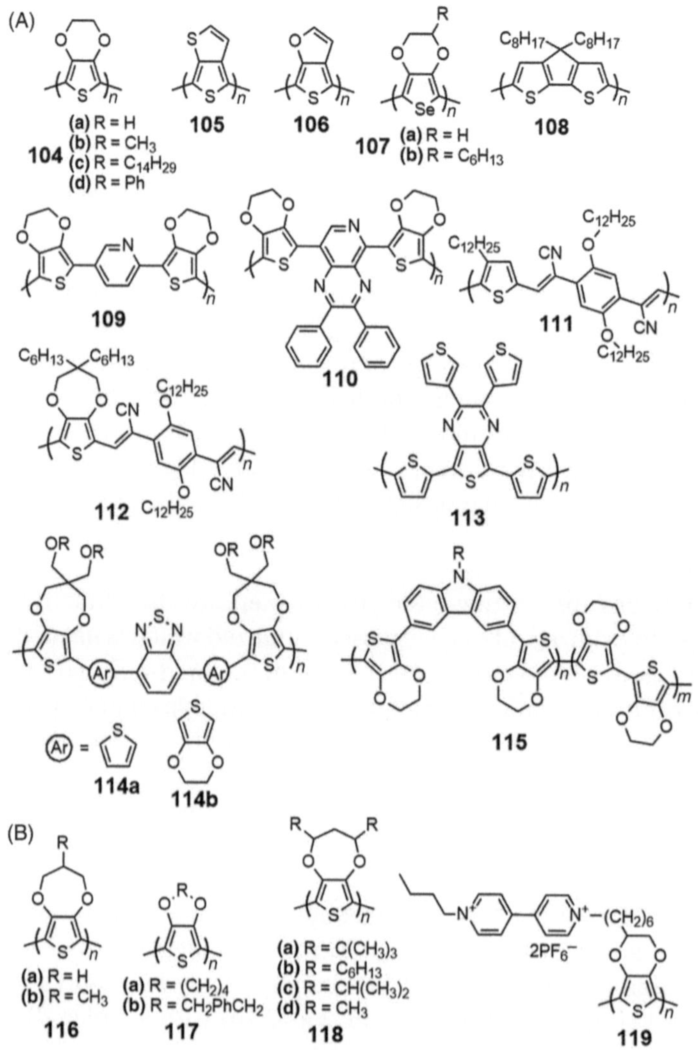

FIGURE 2.7 (A) Chemical structures of **104–115**. (B) Chemical structures of **116–119** for PECDs.

In 2008, Bendikov group reported a new synthetic method for 3,4-ethylenedioxyselenophene (EDOS) and successfully polymerized 2,5-dibromo-3,4-ethylenedioxyselenophene by a solid-state polymerization technique to obtain PEDOS **107a** (Fig. 2.7A) with high yields (Patra et al., 2008). These methods effectively reduced the defects in the conjugation of the polymer chain that formed during conventional oxidative polymerization, so the resulting polyselenophene derivatives showed high electrical conductivities. EDOS is more electron rich than the parent sulfur-based heterocycle, leading to a narrow bandgap of **107a** (1.42 eV) compared to that of **104a**. Similarly, polymer **107a** possessed a highly transmissive oxidized state and a blue neutral state. However, the electrochromic property of **107a** is better, which showed an optical contrast of 55% at 666 nm and a coloration efficiency of 212 $cm^2 \cdot C^{-1}$. The bleaching and coloring times were measured to be 0.4 s and 0.6 s, respectively.

Another common method to obtain a low-bandgap polymer is to tie and rigidify adjacent aromatic units through covalent chemical bonding. Wu et al. reported an electrochromic polymer, poly(4,4′-dioctyl-cyclopenta[2,1-*b*:3,4-*b*′]-dithiophene) **108** (Fig. 2.7A), which had a narrow bandgap of 1.73 eV compared to that of PT (~2.0 eV) (Wu et al., 2007). Polymer **108** was deep blue in its neutral state and turned to transparent bluish after being oxidized. Its film exhibited high coloration efficiency as high as 932 $cm^2 \cdot C^{-1}$ at 580 nm, low response time (0.75 s) and high optical contrast (60%) in the visible light region. Furthermore, this polymer also showed high-level stability for long-term switch with less than 8% contrast loss after 1000 cycles.

To reduce the bandgap, donor–acceptor approach has been proven to be an especially powerful tool. The incorporating of donor and acceptor units with different strengths, along with varying the extent of conjugation in each sub-unit, yields fine control of the bandgap and subsequently the absorption, emission, and absolute values for the redox potentials. In 2002, Reynolds et al. synthesized and characterized two polymers **109** and **110** (Fig. 2.7A) using this approach (DuBois and Reynolds, 2002; DuBois et al., 2004). The use of pyridine-based acceptors showed promising results for manipulation of bandgaps to achieve control over both p- and n-type doping and optical properties in the polymers. **109** and **110** had a bandgap of 1.2 and 1.9 eV, respectively. Interestingly, both of them possessed multicolor electrochromism upon oxidation and reduction. It is worth mentioning that multichromic materials are promising candidates for display device applications.

Another set of D–A polymers is based on cyanovinylene as an acceptor used in concert with electron-rich aromatics. A family of soluble narrow band gap D–A polymers (**111** and **112** in Fig. 2.7A) based on propylenedioxythiophenes and cyanovinylenes have been reported by Reynolds et al. **111** and **112** showed three accessible color states changing from an absorptive blue or purple in the neutral state to a transmissive sky-blue or gray in the oxidized and reduced forms (Thompson et al., 2006). The electrochromic properties of these polymers are similar to those of **104a**. However, these polymers are more advantageous with a highly air-stable neutral state owing to their lower HOMO levels. These polymers possess reduced states that were colored, indicating *n*-type doping abilities and multichromism.

In the design of green EC polymer, donor-acceptor approach has played a major role. Generally speaking, two discrete absorption bands in specific regions of the visible spectrum are necessary for a polymer to display a green color. These absorption bands should be simultaneously depleted upon oxidation. However, it is difficult to create and deplete two absorption bands in the visible region by a single polymer, thereby neutral state green polymers had been perceived as impossible to attain. The previous problems were addressed by Wudl et al. in 2004 with the synthesis of a D–A polymer **113** (Fig. 2.7A) (Sonmez et al., 2004a). **113** was designed to combine an alternating D–A molecular structure with two distinct chromophores: D–A main chain results in absorption of the blue light at wavelengths longer than 600 nm, and the conjugated entity present on each repeat unit absorbs in the red at wavelengths below 500 nm, yet in broken conjugation with the main chain. This polymer can be acclaimed as a breakthrough in the era since it had high green color saturation in the neutral state, fast switching and outstanding stability upon electrochemical switching between neutral and fully oxidized states after 10,000 cycles. Combining with the color-mixing theory, Sonmez et al. rapidly introduced the idea of overlaying RGB colored ECPs to generate other colors (Sonmez et al., 2004b). The electrochromic polymers were chosen from the P3ATs for the red, PEDOT as the blue, and the D–A polymer **113** as the green. Mixtures of any two of the three primary colors in various oxidation levels can produce thousands of colors resulting from possible tones of these polymers that can be obtained at different oxidation levels. Some of the colors can even be obtained from only two redox states of these three polymers (fully oxidized and fully reduced).

It should be noticed that the electrochromic contrasts of polymer **113** were low in the visible region (23% at 370 nm and 12% at 725 nm) and its oxidized state had a persistent yellowish-brown hue with relatively moderate transmissivity. To overcome these obstacles, pentameric oligomers with especially low oxidation potentials were synthesized and chemically polymerized by Reynolds and coworkers to yield two polymers **114a** and **114b** (Fig. 2.7A) (Beaujuge et al., 2008). In these two polymers, alkoxy-derivatized 3,4-propylenedioxythiophene moieties were incorporated to provide the desired solution processability, fine-bandgap-tuning, switching performance, and high color contrasts (45 and 40–44% optical contrast in the visible and in the near infrared region for **114b**, respectively). In particular, these monomers did not involve the presence of a second chromophore in broken conjugation with the main chain, such as that in polymer **113**. As a result, the higher energy transitions of the polymers were depleted effectively on full oxidation. Hence, **114a** and **114b** only exhibited highly transmissive oxidized states with light blue color. Meanwhile, they also possessed small potential windows for operation, fast switching, and long-term stability.

Besides the aforementioned approaches, an alternative method to well tune the neutral-state color of existed polymers is random electrochemical copolymerization of monomers which have similar low oxidation potentials and introduction of new color palettes, obviating synthetic effort at monomer level. For instance, the 3,6-bis-(2-(3,4-ethylenedioxy)thienyl)-*N*-methylcarbazole monomer was copolymerized with 2,2′-bis(3,4-ethylenedioxythiophene) in different feed ratios to produce copolymers **115** (Fig. 2.7A) (Gaupp and Reynolds, 2003). The absorbance maxima and colors of these neutral copolymers ranged from blue (587 nm) to red (464 nm), to orange (429 nm), and finally to yellow (420 nm) with increasing concentration of carbazole-based monomer. Interestingly, all the copolymers passed through a green intermediate state before reaching a blue colored oxidized state, regardless of the polymer composition. These results clearly reveal that tunable multichromism can be further exploited by careful choice of other suitable comonomers.

2.2.4.2 Side Chains

It is well known that side chains are always used to impart the solubility to conjugated polymers. Not only that, they have been discovered to play important roles in certain properties of conjugated polymers, such as inter- and intramolecular interactions, charge transport, and active layer morphology, leading to different performances of PSCs, as discussed in Section 2.2.2.2. For electrochromic polymers, the situation is the same.

Reynolds et al. synthesized a series of alkyl-substituted and unsubstituted poly (3,4-alkylenedioxythiophene)s **104** (Fig. 2.7A), **116** and **117** (Fig. 2.7B) where either the size of the alkylenedioxy ring or the nature of the pendent group was varied (Kumar et al., 1998). Optoelectrochemical experiments revealed that the nature of the substituent groups have little effect on the extent of conjugation of the polymer backbone. Thereby all the polymers switched from a relatively transmissive light green in the oxidized form to an opaque dark blue in the reduced form. However, increasing the ring size or size of the substituent on the alkylenedioxy bridge has an obvious effect to enhance the electrochromic contrast and reduce the switching time. That is to say, these polymers with additional substituent on the alkylenedioxy bridge perform significantly better than their parent polymer **104a.**

Similar results were reported by Bendikov group. They introduced a hexyl chain into the EDOS monomer which is mentioned previously. The subsequent polymerization of this new monomer resulted in a high-performance electrochromic polymer **107b** (Li et al., 2009a). This electrochromic polymer switched in color between a highly absorbing pure-blue and a nearly colorless states accompanying with a remarkable stability. In comparison with **107a** (Patra et al., 2008), **107b** reveals a higher contrast ratio of 88–89%, a higher CE of up to 773 $cm^2 \cdot C^{-1}$ and a faster switching. Later, they performed a systematic study on the electrochromic properties of poly(alkyl-3,4-ethylenedioxyselenophene), where the alkyl chain was changed from a ethyl to dodecyl unit by increments of two carbons (Li et al., 2009b). The results reveal that the electrochromic property reaches the maximum for PEDOS with a hexyl unit, whereas other polymers with shorter and longer alkyl chains possess poorer property.

Sotzing group reported another promising approach for the color control in electrochromic polymers (Dey et al., 2011). They utilized 1,3-substitution on 3,4-propylenedioxythiophene monomer compared to the common 2,2-substitution pattern and synthesized four different polymers (**118a–d** in Fig. 2.7B) where different alkyl groups were used as the substituent. A clear structure-property relationship was observed for these polymers. **118d** had a dark blue-purple color in the neutral state where upon oxidation the polymer switches to transmissive light blue color. A yellow neutral state was observed for **118a** which switched to a green oxidized state upon oxidation. These large differences possibly should originate from the size of the substituents. Obviously, the distortion of backbone planarity of **118** is larger than that in **116**, which gives rise to the distinction of the bandgap as large as 0.73 eV (experimentally measured). This work proves that well tuning of color can be achieved by altering the size of substituents at the 1,3 positions.

Another rather special strategy to obtain electrochromic polymers with unexpected performance is covalently connecting electrochromic units to the conjugated polymer backbone. Lee and coworkers have reported a new polymer **119** (Fig. 2.7B) that is a PEDOT derivative containing a pendant viologen (Ko et al., 2004). Electrochemical and spectroelectrochemical studies showed that this polymer had four colored oxidation states, derived from the combination of the characteristics of both **104a** and viologen. Due to this cooperative electrochromic action of both **104a** and viologen, its optical contrast was greatly enhanced. The maximal transmittance change of this polymer was up to 65% at 610 nm, which is higher than that observed in **104a** (44%) (Kumar et al., 1998). What is interesting is that it has an obvious transmittance change of 72% at ultraviolet region (393 nm), which is rare in the case of the simple alkyl-substituted PEDOT derivatives.

2.2.5 Conclusions

Conjugated polymers have been utilized for the fabrication of a variety of optoelectronic devices, including PSCs, PLEDs, PECDs, and others. To improve the performances of the conjugated polymers, chemical tool is essential. Actually, tremendous progress has been realized in the past two decades attributed to the rational design of novel conjugated polymers. It can be expected that the molecular design of conjugated polymers will continue to play a key role in promoting the application of optoelectronic devices.

References

Alberico, D., Scott, M.E., Lautens, M., 2007. Aryl–aryl bond formation by transition-metal-catalyzed direct arylation. Chem. Rev. 107, 174–238.

Amb, C.M., Dyer, A.L., Reynolds, J.R., 2011. Navigating the color palette of solution-processable electrochromic polymers. Chem. Mater. 23, 397–415.

Babudri, F., Colangiuli, D., Farinola, Gianluca, M., Naso, F., 2002. A general strategy for the synthesis of conjugated polymers based upon the palladium-catalysed cross-coupling of Grignard reagents with unsaturated halides. Eur. J. Org. Chem. 2002, 2785–2791.

Babudri, F., Farinola, G.M., Naso, F., 2004. Synthesis of conjugated oligomers and polymers: the organometallic way. J. Mater. Chem. 14, 11–34.

Baldo, M.A., O'brien, D.F., You, Y., Shoustikov, A., Sibley, S., Thompson, M.E., Forrest, S.R., 1998. Highly efficient phosphorescent emission from organic electroluminescent devices. Nature 395, 151–154.

Beaujuge, P.M., Reynolds, J.R., 2010. Color control in pi-conjugated organic polymers for use in electrochromic devices. Chem. Rev. 110, 268–320.

Beaujuge, P.M., Ellinger, S., Reynolds, J.R., 2008. Spray processable green to highly transmissive electrochromics via chemically polymerizable donor-acceptor heterocyclic pentamers. Adv. Mater. 20, 2772–2776.

Becker, H., Spreitzer, H., Ibrom, K., Kreuder, W., 1999. New insights into the microstructure of GILCH-polymerized PPVs. Macromolecules 32, 4925–4932.

Berrouard, P., Najari, A., Pron, A., Gendron, D., Morin, P.-O., Pouliot, J.-R., Veilleux, J., Leclerc, M., 2012. Synthesis of 5-alkyl[3,4-c]thienopyrrole-4,6-dione-based polymers by direct heteroarylation. Angew. Chem. Int. Ed. 51, 2068–2071.

Beverina, L., Pagani, G.A., Sassi, M., 2014. Multichromophoric electrochromic polymers: colour tuning of conjugated polymers through the side chain functionalization approach. Chem. Commun. 50, 5413–5430.

Bijleveld, J.C., Zoombelt, A.P., Mathijssen, S.G.J., Wienk, M.M., Turbiez, M., De Leeuw, D.M., Janssen, R.A.J., 2009. Poly(diketopyrrolopyrrole-terthiophene) for ambipolar logic and photovoltaics. J. Am. Chem. Soc. 131, 16616–16617.

Bijleveld, J.C., Gevaerts, V.S., Di Nuzzo, D., Turbiez, M., Mathijssen, S.G.J., De Leeuw, D.M., Wienk, M.M., Janssen, R.A.J., 2010. Efficient solar cells based on an easily accessible diketopyrrolopyrrole polymer. Adv. Mater. 22, E242–E246.

Bowley, H.J., Gerrard, D.L., Maddams, W.F., 1985. Formation of conjugated polyenes by chemical and thermal degradation of vinyl chloride copolymers and other vinyl polymers. Makromol. Chem. 186, 715–723.

Brisset, H., Thobie-Gautier, C., Gorgues, A., Jubault, M., Roncali, J., 1994. Novel narrow bandgap polymers from sp^3 carbon-bridged bithienyls: poly(4,4-ethylenedioxy-4H-cyclopenta[2,1-b;3,4-b[prime or minute]]dithiophene). J. Chem. Soc., Chem. Commun., 1305–1306.

Bunz, U.H.F., 2000. Poly(aryleneethynylene)s: syntheses, properties, structures, and applications. Chem. Rev. 100, 1605–1644.

Burn, P.L., Bradley, D.D.C., Friend, R.H., Halliday, D.A., Holmes, A.B., Jackson, R.W., Kraft, A., 1992. Precusor route chemistry and electronic-properties of poly(*p*-phenylene-vinylene), poly (2,5-dimethyl-*p*-phenylene)vinylene and poly(2,5-dimethoxy-*p*-phenylene)vinylene. J. Chem. Soc.-Perkin Trans. 1, 3225–3231.

Burroughes, J.H., Bradley, D.D.C., Brown, A.R., Marks, R.N., Mackay, K., Friend, R.H., Burns, P.L., Holmes, A.B., 1990. Light-emitting diodes based on conjugated polymers. Nature 347, 539–541.

Carsten, B., He, F., Son, H.J., Xu, T., Yu, L., 2011. Stille polycondensation for synthesis of functional materials. Chem. Rev. 111, 1493–1528.

Chen, T.A., Rieke, R.D., 1992. The first regioregular head-to-tail poly(3-hexylthiophene-2,5-diyl) and a regiorandom isopolymer: nickel versus palladium catalysis of 2(5)-bromo-5(2)-(bromozincio)-3-hexylthiophene polymerization. J. Am. Chem. Soc. 114, 10087–10088.

Chen, H.-Y., Hou, J., Zhang, S., Liang, Y., Yang, G., Yang, Y., Yu, L., Wu, Y., Li, G., 2009. Polymer solar cells with enhanced open-circuit voltage and efficiency. Nat. photon. 3, 649–653.

Cheng, Y.-J., Yang, S.-H., Hsu, C.-S., 2009. Synthesis of conjugated polymers for organic solar cell applications. Chem. Rev. 109, 5868–5923.

Chiang, C.K., Fincher, C.R., Park, Y.W., Heeger, A.J., Shirakawa, H., Louis, E.J., Gau, S.C., Macdiarmid, A.G., 1977. Electrical-conductivity in doped polyacetylene. Phys. Rev. Lett. 39, 1098–1101.

Chien, C.-H., Liao, S.-F., Wu, C.-H., Shu, C.-F., Chang, S.-Y., Chi, Y., Chou, P.-T., Lai, C.-H., 2008. Electrophosphorescent polyfluorenes containing osmium complexes in the conjugated backbone. Adv. Funct. Mater. 18, 1430–1439.

Chu, T.-Y., Lu, J., Beaupré, S., Zhang, Y., Pouliot, J.-R., Wakim, S., Zhou, J., Leclerc, M., Li, Z., Ding, J., Tao, Y., 2011. Bulk heterojunction solar cells using thieno[3,4-c]pyrrole-4,6-dione and dithieno[3,2-b:2′,3′-d] silole copolymer with a power conversion efficiency of 7.3%. J. Am. Chem. Soc. 133, 4250–4253.

Coffin, R.C., Peet, J., Rogers, J., Bazan, G.C., 2009. Streamlined microwave-assisted preparation of narrow-bandgap conjugated polymers for high-performance bulk heterojunction solar cells. Nat. Chem. 1, 657–661.

Conticello, V.P., Gin, D.L., Grubbs, R.H., 1992. Ring-opening metathesis polymerization of substituted bicyclo[2.2.2]octadienes: a new precursor route to poly(1,4-phenylenevinylene). J. Am. Chem.l Soc. 114, 9708–9710.

Corriu, R.J.P., Masse, J.P., 1972. Activation of Grignard reagents by transition-metal complexes. A new and simple synthesis of *trans*-stilbenes and polyphenyls. J. Chem. Soc., Chem. Commun. 3, 144a–144a.

Dey, T., Invernale, M.A., Ding, Y., Buyukmumcu, Z., Sotzing, G.A., 2011. Poly(3,4-propylenedioxythiophene) s as a single platform for full color realization. Macromolecules 44, 2415–2417.

Dubois, C.J., Reynolds, J.R., 2002. 3,4-Ethylenedioxythiophene–pyridine-based polymers: redox or n-type electronic conductivity? Adv. Mater. 14, 1844–1846.

Dubois, C.J., Abboud, K.A., Reynolds, J.R., 2004. Electrolyte-controlled redox conductivity and n-type doping in poly(bis-EDOT-pyridine)s. J. Phys. Chem. B 108, 8550–8557.

Edwards, J.H., Feast, W.J., 1980. A new synthesis of poly(acetylene). Polymer 21, 595–596.

Evans, N.R., Sudha Devi, L., Mak, C.S.K., Watkins, S.E., Pascu, S.I., Köhler, A., Friend, R.H., Williams, C.K., Holmes, A.B., 2006. Triplet energy back transfer in conjugated polymers with pendant phosphorescent iridium complexes. J. Am. Chem. Soc. 128, 6647–6656.

Facchetti, A., Vaccaro, L., Marrocchi, A., 2012. Semiconducting polymers prepared by direct arylation polycondensation. Angew. Chem. Int. Ed. 51, 3520–3523.

Funke, V.W., Schütze, E.C., 1964. Polykondensationsreaktionen mit xylylendicyaniden. I. Umsetzung von xylylendicyaniden mit phenylendialdehyden. Makromol. Chem. 74, 71–91.

Gadisa, A., Oosterbaan, W.D., Vandewal, K., Bolsee, J.-C., Bertho, S., D'haen, J., Lutsen, L., Vanderzande, D., Manca, J.V., 2009. Effect of alkyl side-chain length on photovoltaic properties of poly(3-alkylthiophene)/PCBM bulk heterojunctions. Adv. Funct. Mater. 19, 3300–3306.

Garay, R., Lenz, R.W., 1989. Anionic-polymerization of *p*-xylenesulfonium salts. Makromol. Chem.-Macromol. Chem. Phys. 15, 1–7.

Garay, R.O., Lenz, R.W., 1992. Effect of the reaction state on the polymerization of *para*-xylenesulfonium salts. J. Polym. Sci. Part A: Polym. Chem. 30, 977–982.

Garay, R.O., Baier, U., Bubeck, C., Mullen, K., 1993. Low-temperature synthesis of poly(*p*-phenylenevinylene) by the sulfonium salt route. Adv. Mater. 5, 561–564.

Gaupp, C.L., Reynolds, J.R., 2003. Multichromic copolymers based on 3,6-bis(2-(3,4-ethylenedioxythiophene))-*N*-alkylcarbazole derivatives. Macromolecules 36, 6305–6315.

Gaupp, C.L., Welsh, D.M., Rauh, R.D., Reynolds, J.R., 2002. Composite coloration efficiency measurements of electrochromic polymers based on 3,4-alkylenedioxythiophenes. Chem. Mater. 14, 3964–3970.

Gilch, H.G., Wheelwright, W.L., 1966. Polymerization of α-halogenated *p*-xylenes with base. J. Polym. Sci. Part A-1: Polym. Chem. 4, 1337–1349.

Gong, S., Yang, C., Qin, J., 2012. Efficient phosphorescent polymer light-emitting diodes by suppressing triplet energy back transfer. Chem. Soc. Rev. 41, 4797–4807.

Gorelsky, S.I., Lapointe, D., Fagnou, K., 2008. Analysis of the concerted metalation-deprotonation mechanism in palladium-catalyzed direct arylation across a broad range of aromatic substrates. J. Am. Chem. Soc. 130, 10848–10849.

Gorelsky, S.I., Lapointe, D., Fagnou, K., 2012. Analysis of the palladium-catalyzed (aromatic) C—H bond metalation–deprotonation mechanism spanning the entire spectrum of arenes. J. Org. Chem. 77, 658–668.

Gorman, C.B., Ginsburg, E.J., Grubbs, R.H., 1993. Soluble, highly conjugated derivatives of polyacetylene from the ring-opening metathesis polymerization of monosubstituted cyclooctatetraenes: synthesis and the relationship between polymer structure and physical properties. J. Am. Chem. Soc. 115, 1397–1409.

Greiner, A., Heitz, W., 1988. New synthetic approach to poly(1,4-phenylenevinylene) and its derivatives by palladium catalyzed arylation of ethylene. Makromol. Chem., Rapid Commun. 9, 581–588.

Grell, M., Bradley, D.D.C., Inbasekaran, M., Woo, E.P., 1997. A glass-forming conjugated main-chain liquid crystal polymer for polarized electroluminescence applications. Adv. Mater. 9, 798–802.

Grimsdale, A.C., Leok Chan, K., Martin, R.E., Jokisz, P.G., Holmes, A.B., 2009. Synthesis of light-emitting conjugated polymers for applications in electroluminescent devices. Chem. Rev. 109, 897–1091.

Guillerez, S., Bidan, G., 1998. New convenient synthesis of highly regioregular poly(3-octylthiophene) based on the Suzuki coupling reaction. Synth. Met. 93, 123–126.

Gunbas, G., Toppare, L., 2012. Electrochromic conjugated polyheterocycles and derivatives-highlights from the last decade towards realization of long lived aspirations. Chem. Commun. 48, 1083–1101.

Guo, X., Qin, C., Cheng, Y., Xie, Z., Geng, Y., Jing, X., Wang, F., Wang, L., 2009. White electroluminescence from a phosphonate-functionalized single-polymer system with electron-trapping effect. Adv. Mater. 21, 3682–3688.

Gustafsson, G., Cao, Y., Treacy, G.M., Klavetter, F., Colaneri, N., Heeger, A.J., 1992. Flexible light-emitting diodes made from soluble conducting polymers. Nature 357, 477–479.

Halliday, D.A., Burn, P.L., Friend, R.H., Bradley, D.D.C., Holmes, A.B., 1993. Determination of the average molecular weigth of poly(*p*-phenylenevinylene). Synth. Met. 55, 902–907.

Havinga, E.E., Ten Hoeve, W., Wynberg, H., 1993. Alternate donor-acceptor small-band-gap semiconducting polymers; polysquaraines and polycroconaines. Synth. Met. 55, 299–306.

Heitz, W., Brügging, W., Freund, L., Gailberger, M., Greiner, A., Jung, H., Kampschulte, U., Nießner, N., Osan, F., Schmidt, H.-W., Wicker, M., 1988. Synthesis of monomers and polymers by the Heck reaction. Makromol. Chem. 189, 119–127.

Henckens, A., Duyssens, I., Lutsen, L., Vanderzande, D., Cleij, T.J., 2006. Synthesis of poly(*p*-phenylene vinylene) and derivatives via a new precursor route, the dithiocarbamate route. Polymer 47, 123–131.

Hörhold, H.H., Helbig, M., 1987. Poly(phenylenevinylene)s—synthesis and redoxchemistry of electroactive polymers. Makromol. Chem. Macromol. Symp. 12, 229–258.

Hörhold, V.H.-H., Opfermann, J., 1970. Poly-*p*-xylyliden. Synthesen und beziehungen zwischen struktur und elektrophysikalischen eigenschaften. Die Makromolekulare Chemie 131, 105–132.

Hou, J.H., Tan, Z.A., Yan, Y., He, Y.J., Yang, C.H., Li, Y.F., 2006. Synthesis and photovoltaic properties of two-dimensional conjugated polythiophenes with bi(thienylenevinylene) side chains. J. Am. Chem. Soc. 128, 4911–4916.

Huang, F., Zhang, Y., Liu, M.S., Cheng, Y.J., Jen, A.K.Y., 2007. High-efficiency and color stable blue-light-emitting polymers and devices. Adv. Funct. Mater. 17, 3808–3815.

Huang, C.-W., Tsai, C.-L., Liu, C.-Y., Jen, T.-H., Yang, N.-J., Chen, S.-A., 2012. Design of deep blue electroluminescent spiro-polyfluorenes with high efficiency by facilitating the injection of charge carriers through incorporation of multiple charge transport moieties. Macromolecules 45, 1281–1287.

Hummelen, J.C., Knight, B.W., Lepeq, F., Wudl, F., Yao, J., Wilkins, C.L., 1995. Preparation and characterization of fulleroid and methanofullerene derivatives. J. Org. Chem. 60, 532–538.

Huo, L., Zhang, S., Guo, X., Xu, F., Li, Y., Hou, J., 2011. Replacing alkoxy groups with alkylthienyl groups: a feasible approach to improve the properties of photovoltaic polymers. Angew. Chem. Int. Ed. 50, 9697–9702.

Iovu, M.C., Sheina, E.E., Gil, R.R., Mccullough, R.D., 2005. Experimental evidence for the quasi-"living" nature of the Grignard metathesis method for the synthesis of regioregular poly(3-alkylthiophenes). Macromolecules 38, 8649–8656.

Iraqi, A., Barker, G.W., 1998. Synthesis and characterisation of telechelic regioregular head-to-tail poly(3-alkylthiophenes). J. Mater. Chem. 8, 25–29.

Ito, T., Shirakawa, H., Ikeda, S., 1974. Simultaneous polymerization and formation of polyacetylene film on the surface of concentrated soluble Ziegler-type catalyst solution. J. Polym. Sci.: Polym. Chem. Ed. 12, 11–20.

Jiang, J.X., Xu, Y.H., Yang, W., Guan, R., Liu, Z.Q., Zhen, H.Y., Cao, Y., 2006. High-efficiency white-light-emitting devices from a single polymer by mixing singlet and triplet emission. Adv. Mater. 18, 1769–1773.

Kim, H.J., Han, A.R., Cho, C.-H., Kang, H., Cho, H.-H., Lee, M.Y., Fréchet, J.M.J., Oh, J.H., Kim, B.J., 2012. Solvent-resistant organic transistors and thermally stable organic photovoltaics based on cross-linkable conjugated polymers. Chem. Mater. 24, 215–221.

King, A.O., Okukado, N., Negishi, E.-I., 1977. Highly general stereo-, regio-, and chemo-selective synthesis of terminal and internal conjugated enynes by the Pd-catalysed reaction of alkynylzinc reagents with alkenyl halides. J. Chem. Soc., Chem. Commun. 19, 683–684.

Kishimoto, Y., Eckerle, P., Miyatake, T., Kainosho, M., Ono, A., Ikariya, T., Noyori, R., 1999. Well-controlled polymerization of phenylacetylenes with organorhodium(I) complexes: mechanism and structure of the polyenes. J. Am. Chem. Soc. 121, 12035–12044.

Kloppenburg, L., Song, D., Bunz, U.H.F., 1998. Alkyne metathesis with simple catalyst systems: poly(*p*-phenyleneethynylene)s. J. Am. Chem. Soc. 120, 7973–7974.

Ko, H.C., Kang, M., Moon, B., Lee, H., 2004. Enhancement of electrochromic contrast of poly(3,4-ethylenedioxythiophene) by incorporating a pendant viologen. Adv. Mater. 16, 1712–1716.

Kumar, A., Welsh, D.M., Morvant, M.C., Piroux, F., Abboud, K.A., Reynolds, J.R., 1998. Conducting poly(3,4-alkylenedioxythiophene) derivatives as fast electrochromics with high-contrast ratios. Chem. Mater. 10, 896–902.

Kumar, A., Buyukmumcu, Z., Sotzing, G.A., 2006. Poly(thieno[3,4-b]furan). A new low band gap conjugated polymer. Macromolecules 39, 2723–2725.

Kumar, A., Bokria, J.G., Buyukmumcu, Z., Dey, T., Sotzing, G.A., 2008. Poly(thieno[3,4-b]furan), a new low band gap polymer: experiment and theory. Macromolecules 41, 7098–7108.

Lafrance, M., Fagnou, K., 2006. Palladium-catalyzed benzene arylation: incorporation of catalytic pivalic acid as a proton shuttle and a key element in catalyst design. J. Am. Chem. Soc. 128, 16496–16497.

Leclerc, M., Diaz, F.M., Wegner, G., 1989. Structural analysis of poly(3-alkylthiophene)s. Makromol. Chem. 190, 3105–3116.

Lee, K., Sotzing, G.A., 2001. Poly(thieno[3,4-b]thiophene). A new stable low band gap conducting polymer. Macromolecules 34, 5746–5747.

Lee, J., Jung, B.J., Lee, S.K., Lee, J.I., Cho, H.J., Shim, H.K., 2005. Fluorene-based alternating polymers containing electron-withdrawing bithiazole units: Preparation and device applications. J. Polym. Sci. Part A: Polym. Chem. 43, 1845–1857.

Lenz, R.W., Handlovits, C.E., 1960. Thermally stable hydrocarbon polymers: polyterephthalylidenes. J. Org. Chem. 25, 813–817.

Lenz, R.W., Han, C.C., Stengersmith, J., Karasz, F.E., 1988. Preparation of poly(phenylene vinylene) from cycloalkylene sulfonium salt monomers and polymers. J. Polym. Sci. Part A: Polym. Chem. 26, 3241–3249.

Li, Y., 2012. Molecular design of photovoltaic materials for polymer solar cells: toward suitable electronic energy levels and broad absorption. Acc. Chem. Res. 45, 723–733.

Li, J., Bo, Z., 2004. "AB2 + AB" approach to hyperbranched polymers used as polymer blue light emitting materials. Macromolecules 37, 2013–2015.

Li, M., Patra, A., Sheynin, Y., Bendikov, M., 2009a. Hexyl-derivatized poly(3,4-ethylenedioxyselenophene): novel highly stable organic electrochromic material with high contrast ratio, high coloration efficiency, and low-switching voltage. Adv. Mater. 21, 1707–1711.

Li, M., Sheynin, Y., Patra, A., Bendikov, M., 2009b. Tuning the electrochromic properties of poly(alkyl-3,4-ethylenedioxyselenophenes) having high contrast ratio and coloration efficiency. Chem. Mater. 21, 2482–2488.

Li, Y., Sun, B., Sonar, P., Singh, S.P., 2012. Solution processable poly(2,5-dialkyl-2,5-dihydro-3,6-di-2-thienyl-pyrrolo[3,4-c]pyrrole-1,4-dione) for ambipolar organic thin film transistors. Org. Electron. 13, 1606–1613.

Li, Y., Sonar, P., Murphy, L., Hong, W., 2013. High mobility diketopyrrolopyrrole (DPP)-based organic semiconductor materials for organic thin film transistors and photovoltaics. Energ. Environ. Sci. 6, 1684–1710.

Liu, J., Zhou, Q.G., Cheng, Y.X., Geng, Y.H., Wang, L.X., Ma, D.G., Jing, X.B., Wang, F.S., 2005. The first single polymer with simultaneous blue, green, and red emission for white electroluminescence. Adv. Mater. 17, 2974–2978.

Liu, J., Zhou, Q.G., Cheng, Y.X., Geng, Y.H., Wang, L.X., Ma, D.G., Jing, X.B., Wang, F.S., 2006. White electroluminescence from a single-polymer system with simultaneous two-color emission: polyfluorene as the blue host and a 2,1,3-benzothiadiazole derivative as the orange dopant. Adv. Funct. Mater. 16, 957–965.

Liu, J., Lam, J.W.Y., Tang, B.Z., 2009. Acetylenic polymers: syntheses, structures, and functions. Chem. Rev. 109, 5799–5867.

Loewe, R.S., Khersonsky, S.M., Mccullough, R.D., 1999. A simple method to prepare head-to-tail coupled, regioregular poly(3-alkylthiophenes) using Grignard metathesis. Adv. Mater. 11, 250–253.

Loewe, R.S., Ewbank, P.C., Liu, J., Zhai, L., Mccullough, R.D., 2001. Regioregular, head-to-tail coupled poly(3-alkylthiophenes) made easy by the GRIM method: investigation of the reaction and the origin of regioselectivity. Macromolecules 34, 4324–4333.

Louwet, F., Vanderzande, D., Gelan, J., 1992. The synthesis of poly(1,4-phenylene-1,2-ethanediyl) derivatives—an adaptation of the Wessling route. Synth. Met. 52, 125–130.

Louwet, F., Vanderzande, D., Gelan, J., 1995. A general synthetic route to high-molecular-weight poly(*p*-xylylene)-derivatives—a new route to poly(*p*-phenylene vinylene). Synth. Met. 69, 509–510.

Lu, W., Kuwabara, J., Kanbara, T., 2012. Synthesis of 4,4'-dinonyl-2,2'-bithiazole-based copolymers via Pd-catalyzed direct C—H arylation. Polym. Chem. 3, 3217–3219.

Marzano, G., Ciasca, C.V., Babudri, F., Bianchi, G., Pellegrino, A., Po, R., Farinola, G.M., 2014. Organometallic approaches to conjugated polymers for plastic solar cells: from laboratory synthesis to industrial production. Eur. J. Org. Chem. 2014, 6583–6614.

Mccullough, R.D., Lowe, R.D., 1992. Enhanced electrical-conductivity in regioselectively synthesized poly(3-alkylthiophenes). J. Chem. Soc., Chem. Commun. 1, 70–72.

Mccullough, R.D., Lowe, R.D., Jayaraman, M., Anderson, D.L., 1993. Design, synthesis, and control of conducting polymer architectures—structurally homogeneous poly(3-alkylthiophenes). J. Org. Chem. 58, 904–912.

Mcdonald, R.N., Campbell, T.W., 1960. The wittig reaction as a polymerization method1a. J. Am. Chem. Soc. 82, 4669–4671.

Mei, J., Bao, Z., 2014. Side chain engineering in solution-processable conjugated polymers. Chem. Mater. 26, 604–615.

Mercier, L.G., Leclerc, M., 2013. Direct (hetero)arylation: a new tool for polymer chemists. Acc. Chem. Res. 46, 1597–1605.

Miao, Y.J., Bazan, G.C., 1994. Paracyclophene route to poly(*p*-phenylenevinylene). J. Am. Chem. Soc. 116, 9379–9380.

Miyaura, N., Suzuki, A., 1979. Stereoselective synthesis of arylated (E)-alkenes by the reaction of alk-1-enylboranes with aryl halides in the presence of palladium catalyst. J. Chem. Soc., Chem. Commun., 866–867.

Miyaura, N., Yamada, K., Suzuki, A., 1979. New stereospecific cross-coupling by the palladium-catalyzed reaction of 1-alkenylboranes with 1-alkenyl or 1-alkynyl halides. Tetrahedron Lett. 20, 3437–3440.

Neef, C.J., Ferraris, J.P., 2000. MEH-PPV: improved synthetic procedure and molecular weight control. Macromolecules 33, 2311–2314.

Ohmori, Y., Uchida, M., Muro, K., Yoshino, K., 1991. Blue electroluminescent diodes utilizing poly(alkylfluorene). Jpn J. Appl. Phys. Part 2-Lett. 30, L1941–L1943.

Okamoto, K., Housekeeper, J.B., Michael, F.E., Luscombe, C.K., 2013a. Thiophene based hyperbranched polymers with tunable branching using direct arylation methods. Polym. Chem. 4, 3499–3506.

Okamoto, K., Zhang, J., Housekeeper, J.B., Marder, S.R., Luscombe, C.K., 2013b. C—H arylation reaction: atom efficient and greener syntheses of π-conjugated small molecules and macromolecules for organic electronic materials. Macromolecules 46, 8059–8078.

Osaka, I., Mccullough, R.D., 2008. Advances in molecular design and synthesis of regioregular polythiophenes. Acc. Chem. Res. 41, 1202–1214.

Park, S.H., Roy, A., Beaupre, S., Cho, S., Coates, N., Moon, J.S., Moses, D., Leclerc, M., Lee, K., Heeger, A.J., 2009. Bulk heterojunction solar cells with internal quantum efficiency approaching 100%. Nat. photon. 3, 297–302.

Patra, A., Wijsboom, Y.H., Zade, S.S., Li, M., Sheynin, Y., Leitus, G., Bendikov, M., 2008. Poly(3,4-ethylenedioxyselenophene). J. Am. Chem. Soc. 130, 6734–6736.

Peng, Q., Xu, J., Li, M., Zheng, W., 2009. Blue emitting polyfluorenes containing dendronized carbazole and oxadiazole pendants: synthesis, optical properties, and electroluminescent properties. Macromolecules 42, 5478–5485.

Pomerantz, M., Gu, X.M., 1997. Poly(2-decylthieno 3,4-*b* thiophene), a new soluble low-bandgap conducting polymer. Synth. Met. 84, 243–244.

Pomerantz, M., Chalonergill, B., Harding, L.O., Tseng, J.J., Pomerantz, W.J., 1993. New processable low-band-gap, conjugated polyheterocycles. Synth. Met. 55, 960–965.

Price, S.C., Stuart, A.C., Yang, L., Zhou, H., You, W., 2011. Fluorine substituted conjugated polymer of medium band gap yields 7% efficiency in polymer–fullerene solar cells. J. Am. Chem. Soc. 133, 4625–4631.

Pron, A., Gawrys, P., Zagorska, M., Djurado, D., Demadrille, R., 2010. Electroactive materials for organic electronics: preparation strategies, structural aspects and characterization techniques. Chem. Soc. Rev. 39, 2577–2632.

Remy, M.S., Cundari, T.R., Sanford, M.S., 2010. Computational and experimental studies of methyl group exchange between palladium(II) centers. Organometallics 29, 1522–1525.

Reynolds, J.R., 2014. Pi-conjugated polymers: the importance of polymer synthesis. Conjugated Polymers: A Practical Guide to Synthesis. The Royal Society of Chemistry, Cambridge.

Roncali, J., 1997. Synthetic principles for bandgap control in linear π-conjugated systems. Chem. Rev. 97, 173–206.

Roncali, J., 2007. Molecular engineering of the band gap of π-conjugated systems: facing technological applications. Macromol.Rapid Commun. 28, 1761–1775.

Roncali, J., Thobie-Gautier, C., 1994. An efficient strategy towards small bandgap polymers: The rigidification of the π-conjugated system. Adv. Mater. 6, 846–848.

Rudenko, A.E., Thompson, B.C., 2015. Optimization of direct arylation polymerization (DArP) through the identification and control of defects in polymer structure. J Polym. Sci. Part A: -Polym. Chem. 53, 135–147.

Rudenko, A.E., Latif, A.A., Thompson, B.C., 2014. Influence of beta-linkages on the morphology and performance of DArP P3HT-PC61BM solar cells. Nanotechnology 25, 014005.

Sakamoto, J., Rehahn, M., Wegner, G., Schlüter, A.D., 2009. Suzuki Polycondensation: Polyarylenes à la Carte. Macromol Rapid Commun. 30, 653–687.

Schulz, G.L., Chen, X., Chen, S.-A., Holdcroft, S., 2006. Enhancement of phosphorescence of Ir complexes bound to conjugated polymers: increasing the triplet level of the main chain. Macromolecules 39, 9157–9165.

Sévignon, M., Papillon, J., Schulz, E., Lemaire, M., 1999. New synthetic method for the polymerization of alkylthiophenes. Tetrahedron Lett. 40, 5873–5876.

Seyler, H., Jones, D.J., Holmes, A.B., Wong, W.W.H., 2012. Continuous flow synthesis of conjugated polymers. Chem. Commun. 48, 1598–1600.

Sheina, E.E., Liu, J., Iovu, M.C., Laird, D.W., Mccullough, R.D., 2004. Chain growth mechanism for regioregular nickel-initiated cross-coupling polymerizations. Macromolecules 37, 3526–3528.

Son, S., Lovinger, A.J., Galvin, M.E., 1995. A simple organic-soluble precursor route to PPV for highly efficient and pin-hole free LED devices. Polym. Mater. Sci. Eng. 72, 567–568.

Son, H.J., Wang, W., Xu, T., Liang, Y., Wu, Y., Li, G., Yu, L., 2011. Synthesis of fluorinated polythienothiophene-*co*-benzodithiophenes and effect of fluorination on the photovoltaic properties. J. Am. Chem. Soc. 133, 1885–1894.

Sonmez, G., Shen, C.K.F., Rubin, Y., Wudl, F., 2004a. A red, green, and blue (RGB) polymeric electrochromic device (PECD): the dawning of the PECD era. Angewa. Chem. Int. Ed. 43, 1498–1502.

Sonmez, G., Sonmez, H.B., Shen, C.K.E., Wudl, F., 2004b. Red, green, and blue colors in polymeric electrochromics. Adv. Mater. 16, 1905–1908.

Sonogashira, K., Tohda, Y., Hagihara, N., 1975. A convenient synthesis of acetylenes: catalytic substitutions of acetylenic hydrogen with bromoalkenes, iodoarenes and bromopyridines. Tetrahedron Lett. 16, 4467–4470.

Sotzing, G.A., Lee, K., 2002. Poly(thieno[3,4-b]thiophene): a *p*- and *n*-dopable polythiophene exhibiting high optical transparency in the semiconducting state. Macromolecules 35, 7281–7286.

Stille, J.K., 1986. The palladium-catalyzed cross-coupling reactions of organotin reagents with organic electrophiles [new synthetic methods (58)]. Angew. Chem. Inter. Edition 25, 508–524.

Suzaki, Y., Osakada, K., 2003. Cyclization of dinuclear aryl- and aroylpalladium complexes with the metal centers tethered by an oligo(ethylene oxide) chain. Intramolecular transmetalation of the cationic dinuclear arylpalladium complexes. Organometallics 22, 2193–2195.

Svensson, M., Zhang, F., Veenstra, S.C., Verhees, W.J.H., Hummelen, J.C., Kroon, J.M., Inganäs, O., Andersson, M.R., 2003. High-performance polymer solar cells of an alternating polyfluorene copolymer and a fullerene derivative. Adv. Mater. 15, 988–991.

Tamao, K., Sumitani, K., Kumada, M., 1972. Selective carbon bond formation by cross-coupling of Grignard-reagents with organic halides-catalysis by nickel-phosphine complexes. J. Am. Chem. Soc. 94, 4374–4376.

Tang, C., Liu, X.-D., Liu, F., Wang, X.-L., Xu, H., Huang, W., 2013. Recent progress in polymer white light-emitting materials and devices. Macromol. Chem. Phys. 214, 314–342.

Thompson, B.C., Kim, Y.-G., Mccarley, T.D., Reynolds, J.R., 2006. Soluble narrow band gap and blue propylenedioxythiophene-cyanovinylene polymers as multifunctional materials for photovoltaic and electrochromic applications. J. Am. Chem. Soc. 128, 12714–12725.

Tokito, S., Momii, T., Murata, H., Tsutsui, T., Saito, S., 1990. Polyarylenevinylene films prepared from precursor polymers soluble in organic-solvents. Polymer 31, 1137–1141.

Toshima, N., Hara, S., 1995. Direct synthesis of conducting polymers from simple monomers. Prog. Polym. Sci. 20, 155–183.

Tu, G.L., Mei, C.Y., Zhou, Q.G., Cheng, Y.X., Geng, Y.H., Wang, L.X., Ma, D.G., Jing, X.B., Wang, F.S., 2006. Highly efficient pure-white-light-emitting diodes from a single polymer: polyfluorene with naphthalimide moieties. Adv. Funct. Mater. 16, 101–106.

Wagaman, M.W., Grubbs, R.H., 1997. Synthesis of organic and water soluble poly(1,4-phenylenevinylenes) containing carboxyl groups: living ring-opening metathesis polymerization (ROMP) of 2,3-dicarboxybarrelenes. Macromolecules 30, 3978–3985.

Waltman, R.J., Bargon, J., 1986. Electrically conducting polymers—a review of the electropolymerization reaction, of the effects of chemical-structure on polymer film properties, and of applications towards technology. Canadian J. Chem. 64, 76–95.

Wang, E., Wang, L., Lan, L., Luo, C., Zhuang, W., Peng, J., Cao, Y., 2008. High-performance polymer heterojunction solar cells of a polysilafluorene derivative. Appl. Phys. Lett. 92, 033307.

Wang, E., Wang, M., Wang, L., Duan, C., Zhang, J., Cai, W., He, C., Wu, H., Cao, Y., 2009. Donor polymers containing benzothiadiazole and tour thiophene rings in their repeating units with improved photovoltaic performance. Macromolecules 42, 4410–4415.

Wang, Q., Takita, R., Kikuzaki, Y., Ozawa, F., 2010. Palladium-catalyzed dehydrohalogenative polycondensation of 2-bromo-3-hexylthiophene: an efficient approach to head-to-tail poly(3-hexylthiophene). J. Am. Chem. Soc 132, 11420–11421.

Wessling, R., Zimmerman, R., 1968. Patent 3,401,152, 19,661,103. Chem. Abstr. 87735.

Wienk, M.M., Kroon, J.M., Verhees, W.J.H., Knol, J., Hummelen, J.C., Van Hal, P.A., Janssen, R.A.J., 2003. Efficient methano 70 fullerene/MDMO-PPV bulk heterojunction photovoltaic cells. Angew. Chem. Internat. Ed. 42, 3371–3375.

Wu, C.G., Lu, M.I., Chang, S.J., Wei, C.S., 2007. A Solution-processable high-coloration-efficiency low-switching-voltage electrochromic polymer based on polycyclopentadithiophene. Adv. Funct. Mater. 17, 1063–1070.

Wu, H., Ying, L., Yang, W., Cao, Y., 2009. Progress and perspective of polymer white light-emitting devices and materials. Chem. Soc. Rev. 38, 3391–3400.

Wudl, F., Kobayashi, M., Heeger, A.J., 1984. Poly(isothianaphthene). J. Org. Chem. 49, 3382–3384.

Yagyu, T., Hamada, M., Osakada, K., Yamamoto, T., 2001. Cationic arylpalladium complexes with chelating diamine ligands, [PdAr(N—N)(solv)]BF4 (N—N=N,N,N′,N′-tetramethylethylenediamine, 2,2′-bipyridine, 4,4′-dimethyl-2,2′-bipyridine). Preparation, intermolecular coupling of the aryl ligands, and insertion of alkyne and allene into the Pd—C bond. Organometallics 20, 1087–1101.

Yamamoto, T., Morita, A., Miyazaki, Y., Maruyama, T., Wakayama, H., Zhou, Z.H., Nakamura, Y., Kanbara, T., Sasaki, S., Kubota, K., 1992. Preparation of π-conjugated poly(thiophene-2,5-diyl), poly(*p*-phenylene), and related polymers using zerovalent nickel complexes. Linear structure and properties of the π-conjugated polymers. Macromolecules 25, 1214–1223.

Yang, L., Zhou, H., You, W., 2010. Quantitatively analyzing the influence of side chains on photovoltaic properties of polymer-fullerene solar cells. J. Phys. Chem. C 114, 16793–16800.

Ying, L., Ho, C.-L., Wu, H., Cao, Y., Wong, W.-Y., 2014. White polymer light-emitting devices for solid-state lighting: materials, devices, and recent progress. Adv. Mater. 26, 2459–2473.

Yokoyama, A., Miyakoshi, R., Yokozawa, T., 2004. Chain-growth polymerization for poly(3-hexylthiophene) with a defined molecular weight and a low polydispersity. Macromolecules 37, 1169–1171.

Yu, C.Y., Turner, M.L., 2006. Soluble poly(*p*-phenylenevinylene)s through ring-opening metathesis polymerization. Angew. Chem. Int. Ed. 45, 7797–7800.

Yu, W.L., Pei, J., Huang, W., Heeger, A.J., 2000. Spiro-functionalized polyfluorene derivatives as blue light-emitting materials. Adv. Mater. 12, 828–831.

Zhang, Q.T., Tour, J.M., 1998. Alternating donor/acceptor repeat units in polythiophenes. intramolecular charge transfer for reducing band gaps in fully substituted conjugated polymers. J. Am. Chem. Soc. 120, 5355–5362.

Zhang, K., Chen, Z., Zou, Y., Gong, S., Yang, C., Qin, J., Cao, Y., 2009. Effective suppression of intra- and interchain triplet energy transfer to polymer backbone from the attached phosphor for efficient polymeric electrophosphorescence. Chem. Mater. 21, 3306–3314.

Zhang, X., Richter, L.J., Delongchamp, D.M., Kline, R.J., Hammond, M.R., Mcculloch, I., Heeney, M., Ashraf, R.S., Smith, J.N., Anthopoulos, T.D., Schroeder, B., Geerts, Y.H., Fischer, D.A., Toney, M.F., 2011. Molecular packing of high-mobility diketo pyrrolo-pyrrole polymer semiconductors with branched alkyl side chains. J. Am. Chem. Soc. 133, 15073–15084.

Zhen, H., Xu, W., Yang, W., Chen, Q., Xu, Y., Jiang, J., Peng, J., Cao, Y., 2006. White-light emission from a single polymer with singlet and triplet chromophores on the backbone. Macromol. Rapid Commun. 27, 2095–2100.

Zhou, H., Yang, L., Price, S.C., Knight, K.J., You, W., 2010. Enhanced photovoltaic performance of low-bandgap polymers with deep LUMO levels. Angew. Chem. Int. Ed. 49, 7992–7995.

Zhou, H., Yang, L., Stuart, A.C., Price, S.C., Liu, S., You, W., 2011. Development of fluorinated benzothiadiazole as a structural unit for a polymer solar cell of 7% efficiency. Angew. Chem. Int. Ed. 50, 2995–2998.

Zhou, H., Yang, L., You, W., 2012. Rational design of high performance conjugated polymers for organic solar cells. Macromolecules 45, 607–632.

Zhu, R., Wen, G.-A., Feng, J.-C., Chen, R.-F., Zhao, L., Yao, H.-P., Fan, Q.-L., Wei, W., Peng, B., Huang, W., 2005. Di-channel polyfluorene containing spiro-bridged oxadiazole branches. Macromol. Rapid Commun. 26, 1729–1735.

3

Structure and Property of Electronic Polymers

3.1 Introduction

The conception of electronic polymers refers to polymers that are used in energy conversion and storage systems including solar cells, rechargeable batteries and supercapacitors, and the other electronic devices. Polymers used for electronics are often appreciated for these merits: low cost (Zhang et al., 2015b), environmental compatibility (Weng et al., 2015), mechanical flexibility (Granero et al., 2011), large scale processability (Li et al., 2014), as well as high electrical and optical properties. As the performances of the electronic devices strongly rely on the polymers therein, a deep understanding and a clear picture of the correlation between the chemical components, molecular structure and aggregation morphology of polymers, and their electronic properties will definitely underpin the development of devices. The structure-property relations of various polymers used in electronic devices are described accordingly.

3.2 Aggregation Structure

Containing hundreds and thousands of monomeric units, the hierarchical structure of polymers also gives rise to their diversity in aggregations—polymers are not distinctly amorphous or crystalline. The aggregation of polymers often has amorphous subdivisions and crystalline parts which make them exhibit combined features of amorphisms as well as crystals.

Electronic structure of conjugated polymers is of current interest because of the wide range of potential applications for such materials in optoelectronic devices. It is increasingly clear that the electronic properties of conjugated polymers depend sensitively on the physical conformation of the polymer chains and the way the chains pack together in films (Isoniemi et al., 2015; Massonnet et al., 2015; Scrosati et al., 2011; Ugur et al., 2015; Zhang et al., 2015a; Shi et al., 2015). Here we mainly cover the evidence that interchain (adjacent chains) electronic species do form in conjugated polymer films, and that their number and chemical nature depend on processing conditions; the chain conformation, degree of interchain contact, and rate of energy transfer can be controlled by factors, such as, choice of solvent, polymer concentration, thermal annealing, and presence of electrically charged side groups. Taken together, the results reconcile many contradictions in the literature and provide a prescription for the optimization of conjugated polymer film morphology for device applications.

3.2.1 Amorphous

It is believed that the charge transport behavior in an amorphous material can be described as a series of incoherent hopping events. The states are localized by the electrostatic or conformational disorder of the polymer and can be further localized by charge–phonon couplings (Mott and Davis, 2012). It has been demonstrated that the electronic structure of polymers is dictated by the conformational disorder of the individual chains. The effect of electrostatic disorder is modest, and the effect of electronic coupling between chains is even less important. These findings suggest that the charge transport along the polymer chain and across chains may follow different mechanisms, as the interactions between states localized within the same chain or in different chains are different. Variable localization length and difference between interchain and intrachain (along a single backbone) transports increase the intricacy which makes results from theoretical calculations less compatible to any established transport models (Tessler et al., 2009), but instead, some more general models should be built to correctly account for realistic electronic structures (Rivnay et al., 2012). In particular, if the actual structure of low-molecule polymers, such as, poly(*p*-phenylene vinylene) (PPV) is slightly more ordered as a result of very slow equilibration, this will not affect the relation between conformation and electronic structure (Qin and Troisi, 2013). A model is necessary to simulate charge mobility in conjugated polymer. Rühle et al. investigated the charge mobilities in six different assemblies of doped and neutral states of polypyrrole (PPy). It had been found that for noncrystalline conjugated PPy, charge carrier mobility was insusceptible to the molecular conformation. In fact, the key morphological factor, as far as mobility is concerned, is not the regularity of polymer chains, but the local packing. The obtained mobilities, when converted to electrical conductivities, are of the same order of magnitude as the experimentally measured values (Rühle, 2010).

3.2.2 Crystalline

When we say a polymer is crystalline, it does not mean the polymer has periodically translational atomic or molecular structure as small molecules. In fact, the crystalline regions in polymer where chains are folded and arranged regularly, mingles with amorphous regions where the randomness of chains interrupts the ordering (Figs. 3.1A and B).

The charge mobility of electronic polymer, exemplified here by poly(3-hexylthiophene) (P3HT), between adjacent chains (interchain) along the stacking direction would be some 2–3 orders of magnitude lower than along a single backbone (intrachain) (Lan et al., 2010; Crossland et al., 2012). The boundaries formed where two crystal growth fronts meet (i.e., interspherulite boundaries) might be expected to represent a greater barrier to the passage of charge because of the significantly reduced likelihood of tie-chains bridging adjacent ordered regions compared to boundaries between quasiparallel lamellae within the spherulite (Fig. 3.1B) (Street et al., 2005).

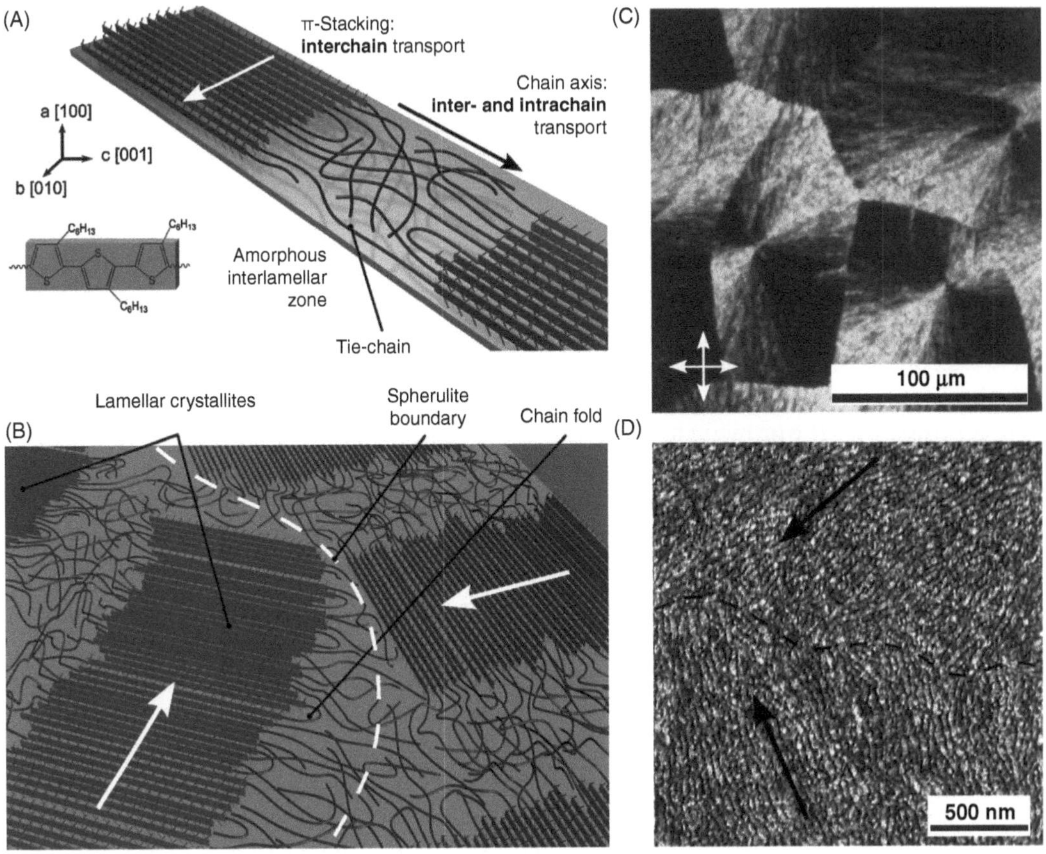

FIGURE 3.1 Molecular and crystalline morphology in P3HT film. (A) Edge-on orientation of chains with side chain *(a-axis)* standing parallel to the substrate. Quasi-parallel nanocrystalline lamellae consists of π-stacked aggregates separated by amorphous zones with occasional tie-molecules spanning the adjacent lamellae. (B) Ordered regions at interspherulite boundaries being far less likely to be bridged by tie-molecules. (C) POM image of a spherulitic morphology at the micron scale in a 25 nm thick P3HT film crystallized at 83%. (D) Tapping-mode AFM image (phase contrast) of a meeting point of two spherulites. Radial growth directions and the approximate location of a spherulite boundary are highlighted in (B) and (D). *Reproduced with permission from reference Crossland, E.J., Tremel, F.K., Fischer, F.F., Rahimi, F.K., Reiter, F.G., Steiner, F.U., Ludwigs, F.S., 2012. Anisotropic charge transport in spherulitic poly (3-hexylthiophene) films. Adv. Mater. 24, 839–844. Copyright 2012, Wiley-VCH.*

Fig. 3.1C shows a typical spherulitic morphology in the film after recrystallization using polarized optical microscopy (POM). The molecular-scale P3HT crystalline surface morphology at the boundary between adjacent spherulites is visualized by atomic force microscopy (AFM) in Fig. 3.1D. The alternating bright and dark stripes are attributable to crystalline lamellae and amorphous interlamellar zones, respectively. Wide-angle grazing incidence X-ray and transmission electron diffraction (not shown) are consistent with the edge-on orientation of the P3HT molecules ([100] alkyl chains standing parallel to the substrate). The b–c plane lies in the film with the b-axis [010] (π-stacking) being parallel to the radial growth direction and c-axis [001] (chain direction) spanning the lamellar width. The position and orientation of the spherulite boundaries in AFM correspond precisely to those visible in POM, proving that the in-plane molecular orientation implied by optical microscopy is present at the free surface of the film (Crossland et al., 2012).

The clarified poly(3,4-ethylenedioxythiophene):polystyrene sulfonate (or PEDOT:PSS) structure indicates that a nanocrystal of PEDOT surrounded by PSS is grown in the solid film from randomly oriented PEDOT in a micelle dispersed in water during the course of film fabrication. The addition of ethylene glycol (EG) to the PEDOT:PSS water dispersion and posttreatment of the pristine film with EG both provide similar improvements in PEDOT crystallinity (Kim et al., 2002; Ashizawa et al., 2005). The crystallite size of PEDOT increases up to a comparable size (~4.8 nm) of the hydrophobic PEDOT core region of the micelle. The electrical conductivity of the solid film is concurrently enhanced by 2 orders of magnitude with the growing nanocrystal of PEDOT (Ouyang et al., 2004). It is found that slow crystal growth conditions give high quality crystals. In order to control this process, tuning the evaporation speed by temperature, time, and humidity may be important, in addition to the selection of the solvent. It is noted that posttreatment with water can also improve the crystallinity and conductivity (Takano et al., 2012).

For both amorphous and crystalline structures in electronic polymers, an anisotropic aggregation often appears and proves important to study their structure-property relationships. In the case of polythiophene, two typical orientations with respect to the substrate surface are known; these are classified as edge-on and face-on orientations. The alkyl side chain (a-axis) is normal to the film plane and both the chain backbone (c-axis) and pi-stacking direction (b-axis) are parallel to the film plane for edge-on orientation, whereas the pi-stacking direction is normal to the film plane and both the alkyl side chain and the backbone are parallel to the film plane in face-on orientation (Hartmann et al., 2011). Recently, friction-transfer (Hosokawa et al., 2012), mechanical rubbing (Biniek et al., 2013), strain-alignment (O'Connor et al., 2011), and flow coating (Schuettfort et al., 2012) technique are verified to be promising methods to prepare oriented thin polythiophene films. All of these methods exploit shear forces to align polymer chains along the direction of rubbing or friction. In some cases, the change of chain conformation is induced from coiled to stretched as for instance in the generation of shish–kebab structures of polyolefins in the presence of elongational flow.

Epitaxy of semicrystalline polymers on aromatic organic crystals is an original and elegant method to grow highly crystalline and oriented polymer thin films with a controlled and regular organization of crystalline domains on a surface. Epitaxy of semicrystalline polymers can be achieved on a large variety of substrates including inorganic substrates, such as, inorganic or infusible aromatic salts, aromatic molecular crystals, and polymeric alignment layers (Rickert and Baer, 1978; Wittmann et al., 1983; Da Costa et al., 1998). Besides material design, epitaxy is a unique method to prepare semiconducting polymer films with high levels of crystallinity and orientation. Such films are ideally suited for fine structural investigations using electron diffraction, especially for the present polymers that seldom crystallize in the form of extended single crystals.

3.3 Morphology of Assembly

So far, many synthetic strategies have been developed for the preparation of a variety of morphologies of conducting polymer materials, for example, nanoparticles, nanofibers, continuous films, porous films, and polymer gels. With a rational synthetic design, structures with different diameters/widths, thicknesses, and lengths can be fabricated with the "morphology control" and "size control" of conducting polymers.

3.3.1 Nanoparticle

Nanoparticles can be regarded as zero-dimensional nanomaterials, the size (typically 1–100 nm) of which is larger than the atomic clusters but smaller than the granules. Nanoparticles share many unique properties linked to its quantum size and surface which has wide application prospects, such as, in physics, chemistry, engineering, and medical science. PEDOT is first discussed as a demonstration. A spectrum of synthetic methods has been developed to prepare high-quality PEDOT nanoparticles, generally, in a solid sphere format. For instance, they were effectively prepared with a good control on the morphology and stability in presence of iron (III) dodecylbenzenesulfonate and ω-functionalized polyisoprenes as costabilizers. The electrical conductivity of the resulting PEDOT bulk material reached 1 S cm^{-1} or even higher.

Besides the solid structure, core-shell and hollow nanoparticles also attract increasing attentions in the recent years. Polymer-coated hollow sulfur particles were prepared through an in situ polymerization (Fig. 3.2) and then used as promising cathode materials in lithium sulfur batteries (Li et al., 2013). The polymer coatings transferred the electrons and prevented the leakage of polysulfides. Poly[(4-styrenesulfonic acid)-*co*-(maleic acid)] (PSS-*co*-MA) anions had been used as active sites for polymerizing 3,4-ethylenedioxythiophene (EDOT) and as soft templates to form core/shell n-PEDOT:PSS-*co*-MA nanoparticles that were dispersible in water. The electrical conductivity was affected by the doping level of PSS-*co*-MA and the core/shell structure. The electrical conductivities of the n-PEDOT and n-PEDOT:PSS-*co*-MA were measured as 3.2 and 0.29 S cm^{-1}, respectively. The core/shell nanoparticle consisted of an insulating PSS-*co*-MA layer wrapped around an n-PEDOT core, which increased the interface dipole and the hole-injection barrier, thereby decreasing the electrical conductivity compared with the n-PEDOT. Moreover, the morphologies of these nanoparticles could be differed by varying the concentration and molecular weight of the PSS-*co*-MA stabilizer (Han et al., 2011).

For a variety of other conjugated polymers that are soluble in organic solvents, the suspensions of their nanoparticles are typically prepared by adding a conjugated polymer solution in a water-miscible organic solvent to water. This preparation does not involve emulsion polymerization or use any surfactant. The suspended nanoparticles reveal a reduction in the mean conjugation length associated with bending or kinking of the polymer backbone as well as energy transfer at intrachain aggregation states. Among them, polyaniline (PANI) has been mostly investigated for electrochemical, electrorheological, and

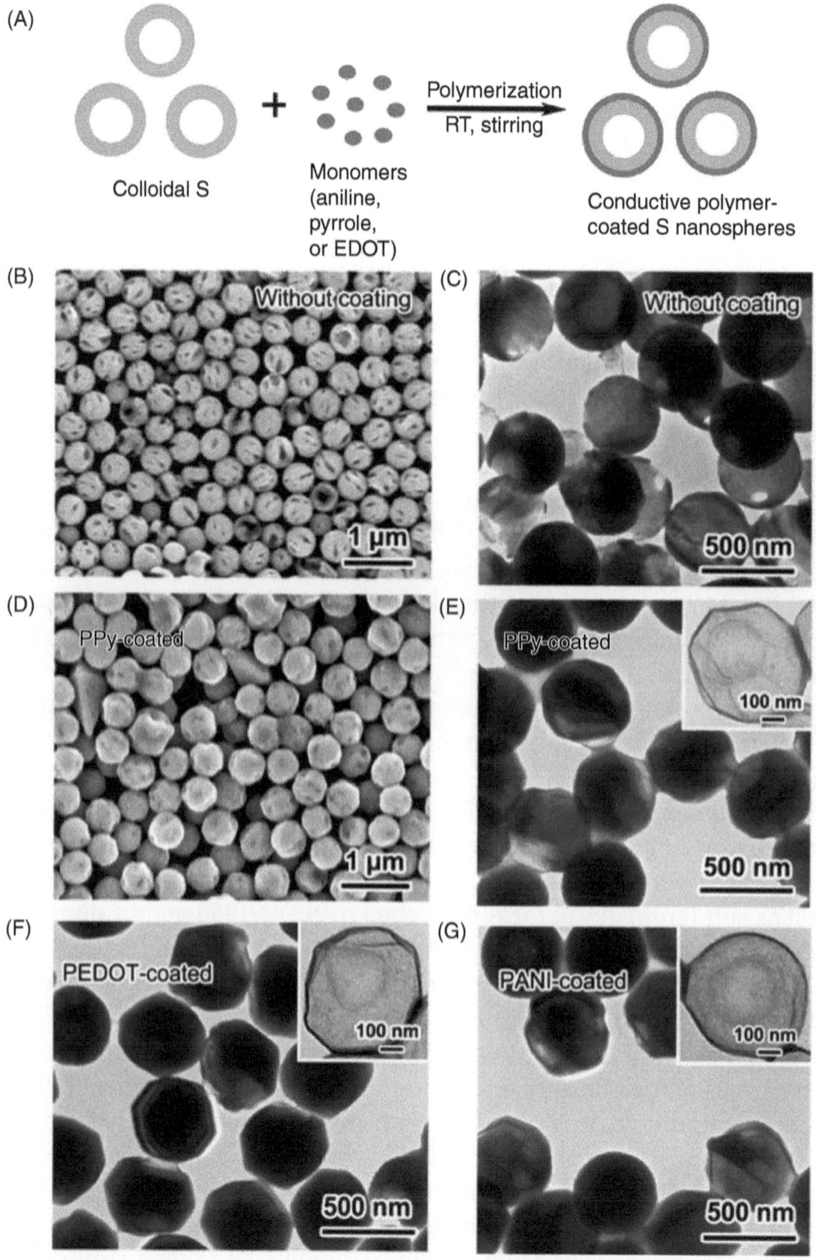

FIGURE 3.2 (A) Schematic illustration of the fabrication process of conductive polymer-coated hollow sulfur nanospheres. *RT* means room temperature. (B) and (D) Scanning electron microscopy (SEM) and (C) and (E) transmission electron microscopy (TEM) images of the hollow sulfur nanospheres before and after coating with polypyrrole *(PPy)*. (F) and (G) TEM images of the poly(3,4-ethylenedioxythiophene) *(PEDOT)* and polyaniline *(PANI)*-coated hollow sulfur nanospheres, respectively. Insets in E, F, G: TEM images of the *PPy, PEDOT,* and *PANI* shell after dissolving sulfur with toluene, respectively. *Reproduced with permission from reference Li, W., Zhang, Q., Zheng, G., She, Z.W., Yao H., Cui, Y., 2013. Understanding the role of different conductive polymers in improving the nanostructured sulfur cathode performance. Nano Lett. 13, 5534–5540. Copyright 2013, American Chemical Society.*

electronic applications (Park et al., 2004) due to a facile synthetic process, tailorability by nonredox acid/base doping, high environmental stability, easy control in electrical conductivity, and possibility for a large-scale production (Cho et al., 2004). Monodisperse PANI nanoparticles were synthesized from an oxidative dispersion polymerization using poly(sodium 4-styrenesulfonate) (PSSA) as both a polymeric stabilizer and a dopant agent due to its acidity. As expected, the molecular weight of the PSSA was found to obviously affect the sizes of PANI nanoparticles, while their electrical conductivities were slightly varied (Park et al., 2004).

The water-soluble stabilizers at the outer surfaces of PANI nanoparticles typically include PSSA and PVA. For both cases, the resulting particles show a low diameter distribution of with a uniform spherical shape. The diameters are appropriately 40 nm with the PSSA and are varied from 100 to 150 nm for the poly(vinyl alcohol) (PVA). The electrical conductivities of the PANI nanoparticles with the use of the PSSA exceed those prepared from the PVA (Cho et al., 2004). Some modified polymerizations have been further developed to enhance the properties of PANI nanoparticles. For instance, polymerization was carried out in a thermostated bath with the assistance of dodecyl benzne sulfonic acid (DBSA) to produce electrical conductivities of 15 ± 3 S cm^{-1} at room temperature (Cho et al., 2004). These PANI nanoparticles were found to be particularly useful for electronic textiles (Moulton et al., 2004).

3.3.2 Nanofiber

Compared with zero-dimensional nanoparticles, the one-dimensional nanofibers are extensively studied for structure characteristics of large aspect ratios and high curvatures that offer high specific surface areas with rapid charge transport along the length direction and remarkable electrochemical activities. The aggregation structure may be further controlled to form networks or assemble into arrays to meet the different requirements in a broad spectrum of fields, such as, composite materials, filtering membranes, and biological materials besides for various electronic materials.

Regioregular P3HTs are the mostly explored organic semiconductors in electronic devices, such as, field-effect transistors, sensors, and solar cells. P3HT nanofibers have been prepared by various routes, among which electrospinning and self-assembly have been widely investigated. Electrospinning represents a general and effective method in producing long and continuous fibers with diameters ranged from 10 nm to submicron. Due to absence of chain entanglement of rigid rod-like P3HT in a solution, the continuous spinning of uniform P3HT nanofibers requires either mixture with high-molecular-weight insulating polymers or use of highly concentrated solutions. The mixture of conductive P3HT with a nonconductive polymer lowers the electrical conductivity compared to the bare P3HT. To avoid this drawback, core-shell P3HT nanofibers have been designed by a coaxial electrospinning process. The shell less conductive or nonconductive shell layers need to be removed for high electronic properties. As shown in Fig. 3.3A, highly aligned core/shell P3HT/poly(methyl methacrylate) (PMMA) nanofibers were first prepared by the electrospinning, and the following removal of the PMMA shell produced bare P3HT nanofibers

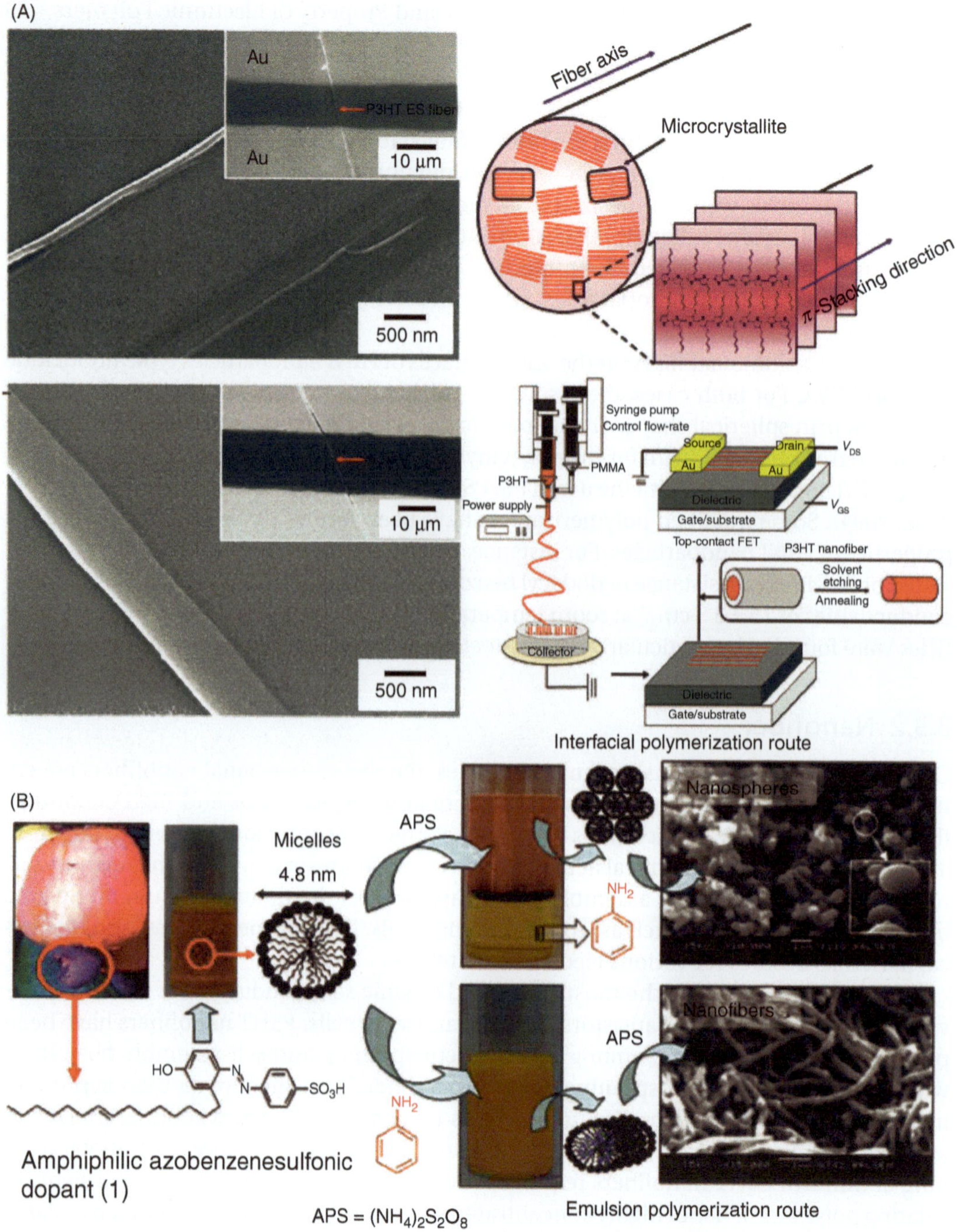

FIGURE 3.3 (A) SEM images of PHTNF-2 (left top) and PHTNF-4 (left bottom), schematics of the coaxial electrospinning setup and process to fabricate the aligned P3HT nanofiber-based OFET (right top), and schematics of the inner microstructure of single P3HT nanofiber, (Chen et al., 2011). (B) Synthesis of PANI nanomaterials using a renewable resource of amphiphilic azobenzenesulfonic acid as a dopant. *(A) Reproduced with permission from reference Chen, J.-Y., Kuo, C.-C., Lai, C.-S., Chen, W.-C., Chen, H.-L., 2011. Manipulation on the morphology and electrical properties of aligned electrospun nanofibers of poly (3-hexylthiophene) for field-effect transistor applications. Macromolecules 44, 2883–2892. Copyright 2011, American Chemical Society. (B) Reproduced with permission from reference Anilkumar, P., Jayakannan, M., 2007. Macromolecules. Single-molecular-system-based selective micellar templates for polyaniline nanomaterials: control of shape, size, solid state ordering, and expanded chain to coillike conformation. Macromolecules 40, 7311–7319. Copyright 2007, American Chemical Society.*

that had been fabricated into organic field-effect transistors with high performances (Chen et al., 2011). Recently, pure P3HT nanofibers were directly prepared through the use of a concentrated P3HT solution, and the physical entanglement among P3HT backbones in highly concentrated solutions improved the spinnability. The strong shear force and the geometrical confinement induce the orientation of polymer chains along the fiber axis.

PANI has also received a great deal of attention for the preparation into nanofibers. PANI nanofibers were typically synthesized through a rapid mixing method with the use of HCl and m-chlorobenzoic acid (m-CPBA) aqueous solution. The previous mixture was shaken vigorously and the reaction occurred under stirring at room temperature. The resulting PANI nanofibers exhibit a relatively large specific capacitance and high columbic efficiency due to the rapid diffusion of H^+ ions through the PANI during the redox process (Mahmood et al., 2013). It was also demonstrated to synthesize large quantities of uniform PANI nanofibers by a template-guided process in a dilute solution of sodium alginate. The diameters of the PANI nanofibers were easily controlled by the concentration of aniline and sodium alginate (Yu et al., 2006).

A strategy to use renewable resource for PANI nanofibers was developed from a raw material, cardanol, which is an industrial waste and pollutant from the cashew nut industry. A new amphiphilic molecule, 4-[4-hydroxy-2-((*Z*)-pentadec-8-enyl) phenylazo] benzenesulfonic acid, was studied for cardanol and employed as a dopant to produce PANI nanomaterials with a variety of morphologies including microspheres and nanotubes besides linear and dendritic nanofibers (Anilkumar and Jayakannan, 2007). For the PANI nanofibers, they displayed ~5 μm in length and 200 nm in diameter (Fig. 3.3B).

A facile chemical route was reported to synthesize high-quality PANI nanofibers under ambient conditions using aqueous/organic interfacial polymerization. The synthesis is based on the well-known chemical oxidative polymerization of aniline in a strongly acidic environment, with ammonium peroxydisulfate as the oxidant. The nanofibers showed uniform diameters between 30 and 50 nm with lengths varied from 500 nm to several micrometers (Huang et al., 2003). This aqueous/organic interfacial synthesis offers a few promising advantages: (1) the synthesis and purification are simple without any template, and it can be also easily scaled up with a good reproducibility; (2) a high production of >95% uniform nanofibers is achieved; (3) the PANI nanofibers can be well dispersed in water, which facilitates an environmentally friendly processing and biological applications.

PANI nanofibers could be also synthesized from hard templates like alumina and zeolites, soft templates like gels, and liquid crystal, emulsion, dispersion, interfacial seeding, and dilute polymerizations methods. Among them, the emulsion method has attracted increasing interests due to the formation of high-quality nanomaterials and the simplicity in adapting synthetic conditions. The synthesis was susceptible to the dopant/aniline ratio, and the uniform nanofibers were available only for specific compositions (Anilkumar and Jayakannan, 2009). The use of novel amphiphilic sulfonic acid dopants as structure directing agents is underway to further enhance the quality of PANI nanofibers while with low cost.

3.3.3 Solid Film

Organic polymers have been extensively studied to be made into films with both high strength and toughness. In the case of the electronic applications, the conducting polymers films attract considerable interests as they are lightweight and flexible compared with their inorganic counterparts (Grabowski et al., 2014).

PEDOT:PSS represents one of the most explored conductive polymers and is available for some commercial applications in the fabrication of low-cost, flexible, and printable electronic devices. Although the PEDOT:PSS films are also widely investigated for high electrical conductivities (Liu et al., 2015), they remain obviously lower than their inorganic counterparts (Döbbelin et al., 2007). Even worse, the performances, such as efficiency and lifetime of the electronic devices are deteriorated as PSS is strongly acidic and hygroscopic. Therefore, it is critical to further improve the electrical conductivity and stability of the PEDOT:PSS film.

It proves effective to introduce new reactive polyisoprene stabilizers for high-performances PEDOT:PSS films. The electrical conductivity of the PEDOT:PSS film was increased from 14 to 136 S cm^{-1} with the assistance of ionic liquids (Döbbelin et al., 2007). Similarly, with the addition of dimethyl sulfate, the electrical conductivity had been enhanced on the basis of preserving the high flexibility and visible-light transparency in the PEDOT:PSS film. This enhancement is produced, principally, by a decrease in the pH of the solution and the replacement of some PSS^- units with SO_4^{2-} ions. The morphology and size of the conductive domains, and the partial PSS phase separation are another two key factors that affect the electrical conductivity (Reyes-Reyes et al., 2010). A conclusive explanation for the origin of the high electrical conductivity was revealed by growing PEDOT nanocrystals in different solvents (Takano et al., 2012). A series of PEDOT:PSS films was prepared by mixing a PEDOT:PSS aqueous solution with EG, poly(ethylene glycol) (PEG), methanol, or formic acid. The maximal electrical conductivity had been enhanced to 1900 S cm^{-1} with the use of formic acid.

For the PEDOT:PSS film, PSS is used as the counter ion, charge compensator, and template for polymerization of PEDOT; PSS also makes it easily dispersible in water. However, PSS, a nonionized and insulating dopant, is present in excess and has been recognized as the main reason for the relatively low electrical conductivity of the commercial PEDOT:PSS material. Selective removal of the PSS has been thus studied as an effective strategy to enhance the electrical conductivity. The treatments using secondary dopants affect the structures of PEDOT:PSS films, which in turn also affect their electrical conductivities. After treatments, the grain sizes of particles are increased to reduce their boundaries in a given volume or area with lower energy barriers for charge conduction, which facilitates the charge hopping among polymer chains. The charge hopping is believed to be the dominant conduction mechanism in conducting polymers (Mengistie et al., 2014). Therefore, the increase in electrical conductivity after addition of diethylene glycol (DEG) was demonstrated and studied by using the PEDOT:PSS film as the anode to fabricate polymer light-emitting diodes (PLEDs). Poly[2-methoxy-5-(2-ethylhexyloxy)-1,4-phenylenevinylene], or called MEH–PPV, was coated at the top of the PEDOT:PSS film without or with DEG (Crispin et al., 2006). The electrical conductivity of a pristine PEDOT:PSS was ~0.006 S cm^{-1}, and it sharply increased to 10 S cm^{-1} at a DEG weight concentration of 0.3%. This great increase in electrical conductivity was not produced by the screening effect of the interaction between ions and charge carriers by the solvent, which would be progressive, but rather a phase separation (Crispin et al., 2006).

To investigate whether and how the phase separation occurred, the surface morphologies of PEDOT and PSS were traced. For a polymer material studied by AFM at tapping model, a bright area in the phase image is assigned to a relatively hard phase. In other words, the regions with lower phase signals correspond to softer materials with respect to the surrounding. In the PEDOT:PSS film, the regions or islands are thus attributed to the excess PSS component. The surface areas rich with this hygroscopic polyelectrolyte are expected to swell and soften due to the ambient humidity and/or the added DEG. Therefore, the added DEG dopant induced a phase separation at nanoscale (Crispin et al., 2006). The enhanced electrical conductivity of the PEDOT:PSS film with DEG was explained later. During the polymerization of EDOT monomers with an oxidant in the PSS water solution, a lot of PEDOT chains are produced. However, a stable PEDOT:PSS emulsion is only available under vigorous shaking to increase the contact area between PEDOT and PSS that are assembled into particles in the PSS water solution. For the aforementioned process, the mechanical energy is transformed to interfacial energy between the PEDOT and PSS phases, and a stable emulsion is verified by a positive Gibbs energy. Spin coating such an emulsion produces the PEDOT:PSS film with PEDOT:PSS particles surrounded by the excess PSS. The PEDOT:PSS particles are barely touched with each other, so the electrical conductivity of the resulting film is relatively low.

A lot of efforts are also made to study the morphology and charge transport of regioregular P3HT. Weight-average molecular weight (M_W) is a key to them. As expected, a lower M_W displays a higher degree of crystallinity. Surprisingly, the charge carrier mobility is decreased by 4 orders of magnitude when the M_W is reduced by 1 order of magnitude. The neighboring backbones are more likely to show similar orientations, and a better intergrain transport may be achieved, while the increase in the in-plane pi-stacking results in both higher intragrain transport and mobility. Furthermore, the poor connectivity and insulating grain boundaries among nonoriented neighboring crystals limit the charge carrier mobility of as-spun low-M_W films. Allowing more time for crystallization or modifying the surface treatment facilitates the crystals to preferentially orient with their insulating side chains being normal to the substrate, thus reducing the number of in-plane insulating grain boundaries and increasing the charge carrier mobility. The chains of the medium and high M_W films are longer than the domains and minimize the effects of the grain boundaries by bridging neighboring grains (Kline and McGehee, 2006).

P3HT films with controlled thicknesses, typically from 1 to 19 nm, are spin-cast from P3HT-chlorobenzene solutions. By scanning the potential positively, the polymer film can be oxidized with more positive potentials that correspond to higher percentages of oxidized thiophene species. The extent of electrochemical oxidation is determined by integrating the oxidative scan of the cyclic voltammetry. After the initial voltammetric characterization, P3HT films are doped to desired levels using potential steps (choronoamperometry); the film is stepped to the desired potential and then rapidly removed from the solution under potential control. Electrochemically doped polymers provide a unique methodology for quantifying the extent of doping. The charge exchange is traced by studying the complementary fraction of reduced species in the contact material with the P3HT. A contact material shows a well-defined spectral feature associated with a change in oxidation state (e.g., molybdenum oxide). Finally, the difference in chemical composition between the contact material/P3HT interface and bulk film can be investigated using angle-resolved X-ray photoelectron spectroscopy, providing a facile way to determine whether more stringent thickness-dependent measurements are necessary to describe a specific substrate/semiconductor interaction (Shallcross et al., 2015).

3.3.4 Porous Film

Porous conducting polymers are of great interest because of the huge potential to enhance their physical properties on the basis of high specific surface areas. Currently, the in situ polymerization has been mostly used in the preparation of porous conductive films, mainly due to the advantages of mild conditions in reaction and availability in effectively controlling their structures and compositions. PEDOT:PSS is first studied as a demonstration here. For a solid PEDOT:PSS film, it shows a relatively low conductivity, low coatability on hydrophobic substrate, and decreased conductivity at large strains. To this end, a general and effective strategy is developed through the addition of a small amount of nonionic surfactant (Fig. 3.4A) (Oh et al., 2014). This surfactant can considerably reduce the surface tension of the PEDOT:PSS solution, thus permitting conformal coatings of PEDOT:PSS thin film on a diverse range of hydrophobic substrates. It also induces the formation of PEDOT nanofibrils during coating, which leads to a high conductivity and mechanical stability at large strains. A pristine PEDOT:PSS film displays a low electrical conductivity of 0.24 S cm^{-1}, whereas a much higher electrical conductivity of 100 S cm^{-1} has been achieved through the use of a surfactant. The electrical conductivity may be further increased to (347 ± 75) S cm^{-1} by washing with methanol on a glass substrate.

P3HT has been extensively employed in polymer-based organic devices including thin-film transistors, light-emitting diodes, and solar cells. The self-assembled one-dimensional nanostructures are found to largely enhance the crystallinity, mechanical flexibility, and the speed of charge transport, compared with the solid thin film. For a typical preparation, crystal seeds are prepared in solution through a cycle of cooling and heating treatments to a P3HT solution. The blend solution is then spin-coated onto a substrate, followed by drying in air to remove the solvent. The electronic properties of P3HT nanofibrils are compared with the conventional solid P3HT film. The power conversion efficiency of the solar cell based on the conventional coating for P3HT is typically low; in contrast, for the solar cell fabricated from the P3HT after a cooling/heating cycle, the power conversion efficiency is largely enhanced (Oh et al., 2012).

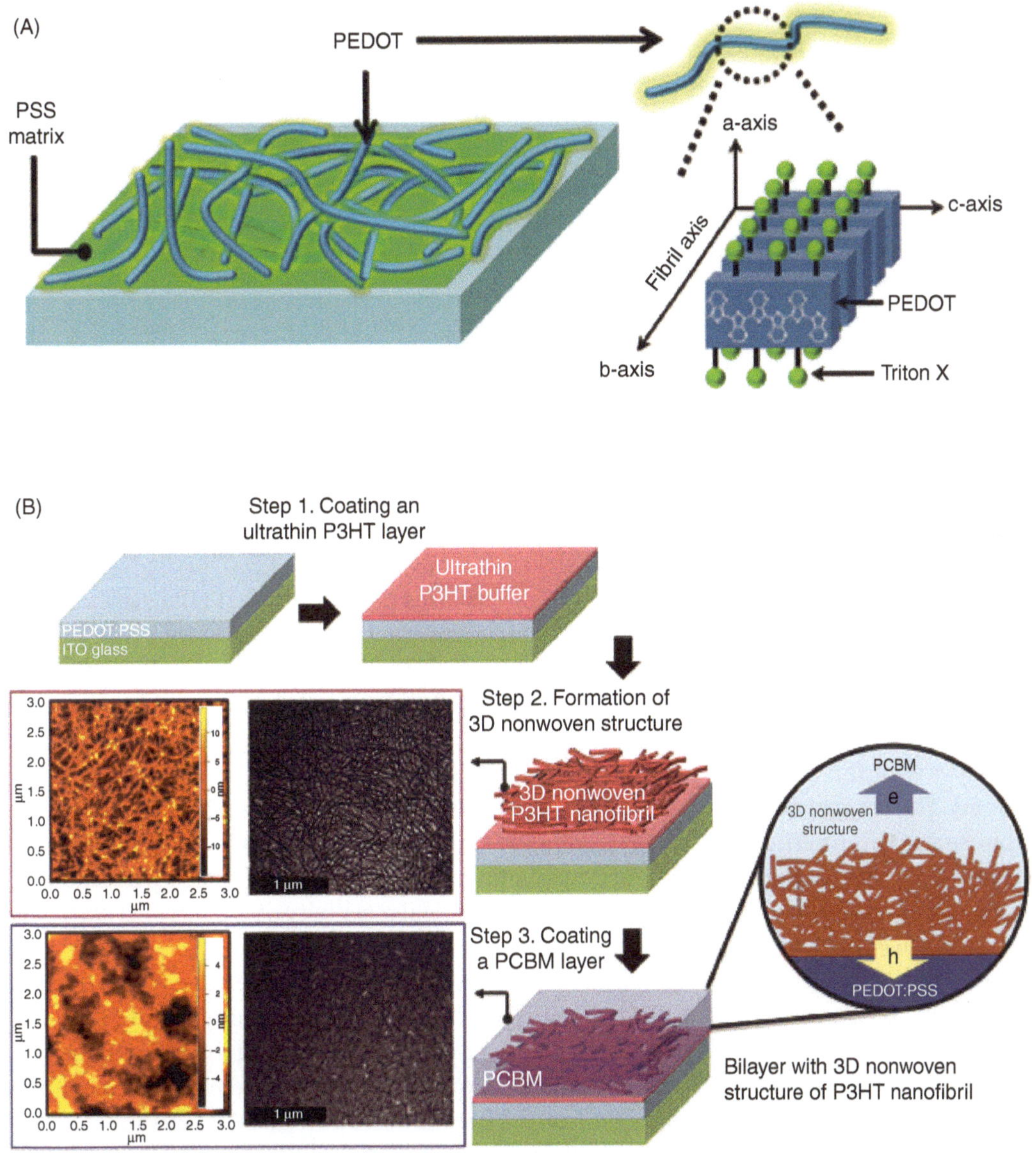

FIGURE 3.4 (A) Fabrication of polymer solar cell from PEDOT:PSS thin film. (B) Fabrication of a three-dimensional (3D) nonwoven nanofabric-based organic solar cell. *(A) Reproduced with permission from reference Oh, J.Y., Shin, M., Lee, J.B., Ahn, J.-H., Baik, H.K., Jeong, U., 2014. Effect of PEDOT nanofibril networks on the conductivity, flexibility, and coatability of PEDOT: PSS films. ACS Appl. Mater. Interfaces 6, 6954–6961. Copyright 2014, Royal Society of Chemistry. (B) Reproduced with permission from reference Oh, J.Y., Lee, T.I., Jang, W.S., Chae, S.S., Park, J.H., Lee, H.W., Myoung, J.-M., Baik H.K., 2013. Mass production of a 3D non-woven nanofabric with crystalline P3HT nanofibrils for organic solar cells. Energy Environ. Sci. 6, 910–917. Copyright 2013, American Chemical Society.*

The mechanism for the formation of P3HT nanofibrils by solvent annealing is summarized later. The monomers of P3HT that randomly aggregate in the amorphous film obtain enough energy to migrate and crystallize because they possess high mobility at a partially dissolved state through the diffusion of solvent molecules into the amorphous film. The activated monomers spontaneously assemble and form nanoscaled crystalline aggregates, that is, nanofibrils. As shown in Fig. 3.4B, a three-dimensional nonwoven structure for the P3HT nanofibrils is further maintained after coating a second material such as phenyl-C61-butyric acid methyl ester (PCBM), which is important for the electronic application, such as organic solar cells.

An in situ fabrication is thus developed for a large-scale production of organic solar cells from the nonwoven fabric based on crystalline P3HT nanofibrils. A P3HT solution is injected into a cooling double jacket by a syringe pump and then dispensed onto the PEDOT:PSS-coated substrate. After drying, a PCBM layer is spin or spray-coated onto the nonwoven fabric. The density and thickness of the P3HT nanofibril layer are controlled by varying the P3HT concentration. The bilayer solar cell is invented from the hybrid P3HT/PCBM fabric with power conversion efficiency near 4%, which represents the state-of-the-art value in bilayer based solar cell (Oh et al., 2013).

Pyrrole (Py) monomers can be diffused onto porous polyurethane (PU) elastomers and then polymerized in situ to synthesize a stretchable electric material with a thin layer of PPy strongly anchored at the top surface of PU. The structure and electrical conductivity of porous PU and PPy/PU elastomers are carefully investigated. After the polymerization of pyrrole monomers, the surfaces of both sides of the PU elastomer are smooth, and a PPy layer with the thickness of ~40 μm on and inside the porous PU elastomer has been formed. The large specific surface area of the porous PU promotes the diffusion of pyrrole monomers. This interpenetrating layer helps to strengthen the interface adhesion between PU and PPy. The strong interface effectively prevents the detachment of PPy from the PU framework under bending and stretching. The electrical resistivity of the PPy/PU elastomer reaches 8.364 Ω·cm. The structure and electrical property are further investigated during and after stretching and releasing. A lot of cracks on a micrometer scale appear at an elongation of 10%; the numbers of cracks are increased obviously at larger elongations of 30–50%. Simultaneously, the electrical resistances are increased sharply during the above stretching process. Importantly, the electrical resistances can be reversed after releasing, indicating a high stretchability and repeatability for human breath detection (Li et al., 2014).

PANI has been generally synthesized by electropolymerization, and the preparation can be well controlled by varying the polymerization time, applied potential, and current. The synthesis of nanostructured PANI materials is summarized later. Oppositely charged polystyrene (PS) nanoparticles are assembled onto the surface of a glassy carbon electrode to form the PS template on a conducting substrate. The electropolymerization of aniline is then carried out in an acidic solution containing aniline. As the sulfate-PS nanoparticles are negatively charged, they can absorb the positively charged aniline monomers, which polymerize around the PS nanoparticles to form PS/PANI core/shell nanoparticles.

The following removal of the PS nanoparticles produces a nanoporous PANI film. Here the size and shape of the PANI can be controlled by varying the electropolymerization time (Luo et al., 2007).

Three-dimensional highly ordered macroporous conducting polymers may be synthesized from a self-assembled colloidal template based on PS latex spheres. A working electrode is obtained by sequentially depositing a chromium adhesion layer and a gold layer onto a thin glass slide. The PS latex sphere template is then prepared by coating PS spheres onto the gold electrode for electropolymerization. After electrochemical deposition, the PS template is removed by soaking the conducting polymer film in toluene. The resulting macroporous conducting polymer films are electrically conductive, similar to the solid counterparts. The pore size and uniformity of the conducting polymer film can be well controlled by using the specific PS latex spheres. The ability to control the porosity and morphology shows many implications in terms of kinetic studies, property improvements, and electronic applications (Bartlett et al., 2001).

Increasing interests are also attracted to create macroscopically periodic materials for a range of applications. Some dielectric structures with a three-dimensional periodicity or called photonic crystals are demonstrated to be promising in low-loss waveguides, low-threshold lasers, optical switching elements, and radically improved photochemical reactors. They are typically prepared by template-directed electrodeposition. The used spheres in a template are generally arranged in a hexagonal format, while the formed voids among them appear in a triangular lattice. The ease to remove the template by solvent/acid etching or burning through the void is a key consideration in designing the template (Braun and Wiltzius, 2001).

3.3.5 Polymer Gel

Three-dimensionally nanostructured hydrogels based on conducting polymers provide intrinsic conducting frameworks that promote the transport of charges, ions, and molecules. They also provide an ideal interface between electronically and ionically transporting phases, between biological and synthetic phases, and between soft and hard phases. Therefore, such hydrogels are proposed as ideal candidates for energy storage, molecular electronics, and bioelectronics. The microstructure and mechanical and electrochemical properties can be controlled by varying the ratios of additives to monomers during synthesis (Shi et al., 2014).

Theoretically, the inherent rigidity of most conducting polymers leads to the generation of micropores (diameters of <2 nm) with high specific surface areas, for example, above 1000 $m^2\ g^{-1}$ for microporous poly(arylene ethynylene) (PAE) materials. Mesopores (diameters of 2–50 nm) and macroporous (diameters of >50 nm) are also produced for some conducting polymers, such as PANI and PPy by chemical oxidation or electropolymerization of the corresponding monomers in the presence of organic molecules or colloid particles as templates. The importance of the porous structure has inspired people to develop functional conducting polymer hydrogels, such as, PEDOT:PSS that are well-known as nonporous conducting polymers. The PEDOT:PSS hydrogel is prepared by an ice-segregation-induced self-assembly. Briefly, EDOT monomers are diluted with ethanol and NaPSS aqueous solution to form a milky dispersion, followed by stirring with oxidizing agent to form cryogel. The PEDOT:PSS cryogel includes considerable interconnected open macropores, which serve as effective channels for the transport of various ions and molecules.

The concentration plays a significant role in determining the structure of the cryogel. For example, a concentration of 5 wt.% produces brushy macropores, while the concentration of 10 wt.% results in smooth macropores. Importantly, each domain in a the monolithic cryogel is composed of aligned macropores, and the whole cryogel can be regarded as many similar domains integrated along the mutually longitudinal direction. This longitudinally aligned structure demonstrates a significant influence on the electrical conductivity of the resulting cryogel, for example, 10^{-3} S cm^{-1} along the aligned direction while 10^{-4}–10^{-6} S cm^{-1} in the radial direction for a PEDOT:PSS cryogel prepared from a 10 wt.% PEDOT:PSS dispersion precursor (Zhang et al., 2011).

Conducting polymers can be mixed with or electrochemically deposited onto transition metal oxides to achieve better performances as electrode materials. However, the wrapping of conducting polymers on metal oxides may reduce the effective contact area between metal oxides and electrolytes. This will further impede insertion of alkali ions into metal oxides and result in poor rate performance with low specific capacitance. A rational strategy is to use conducting polymers as active electrode materials directly and optimize their structures. To this end, a hierarchically nanostructured conductive polymer hydrogel was synthesized by a facile interfacial polymerization (Shi et al., 2014).

For the conducting polymer hydrogel, the added second phase plays a critical role on the performance. In the case of a PPy hydrogel, the ratio of pyrrole to phytic acid largely impacts the microstructure and electrical property (Fig. 3.5). When the ratio increases, the PPy structures are changed from interconnected microstructured hollow spheres to smaller plate-like particles. Accordingly, more pores have been formed, but the electrical conductivities are decreased (Shi et al., 2014).

To enhance the physical properties, such as electrical conductivities of conjugated polymers, a lot of studies are available to make polymer chains to adopt aligned, straight-chain conformations, which are realized by increasing the solid-state crystallinity, for example, conjugated polymers are incorporated into rigid matrices or packed more efficiently by molecular self-assembly. An amphiphilic conjugated polyelectrolyte, poly(fluorene-*alt*-thiophene) (PFT) that assembles into cylindrical micelles for high capability in conducting electrons efficiently over long distances, is demonstrated in this book. At an enough high concentration, PFT reversibly forms entangled hydrogels. Strain-controlled oscillatory rheological measurements indicate that these hydrogels display solid-like properties, for example., a high degree of strain hardening. This network of semiflexible conjugated polymer micelles is electrically conductive in three dimensions, making it a promising material for optoelectronic applications.

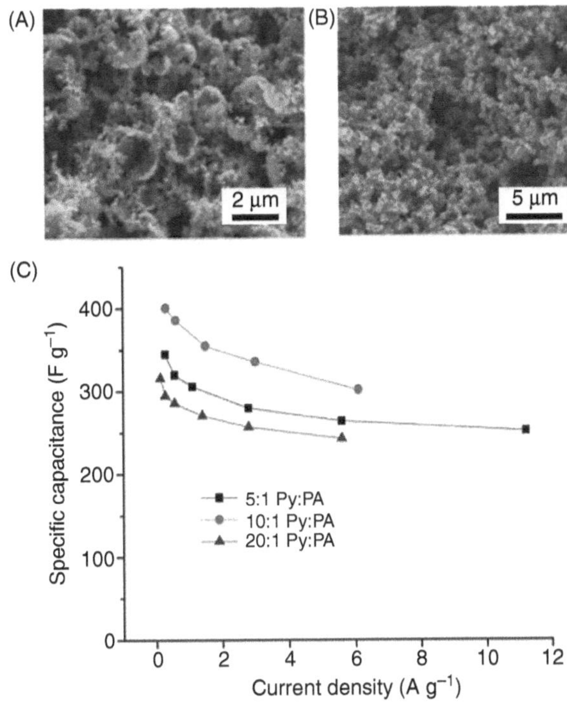

FIGURE 3.5 (A) and (B) SEM images of a PPy hydrogels with the ratio of pyrrole *(Py)* to phytic acid *(PA)* as 5:1 and 10:1, respectively. (C) Specific capacitance versus current density profiles of PPy hydrogel electrodes with Py/PA ratios of 5:1, 10:1, and 20:1. *Reproduced with permission from reference Shi, Y., Pan, L., Liu, B., Wang, Y., Cui, Y., Bao, Z., Yu, G., 2014. Nanostructured conductive polypyrrole hydrogels as high-performance, flexible supercapacitor electrodes. J. Mater. Chem. A 2, 6086–6091. Copyright 2014, Royal Society of Chemistry.*

3.4 Properties

3.4.1 Mechanical Properties

The charge-transport properties are the core of electronic polymers. However, the mechanical properties likewise occupy a range of values that are associated with the stability of electronic devices. The mechanical response of the polymer upon indentation depends on its elastic properties, which are mainly related to the local density, degree of crystallinity, and arrangement of the polymer molecule (Camposeo et al., 2013). The all-sp^2 hybridization of the carbon atoms in the main chain of a conjugated polymer is responsible for semiconducting behavior but also produces high stiffness and frangibility (Seitz, 1993). Therefore, continuous films made of conjugated polymers show a low mechanical integrity, associated with adhesive and cohesive debonding of the layers (Dupont et al., 2014). Moreover, the mechanical behavior highly depends on the molecular structure, such as, the presence of fused or isolated rings in the main chain, length and composition of pendant groups, propensity to form crystallites, and microstructural order (Savagatrup et al., 2014b). Polythiophenes are greatly different in tensile modulus due to the different

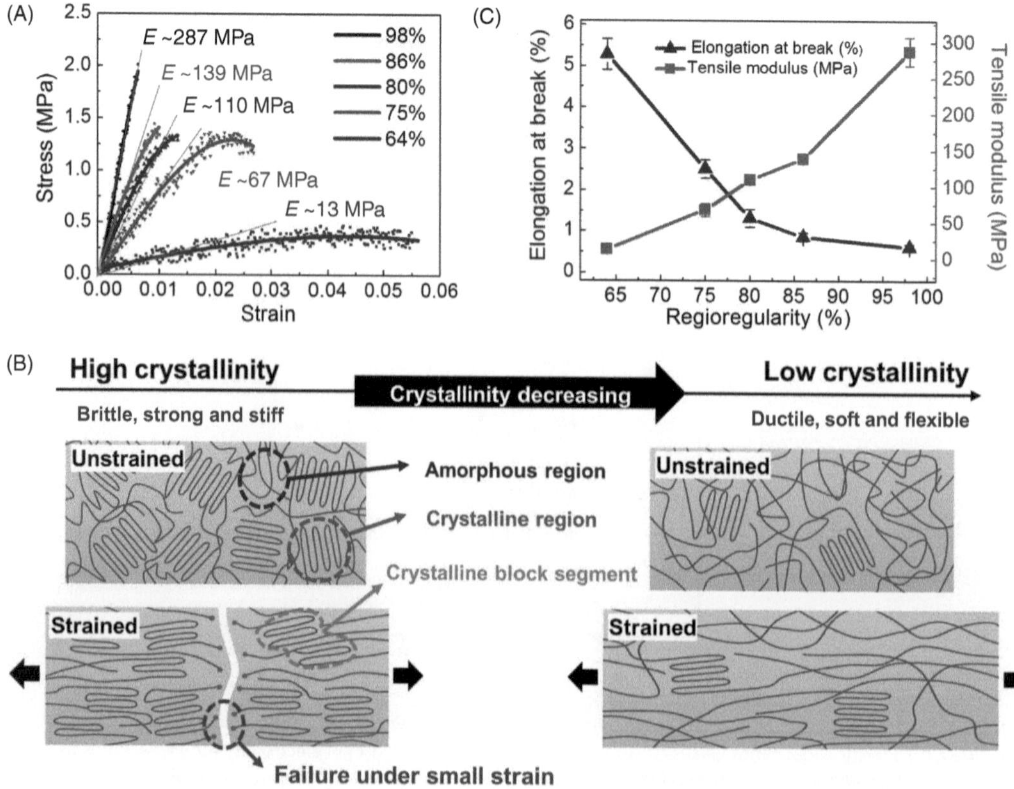

FIGURE 3.6 Mechanical properties and deformation mechanism of P3HT thin films as a function of RR. (A) Stress–strain curves of the P3HT thin films with the RR of 98, 86, 80, 75, and 64%. (B) Tensile modulus and elongation at breaking of the P3HT thin films. (C) Strain-evolved deformation mechanism of conjugated polymer thin films under tensile strain with respect to the degree of crystallinity. *Reproduced with permission from reference Kim, J.-S., Kim, J.-H., Lee, W., Yu, H., Kim, H.J., Song, I., Shin, M., Oh, J.H., Jeong, U., Kim, T.-S., 2015. Tuning mechanical and optoelectrical properties of poly(3-hexylthiophene) through systematic regioregularity control. Macromolecules 48, 4339–4346. Copyright 2015, American Chemical Society.*

lengths of alkyl side chains. The tensile modulus decreases with the increasing length of the alkyl side chain, from 1.87 GPa for a butyl side chain to 0.16 GPa for a dodecyl chain (Savagatrup et al., 2014c). Meanwhile, the effect of regioregularity on intrinsic mechanical properties of P3HT thin films is investigated by lowering the degree of crystallinity from 98% to 65% (Figs. 3.6A and B). Although P3HT with a lower regioregularity reduces electronic properties, it shows advantages in increasing the mechanical properties, that is, the decreased tensile modulus at a higher elongation enhance its mechanical stability (Kim et al., 2015a).

Molecular weight also plays an important role in determining the mechanical performance. The relationship between M_w and mechanical property is demonstrated for P3HT films. The simulations suggest that high mechanical properties occur for chain lengths to be greater than 200 monomers or M_w to be higher than 35 kDa (Tummala et al., 2015).

A series of cross-linked, monodispersed PS nanoparticles with increasing diameters of 60, 80, and 100 nm are used to improve the mechanical and ambient stabilities of polymer solar cells. These PS nanoparticles acted as not only the binders in the PEDOT:PSS domains, but also the interfacial modifiers between PEDOT:PSS and the active layer in the solar cell (Kang et al., 2015). In addition to the intrinsic stiffness of conjugated materials, the addition of small molecules, such as fullerenes to organic solar cells, often displays deleterious effects on the compliance of the active layers and the adhesion of the active layers with other layers in the device (Dupont et al., 2012). The addition of PCBM to P3HT produces a stiffening effect, for example, with an increase of the tensile strength by a factor of five (Lipomi et al., 2012). The tensile modulus and brittleness of regioregular poly(3-alkylthiophene) (P3AT) and its blends with PCBM are decreased with the increasing length of the alkyl side chain, which affects the long-range ordering, that is, intracrystalline ordering. In addition, a significant influence of the presence of processing additives and the adhesion of the semiconductor film to the substrate is found for the mechanical properties of conjugated polymers and their composites (Savagatrup et al., 2014a). These conjugated polymers have been also made into a fiber shape, and a core-shell architecture is found to be an effective design. For instance, for the MEH–PPV nanofiber prepared via electrospinning, the high-density core provides a high mechanical strength, whereas the porous shell becomes a suitable soft substrate.

3.4.2 Electronic Properties

Conducting polymers also possess promising electronic properties. Since the intrinsic density of free charge carriers in most conjugated polymers is negligible, and the injection barriers in most devices are sufficiently low, the space charge-limited conduction is assumed, that is, the electric field from neighboring carriers dominates under the applied bias. In some modified models, there are distributions of traps that lie lower in energy than the carrier transport state. Such traps might be created by kinks in the polymer chains, which interrupt the conjugation along the backbone. Minimizing the number of kinks in the polymer backbone therefore maximizes the electrical conductivity. Similarly, in hopping transport models, the conduction is facilitated by charge carriers that hop from a chromophore to neighboring one. This process strongly depends on the orientation and distance between two neighboring chromophores, implying that optimizing the geometry and packing density of adjacent polymers may also maximize the electrical conductivity.

Although there are still debates on which model is more accurate, the measurements on the field dependence of the hole mobility in conducting polymers have revealed that higher mobilities correlate with improved packing of polymer chains in coplanar geometries, and that lower mobilities are derived from disruption of the packing geometry and the conjugation length. The conductivity in the resulting electronic devices can therefore be improved by making the polymer chains aligned to enhance their packing (Yin and Zheng, 2012).

Ordered polymer conformations have been demonstrated to increase the charge carrier mobility. However, it does not mean that the charge transport property can be enhanced by increasing the solid-state crystallinity. As molecular chains are longer than the crystalline domains, they can form bridges between neighboring grains, reducing the grain boundary resistance and creating more continuous conductive pathways (Clark et al., 2013). The charge mobility is significantly higher along the polymer backbone direction than the other crystallographic directions (O'Connor, 2013). In the context of poor electrical conductivity in pristine conducting polymer, doping represents a popular method to enhance its electronic properties.

PPy doped with protonic acid shows a higher electrical conductivity compared to that doped with organic sulfonic acid (Lee et al., 2014), and a PPy film with electrical conductivity up to 2000 S cm^{-1} was prepared (Qi et al., 2012). For PEDOT:PSS, many secondary dopants, commonly called conductivity enhancement agents, such as, polyols, alcohols, surfactants, salts, acids, and organic solvents, have been widely used to enhance the electrical conductivity (McCarthy et al., 2014). It reached 2084 S cm^{-1} with the use of a nonvolatile ionic liquid (Badre et al., 2012) and 3065 S cm^{-1} or even higher after H_2SO_4 treatment at high temperature (Xia et al., 2012).

3.4.3 Electrochemical Properties

Electrochemical reaction usually consists of a blend of two materials: an electron–donor π-conjugated polymer (donor, D) and an electron–acceptor fullerene derivative (acceptor, A). Polymers with electrochemical properties have attracted considerable attention over past decades due to potential applications in various fields including low-cost, lightweight, and flexible electrode materials in photovoltaic devices, such as, solar cells and energy storage devices like supercapacitors (Ripolles-Sanchis et al., 2013; Gelinck et al., 2010; Snook et al., 2011).

The development of organic semiconductor-based optoelectronics requires a much better understanding to the nature of electronic structure and charge transport properties (Facchetti, 2010). Polymeric semiconductors for organic thin-film transistors (OTFTs) and bulk-heterojunction photovoltaic applications must present two essential structural features. As shown in Fig. 3.7A, the first is a π-conjugated backbone composed of linked unsaturated units, resulting in extended π orbitals along the polymer chain, thus enabling proper charge transport and optical absorption (Facchetti et al., 2005). The second is the functionalization of the polymer core with solubilizing substituents, which is essential for inexpensive manufacture by solution methods as well as to enhance solid state core interactions (Allard et al., 2008).

Among the most common unsaturated units, there are mono (poly) cyclic aromatic hydrocarbons, heterocycles, benzofused systems, and olefinic and acetylenic groups, typically paired with various fullerene derivatives. The extent of conjugation/interaction between these units determine the polymer solution/solid-state electronic structure, which in turn control polymer properties, such as, optical absorption/emission, redox

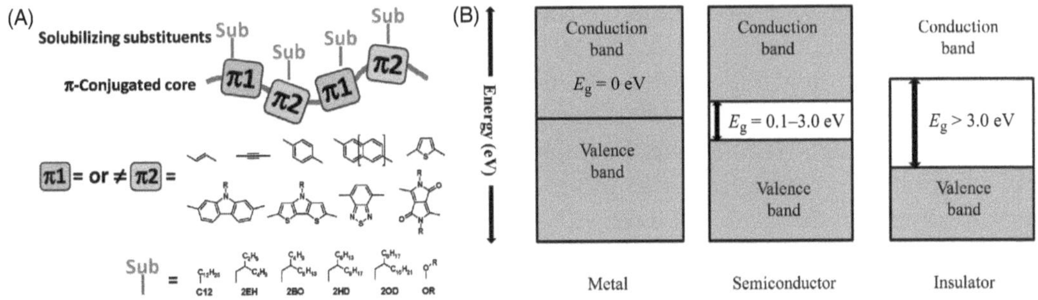

FIGURE 3.7 (A) Schematic representation of a polymer chain showing unsaturated (π) and solubilizing *(sub)* units. (B) Energy band diagram demonstrating band gaps. *(A) Reproduced with permission from reference Facchetti, A., 2010. π-Conjugated polymers for organic electronics and photovoltaic cell applicationst. Chem. Mater. 23, 733–758. Copyright 2010, American Chemical Society.*

characteristics, and frontier molecular orbital energy levels (Facchetti, 2010). A series of linear and soluble fullerene/oligothiophene alternating copolymers by polycondensation were synthesized to demonstrate enhancement of the charges transfer and transport during the photovoltaic process (Perrin et al., 2015).

The power conversion efficiency in photovoltaics has largely increased as a result of the advancement in synthesizing new materials and designing novel device structures, but it is still much below the theoretical value (Dou et al., 2013). The mechanisms that limit the power conversion efficiency have been already well understood. They include the energy losses associated with exciton dissociation at the bulk heterojunction interface, optical absorption losses by nonoptimal absorber bandgaps, and recombination losses through various processes (Street et al., 2014). The energy band that results from the bonding orbitals of a molecule is known as the valence band (VB), while the conduction band (CB) is a result of the antibonding orbitals of the molecule. The width of individual bands across the range of energy levels is called band width. As shown in Fig. 3.7B, VB represents the highest occupied molecular orbital (HOMO) and CB represents the lowest unoccupied molecular orbital (LUMO) (Cheng et al., 2009). The gap between the highest filled energy level and lowest unfilled energy level is called bandgap (E_g). This bandgap represents a range of energies which is not available to electrons, and this gap is known variously as "the fundamental energy gap," the "band gap," the "energy gap," or the "forbidden gap". The level of electrons in a system which is reached at absolute zero is called the Fermi level (F_g) (Martin, 2010). It has been demonstrated that, in order to allow the formation of delocalized electronic states, conducting polymer (CP) molecular arrangement must be conjugated (Heeger, 2001). The delocalization of the electronic states relies on the resonance-stabilized structure of the polymer. The size of the energy bandgap depends on the extending of delocalization and the alternation of double and single bonds (Bolto et al., 1963). Moreover, the size of the energy bandgap will determine whether the CP is metal, semiconductor, or insulator (Cheng et al., 2009). Combining the concepts explained in both atomic and molecular orbital theory, the electronic properties of metals, semiconductors, and insulators can be differentiated with reference to the energy band gap (Molapo et al., 2012).

Bandgap of electronic polymer is not the only parameter for photovoltaic applications. Proper alignment of the HOMO and LUMO energy levels are also critical parameters. The common strategy to enable a good hole-transporting polymer for conjugate polymer is made by a delicate balance of the HOMO energy level which should lie somewhere around −5 to −5.5 eV. When HOMO is too high (low ionization potential), facile oxidation by air and acceptor sites dramatically compromise thin-film transistor (TFT) ambient stability and on/off current ratio. For HOMO of −5.5 eV (high ionization potential), very large on/off current ratio and good mobility can be achieved. The performance of P3HT and its derivatives have been investigated by various groups considering the effects of P3HT molecular structure, solvent, film morphology, film thickness, and fabrication process, as well as humidity and core substituent (alkyl chain) length (Lücke et al., 2015; Zen et al., 2006). These studies have deepened our understanding of the charge transport properties of polymeric semiconductors as a whole (Facchetti, 2010). The role of the charge carrier recombination kinetics on the achievable open-circuit voltage of polymer/fullerene solar cells has been evaluated. Small electronic coupling and large reorganization energy between initial and final states are needed. Benzotriazole-based donor–acceptor-type conjugated polymers were synthesized with high optical contrast in near-infrared (NIR) region and less than 1s switching time (Cevher et al., 2013). Two classes of polymers based on fluorene and indenofluorene copolymerized with triarylamine and indacenodithiophene copolymerized with benzothiadiazole and thienothiophene can be used to fabricate transistors with maximal hole mobilities of 0.8–1.2 $cm^2 V^{-1} S^{-1}$, current on/off ratio of $\sim 10^4$ and a threshold voltage of ~ -30 V (Zhang et al., 2010). Introducing flexible conjugation-break spacers would open a venue to design future high-performance melt-processable semiconducting polymers. Because melt-processed thin films present higher crystallinity and more ordered microstructures than their solution-processed counterparts, the transport properties are thus enhanced (Zhao et al., 2015).

Regarding the application in supercapacitor, the flexible PANI nanotube arrays with high and stable specific capacitances were prepared via electrochemical polymerization. The specific capacitances increased with decreasing wall thicknesses (Wang et al., 2012). A three-dimensionally hierarchical porous PPy hydrogel with various mechanical and electrochemical properties are fabricated by controlling the ratio of phytic acid to pyrrole monomer in the synthetic process. The solid-state supercapacitor showed a weight specific capacitance of $\sim$380 F g^{-1} and areal specific capacitance of $\sim$6.4 F cm^{-2} at a mass load of 20 mg cm^{-2} (Shi et al., 2014).

However, electronically active polymers are not limited to π-conjugated polymers. Nonconjugated redox-active polymers, particularly radical polymers, are emerging as a family of functional materials with a nonconjugated backbone and stable radical groups that are pendant to this nonconjugated backbone. There are a few promising advantages. (1) They can be readily synthesized at relatively large quantities using the widely studied, often controlled polymerization processes (e.g., atom transfer radical polymerization, reversible addition-fragmentation chain transfer polymerization, and anionic polymerization). (2) The radical polymers can be generated with high electronic properties and device stability. (3) It is available to produce p-type and n-type radical polymers by selectively tuning the functional pendant groups (Tomlinson et al., 2014). In principle, electroactive polymers (EAP) show similar wide potentials to and even higher redox capacity than inorganic electrode materials. Particularly, a conducting polymer could be p-doped at high potential or n-doped at low potential, serving as both cathodic and anodic materials at a totally organic battery. Such a bipolar polymer could be developed with a large potential gap and sufficiently high capacity for its n- and p-type redox reactions. A bipolar polyparaphenylene was synthesized and investigated as both the n- and p-type redox-active material for electrochemical energy storage. Such a polyparaphenylene-based organic battery can work at 3.0 V with considerably high capacity and cycling stability (Zhu et al., 2013).

3.4.4 Electroluminescent Properties

Electroluminescent (EL)-conjugated polymers consist of triplet energy of organic hosts (D) and phosphorescent dopant acceptor (A), and they are widely used for a variety of applications including light-emitting devices and photovoltaic devices, as well as emerging applications, such as, molecular electronics and fluorescent nanoparticles (Kim et al., 2009; Wu et al., 2010). They display several advantages of good processability, high flexibility, low operating voltage, fast response times, and facile color tunability over the full visible range. As shown in Fig. 3.8, among the variety of electronic polymers, PPV, PT, and poly(alkylfluorene) (PF) are three typical p-type electron–donor materials that have been mostly explored. PPV and its derivatives are known as promising materials for

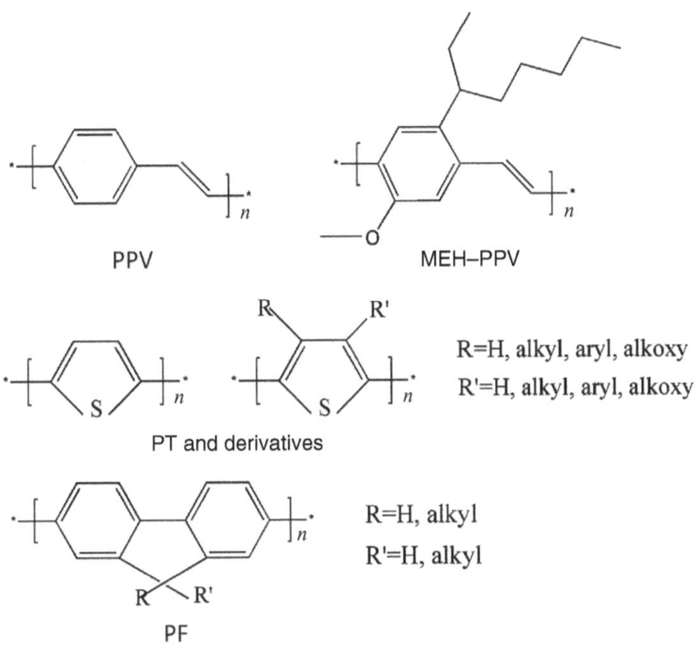

FIGURE 3.8 Chemical structures of electroluminescent polymers.

PLEDs owing to their relatively high photoluminescence and EL quantum efficiencies as well as their good mechanical strength, high thermal stability, and the possibility to vary the colors by changing the substituents at the aromatic ring. Functionalization of PPV with alkoxy groups reduces the energy bandgap, leading to a distinct red-shift in emission (Peng et al., 1999; Grimsdale et al., 2009).

It is easy to control the emission colors using organic materials by controlling the intra- and intermolecular interactions at molecular level through micro- and nanopatterns and thus realize either pure color or natural white for household lighting purposes (Nie et al., 2014). After nearly 2 decades of efforts, light-emitting polymers with three primary colors (blue, green, and red) have been developed. Devices based on them exhibit not only high efficiency and brightness, but also improved operational lifetime. However, white light-emitting polymers, which are very desirable for full color displays with color filters, backlights, and lighting applications, have been continually explored (Liu et al., 2008). The basic molecular design of these single white-emitting polymers (SWPs) involves covalent tethering of chromophores with either two complementary colors (blue and yellow; named as two-color SWPs) or three primary colors (red, green, and blue; named as three-color SWPs) to the main or side chain of a polymer host at the same time. In other words, one macromolecule contains two or three chromophores, forming an intramolecular dopant/host system. Through fine tuning of their doping contents at a very low level to manage the energy transfer and charge trapping, individual emissions from all of the incorporated chromophores can be achieved simultaneously, resulting in a continuous EL spectrum close to that of standard white light (Shao et al., 2012). The introduction of transition-metal phosphors in organic light-emitting diodes (OLEDs), which allow the harnessing of luminescence from triplet excitons, was a major breakthrough in the efficiency (Yersin, 2008). Utilizing triplet excited states in radiative processes is of utmost importance, and the ability to harvest the energy of triplet excited states is important for obtaining the maximum light output in OLED. The triplet energy of organic hosts (donor) may be harvested by a phosphorescent dopant acceptor dispersed in a host. The efficacy of the triplet-energy transfer depends on the alignment of triplet-energy levels of the components.

Similar to an electron-transfer process, approximate matching between the energy levels of the donor and the bridge is crucial for promoting the molecular wire behavior. The instability phenomenon of a liquid film was addressed to obtain monochromatic (OLED) with positive or negative pixelated structures over a large area by modulating the solvent polarity. A bicolor micro pixel OLED array was realized by a single luminescent molecular material, using a single-step lithographic approach based on a surface-tension-driven lithography (Viola et al., 2007; Viola et al., 2010). Hydrogen-bonded supramolecular conjugated polymer, poly{9-[4-(octyloxy)phenyl]fluoren-9-ol-2,7-diyl} (PPFOH) nanoparticles were obtained via reprecipitation which exhibited fascinating size-dependent luminescent properties, affording a flexible method to tune their emission color from blue-green and yellow to white. PPFOH nanoparticles with size of c. 80 nm were observed to emit nearly white light (Lin et al., 2014).

Strongly emissive and vacuum-sublimable neutral Re (I) diimine complexes were designed by using the anionic isocyanotris (pentafluorophenyl) borate ligand with exceptionally strong π-accepting ability (Chu et al., 2014). The triplet-energy transfer occurred in an ultra-fast time scale ($K_{TTET} = 4–6 \times 10^{10}\ S^{-1}$). Finally, it requires a careful energy alignment of the components ($\Delta E \geq 0.1$ eV) to prevent endothermic energy transfer and severe quenching of the electroluminescence (Montes et al., 2007). A yellow OLED based on the star-shaped donor compound tri(9-hexylcarbazol-3-yl) amine was fabricated, which provides formation of the interface exciplexes with the iridium(III) bis[4,6-difluorophenyl]-pyridinato-*N*,C2]-picolinate. The fabricated exciplex-type device exhibits a high brightness of 38,000 cd m^{-2} and high external quantum efficiency (Cherpak et al., 2015). Devices based on poly(*p*-phenylene-ethynylene)-poly(*p*-phenylene vinylene) (PPE–PPV) copolymers with various statistical side-chain distributions showed low turn-on voltage of ~2 V and exhibited red emissions with maximal peaks at between 608 and 620 nm, a maximum luminance of 2070 cd m^{-2} was produced at 5 V (Usluer et al., 2012). Highly efficient and fully solution-processed white OLEDs were fabricated based on fluorescent small molecules [4,9-bis(9,9,9′,9′-tetrakis(6-(9H-carbazol-9-yl)hexyl)-9H,9′H-[2,2′-bifluoren]-7-yl) naphtho[2,3-c][1,2,5] thiadiazole (OCNzC),4,7-bis(9,9,9′,9′-tetrakis(6-(9H-carbazol-9-yl)hexyl)-9H,9′H-[2,2′-bifluoren]-7-yl) benzo[c][1,2,5] thiadiazole (OCBzC),2‴,7‴-bis(9,9,9′,9′-tetrakis(6-(9H-carbazol-9-yl)hexyl)-9H,9′H-[2,2′-bifluoren]-7yl)-9,9‴-spirobi [fluorene] (OCPC),] and a methanol/water soluble poly[(9,9-bis(30-(*N*,*N*-dimethylamino)propyl)-2,7-fluorene)-*alt*-2,7-(9,9-dioctylfluorene) with higher luminous efficiency (9.2 cd A^{-1}), high power efficiency (6.1 l mW^{-1}), and good luminance (16,000 cd m^{-2}) by solution-processing techniques (Xue et al., 2012). Interestingly, some groups found that molecules are practically nonfluorescent when dissolved in good solvents as isolated species, but become highly emissive when aggregated into nanoparticles in poor solvents (e.g., water) or fabricated into thin films in the solid state (Luo et al., 2001). A group of benzene-cored luminogens was synthesized with high solid-state fluorescence quantum yields and enjoys high thermal stabilities. The fabrication produced combined high luminance, current, and external quantum efficiency up to 10,800 cd m^{-2}, 5.8 cd A^{-1}, and 2.7%, respectively (Chan et al., 2012).

3.4.5 Electrochromic Properties

Electrochromic polymers (ECPs) materials exhibit a change in transmittance and/or reflectance of electromagnetic radiation induced by an electrochemical oxidation–reduction reaction (called electrochromism) (Rosseinsky and Mortimer, 2001). The ECPs can be used as electrochromic displays, glasses, and even fabrics (Remmele et al., 2015). A certain set of properties, such as, desired absorption profile, easily accessible redox process and ease of processing are of essential importance for materials to serve as effective electrochromes (Dyer and Reynolds, 2007).

Common magnitudes that define the performance of an electrochromic device include optical contrast, color change, color uniformity, switching speed, and the required voltage to power the device, along with long-term switching and environmental stability. Toward ECPs with these performances, numerous approaches have been employed to control the color of the polymer, including increasing the conjugation length, utilizing donor–acceptor effect, modifying the HOMO–LUMO levels, and using sterically hindered substituent. Also, side chain substituents provide a viable method to tune optical properties of the polymer (Steckler et al., 2014). The resulting effect is reflected by a difference in absorption shifts, resulting in changes in colors. Typically, the clear-to-colored (light-to-dark) transition is faster and typically occurs in 10–60 s, whereas the colored-to-clear transition needs up to several minutes.

Electrochromic devices based on solution-processable polymers enable the utilization of printing and coating methods for processing of the functional layers. Similar to all solution-based deposition methods, the preparation includes the dissolution of the polymer in an appropriate solvent, deposition onto a substrate, and evaporation of the solvent to form a film. Three typical coating methods, that is, spray, blade, and spin coating, are compared for the optical contrast in solution-processable ECPs. It was found that spin-coated films tended to show the highest contrast, closely followed by blade-coated films and finally spray-coated films (Padilla et al., 2015). Solution-processable ECPs are an important class of ECPs owing to their fast response time, ease of color tunability, high contrast ratio and good processability (Hassab and Padilla, 2014).

The discovery of neutral state green color based on a donor–acceptor design represents a great advancement in ECPs (Sonmez et al., 2004). A colorless polymer was thus obtained at an oxidized state in the green ECPs (Durmus et al., 2007). Recently, all red, green and blue (RGB) and cyan, magenta, yellow and black (CMYK) colors at neutral state were reported to make ECPs colorless at their oxidized state (Dyer et al., 2011). A series of styrene or acrylic based polymers containing carbazole-electro-active moiety as pendant unit were synthesized and then cross-linked electrochemically. Thereinto, ECP with benzyloxy spacer exhibits the best electrochemical performance: contrast ratio of 55 at 690 nm, response time of 2.1 s, coloration efficiency of 331 $cm^2 C^{-1}$ and stability by maintaining 94.3% performance even after 1000 cycles (Gulfidan et al., 2014).

Some new ECPs are synthesized via Suzuki polycondensation. They are yellow and near-yellow at neutral states with varying oxidation potentials. The phenylene-type propylenedioxythiophene in a polymer chain leads to absorption of high-energy light in the visible, yielding vibrant yellow neutral states. The designed rings are highly aromatic and cause the polymers to oxidize at elevated potentials (Steckler et al., 2014). Electrochromic windows prepared using this in situ method have a higher contrast than those prepared using the corresponding conventional ex situ electrochemical polymerization due to a more transparent bleached state attributed to a lower absorption in the IR region with higher interchain distances. Another advantage of the in situ process is that complex patterns can be rapidly formed with relatively high resolutions. Recently, a high-throughput screening method was developed for rapid color selection for electrochromic devices using the in situ method (Alhasmi Alamar, 2014).

Polyselenophenes show some advantages over their polythiophene analogs, such as, lower aromaticity, lower redox potentials, lower band gaps, more rigid, robust, and quinoid structures, and better interchain charge transfers (Salzner et al., 1998). A lot of efforts are thus made to explore effective synthetic methods. A series of EDOS derivatives substituted by alkyl chains with various lengths to the bridge have been reported (Li et al., 2009).

Among them, hexylsubstituted PEDOS exhibits unprecendented properties, for example, it displays a maximum of 88–89% in transmittance change, which is one of the highest reported transmittance change for ECPs. Furthermore, the film is highly robust and stable. It exhibits 84% of its optical contrast ratio when switched between redox states for 50,000 cycles in the presence of air (Karabay et al., 2015). The design of random copolymers based on thiophenes with different monomer ratios also improves the switching speed with a relatively high full optical contrast from a red neutral state (Chen et al., 2015b). Thiadiazolo [3,4-*c*]pyridine (PT) represents a promising alternative to benzothiadiazole (BT) as donor–acceptor-type polymer with higher efficiency, favorable optical memory, lower bandgap, and rapider response (Ming et al., 2015). The electrochromic device exhibits a color uniformity and high switch speed. In addition, this method can be used to fabricate a large-area flexible electrochromic device (Otley et al., 2014).

Through the variation on a poly(3,4-propylenedioxy-thiophene) (ProDOT) backbone, a family of ECPs spanning the breadth of the color palette becomes available, and they exhibit both vibrantly colored and highly transmissive states (Bhuvana et al., 2013). Base on this discovery, it has been widely made into electrochromic devices. A conjugated electroactive copolymer based on an alkoxy-substituted ProDOT and an acrylate-substituted propylenedioxythiophene ProDOT has been presented. It allows solution processability of an electrochromic material from common organic solvents, followed by photocross-linking to make the resulting film insoluble while maintain electrochemical, spectroscopic, and colorimetric properties. It has also been shown for printing and patterning by utilizing photolithographic methods to create a patterned surface with millimeter-sized dimensions (Jensen et al., 2013). A segmented ECP display is fabricated from solution-based processes using PProDOT-$(CH_2OEtHx)_2$, a kind of propylenedioxy-thiophene derivatives. This device architecture, combined with the inherently stable polymers used, afforded a rapid segment switching within a narrow voltage window as well as exceptional long-term stability over 200,000 cycles (Chandrasekhar et al., 2014). High-throughput screening method was provided for preparing a set of conjugated copolymers exhibiting a single absorption peak within the visible region in the colored state, and a highly transmissive sky blue color in the bleach state has been established. This method has further been demonstrated using other two monomer combinations, such as, EDOT/bithiophene and ProDOT/bithiophene. By understanding the diffusion of monomer in the polymer gel electrolyte, the exact composition of the monomer feed can be determined and matched to a specific color that can fit the application need (Alamer et al., 2013). A method was developed to prepare carbazole-protected amino acid fibrous hydrogels (Kubiak et al., 2015). The polymers were electrochromic (colorless to green) and retained the alanine functionality, opening the possibility that a wide range of protected peptides could be fully or partially polymerized to form polymers with unique and useful microstructures.

3.4.6 Electromechanical Properties

The electromechanical devices that convert electric energy into mechanical energy are generally made from ferroelectric ceramics, shape memory alloys, and EAPs, typically conducting polymers. The electromechanical actuators have recently attracted increasing attentions due to their promising applications, such as, artificial muscles, microrobots, sensors, and biomimetic devices (Chen et al., 2015a). Considerable attention has been given to EAPs that show a mechanical response to an applied electric field, aiming at applications of actuators and sensors (Shankar et al., 2007). Actuating functionalities of EAPs are particularly fascinating because they can be driven under low-voltage conditions (<5 V), paving the way for the development of low-power electrochemical devices (Takashima et al., 2012). The inherent features of polymers, such as, flexibility, light weight, and ease of manufacturing into various shapes have also been highlighted as desirable properties of cutting-edge technologies to be integrated into microelectromechanical systems (Liu et al., 2013). Ionic polymers are of increasing interest for EAP actuators, in which bending motions are caused by ion migration in the direction of the oppositely charged electrode surface under an applied electric field (Kiyohara et al., 2009). One of the most commonly employed ionic polymers for EAP actuators is a perfluorosulfonic acid ionomer (commercialized under the trademark Nafion), which has the advantage of high conductivity in the presence of moisture (Liu et al., 2012. As a class of conventional EAPs, their drawbacks discouraged their widespread use in various applications. Then ionic liquids were gradually widely used in actuators. When doped with ionic liquid, Nafion is permeable to both cations and anions of the ionic liquid; and the electromechanical response of EAP actuators is directly proportional to concentration of the ions from dopant. The exchange of the H^+ counterion of Nafion with a cation results in the generation of enhanced cationic strain (Hong et al., 2014). Recently, some attempts are made to employ new ionic polymers in EAP actuators to attain synergy between the polymer matrix and the ionic liquid (Gao et al., 2012). It is suggested that, for applications where high displacements are desired, increasing concentrations of ionic liquids may prove advantageous.

Microphase-separated ionic polymers are also incorporated for EAP actuators (Gao et al., 2012. The actuation properties (good mechanical integrity and fast actuation response) are largely increased for sulfonated pentablock copolymers, imidazolium-tethered block copolymers, and zwitterionomers with tunable ionic interactions. Phase-separated sulfonated pentablock copolymers are introduced to EAP actuators for enhancing the actuation performance. The integration of ionic liquids, in conjunction with surface-modified inorganic fillers, into nanoscale ionic domains eliminated back relaxation by allowing both fast ion transport and good mechanical strength. Key factors affecting the electromechanical properties of ionic polymer actuators are determined by employing a set of linear sulfonated di- and triblock copolymers and imidazolium ionic liquids. The ionic liquids with sulfonated block copolymers displayed well defined hexagonal cylindrical morphologies where the ionic liquid was selectively incorporated into PSS phases (Kim et al., 2014). CPs, such as PPy, undergoes volume changes due to ion/solvent diffusion in and out of the polymer matrix upon electrochemical switching between oxidized and reduced states (Smela, 2003). Depending on whether anion or cation exchange predominates, the active layer will either expand upon oxidation due to anion insertion and contract upon reduction due to anion expulsion, or it contract for oxidation due to cation expulsion and expand for reduction due to cation insertion. Actuators based on CPs can be electrically controlled at low operating voltages (typically 1–3 V), continuously switched between expanded/contracted states, and operated well in liquid electrolytes. Films composed of an interpenetrating network of silk fibroin and PPy can function as electromechanical actuators in aqueous electrolytes (Romero et al., 2014).

Self-healing materials based on electromechanical properties have attracted increasing interests after the discovery of an electrically triggered self-healing polymeric material based on the complexation between ferrocene-modified poly(glycidyl methacrylate) and β-cyclodextrin groups (Zhong and Post, 2015). For instance, single-layer fabric ribbons with tunable electromechanical actuations are developed by integrating spring-like carbon nanotube (CNT) fibers through a sewing method. They are flexible and mechanically robust. The fabric ribbons have been further designed into helically three-dimensional superstructures to generate lengthwise contraction or elongation besides the typical bending/unbending deformation in planar ribbons. The programmable actuations are rapid, reversible, operated by low voltage ($<10\,V\,cm^{-1}$), and can be well repeated for thousands of cycles without fatigue (Chen et al., 2015a).

3.4.7 Stability

Despite the high electrical performance, polymer-based electronic devices are highly sensitive to reaction conditions, such as temperature, humidity, and impurities, which often result in heterogeneous surface characteristics and critically affect the electrical performance and reproducibility. Thus, there are raising interests to explore the stability and durability of electronic polymers in various environments. The widely used solar cell layer made of electronic polymers, such as PEDOT:PSS, suffers from short lifetime due to its acidic and hygroscopic nature. For a long-term stability of polymer solar cells, interfaces with matched surface energy are required to prevent interfacial dewetting and delamination. Over the past years, enormous progress has been achieved on improving the performance of organic solar cells with power conversion efficiencies of 8–9% and lifetime over 6 years (Peters et al., 2011). The active layer of a polymer solar cell is mainly based on a blend of two components: a semiconducting polymer (electron donor) and a fullerene derivative (electron acceptor) to form the bulk heterojunction (Wantz et al., 2014). However, a significant drawback lies in the susceptibility to degradation upon attack by nucleophiles, especially in the case of conjugated organoboranes with high Lewis acidity (Ashley et al., 2011). The low stability becomes an obstacle in the further progress of organic solar cells (Yin et al., 2014).

There are multiple techniques under development to stabilize the bulk heterojunction and consequently improve performance stability: the use of ternary blends with compatibilizers, such as, block copolymers, amorphous fullerene derivatives, high glass transition temperature polymers, thermos-cleavable polymers, specifically designed donor–acceptor systems with enhanced interactions, functionalized side chains on the polymer and cross-linkable materials (Mulherin et al., 2011). The cross-linking materials are described as a demonstration. There are three methods considering that the bulk heterojunction is composed of an electron-donating polymer (D) and electron acceptor (A). We can cross-link the donor with the donor (D–D), the donor with the acceptor (D–A) or the acceptor with the acceptor (A–A). In the D–D method, the degree of reticulation can be varied to control the structure of the polymer. Various functions are possible to promote this reaction in the solid state using thermal or UV treatment. The latter is more interesting as the cross-linking step can be disconnected from the thermal annealing that is necessary to optimize the morphology. However, in most cases, PCBM crystallization cannot be completely eliminated. In the D–A method, the stabilization of the morphology is improved, and the link to the acceptor can be controlled via the amount of introduced functions. However, this method is limited by the choice of functions available to enable reactions of donors with fullerene. The A–A method appears to be the most efficient for inhibiting PCBM crystallization. In any case, the cross-linking stimulus must be compatible with the fabrication process of the solar cell without affecting or degrading the other layers (Wantz et al., 2014).

A series of perfectly coplanar arrangement of the thienylborane skeleton is used to facilitate p-π conjugation through the boron atoms. The bulky groups greatly increase the stability, thus making the borane species inert to air, moisture, and even acids and bases (Yin et al., 2014). A Cu:NiO_x-based device revealed that power conversion efficiencies are maintained to be above 90% even after 240 h of storage in air. Cu doping is a simple and effective approach to render NiO_x a promising hole and electron-transporting layers for high-performance perovskite solar cells due to the improved electrical conductivity and favorable perovskite crystallization. As a result, the losses in short-circuit current and fill factor in the pristine NiO_x-based devices are mitigated, and a high power conversion efficiency up to 15.40% with decent environmental stability is achieved (Kim et al., 2015b). For the first time, graphene nanoribbon that possesses proper energy level alignment, good solubility, and high film-forming property serves as an excellent hole extraction material to enhance efficiency and stability of polymer solar cells. Polymer solar cells based on the graphene nanoribbon hole extraction layer outperform their counterparts based on conventional hole extraction materials, including PEDOT:PSS. The power conversion efficiency of the graphene nanoribbon-based device is maintained by 86% after storage for 90 days (Liu et al., 2014). An organic field-effect transistor overcomes this barrier using a solution-processable isoindigo-based polymer semiconductor. More importantly, these organic field-effect transistors are stable in both freshwater and seawater. The organic field-effect transistors are further capable of selectively sensing heavy-metal ions in seawater (Knopfmacher et al., 2014).

It is well acknowledged that nonohmic contact will impede the charge-collecting efficiency of electrodes and cause undesirable charge accumulation at the organic/electrode interfaces, leading to decreased device performances. To alleviate these interfacial energy barriers, new interfacial materials with desired charge selectivity and compatibility are developed for all-solution-processed multilayer devices, and proper integration of the interfacial layer with new active materials are important to further improve the efficiency and stability of organic photovoltaic (OPV) devices (Yip and Jen, 2012). Previously, interfacial layers have been introduced to improve adhesion between the photoactive layer and the electrode and to enhance the durability and stability of device, particularly, for flexible polymer solar cell. In addition, the interlayer can also serve as an additional protecting layer for the bulk heterojunction to impede the permeation of oxygen and water. At present, PEG derivatives are the most widely used functional materials. Usually, PEG derivatives are blended into photoactive layers to form in situ self-organized PEG-based interlayers during the morphological evolution in bulk heterojunction because they tend to migrate to the surface spontaneously. These results show that the combination of proper interfacial engineering and encapsulation will enhance long-term stability of polymer solar cells (Chueh et al., 2015).

3.4.8 Biocompatibility

Biocompatibility refers to the ability of a biomaterial to perform with an appropriate host in response in a human body or other biological systems. The softness and flexibility of organic bioelectronics must be improved to increase its biocompatibility because the flexible tissues in the biological systems contain various kinds of movements in their physiological functions. On the other hand, flexible bioelectronic devices can be easily integrated with the textiles to fabricate the wearable smart devices (Zeng et al., 2014). As a result, systems capable of biomimetic movements are highly sought for biomedical devices, such as, steerable catheters, valves, blood vessel sutures, cochlear implants, and controlled drug release devices (Liao et al., 2014). Nowadays, bioelectronic devices based on organic materials, such as OTFTs for the sensitive detection of biological analytes, organic electrochemical ion pumps (OEIPs) for the dynamic control of drug delivery, and polymer electrodes for recording/stimulation of cells and neural, have all been successfully realized through easy fabrication processes (Irimia-Vladu, 2014). For example, PEDOT has been electrochemically polymerized around neurons and maintained high cell viability, and organic electrochemical transistors (OECTs) based on PEDOT:PSS have been successfully used as cell-based biosensors (Richardson-Burns et al., 2007). A bioelectronic device normally consists of several essential parts, such as active elements to transduce the signals across the electronics/biology interface, electronics to control and record corresponding signals, and power sources, among which the used materials play the most significant role (Liao et al., 2014). As one of the simplest organic electronic devices, polymer-based electrodes have emerged as a viable platform for various stimulation and sensing applications (Table 3.1), including medical diagnose, cell monitoring, and drug delivery.

Regarding the polymer backbones, the most investigated conducting polymers for flexible organic bioelectronics are PT, PPy, and PANI. Considering the versatility of organic chemistry toolbox, polythiophene backbone-based conducting polymers are poised to have a widespread use in organic bioelectronics. Significant effort has been devoted to investigate the polythiophene derivative, such as, PEDOT, which is

Table 3.1 Summary of Flexible Polymer Electrodes for Bioelectronic Applications

Types	Substrate	Active Organic Material	Application	Performance	References
Sensing	PI	PVC	pH, glucose sensor	Glucose: 5 mM to 30 mM	Urban et al. (1992)
	PET	PEDOT:PSS	Glucose sensor	Glucose: 100 μM to 1 M	Chiu et al. (2009)
	Polymer film	PANI	pH sensor	pH: 1–13	Kaempgen and Roth (2006)
	Carbon fiber	Doped PPy	Nitrate ion sensor	Nitrate ion: c. 10^{-6} M	Bendikov et al. (2007)
Neural recording/ stimulation	Polymer substrates	PPy	Neural prosthetics	/	George et al. (2005)
	Flexible elastomer	PPy	Neural communication	/	(Nyberg et al., 2002)
	Parylene C	PEDOT:PSS	ECoG	Recording of brain activity	Khodagholy et al. (2011)
	PI		EEG	Recording of brain activity	Leleux et al. (2014a)
	Kapton PI		Cutaneous electrophysiological recordings	Long-term recording of cutaneous nerve	Leleux et al. (2014b)
Drug delivery	Plastic coverslips	PPy	Controlled release of dexamethasone	Localized delivery of dexamethasone	Wadhwa et al. (2006)
	Kapton PI	PEDOT:PSS-CNT composite	Drug delivery	/	Honda et al. (2014)
Cell	PET	PEDOT:PSS	Controlled cell activity	Control of epithelial cell adhesion	Svennersten et al. (2009)
	Hydrogel	PEDOT		Induce contraction of muscle cells	Sekine et al. (2010)

ECoG, Electrocorticography; EEG, electroencephalography.

electronically stable under physiological conditions (Groenendaal et al., 2000). A successful polymerization of PEDOT around living neuronal cells is performed without introducing any toxic effects (Richardson-Burns et al., 2007). In addition, PEDOT doped with *p*-toluenesulfonate also showed excellent properties for organic bioelectronics (Bolin et al., 2009). PPy shows high biocompatibility both in vitro and in vivo. For example, a PPy film electrochemically deposited on an electrode could support the secretory function of the cultured chromaffin cells. Cells seeded on PPy in vivo demonstrated a higher survival and proliferation rate than that of control groups (Wang et al., 2004). PPy also can modulate the cellular functions using the electrical stimulations. Conducting

PANI would be typically functionalized with selected dopants via either noncovalent or covalent approaches. In addition, nanostructured PANI materials, such as nanorods, nanowires and nanofibers, offer the possibilities to improve the performance of the PANI-based devices (Huang et al., 2003). PANI has demonstrated its biocompatibility in vivo and sparked great interests in tissue engineering. The biocompatibility of PANI can be further improved by the introduction of biocompatible elements without sacrificing its electric conductivity.

Several strategies have been proposed to improve the biocompatibility of conducting polymers. First, the organic materials are functionalized with biomolecules, including peptides, proteins, and polysaccharides, either by chemically covalent methods or physical entrapment. Second, the naturally occurring materials, including indigo derivatives, carotenoid polyenes, and hydrogen-bonded analogs of linear acenes, are explored as the active organic materials for bioelectronics. Glucose sensors fabricated with enzyme entrapped-PEDOT nanofibers offer higher sensitivity and increased lifetime in comparison to the devices based on planar PEDOT films. The PEDOT nanofibers coated on the microelectrode provide a suitable nanoscale matrix to entrap GOx and reduce the impedance, and thus can significantly enhance the performance in sensitivity, detection limit, and longevity (Yang et al., 2014).

Overall, much effort has been made to develop biocompatible organic materials, which allows for the ultimate integration between the electronic device and biological system. The possibility of fabricating memory devices on biodegradable substrates, such as, rice paper and chitosan is also demonstrated. Biocompatible and flexible resistive switching memory devices are made on the basis of Ag-doped chitosan as the natural solid polymer electrolyte layer on the transparent and bendable substrate. Decomposable devices, where chitosan layer serves as the substrate while Mg as the electrode, have been also achieved (Hosseini and Lee, 2015). A comparison of the biocompatible material-based resistive switching memory devices is made in Table 3.2.

3.5 Summary

Structures and morphologies are the key factors to determine the applications of conducting polymers and have been mainly studied in the recent years. In particular, the combination with nanotechnology offers the possibility to more effectively tune and utilize their excellent electrical, optoelectronic, and mechanical properties. For instance, fabricating conducting polymers into well-defined one-dimensional nanostructured materials demonstrate the emerging opportunities for interdisciplinary research in a variety of fields as well as a lot of promising applications. Many efforts are also made to develop new, low-cost and environmentally friendly conducting polymers. It is concentrated to precisely control the composition, size, and structure of the polymer.

Table 3.2 Comparison of Biocompatible Material-Based Resistive Random Access Memory Devices

The Structure of Memory Device	Switching Material	Substance	Flexible Substance	Transient Electrode	Degradable Substance	Voltage (V)	Current Level (A)	Retention	On/Off Ratio	References
$Si/SiO_2/Ti/Pt/Ag$-CNP/Ag	Ag-CNP	Si/SiO_2	Yes	No	No	±0.5	10^{-3}	10^5	10^6	Nagashima et al. (2014)
Al foil/Ag-CNP/Ag		Al foil								
$Si/SiO_2/Ti/Au/Ag$		Si/SiO_2	Yes	No	No	3, −1.5	10^{-2}, 10^{-3}, 10^{-4}	10^3	10^6	Wang et al. (2013)
PET/ITO/Ag		PET/ITO								
Glass/ITO/albumene/Al	Al	Glass/ITO	No	No	No	4, −3	10^{-3}, 10^{-1}	10^4	10^3	Chen et al. (2015c)
Glass/ITO/Silk/Al	Silk fibroin	Glass/ITO	No	No	No	15, −15	0.04	800	10	Hota et al. (2012)
ITO/DNA-CTMA NPs/Ag	DNA-CTMA NPs	ITO	No	No	No	±6	10^{-1}	10^5	>10	Hung et al. (2011)
PET/ITO/Au NPs-silk/Al	Au NPs-silk	PET/ITO	Yes	No	No	±2	10^{-2}	10^2	10^6	Gogurla et al. (2013)
$Si/SiO_2/Ti/Pt/Ag$-chitosan/Ag	Ag-chitosan	Si/SiO_2	Yes	No	No	±1.5	10^{-4}	10^4	10^5	Raeis Hosseini and Lee (2014)
PES/Ti/Pt/Ag-chitosan/Ag		PES								
PET/ITO/Mg/Ag-chitosan/Mg	Ag-chitosan	PET/ITO Rice paper Chitosan	Yes	Yes	Yes	2, −3	10^{-4}	10^4	$>10^2$	Hosseini and Lee (2015)

References

Alamer, F.A., Otley, M.T., Ding, Y., Sotzing, G.A., 2013. Solid-state high-throughput screening for color tuning of electrochromic polymers. Adv. Mater. 25, 6256–6260.

Alhasmi Alamar, F., 2014. Solid-state electrochromic devices: relationship of contrast as a function of device preparation parameters. J. Mater. Chem. C 2, 2510–2516.

Allard, S., Forster, M., Souharce, B., Thiem, H., Scherf, U., 2008. Organic semiconductors for solution-processable field-effect transistors (OFETs). Angew. Chem. Int. Ed. 47, 4070–4098.

Anilkumar, P., Jayakannan, M., 2007. Single-molecular-system-based selective micellar templates for polyaniline nanomaterials: control of shape, size, solid state ordering, and expanded chain to coil like conformation. Macromolecules 40, 7311–7319.

Anilkumar, P., Jayakannan, M., 2009. Self-assembled cylindrical and vesicular molecular templates for polyaniline nanofibers and nanotapes. J. Phys. Chem. B 113, 11614–11624.

Ashizawa, S., Horikawa, R., Okuzaki, H., 2005. Effects of solvent on carrier transport in poly (3, 4-ethylenedioxythiophene)/poly (4-styrenesulfonate). Synth. Met. 153, 5–8.

Ashley, A.E., Herrington, T.J., Wildgoose, G.G., Zaher, H., Thompson, A.L., Rees, N.H., KräMer, T., O'hare, D., 2011. Separating electrophilicity and Lewis acidity: the synthesis, characterization, and electrochemistry of the electron deficient *tris*(aryl)boranes $B(C_6F_5)_{3-n}(C_6C_{l5})_n$ (n = 1–3). J. Am. Chem. Soc. 133, 14727–14740.

Badre, C., Marquant, L., Alsayed, A.M., Hough, L.A., 2012. Highly Conductive poly(3, 4-ethylenedioxythiophene): poly(styrenesulfonate) films using 1-ethyl-3-methylimidazolium tetracyanoborate ionic liquid. Adv. Funct. Mater. 22, 2723–2727.

Bartlett, P.N., Birkin, P.R., Ghanem, M.A., Toh, C.-S., 2001. Electrochemical syntheses of highly ordered macroporous conducting polymers grown around self-assembled colloidal templates. J. Mater. Chem. 11, 849–853.

Bendikov, T.A., Miserendino, S., Tai, Y.-C., Harmon, T.C., 2007. A parylene-protected nitrate selective microsensor on a carbon fiber cross section. Sensor. Actuat. B: Chem. 123, 127–134.

Bhuvana, T., Kim, B., Yang, X., Shin, H., Kim, E., 2013. Reversible full-color generation with patterned yellow electrochromic polymers. Angew. Chem. Int. Ed. 52, 1180–1184.

Biniek, L., Leclerc, N., Heiser, T., Bechara, R., Brinkmann, M., 2013. Large scale alignment and charge transport anisotropy of pBTTT films oriented by high temperature rubbing. Macromolecules 46, 4014–4023.

Bolin, M.H., Svennersten, K., Nilsson, D., Sawatdee, A., Jager, E.W., Richter-Dahlfors, A., Berggren, M., 2009. Active control of epithelial cell-density gradients grown along the channel of an organic electrochemical transistor. Adv. Mater. 21, 4379–4382.

Bolto, B.A., Mcneill, R., Weiss, D., 1963. Electronic conduction in polymers. III. Electronic properties of polypyrrole. Aust. J. Chem. 16, 1090–1103.

Braun, P.V., Wiltzius, P., 2001. Electrochemical fabrication of 3D microperiodic porous materials. Adv. Mater. 13, 482–485.

Camposeo, A., Greenfeld, I., Tantussi, F., Pagliara, S., Moffa, M., Fuso, F., Allegrini, M., Zussman, E., Pisignano, D., 2013. Local mechanical properties of electrospun fibers correlate to their internal nanostructure. Nano lett. 13, 5056–5062.

Cevher, S.C., Unlu, N.A., Ozelcaglayan, A.C., Apaydin, D.H., Udum, Y.A., Toppare, L., Cirpan, A., 2013. Fused structures in the polymer backbone to investigate the photovoltaic and electrochromic properties of donor–acceptor-type conjugated polymers. J. Polym. Sci. Part A: Polym. Chem. 51, 1933–1941.

Chan, C.Y., Zhao, Z., Lam, J.W., Liu, J., Chen, S., Lu, P., Mahtab, F., Chen, X., Sung, H.H., Kwok, H.S., 2012. Efficient light emitters in the solid state: synthesis, aggregation-induced emission, electroluminescence, and sensory properties of luminogens with benzene cores and multiple triarylvinyl peripherals. Adv. Funct. Mater. 22, 378–389.

Chandrasekhar, P., Zay, B.J., Cai, C., Chai, Y., Lawrence, D., 2014. Matched-dual-polymer electrochromic lenses, using new cathodically coloring conducting polymers, with exceptional performance and incorporated into automated sunglasses. J. Appl. Polym. Sci. 131, 547–557.

Chen, J.-Y., Kuo, C.-C., Lai, C.-S., Chen, W.-C., Chen, H.-L., 2011. Manipulation on the morphology and electrical properties of aligned electrospun nanofibers of poly (3-hexylthiophene) for field-effect transistor applications. Macromolecules 44, 2883–2892.

Chen, P., He, S., Xu, Y., Sun, X., Peng, H., 2015a. Electromechanical actuator ribbons driven by electrically conducting spring-like fibers. Adv. Mater. 27, 4982–4988.

Chen, X., Xu, Z., Mi, S., Zheng, J., Xu, C., 2015b. Spray-processable red-to-transmissive electrochromic polymers towards fast switching time for display applications. New J. Chem. 39, 5389–5394.

Chen, Y.-C., Yu, H.-C., Huang, C.-Y., Chung, W.-L., Wu, S.-L., Su, Y.-K., 2015c. Nonvolatile bio-memristor fabricated with egg albumen film. Sci. Rep. 5, 10022.

Cheng, Y.-J., Yang, S.-H., Hsu, C.-S., 2009. Synthesis of conjugated polymers for organic solar cell applications. Chem. rev. 109, 5868–5923.

Cherpak, V., Stakhira, P., Minaev, B., Baryshnikov, G., Stromylo, E., Helzhynskyy, I., Chapran, M., Volyniuk, D., Hotra, Z., Dabuliene, A., 2015. Mixing of phosphorescent and exciplex emission in efficient organic electroluminescent devices. ACS Appl. Mater. Interfaces 7, 1219–1225.

Chiu, J.-Y., Yu, C.-M., Yen, M.-J., Chen, L.-C., 2009. Glucose sensing electrodes based on a poly (3, 4-ethylenedioxythiophene)/Prussian blue bilayer and multi-walled carbon nanotubes. Biosens. Bioelectron. 24, 2015–2020.

Cho, M.S., Park, S.Y., Hwang, J.Y., Choi, H.J., 2004. Synthesis and electrical properties of polymer composites with polyaniline nanoparticles. Mater. Sci. Eng. C 24, 15–18.

Chu, W.-K., Ko, C.-C., Chan, K.-C., Yiu, S.-M., Wong, F.-L., Lee, C.-S., Roy, V., 2014. A simple design for strongly emissive sky-blue phosphorescent neutral rhenium complexes: synthesis, photophysics, and electroluminescent devices. Chem. Mater. 26, 2544–2550.

Chueh, C.-C., Li, C.-Z., Jen, A.K.-Y., 2015. Recent progress and perspective in solution-processed Interfacial materials for efficient and stable polymer and organometal perovskite solar cells. Energy Environ. Sci. 8, 1160–1189.

Clark, A.P.-Z., Shi, C., Ng, B.C., Wilking, J.N., Ayzner, A.L., Stieg, A.Z., Schwartz, B.J., Mason, T.G., Rubin, Y., Tolbert, S.H., 2013. Self-assembling semiconducting polymers-rods and gels from electronic materials. ACS Nano 7, 962–977.

Crispin, X., Jakobsson, F., Crispin, A., Grim, P., Andersson, P., Volodin, A., Van Haesendonck, C., Van Der Auweraer, M., Salaneck, W.R., Berggren, M., 2006. The origin of the high conductivity of poly (3, 4-ethylenedioxythiophene)-poly (styrenesulfonate)(PEDOT-PSS) plastic electrodes. Chem. Mater. 18, 4354–4360.

Crossland, E.J., Tremel, K., Fischer, F., Rahimi, K., Reiter, G., Steiner, U., Ludwigs, S., 2012. Anisotropic charge transport in spherulitic poly (3-hexylthiophene) films. Adv. Mater. 24, 839–844.

Da Costa, V., Le Moigne, J., Oswald, L., Pham, T., Thierry, A., 1998. Thin film orientation by epitaxy of carbazolyl polydiacetylenes: Guest-host interaction on a crystal surface. Macromolecules 31, 1635–1643.

Döbbelin, M., Marcilla, R., Salsamendi, M., Pozo-Gonzalo, C., Carrasco, P.M., Pomposo, J.A., Mecerreyes, D., 2007. Influence of ionic liquids on the electrical conductivity and morphology of PEDOT: PSS films. Chem. mater. 19, 2147–2149.

Dou, L., You, J., Hong, Z., Xu, Z., Li, G., Street, R.A., Yang, Y., 2013. 25th anniversary article: a decade of organic/polymeric photovoltaic research. Adv. Mater. 25, 6642–6671.

Dupont, S.R., Oliver, M., Krebs, F.C., Dauskardt, R.H., 2012. Interlayer adhesion in roll-to-roll processed flexible inverted polymer solar cells. Sol Energy Mater. Sol. Cells 97, 171–175.

Dupont, S.R., Novoa, F., Voroshazi, E., Dauskardt, R.H., 2014. Decohesion kinetics of PEDOT:PSS conducting polymer films. Adv. Funct. Mater. 24, 1325–1332.

Durmus, A., Gunbas, G.E., Camurlu, P., Toppare, L., 2007. A neutral state green polymer with a superior transmissive light blue oxidized state. Chem. Commun. 31, 3246–3248.

Dyer, A.L., Reynolds, J.R., 2007. Electrochromism of conjugated conducting polymers. Handbook of Conducting Polymers, 1. pp. 1–63, CRC Press, Boca Raton.

Dyer, A.L., Thompson, E.J., Reynolds, J.R., 2011. Completing the color palette with spray-processable polymer electrochromics. ACS Appl. Mater. Interfaces 3, 1787–1795.

Facchetti, A., 2010. π-conjugated polymers for organic electronics and photovoltaic cell applications†. Chem. Mater. 23, 733–758.

Facchetti, A., Yoon, M.H., Marks, T.J., 2005. Gate dielectrics for organic field-effect transistors: new opportunities for organic electronics. Adv. Mater. 17, 1705–1725.

Gao, R., Wang, D., Heflin, J.R., Long, T.E., 2012. Imidazolium sulfonate-containing pentablock copolymer–ionic liquid membranes for electroactive actuators. J. Mater. Chem. 22, 13473–13476.

Gelinck, G., Heremans, P., Nomoto, K., Anthopoulos, T.D., 2010. Organic transistors in optical displays and microelectronic applications. Adv. mater. 22, 3778–3798.

George, P.M., Lyckman, A.W., Lavan, D.A., Hegde, A., Leung, Y., Avasare, R., Testa, C., Alexander, P.M., Langer, R., Sur, M., 2005. Fabrication and biocompatibility of polypyrrole implants suitable for neural prosthetics. Biomaterials 26, 3511–3519.

Gogurla, N., Mondal, S.P., Sinha, A.K., Katiyar, A.K., Banerjee, W., Kundu, S.C., Ray, S.K., 2013. Transparent and flexible resistive switching memory devices with a very high ON/OFF ratio using gold nanoparticles embedded in a silk protein matrix. Nanotechnology 24, 345202.

Grabowski, C.A., Koerner, H., Meth, J.S., Dang, A., Hui, C.M., Matyjaszewski, K., Bockstaller, M.R., Durstock, M.F., Vaia, R.A., 2014. Performance of dielectric nanocomposites: matrix-free, hairy nanoparticle assemblies and amorphous polymer–nanoparticle blends. ACS Appl. Mater. Interfaces 6, 21500–21509.

Granero, A.J., Wagner, P., Wagner, K., Razal, J.M., Wallace, G.G., 2011. Highly stretchable conducting SIBS-P3HT fibers. Adv. Funct. Mater. 21, 955–962.

Grimsdale, A.C., Leok Chan, K., Martin, R.E., Jokisz, P.G., Holmes, A.B., 2009. Synthesis of light-emitting conjugated polymers for applications in electroluminescent devices. Chem. Rev. 109, 897–1091.

Groenendaal, L., Jonas, F., Freitag, D., Pielartzik, H., Reynolds, J.R., 2000. Poly (3, 4-ethylenedioxythiophene) and its derivatives: past, present, and future. Adv. Mater. 12, 481–494.

Gulfidan, D., Sefer, E., Koyuncu, S., Acar, M.H., 2014. Neutral state colorless electrochromic polymer networks: Spacer effect on electrochromic performance. Polymer 55, 5998–6005.

Han, Y.K., Yih, J.N., Chang, M.Y., Huang, W.Y., Ho, K.S., Hsieh, T.H., Lou, J.G., 2011. Facile synthesis of aqueous-dispersible nano-PEDOT:PSS-*co*-MA core/shell colloids through spray emulsion polymerization. Macromol. Chem. Phys. 212, 361–366.

Hartmann, L., Tremel, K., Uttiya, S., Crossland, E., Ludwigs, S., Kayunkid, N., Vergnat, C., Brinkmann, M., 2011. 2D versus 3D crystalline order in thin films of regioregular poly(3-hexylthiophene) oriented by mechanical rubbing and epitaxy. Adv. Funct. Mater. 21, 4047–4057.

Hassab, S., Padilla, J., 2014. Use of ionic liquids in electrochromic devices. Ionic Liq. Sep. Technol., 301.

Heeger, A.J., 2001. Semiconducting and metallic polymers: the fourth generation of polymeric materials. J. Phys. Chem. B 105, 8475–8491.

Honda, W., Harada, S., Arie, T., Akita, S., Takei, K., 2014. Wearable, human-interactive, health-monitoring, wireless devices fabricated by macroscale printing techniques. Adv. Funct. Mater. 24, 3299–3304.

Hong, W., Almomani, A., Montazami, R., 2014. Influence of ionic liquid concentration on the electromechanical performance of ionic electroactive polymer actuators. Org. Electron. 15, 2982–2987.

Hosokawa, Y., Misaki, M., Yamamoto, S., Torii, M., Ishida, K., Ueda, Y., 2012. Molecular orientation and anisotropic carrier mobility in poorly soluble polythiophene thin films. Appl. Phys. Lett. 100, 203305.

Hosseini, N.R., Lee, J.S., 2015. Biocompatible and flexible chitosan-based resistive switching memory with magnesium electrodes. Adv. Funct. Mater. 25, 5586–5592.

Hota, M.K., Bera, M.K., Kundu, B., Kundu, S.C., Maiti, C.K., 2012. A natural silk fibroin protein-based transparent bio-memristor. Adv. Funct. Mater. 22, 4493–4499.

Huang, J., Virji, S., Weiller, B.H., Kaner, R.B., 2003. Polyaniline nanofibers: facile synthesis and chemical sensors. J Am. Chem. Soc. 125, 314–315.

Hung, Y.-C., Hsu, W.-T., Lin, T.-Y., Fruk, L., 2011. Photoinduced write-once read-many-times memory device based on DNA biopolymer nanocomposite. Appl. Phys. Lett. 99, 253301.

Irimia-Vladu, M., 2014. "Green" electronics: biodegradable and biocompatible materials and devices for sustainable future. Chem. Soc. Rev. 43, 588–610.

Isoniemi, T., Tuukkanen, S., Cameron, D.C., Simonen, J., Toppari, J.J., 2015. Measuring optical anisotropy in poly (3, 4-ethylene dioxythiophene): poly(styrene sulfonate) films with added graphene. Org. Electron. 25, 317–323.

Jensen, J., Dyer, A.L., Shen, D.E., Krebs, F.C., Reynolds, J.R., 2013. Direct photopatterning of electrochromic polymers. Adv. Funct. Mater. 23, 3728–3737.

Kaempgen, M., Roth, S., 2006. Transparent and flexible carbon nanotube/polyaniline pH sensors. J. Electroanal. Chem. 586, 72–76.

Kang, D.J., Cho, H.-H., Lee, I., Kim, K.-H., Kim, H.J., Liao, K., Kim, T.-S., Kim, B.J., 2015. Enhancing mechanical properties of highly efficient polymer solar cells using size-tuned polymer nanoparticles. ACS Appl. Mater. Interfaces 7, 2668–2676.

Karabay, B., Pekel, L.C., Cihaner, A., 2015. A pure blue to highly transmissive electrochromic polymer based on poly(3, 4-propylenedioxyselenophene) with a high optical contrast ratio. Macromolecules 48, 1352–1357.

Khodagholy, D., Doublet, T., Gurfinkel, M., Quilichini, P., Ismailova, E., Leleux, P., Herve, T., Sanaur, S., Bernard, C., Malliaras, G.G., 2011. Highly conformable conducting polymer electrodes for in vivo recordings. Adv. Mater. 23, H268–H272.

Kim, J., Jung, J., Lee, D., Joo, J., 2002. Enhancement of electrical conductivity of poly(3, 4-ethylenedioxythiophene)/poly (4-styrenesulfonate) by a change of solvents. Synth. Met. 126, 311–316.

Kim, H.-C., Park, S.-M., Hinsberg, W.D., 2009. Block copolymer based nanostructures: materials, processes, and applications to electronics. Chem. Rev. 110, 146–177.

Kim, O., Kim, S.Y., Park, B., Hwang, W., Park, M.J., 2014. Factors affecting electromechanical properties of ionic polymer actuators based on ionic liquid-containing sulfonated block copolymers. Macromolecules 47, 4357–4368.

Kim, J.-S., Kim, J.-H., Lee, W., Yu, H., Kim, H.J., Song, I., Shin, M., Oh, J.H., Jeong, U., Kim, T.-S., 2015a. Tuning mechanical and optoelectrical properties of poly(3-hexylthiophene) through systematic regioregularity control. Macromolecules 48, 4339–4346.

Kim, J.H., Liang, P.W., Williams, S.T., Cho, N., Chueh, C.C., Glaz, M.S., Ginger, D.S., Jen, A.K.Y., 2015b. High-performance and environmentally stable planar heterojunction perovskite solar cells based on a solution-processed copper-doped nickel oxide hole-transporting layer. Adv. Mater. 27, 695–701.

Kiyohara, K., Sugino, T., Takeuchi, I., Mukai, K., Asaka, K., 2009. Expansion and contraction of polymer electrodes under applied voltage. J. Appl. Phys. 105, 063506.

Kline, R., Mcgehee, M., 2006. Morphology and charge transport in conjugated polymers. J. Macromol. Sci. Part C: Polym. Rev. 46, 27–45.

Knopfmacher, O., Hammock, M.L., Appleton, A.L., Schwartz, G., Mei, J., Lei, T., Pei, J., Bao, Z., 2014. Highly stable organic polymer field-effect transistor sensor for selective detection in the marine environment. Nat. Commun. 5, 149–168.

Kubiak, P., Awhida, S., Hotchen, C., Deng, W., Alston, B.M., Mcdonald, T.O., Adams, D.J., Cameron, P.J., 2015. Polymerization of low molecular weight hydrogelators to form electrochromic polymers. Chem. Commun. 51, 10427–10430.

Lan, Y.K., Yang, C.H., Yang, H.C., 2010. Theoretical investigations of electronic structure and charge transport properties in polythiophene-based organic field-effect transistors. Polym. Int. 59, 16–21.

Lee, S.H., Lee, S., Ryu, H.W., Park, H., Kim, Y.S., Kim, J.H., 2014. Synthesis and in situ doping of highly conductive polypyrrole nanocomplexes with binary acids. J. Polym. Sci. Part A: Polym. Chem. 52, 2329–2336.

Leleux, P., Badier, J.M., Rivnay, J., B Nar, C., Herv, T., Chauvel, P., Malliaras, G.G., 2014a. Conducting polymer electrodes for electroencephalography. Adv. Healthc. Mater. 3, 490–493.

Leleux, P., Johnson, C., Strakosas, X., Rivnay, J., Herv, T., Owens, R.M., Malliaras, G.G., 2014b. Ionic liquid gel-assisted electrodes for long-term cutaneous recordings. Adv. Healthc. Mater. 3, 1377–1380.

Li, M., Patra, A., Sheynin, Y., Bendikov, M., 2009. Hexyl-derivatized poly(3, 4-ethylenedioxyselenophene): novel highly stable organic electrochromic material with high contrast ratio, high coloration efficiency, and low-switching voltage. Adv. Mater. 21, 1707–1711.

Li, W., Zhang, Q., Zheng, G., Seh, Z.W., Yao, H., Cui, Y., 2013. Understanding the role of different conductive polymers in improving the nanostructured sulfur cathode performance. Nano Lett. 13, 5534–5540.

Li, M., Li, H., Zhong, W., Zhao, Q., Wang, D., 2014. Stretchable conductive polypyrrole/polyurethane (PPy/PU) strain sensor with netlike microcracks for human breath detection. ACS Appl. Mater. Interfaces 6, 1313–1319.

Liao, C., Zhang, M., Yao, M.Y., Hua, T., Li, L., Yan, F., 2014. Flexible organic electronics in biology: materials and devices. Adv. Mater. 27, 7493–7527.

Lin, J.Y., Wong, J., Xie, L.H., Dong, X.C., Yang, H.Y., Huang, W., 2014. Hydrogen-bonded supramolecular conjugated polymer nanoparticles for white light-emitting devices. Macromol. Rap. Commun. 35, 895–900.

Lipomi, D.J., Chong, H., Vosgueritchian, M., Mei, J., Bao, Z., 2012. Toward mechanically robust and intrinsically stretchable organic solar cells: evolution of photovoltaic properties with tensile strain. Sol. Energy Mater. Sol. Cells 107, 355–365.

Liu, J., Gao, B., Cheng, Y., Xie, Z., Geng, Y., Wang, L., Jing, X., Wang, F., 2008. Novel white electroluminescent single polymer derived from fluorene and quinacridone. Macromolecules 41, 1162–1167.

Liu, Y., Ghaffari, M., Zhao, R., Lin, J.-H., Lin, M., Zhang, Q., 2012. Enhanced electromechanical response of ionic polymer actuators by improving mechanical coupling between ions and polymer matrix. Macromolecules 45, 5128–5133.

Liu, Y., Lu, C., Twigg, S., Ghaffari, M., Lin, J., Winograd, N., Zhang, Q., 2013. Direct observation of ion distributions near electrodes in ionic polymer actuators containing ionic liquids. Sci. Rep. 3, 973.

Liu, J., Kim, G.H., Xue, Y., Kim, J.Y., Baek, J.B., Durstock, M., Dai, L., 2014. Graphene oxide nanoribbon as hole extraction layer to enhance efficiency and stability of polymer solar cells. Adv. Mater. 26, 786–790.

Liu, J., Wang, X., Li, D., Coates, N.E., Segalman, R.A., Cahill, D.G., 2015. Thermal conductivity and elastic constants of PEDOT: PSS with high electrical conductivity. Macromolecules 48, 585–591.

Lücke, A., Schmidt, W.G., Rauls, E., Ortmann, F., Gerstmann, U., 2015. Influence of structural defects and oxidation onto hole conductivity in P3HT. J. Phys. Chem. B 119, 6481–6491.

Luo, J., Xie, Z., Lam, J.W., Cheng, L., Chen, H., Qiu, C., Kwok, H.S., Zhan, X., Liu, Y., Zhu, D., 2001. Aggregation-induced emission of 1-methyl-1, 2, 3, 4, 5-pentaphenylsilole. Chem. Commun., 1740–1741.

Luo, X., Killard, A.J., Morrin, A., Smyth, M.R., 2007. Electrochemical preparation of distinct polyaniline nanostructures by surface charge control of polystyrene nanoparticle templates. Chem. Commun., 3207–3209.

Mahmood, I., Ahmad, I., Huizhou, L., Chen, G., 2013. A new strategy for the synthesis of polyaniline nanostructures using *m*-CPBA as an oxidant. J. Mater. Sci.: Mater. Electron. 24, 1181–1186.

Martin, E.A., 2010. A Dictionary of Science. Oxford University Press, Oxford.

Massonnet, N., Carella, A., De Geyer, A., Faure-Vincent, J., Simonato, J.-P., 2015. Metallic behaviour of acid doped highly conductive polymers. Chem. Sci. 6, 412–417.

Mccarthy, J.E., Hanley, C.A., Brennan, L.J., Lambertini, V.G., Gun'ko, Y.K., 2014. Fabrication of highly transparent and conducting PEDOT: PSS films using a formic acid treatment. J. Mater. Chem. C 2, 764–770.

Mengistie, D.A., Chen, C.-H., Boopathi, K.M., Pranoto, F.W., Li, L.-J., Chu, C.-W., 2014. Enhanced thermoelectric performance of PEDOT: PSS flexible bulky papers by treatment with secondary dopants. ACS Appl. Mater. Interfaces 7, 94–100.

Ming, S., Zhen, S., Lin, K., Zhao, L., Xu, J., Lu, B., 2015. Thiadiazolo [3, 4-*c*] pyridine as an acceptor toward fast-switching green donor-acceptor-type electrochromic polymer with low bandgap. ACS Appl. Mater. Interfaces 7, 11089–11098.

Molapo, K., Ndangili, P., Ajayi, R., Mbambisa, G., Mailu, S., Njomo, N., Masikini, M., Baker, P., Iwuoha, E.I., 2012. Electronics of conjugated polymers (I): polyaniline. Int. J. Electrochem. Sci. 7, 11859–11875.

Montes, V.A., Perez-Bolivar, C., Estrada, L.A., Shinar, J., Anzenbacher, P., 2007. Ultrafast dynamics of triplet excitons in Alq3-Bridge-Pt (II) porphyrin electroluminescent materials. J. Am. Chem. Soc. 129, 12598–12599.

Mott, N.F., Davis, E.A., 2012. Electronic Processes in Non-crystalline Materials. Oxford University Press, Oxford.

Moulton, S., Innis, P., Kane-Maguire, L., Ngamna, O., Wallace, G., 2004. Polymerisation and characterisation of conducting polyaniline nanoparticle dispersions. Curr. Appl. Phys. 4, 402–406.

Mulherin, R.C., Jung, S., Huettner, S., Johnson, K., Kohn, P., Sommer, M., Allard, S., Scherf, U., Greenham, N.C., 2011. Ternary photovoltaic blends incorporating an all-conjugated donor–acceptor diblock copolymer. Nano Lett. 11, 4846–4851.

Nagashima, K., Koga, H., Celano, U., Zhuge, F., Kanai, M., Rahong, S., Meng, G., He, Y., De Boeck, J., Jurczak, M., 2014. Cellulose nanofiber paper as an ultra flexible nonvolatile memory. Sci. Rep. 4, 5532.

Nie, W., Chen, Y., Smith, G., Xia, Y., Hewitt, C., Carroll, D., 2014. Nano graphite platelets enhanced blue emission in alternating current field induced polymer based electroluminescence devices using Poly (9,9-dioctylfluorenyl-2,7-diyl) as the emitter. Org. Electron. 15, 99–104.

Nyberg, T., Ingan S, O., Jerreg Rd, H., 2002. Polymer hydrogel microelectrodes for neural communication. Biomed. Microdevices 4, 43–52.

O'connor, B.T., 2013. Using mechanical deformation to elucidate structure-property relationships in polymer semiconductors. 223rd ECS Meeting, May 12–17, 2013.

O'connor, B., Kline, R.J., Conrad, B.R., Richter, L.J., Gundlach, D., Toney, M.F., Delongchamp, D.M., 2011. Anisotropic structure and charge transport in highly strain-aligned regioregular poly(3-hexylthiophene). Adv. Funct. Mater. 21, 3697–3705.

Oh, J.Y., Shin, M., Lee, T.I., Jang, W.S., Min, Y., Myoung, J.-M., Baik, H.K., Jeong, U., 2012. Self-seeded growth of poly(3-hexylthiophene) (P3HT) nanofibrils by a cycle of cooling and heating in solutions. Macromolecules 45, 7504–7513.

Oh, J.Y., Lee, T.I., Jang, W.S., Chae, S.S., Park, J.H., Lee, H.W., Myoung, J.-M., Baik, H.K., 2013. Mass production of a 3D non-woven nanofabric with crystalline P3HT nanofibrils for organic solar cells. Energy Environ. Sci. 6, 910–917.

Oh, J.Y., Shin, M., Lee, J.B., Ahn, J.-H., Baik, H.K., Jeong, U., 2014. Effect of PEDOT nanofibril networks on the conductivity, flexibility, and coatability of PEDOT: PSS films. ACS Appl. Mater. Interfaces 6, 6954–6961.

Otley, M.T., Zhu, Y., Zhang, X., Li, M., Sotzing, G.A., 2014. Color-tuning neutrality for flexible electrochromics via a single-layer dual conjugated polymer approach. Adv. Mater. 26, 8004–8009.

Ouyang, J., Xu, Q., Chu, C.-W., Yang, Y., Li, G., Shinar, J., 2004. On the mechanism of conductivity enhancement in poly (3, 4-ethylenedioxythiophene): poly (styrene sulfonate) film through solvent treatment. Polymer 45, 8443–8450.

Padilla, J., Österholm, A.M., Dyer, A.L., Reynolds, J.R., 2015. Process controlled performance for soluble electrochromic polymers. Sol. Energy Mater. Sol. Cells 140, 54–60.

Park, S., Cho, M., Choi, H., 2004. Synthesis and electrical characteristics of polyaniline nanoparticles and their polymeric composite. Curr. Appl. Phys. 4, 581–583.

Peng, Z., Zhang, J., Xu, B., 1999. New poly(*p*-phenylene vinylene) derivatives exhibiting high photoluminescence quantum efficiencies. Macromolecules 32, 5162–5164.

Perrin, L., Legros, M., Mercier, R.G., 2015. Design of a series of polythiophenes containing C60 groups: synthesis and optical and electrochemical properties. Macromolecules 48, 323–336.

Peters, C.H., Sachs-Quintana, I., Kastrop, J.P., Beaupre, S., Leclerc, M., Mcgehee, M.D., 2011. High efficiency polymer solar cells with long operating lifetimes. Adv. Energy Mater. 1, 491–494.

Qi, G., Huang, L., Wang, H., 2012. Highly conductive free standing polypyrrole films prepared by freezing interfacial polymerization. Chem. Commun. 48, 8246–8248.

Qin, T., Troisi, A., 2013. Relation between structure and electronic properties of amorphous MEH-PPV polymers. J. Am. Chem. Soc. 135, 11247–11256.

Raeis Hosseini, N., Lee, J.-S., 2014. Resistive switching memory based on bioinspired natural solid polymer electrolytes. ACS Nano 9, 419–426.

Remmele, J., Shen, D.E., Mustonen, T., Fruehauf, N., 2015. High performance and long-term stability in ambiently fabricated segmented solid-state polymer electrochromic displays. ACS Appl. Mater. Interfaces 7, 12001–12008.

Reyes-Reyes, M., Cruz-Cruz, I., L Pez-Sandoval, R., 2010. Enhancement of the electrical conductivity in PEDOT: PSS films by the addition of dimethyl sulfate. J. Phys. Chem. C 114, 20220–20224.

Richardson-Burns, S.M., Hendricks, J.L., Foster, B., Povlich, L.K., Kim, D.-H., Martin, D.C., 2007. Polymerization of the conducting polymer poly (3, 4-ethylenedioxythiophene)(PEDOT) around living neural cells. Biomaterials 28, 1539–1552.

Rickert, S., Baer, E., 1978. Epitaxial crystallization of isotactic polypropylene on alkali halide surfaces. J. Mater. Sci. 13, 451–455.

Ripolles-Sanchis, T., Raga, S.R., Guerrero, A., Welker, M., Turbiez, M., Bisquert, J., Garcia-Belmonte, G., 2013. Molecular electronic coupling controls charge recombination kinetics in organic solar cells of low bandgap diketopyrrolopyrrole, carbazole, and thiophene polymers. J. Phys. Chem. C 117, 8719–8726.

Rivnay, J., Mannsfeld, S.C., Miller, C.E., Salleo, A., Toney, M.F., 2012. Quantitative determination of organic semiconductor microstructure from the molecular to device scale. Chem. Rev. 112, 5488–5519.

Romero, I.S., Bradshaw, N.P., Larson, J.D., Severt, S.Y., Roberts, S.J., Schiller, M.L., Leger, J.M., Murphy, A.R., 2014. Biocompatible electromechanical actuators composed of silk-conducting polymer composites. Adv. Funct. Mater. 24, 3866–3873.

Rosseinsky, D.R., Mortimer, R.J., 2001. Electrochromic systems and the prospects for devices. Adv. Mater. 13, 783–793.

Rühle, V., 2010. Morphology and Charge Transport in Conjugated Polymers. Dissertation. University of Mainz, Germany.

Salzner, U., Lagowski, J., Pickup, P., Poirier, R., 1998. Comparison of geometries and electronic structures of polyacetylene, polyborole, polycyclopentadiene, polypyrrole, polyfuran, polysilole, polyphosphole, polythiophene, polyselenophene and polytellurophene. Synth. Met. 96, 177–189.

Savagatrup, S., Makaram, A.S., Burke, D.J., Lipomi, D.J., 2014a. Mechanical properties of conjugated polymers and polymer-fullerene composites as a function of molecular structure. Adv. Funct. Mater. 24, 1169–1181.

Savagatrup, S., Printz, A.D., O'connor, T.F., Zaretski, A.V., Lipomi, D.J., 2014b. Molecularly stretchable electronics. Chem. Mater. 26, 3028–3041.

Savagatrup, S., Printz, A.D., Rodriquez, D., Lipomi, D.J., 2014c. Best of both worlds: conjugated polymers exhibiting good photovoltaic behavior and high tensile elasticity. Macromolecules 47, 1981–1992.

Schuettfort, T., Watts, B., Thomsen, L., Lee, M., Sirringhaus, H., Mcneill, C.R., 2012. Microstructure of polycrystalline PBTTT films: domain mapping and structure formation. ACS Nano 6, 1849–1864.

Scrosati, B., Hassoun, J., Sun, Y.-K., 2011. Lithium-ion batteries. A look into the future. Energy Environ. Sci. 4, 3287–3295.

Seitz, J., 1993. The estimation of mechanical properties of polymers from molecular structure. J. Appl. Polym. Sci. 49, 1331–1351.

Sekine, S., Ido, Y., Miyake, T., Nagamine, K., Nishizawa, M., 2010. Conducting polymer electrodes printed on hydrogel. J. Am. Chem. Soc. 132, 13174–13175.

Shallcross, R.C., Stubhan, T., Ratcliff, E.L., Kahn, A., Brabec, C.J., Armstrong, N.R., 2015. Quantifying the extent of contact doping at the interface between high work function electrical contacts and poly(3-hexylthiophene) (P3HT). J. Phys. Chem. Lett. 6, 1303–1309.

Shankar, R., Ghosh, T.K., Spontak, R.J., 2007. Dielectric elastomers as next-generation polymeric actuators. Soft Matter 3, 1116–1129.

Shao, S., Ding, J., Wang, L., Jing, X., Wang, F., 2012. White electroluminescence from all-phosphorescent single polymers on a fluorinated poly (arylene ether phosphine oxide) backbone simultaneously grafted with blue and yellow phosphors. J. Am. Chem. Soc. 134, 20290–20293.

Shi, Y., Pan, L., Liu, B., Wang, Y., Cui, Y., Bao, Z., Yu, G., 2014. Nanostructured conductive polypyrrole hydrogels as high-performance, flexible supercapacitor electrodes. J. Mater. Chem. A 2, 6086–6091.

Shi, Y., Peng, L., Ding, Y., Zhao, Y., Yu, G., 2015. Nanostructured conductive polymers for advanced energy storage. Chem. Soc. Rev. 44, 6684–6696.

Smela, E., 2003. Conjugated polymer actuators for biomedical applications. Otero 7, 22.

Snook, G.A., Kao, P., Best, A.S., 2011. Conducting-polymer-based supercapacitor devices and electrodes. J. Power Sources 196, 1–12.

Sonmez, G., Sonmez, H.B., Shen, C.K., Wudl, F., 2004. Red, green, and blue colors in polymeric electrochromics. Adv. Mater. 16, 1905–1908.

Steckler, T.T., Henriksson, P., Mollinger, S., Lundin, A., Salleo, A., Andersson, M.R., 2014. Very low band gap thiadiazoloquinoxaline donor–acceptor polymers as multi-tool conjugated polymers. J. Am. Chem. Soc. 136, 1190–1193.

Street, R., Northrup, J., Salleo, A., 2005. Transport in polycrystalline polymer thin-film transistors. Phys. Rev. B 71, 165202.

Street, R.A., Hawks, S.A., Khlyabich, P.P., Li, G., Schwartz, B.J., Thompson, B.C., Yang, Y., 2014. Electronic structure and transition energies in polymer–fullerene bulk heterojunctions. J. Phys. Chem. C 118, 21873–21883.

Svennersten, K., Bolin, M.H., Jager, E.W., Berggren, M., Richter-Dahlfors, A., 2009. Electrochemical modulation of epithelia formation using conducting polymers. Biomaterials 30, 6257–6264.

Takano, T., Masunaga, H., Fujiwara, A., Okuzaki, H., Sasaki, T., 2012. PEDOT nanocrystal in highly conductive PEDOT: PSS polymer films. Macromolecules 45, 3859–3865.

Takashima, Y., Hatanaka, S., Otsubo, M., Nakahata, M., Kakuta, T., Hashidzume, A., Yamaguchi, H., Harada, A., 2012. Expansion–contraction of photoresponsive artificial muscle regulated by host–guest interactions. Nat. Commun. 3, 1270.

Tessler, N., Preezant, Y., Rappaport, N., Roichman, Y., 2009. Charge transport in disordered organic materials and its relevance to thin-film devices: a tutorial review. Adv. Mater. 21, 2741–2761.

Tomlinson, E.P., Hay, M.E., Boudouris, B.W., 2014. Radical polymers and their application to organic electronic devices. Macromolecules 47, 6145–6158.

Tummala, N.R., Risko, C., Bruner, C., Dauskardt, R.H., Br Das, J.L., 2015. Entanglements in P3HT and their influence on thin-film mechanical properties: insights from molecular dynamics simulations. J. Polym. Sci. Part B: Polym. Phys. 53, 934–942.

Ugur, A., Katmis, F., Li, M., Wu, L., Zhu, Y., Varanasi, K.K., Gleason, K.K., 2015. Low-dimensional conduction mechanisms in highly conductive and transparent conjugated polymers. Adv. Mater. 27, 4604–4610.

Urban, G., Jobst, G., Keplinger, F., Aschauer, E., Tilado, O., Fasching, R., Kohl, F., 1992. Miniaturized multi-enzyme biosensors integrated with pH sensors on flexible polymer carriers for in vivo applications. Biosens. Bioelectron. 7, 733–739.

Usluer, Ö., K Stner, C., Abbas, M., Ulbricht, C., Cimrova, V., Wild, A., Birckner, E., Tekin, N., Sariciftci, N.S., Hoppe, H., 2012. Charge carrier mobility, photovoltaic, and electroluminescent properties of anthracene-based conjugated polymers bearing randomly distributed side chains. J. Polym. Sci. Part A: Polym. Chem. 50, 3425–3436.

Viola, I., Della Sala, F., Piacenza, M., Favaretto, L., Gazzano, M., Anni, M., Barbarella, G., Cingolani, R., Gigli, G., 2007. Bicolor pixels from a single active molecular material by surface-tension-driven deposition. Adv. Mater. 19, 1597–1602.

Viola, I., Piliego, C., Favaretto, L., Barbarella, G., Cingolani, R., Gigli, G., 2010. Bicolor electroluminescent pixels from single active molecular material. ACS Appl. Mater. Interfaces 2, 484–490.

Wadhwa, R., Lagenaur, C.F., Cui, X.T., 2006. Electrochemically controlled release of dexamethasone from conducting polymer polypyrrole coated electrode. J. Controlled Release 110, 531–541.

Wang, X., Gu, X., Yuan, C., Chen, S., Zhang, P., Zhang, T., Yao, J., Chen, F., Chen, G., 2004. Evaluation of biocompatibility of polypyrrole in vitro and in vivo. J. Biomed. Mater. Res. Part A 68, 411–422.

Wang, Z.-L., Guo, R., Li, G.-R., Lu, H.-L., Liu, Z.-Q., Xiao, F.-M., Zhang, M., Tong, Y.-X., 2012. Polyaniline nanotube arrays as high-performance flexible electrodes for electrochemical energy storage devices. J. Mater. Chem. 22, 2401–2404.

Wang, H., Meng, F., Cai, Y., Zheng, L., Li, Y., Liu, Y., Jiang, Y., Wang, X., Chen, X., 2013. Sericin for resistance switching device with multilevel nonvolatile memory. Adv. Mater. 25, 5498–5503.

Wantz, G., Derue, L., Dautel, O., Rivaton, A., Hudhomme, P., Dagron-Lartigau, C., 2014. Stabilizing polymer-based bulk heterojunction solar cells via crosslinking. Polym. Int. 63, 1346–1361.

Weng, Y.-T., Pan, H.-A., Wu, N.-L., Chen, G.Z., 2015. Titanium carbide nanocube core induced interfacial growth of crystalline polypyrrole/polyvinyl alcohol lamellar shell for wide-temperature range supercapacitors. J. Power Sources 274, 1118–1125.

Wittmann, J., Hodge, A., Lotz, B., 1983. Epitaxial crystallization of polymers onto benzoic acid: polyethylene and paraffins, aliphatic polyesters, and polyamides. J. Polym. Sci.: Polym. Phys. Ed. 21, 2495–2509.

Wu, C., Schneider, T., Zeigler, M., Yu, J., Schiro, P.G., Burnham, D.R., Mcneill, J.D., Chiu, D.T., 2010. Bioconjugation of ultrabright semiconducting polymer dots for specific cellular targeting. J. Am. Chem. Soc. 132, 15410–15417.

Xia, Y., Sun, K., Ouyang, J., 2012. Solution-processed metallic conducting polymer films as transparent electrode of optoelectronic devices. Adv. Mater. 24, 2436–2440.

Xue, S., Yao, L., Shen, F., Gu, C., Wu, H., Ma, Y., 2012. Highly efficient and fully solution-processed white electroluminescence based on fluorescent small molecules and a polar conjugated polymer as the electron-injection material. Adv. Funct. Mater. 22, 1092–1097.

Yang, G., Kampstra, K.L., Abidian, M.R., 2014. High performance conducting polymer nanofiber biosensors for detection of biomolecules. Adv. Mater. 26, 4954–4960.

Yersin, H., 2008. Highly Efficient OLEDs With Phosphorescent Materials. John Wiley & Sons, Hoboken.

Yin, Z., Zheng, Q., 2012. Controlled synthesis and energy applications of one-dimensional conducting polymer nanostructures: an overview. Adv. Energy Mater. 2, 179–218.

Yin, X., Chen, J., Lalancette, R.A., Marder, T.B., J Kle, F., 2014. Highly electron-deficient and air-stable conjugated thienylboranes. Angew. Chem. Int. Ed. 53, 9761–9765.

Yip, H.-L., Jen, A.K.-Y., 2012. Recent advances in solution-processed interfacial materials for efficient and stable polymer solar cells. Energy Environ. Sci. 5, 5994–6011.

Yu, Y., Zhihuai, S., Chen, S., Bian, C., Chen, W., Xue, G., 2006. Facile synthesis of polyaniline-sodium alginate nanofibers. Langmuir 22, 3899–3905.

Zen, A., Saphiannikova, M., Neher, D., Grenzer, J., Grigorian, S., Pietsch, U., Asawapirom, U., Janietz, S., Scherf, U., Lieberwirth, I., 2006. Effect of molecular weight on the structure and crystallinity of poly(3-hexylthiophene). Macromolecules 39, 2162–2171.

Zeng, W., Shu, L., Li, Q., Chen, S., Wang, F., Tao, X.M., 2014. Fiber-based wearable electronics: a review of materials, fabrication, devices, and applications. Adv. Mater. 26, 5310–5336.

Zhang, W., Smith, J., Watkins, S.E., Gysel, R., Mcgehee, M., Salleo, A., Kirkpatrick, J., Ashraf, S., Anthopoulos, T., Heeney, M., 2010. Indacenodithiophene semiconducting polymers for high-performance, air-stable transistors. J. Am. Chem. Soc. 132, 11437–11439.

Zhang, X., Li, C., Luo, Y., 2011. Aligned/unaligned conducting polymer cryogels with three-dimensional macroporous architectures from ice-segregation-induced self-assembly of PEDOT-PSS. Langmuir 27, 1915–1923.

Zhang, Q., Tsang, D., Kuwabara, H., Hatae, Y., Li, B., Takahashi, T., Lee, S.Y., Yasuda, T., Adachi, C., 2015a. Nearly 100% internal quantum efficiency in undoped electroluminescent devices employing pure organic emitters. Adv. Mater. 27, 2096–2100.

Zhang, Y.-Z., Wang, Y., Cheng, T., Lai, W.-Y., Pang, H., Huang, W., 2015b. Flexible supercapacitors based on paper substrates: a new paradigm for low-cost energy storage. Chem. Soc. Rev. 44, 5181–5199.

Zhao, Y., Zhao, X., Zang, Y., Di, C.-A., Diao, Y., Mei, J., 2015. Conjugation-break spacers in semiconducting polymers: impact on polymer processability and charge transport properties. Macromolecules 48, 2048–2053.

Zhong, N., Post, W., 2015. Self-repair of structural and functional composites with intrinsically self-healing polymer matrices: a review. Compos. Part A: Appl. Sci. Manuf. 69, 226–239.

Zhu, L., Lei, A., Cao, Y., Ai, X., Yang, H., 2013. An all-organic rechargeable battery using bipolar polyparaphenylene as a redox-active cathode and anode. Chem. Commun. 49, 567–569.

4

Electronic Polymer Composite

4.1 Synthesis

To synthesize polymer composites, the second inorganic phase is usually incorporated into a polymer matrix. The dispersion, content, and even the alignment of inorganic moieties in polymer matrix as well as the interfacial interactions between the two components should be considered for target composites from different synthetic processes, which determine the final structures and properties of polymer composites. Based on the natural properties of polymers and inorganic moieties, a variety of methods have been employed to prepare polymer composites. According to the synthetic sequence, they are divided into four main categories (Fig. 4.1). First, polymer and the second phase are synthesized separately, and the composites are obtained by direct mixing the two components through solution, melting, and infiltration processes. Second, the composites are prepared by in situ polymerization of monomers in the presence of the second inorganic phase. Third, the second phase is in situ generated in the polymer. Last, the two components are in situ synthesized simultaneously. In addition, multicomponent composites can be synthesized by one method or the combination of several methods.

4.1.1 Direct Mixing

For this method, the polymer and second phase are first synthesized separately, and the composites are then obtained by direct mixing the two components. It is simple and favorable for mass production. All kind of available methods for the two components can be employed separately without any influence by each other. Both the size and size distribution of the inorganic components can be carefully controlled during the synthesis. The composites can be produced through melt blending, solution mixing, and infiltrating polymer solutions into the inorganic templates as well as some other complex processes, such as, layer-by-layer assembly and fiber spinning. This method is suitable for polymers that can be molten at certain temperature or soluble in solvents. Various inorganic components have been added to polymers, such as, metals, metallic and nonmetallic oxides, and carbonaceous nanomaterials. Although they are mixed in different temperatures or solvents, they show a similar mixing process, thus here we take carbon nanotube (CNT) as an example.

CNTs can be first synthesized by arc discharge, laser ablation, and chemical vapor decomposition. CNTs are dispersed in a polymer matrix by melt blending under a high temperature and high shear force, which is also compatible with current industrial practices (Moniruzzaman and Winey, 2006). Melt blending is simple and useful for thermoplastic polymers, such as, polyethylene, polypropylene, polycarbonate, poly(methyl methacrylate)

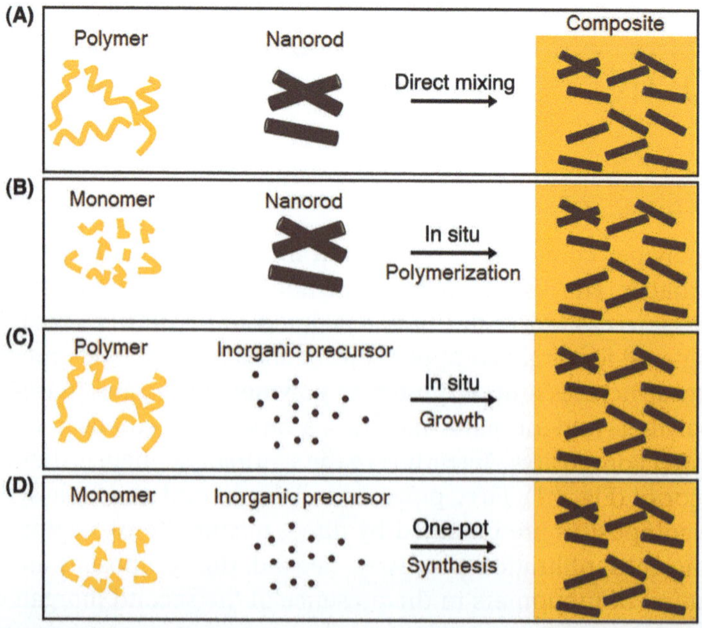

FIGURE 4.1 Schematic illustration of the synthetic process of polymer nanocomposites. (A) Direct mixing of the two components. (B) In situ polymerization of monomers in the presence of the inorganic component. (C) In situ growth of the inorganic component in the polymers. (D). The two components synthesized simultaneously in one-pot approach.

(PMMA), and nylon-6 (Sahoo et al., 2010). The process avoids the use of toxic solvents as in the case of solution mixing. However, it is not suitable for most electronic polymers for a melting process. Moreover, the CNTs in polymers are limited at lower concentrations due to the high viscosities of the composites at higher CNT loadings. Therefore, melt blending is less used in electronic polymer composites.

Solution process involves dispersing CNTs and dissolving polymer in a suitable solvent, and then the evaporation of solvent produces composites by precipitating or casting into film. It is more fitting for electronic polymers. The dispersion of CNTs in solution will influence their distribution in polymer matrix and finally impact the properties of composites. However, it is difficult to disperse CNTs in solvent due to the strong interaction among them even with the assistance of ultrasonication, thus resulting in a low content. For instance, the CNT/PVDF nanocomposite prepared by the solution casting method exhibited CNT contents not higher than 0.3 wt.% (Ning et al., 2013).

CNTs can be further functionalized through noncovalent or covalent interactions for better dispersibility and stability in polymer matrix. The noncovalent functionalization of CNTs normally involves van der Waals, π–π, CH–π, or electrostatic interactions between polymer molecules and CNT surfaces. The advantage of noncovalent functionalization is that it does not destroy the conjugated system of the CNT sidewall, and therefore it does not affect their final properties (Spitalsky et al., 2010). It should be noted that, the graphitic sidewalls of CNTs provide the possibility for π-stacking interactions with conjugated and aromatic group-containing polymers, as well as organic polymers containing heteroatoms with free electron pairs. Therefore, the CNTs can be wrapped with these polymers to form uniform dispersion and composites. Through this interaction, CNT and poly(3-hexylthiophene) were mixed in solvent like chlorobenzene under sonication, and the mixed solution can be spin coated to form thin composite film for the active layer of photovoltaic cells (Berson et al., 2007). The dispersion of CNTs can also be improved by incorporating surfactants. Those amphiphilic molecules have both polar and apolar groups, and are adsorbed at the interface between immiscible bulk phases to reduce the surface tension (Vaisman et al., 2006). The better dispersion the CNT has, the higher performance the composites may have. For instance, with the surfactant as the wetting agent, the addition of only 1 wt.% CNTs in the polymer composite increased the glass transition temperature from 63 to 88°C (Gong et al., 2000). The elastic modulus was also increased by more than 30%. In contrast, the addition of CNTs without the surfactant only had moderate effects on the glass transition temperature and the mechanical properties. However, the content of CNT in the composite from this method is still low, typically in the range of 1–15 wt.%. Moreover, phase segregation may occur during solvent evaporation. Therefore, the improvement of properties is restricted.

CNTs are then covalently modified by grafting polar groups or polymers. The most common and facile method is oxidation of CNTs, which results in the formation of a number of carboxylic acid groups (—COOH or —OH) on the surface of CNTs (Byrne and Gun'ko, 2010). This is usually realized by treating CNTs with HNO_3/H_2O_2 and HNO_3/H_2SO_4. The acid functionalization significantly improves the dispersion and stability of CNTs in polar solvent and the interfacial bonding properties between the CNT and a polymer matrix. For instance, after carboxyl-functionalization, CNTs could be dispersed in water, and incorporated into poly(3,4-ethylenedioxythiophene): poly(styrene sulfonate) (PEDOT:PSS). The resulting transparent film exhibited enhanced conductivity compared to the bare PEDOT:PSS films (Byrne and Gun'ko, 2010). The presence of active functional groups also allows for further covalent functionalization of CNTs with polymer molecules. Therefore, the functionalization of CNTs can be tailored to different polymers. In the case of covalent functionalization, the structures and properties of CNT, such as, electronic and transport will be influenced due to the introduction of defects. The degree of functionalization and the enhancement of properties should be balanced in composites.

The modified and charged CNTs can be composited with polyelectrolytes through layer-by-layer assembly (Mamedov et al., 2002). By alternately dipping a solid substrate into CNT dispersion and polymer solution, a multilayer composite film can be obtained. For instance, active-ester-modified CNT/poly(allylamine hydrochloride) composite thin film has been deposited on an activated surface of a quartz slide through this method (Park et al., 2008). The adsorptions of CNT and polymer are controlled to be monolayer, thus realizing homogeneity at the nanometer level and preventing phase segregation. In addition, the loading of CNT reaches up to 50 wt.%, and the tensile strength of the composites is several times higher than those of composites made by mixing (Srivastava and Kotov, 2008). However, this method is only suitable for thin films, has low efficiency and is hard for mass production.

To avoid the aggregation and phase segregation of CNTs in polymer matrix, porous CNT scaffolds are then developed as the template for polymer infiltration. This method does not involve the dispersion of CNTs in polymer matrix during fabrication, and can offer a higher load of CNTs. The structures of composite materials can be controlled by designing and varying the CNT scaffolds, for example, sprayed or spin-coated CNT film, CNT buckypapers, CNT sponges in a network structure and CNT arrays, sheets and fibers in an aligned structure (Sun et al., 2013b). After the incorporation of polymer solutions or melts, the CNT structures are well maintained with the desired properties. For the film and buckypaper, they still need to disperse CNTs in solvent, while sponges and aligned CNT assemblies can be directly synthesized without a solution process. By this method, PEDOT:PSS/CNT nanocomposite has been fabricated by sequentially spin casting CNT dispersion and PEDOT:PSS solution (Song et al., 2013). The polymer was infiltrated into the CNT network and coated on its surface to form a layered nanostructure. CNT buckypaper has been mainly formed through the filtration of CNT dispersions on filtering membranes (Han et al., 2014b). However, CNTs in buckypaper are densely packed to form small voids among them, which limit the penetration rate and efficiency of relatively large polymer molecules. CNT sponges have been synthesized directly by chemical vapor deposition (CVD) (Gui et al., 2010). They were composed of interconnected CNT skeletons with a porosity of >99% and density close to the lightest aerogel. The highly porous structure provides them as a good template in the synthesis of composite materials. In order to effectively take advantage of the remarkable properties from CNTs, the alignment of CNTs is required; highly aligned CNT/polymer materials have been recently synthesized from aligned CNT scaffolds including arrays, sheets, and fibers (Sun et al., 2013a). Vertically aligned CNT arrays were synthesized by CVD. Aligned CNT sheets and fibers had been continuously pulled out of CNT arrays by a dry spinning process. Composite materials were then fabricated by infiltrating polymers into CNT sponge and array, spin-coating them into CNT sheets, and dip-coating them into CNT fiber (Sun et al., 2013b).

These techniques have been used extensively to incorporate various inorganic nanoparticles, nanosheets, nanorods, nanowires, and nanotubes into polymers. During a mixing process, the compatibility of the two components is an important issue. They should have similar hydrophilic/hydrophobic nature to achieve uniform composites. Otherwise, the inorganic particles will aggregate, which is unfavorable for the high properties of composites. Therefore, most of inorganic particles need surface modifications. The inorganic oxides are usually easily modified due to the generated functional groups on the surface during preparation. For the infiltration process, inorganic components should be controlled to assemble into frameworks with a porous structure, and the connectivity and the pore size are critical. The polymer may not completely fill the too small inner pores of the nanostructured inorganic component due to self-blocking. Therefore, the filtration of monomers followed by in situ polymerization is a better way, which will be introduced in the following part. The layer-by-layer method can generate uniform composites, especially for thin films, but the requirement for materials is high, and it is also difficult for large-scale production. For all mixed composites, the problem is the weak interaction between the two components.

4.1.2 In Situ Polymerization in Inorganic Matrix

In this process, the inorganic components are first synthesized, then monomers are mixed with the already synthesized inorganic components, and the composites are finally fabricated by in situ polymerization of monomers. Inorganic components can be particles which are dispersed in solution or assembled into macroscopic scaffolds with nanostructures. The interaction between two components will be enhanced by controlling polymerization of the polymer onto the surface of the inorganic particles. This method is useful for the polymers that cannot be processed by solution or melt blending, for example, insoluble and thermally unstable polymers. Most of conducting polymers and cross-linked polymer composites are synthesized by this method. Chemical synthesis of conducting polymers is simply achieved by oxidation of corresponding monomers using an oxidizing agent and it is preferred for mass production. For instance, aniline can be chemically oxidized into polymer in aqueous solution by a variety of oxidants, such as, $(NH_4)S_2O_8$, $FeCl_3$, and $KMnO_4$. Therefore, aniline monomer was first added to the inorganic particle dispersion under ultrasonication, and then freshly prepared catalyst solution was slowly added dropwise. After polymerization, the composites were obtained by filtering, rinsing, and drying the reaction mixture (Peng et al., 2008).

As with solution blending, the functionalization of inorganic particles can improve the initial dispersion in the liquid (monomer, solvent) and consequently in the composites. Furthermore, in situ polymerization method enables covalent bonding between functionalized inorganic particles and the polymer matrix using various condensation reactions of monomers. For example, water-soluble, polyelectrolyte-grafted CNTs were synthesized using a "grafting to" route (Han et al., 2010). The composite films were then fabricated by an electrostatic spray technique and could be used as the counter electrode for dye-sensitized solar cells (DSSCs). The sprayed films were uniform over a large area, and the well exfoliated CNTs formed highly interconnected network structures. The covalent approach allows for the formation of a strong interface between the CNT and polymer matrix due to strong chemical bonding of polymer molecules to the CNT surface.

The most effective way is to carry out the in situ polymerization of monomers in the presence of inorganic frameworks. Monomers can infiltrate into the framework easier than polymers, thus form the composite effectively. As mentioned before, porous CNT scaffolds can be prepared in various forms. Monomers are then incorporated and polymerized by in situ chemical, electrochemical, and photochemical polymerizations to form composites. In this process, the inorganic frameworks will reserve the assembled structure. Conducting polymers, such as polyaniline (PANI), have been incorporated into CNTs through mixing monomer with CNTs followed by chemical polymerization, or dissolving monomer in electrolyte followed by electrochemical deposition onto CNTs (Ge et al., 2011; Meng et al., 2010; Chabi et al., 2014). The morphology, uniformity, and amount of conducting polymers could be controlled by using different modes of electrodeposition, such as, potentiodynamic, potentiostatic, and galvanostatic (Benson et al., 2013). The composites could maintain the construction of CNT scaffolds with thin layers of polymers coated on the surface. Electrochemically grown conducting polymers decreased the contact resistance between CNTs, producing composites with improved electrical conductivity. Polydiacetylene (PDA)/CNT composite fiber and liquid crystalline polymer/CNT composite film and fiber have been fabricated by incorporating monomers into CNT sheets and fibers and further photopolymerization (Sun et al., 2012; Peng et al., 2009).

The monomers have much better solubility and permeability than polymers. The interaction and homogeneity are expected to be better in the resulted composite. The coexisting inorganic moieties can act as nuclei, thus the formed insoluble polymer chains are deposited on them. In addition, the cross-linked polymers cannot be dissolved in solvent, so their composites should be prepared through this in situ polymerization process. The composites of epoxy, polydimethylsiloxane (PDMS), and polymer gels have been prepared via this method to improve mechanical and electrical properties (Sun et al., 2013b).

4.1.3 In Situ Growth of Inorganic Component in Polymer Matrix

In this process, polymers are first prepared to form solution or film; inorganic precursors are then incorporated and in situ grown to form inorganic particles in presence of polymer matrix. This method is suitable for the inorganic moieties that can be synthesized through chemical reactions in polymer solutions or electrochemical reactions on conductive polymers. Metals and inorganic oxide components are compatible with this method, while polymer composites with carbonaceous materials, such as, CNTs cannot be prepared via this process due to the high requirements in CNT synthesis. Therefore, the following two sections focus on metal oxides, such as, MnO_2 as an example.

For polymer-based electronic materials, conducting polymers are usually employed as the matrix and some inorganic particles are incorporated to enhance the electrical and electrochemical properties. It is reported that conducting polymers possess several intrinsic oxidation states and the as-synthesized conducting polymer are typically in the intermediate oxidation state between fully oxidized and reduced forms. Therefore, conducting polymers show reduction ability toward $KMnO_4$ in aqueous solution at room temperature for obtaining conducting polymer/MnO_2 composite (Han et al., 2012a). Through a chemical reaction, ultrafine and loosened MnO_2 floccules were uniformly loaded on the surface of PANI nanofibers by simply soaking PANI nanofibers in a $KMnO_4$ solution (Jiang et al., 2012). Other nanostructures, such as, ultrathin nanorods could also be formed on surfaces of conducting polymer nanofibers by using conducting polymer nanofibers and $KMnO_4$ as precursors (Han et al., 2012a). The content of MnO_2 in the composites could easily be controlled by changing the concentration of $KMnO_4$. The conducting polymer films can act as an electrode, and nanostructured metal oxides could also be electrodeposited on conducting polymer films. The structure and content could be tuned by the concentration of the precursor and electrolyte solution, the deposition time, and the deposition potential or applied current. Metal oxide particles, such as, SnO_2, TiO_2, and WO_3, can also be incorporated into conducting polymers, such as, polypyrrole (PPy), PANI, and PEDOT.

Metal particles could also be loaded on conducting polymer matrix. The synthesis of metal nanoparticles by chemical deposition on conducting polymers has been well developed. For instance, platinum nanoparticles were in situ synthesized by using a reduction of hexachloroplatinic acid hydrate with a sodium borohydride solution in the presence of aqueous PEDOT:PSS solution (Woo et al., 2014). Electrochemical deposition is a more common and well-established technique for the fabrication of conducting polymer/metal composite materials. Platinum nanoparticles have been electrodeposited onto the outer surface and even inside the conducting polymer matrix by different routes, such as, potentiostatic and pulsed depositions (Armel et al., 2012). It had been found that the potentiostatic technique favored the deposition of metal particles on the surface of conducting polymer, whereas the pulse potentiostatic technique leaded to an efficient dispersion of platinum particles inside the polymer matrix (Li et al., 2009a). Other metal particles, such as, gold, silver, and copper could also be introduced into polymers through similar processes. It is demonstrated that the initial oxidation state of the conducting polymer layer and its surface, and bulk structure play important roles for the location of the metal nanoparticles. The use of metal anion complexes instead of the corresponding metal cations also presented a helpful tool for affecting the location and number of metal crystals (Tsakova, 2008).

4.1.4 One-Pot Synthesis

In the one-pot process, monomer and inorganic precursor are first mixed together in one solution, and the polymerization of monomer and the growth of inorganic particles are carried out simultaneously to form composites. The main advantage of this approach lies in the simplicity and high efficiency. The resulted composites exhibit a homogeneous distribution of inorganic component in polymer matrix due to the uniform mixing of precursors of the two components. This approach has been exploited to obtain hybrids from conducting polymers including PPy, PEDOT and PANI, and also metal or particularly metal oxide particles, such as Fe_2O_3, MnO_2, MoO_3, NiO_x, and WO_3. However, it has the drawback of limited control over structure and morphology of the resultant composite material.

This method can be made via chemical, electrochemical, and photochemical processes. Chemical approach is widely used but with severe aggregation and random distribution of the inorganic particles, which may hamper the application in electronics to some extent. In this route, all reactants are dispersed homogenously in precursor solution and existed as ions and molecules, and involved reactions of ions and molecules produce polymers and inorganics simultaneously. For instance, a PANI/MnO_2 composite has been synthesized via a simultaneous oxidation route (Zhang et al., 2012). $KMnO_4$ was used as the oxidizing agent to oxidize aniline and $MnCl_2$ to prepare PANI and MnO_2, and meanwhile $KMnO_4$ was also reduced to form MnO_2. PANI and MnO_2 were contacted with each other at a molecular level and arranged alternately in the composite. The intermolecule contact improved the conductivity of the composite. The alternative arrangement of PANI and MnO_2 molecules prevented the aggregation of PANI and cluster of MnO_2 so as to decrease the size of the composite particle.

Electrochemical codeposition is a more efficient and frequently used method for incorporating metal or metal oxide nanoparticles into the conducting polymer films. It usually requires a conducting support where both conducting polymer and inorganic nanoparticle can be simultaneously deposited. In this procedure, various parameters can be tuned to precisely control the structures, such as thickness, porosity and morphology, and properties of the composites toward different applications. PEDOT/MnO_2 coaxial nanowires were prepared through electrochemical codeposition in a porous alumina template in aqueous solution containing manganese acetate and ethylenedioxy thiophene monomers (Liu and Lee, 2008). The structures of coaxial nanowires, such as, PEDOT shell thickness and nanowire length could be controlled by varying the applied potential. The conductive, porous, and flexible PEDOT shell facilitates electron transport and ion diffusion into the MnO_2 cores and protects the electrode from mechanical failure. PANI/MnO_2 composite was also electrochemically codeposited on conductive carbon cloth from solutions of $MnSO_4$ and aniline through potential cycling, producing fibrous structures with larger effective areas. A lot of metal nanoparticles can also be composited with polymer through this process. For instance, PANI/silver composite nanowires were achieved by simultaneous electrochemical oxidative polymerization of aniline and reduction of silver ions to silver nanowires in the nanopores of an anodic aluminum oxide (AAO) template (Drury et al., 2007).

Polymer composites with carbonaceous materials, such as, graphene oxides (GO) and charged CNTs can also be fabricated by electrodeposition along with the electrochemical polymerization of monomers. For example, the relatively large anionic GO served as a weak electrolyte and was entrapped in the PPy nanocomposite during the electropolymerization of pyrrole, and also acted as an effective charge-balancing dopant within the PPy film. The obtained GO/PPy nanocomposites exhibited good electrochemical properties (Zhu et al., 2012). This process was extended to the synthesis of other conducting polymer composites with GO, such as, PEDOT. Recently, it was found that the GO could be electrochemically reduced, thus it is expected that the uniform graphene/conducting polymer composite could be obtained by reduction of GO and polymerization of monomers at the same time.

The photochemical method does not require conducting support and the composite can form on any surface. UV and visible light can both be employed to conduct the reaction, and the reaction can be easily controlled by tuning the wavelength and intensity of light to form homogeneous composites. Conducting polymer/metal nanoparticle composites were produced by dissolving monomers, such as pyrrole, in water and mixing with metal salt solutions, such as silver nitrate, copper sulfate, or gold trichloride, followed by photochemical reaction under a UV light. The polymerization and oxidation of pyrrole and reduction of metal ions occurred simultaneously in this process (Breimer et al., 2001). PPy/silver nanoparticles with a core/shell structure were synthesized by one-step UV-induced polymerization (Yang and Lu, 2005). The photopolymerization was made in quartz tube containing distilled pyrrole and silver nitrate in polyvinylpyrrolidone aqueous solutions. The polyvinylpyrrolidone was used as stabilizer to assist the formation of core-shell structure, while the reaction in aqueous solution without polyvinylpyrrolidone just generated silver–polypyrrole mixture.

4.1.5 Multicomponent Preparation

Most metal oxides and carbon materials can be incorporated into polymers at the same time for optimizing properties by fully taking the advantages of each component. For instance, the composites of conducting polymers, metal oxides and CNTs, or graphene were prepared for high performance supercapacitors. Metal oxide materials show high energy density and low cost but limited electric conductivity and structure controllability. Conducting polymers and metal oxides both suffer from mechanical instability and poor cycle ability. Carbon nanomaterials can supplement these properties but with low capacitance. Therefore, the multicomponent materials may show high performances. The preparation can draw lessons from the binary composites via one or two steps.

The first type is wrapping or coating conducting polymers onto metal oxide/carbon nanomaterial composites. CNT/MnO_2/PEDOT-PSS composite was prepared by this process (Hou et al., 2010). CNTs were first treated by acid to attach carboxylic groups or hydroxyl groups on the sidewall of the outer shell. MnO_2 precursors ($KMnO_4$ and $MnSO_4$) were exposed to ultrasonic irradiation in the presence of modified CNTs, leading to the formation of the hierarchical MnO_2 nanospheres on CNT mesoporous networks within short time. PEDOT–PSS was then wrapped on the composites and functioned as an effective dispersant for MnO_2/CNTs structures and as binder material in improving the adhesion to the substrate and the connection. Graphene/MnO_2 composite has also been three-dimensionally wrapped by conducting polymer, and the resulting ternary composites showed an increased specific capacitance (Yu et al., 2011a; Xia et al., 2012). Mesoporous carbon, was also used to form ternary nanocomposites with PANI and MnO_2. The PANI nanolayer was uniformly deposited on the mesoporous carbon/MnO_2 particles by chemical oxidative polymerization (Yan et al., 2012). The PANI nanolayer in the ternary composite restrained the dissolution of MnO_2 nanoparticles in acidic electrolyte so as to enhance their electrochemical utilization. The synergistic effect among three components may result in enhanced specific capacitance and cycling stability of the ternary composites.

Ternary nanocomposites composed of two kinds of carbon nanomaterials and polymers can also be fabricated by this process. Graphene/CNT/PPy composite has been prepared via in situ chemical polymerization of PPy. The introduction of one-dimensional CNT can inhibit the stacking of nanosheet-like graphene and PPy to form three-dimensional hierarchical architecture for highly conductive pathways and high mechanical strength (Lu et al., 2012). It was found that the ternary composites achieved a higher specific capacitance than that of the composites containing two components.

Another type is to introduce metal oxides to polymer/carbon nanocomposites. Ternary composites of MnO_2 nanorods, PANI, and GO have been prepared by incorporating MnO_2 nanorods to intercalated PANI/GO nanosheet composites (Han et al., 2014a). The PANI layer could assist to anchor MnO_2 nanorods onto the surface of GO, prevent GO sheets and MnO_2 nanorods from stacking/aggregating, improve the charge transfer between GO and MnO_2, and prevent MnO_2 from falling off GO sheets during cycling electrochemistry. The composites exhibited greatly improved capacitive performance than two component composites of both GO/MnO_2 and GO/PANI. MnO_2/PANI/CNT ternary coaxial structures were fabricated via a simple wet chemical method (Li et al., 2011b). A PANI layer with appropriate thickness was first coated onto CNT surface and the nanostructured MnO_2 was chemically generated on PANI. The thickness of PANI coatings was controlled by tuning the aniline/CNT ratio, and the appropriate thickness of PANI layers was important for building ternary coaxial structures without the agglomeration of MnO_2 nanoflakes. The ternary coaxial structures provided large interaction area between the MnO_2 nanoflakes and electrolyte, and decreased the contact resistance between MnO_2 and PANI layer coated with CNTs, leading to intriguing electrochemical properties for the applications in supercapacitors.

For the third process, conducting polymers and metal oxides can be simultaneously deposited onto carbonaceous scaffolds by using CNTs or graphene as frameworks during chemical process or electrodes in the electrochemical procedure. This is similar to the one-pot synthesis of binary composites. Besides sequential compositing, ternary composites can also be fabricated by a one-step electrochemical deposition to reach desirable porous architectures due to its simplicity and reliability. It was proposed to prepare unique PPy/reduced GO/CNT ternary composites, where reduced GO, CNT, and PPy were electrodeposited simultaneously to construct a three-dimensionally highly porous film electrode (Ding et al., 2012).

4.2 Structure

The polymer composites possess different structures due to the introduction of the second phases. The nanostructures are essential for different electronic devices. Nanostructured conductive composite polymers have aroused considerable research interest owing to their unique properties over their bulk counterparts, such as larger surface areas and shorter pathways for charge/mass transport, which make them promising candidates for broad applications in energy conversion and storage, sensing, and actuating (Shi et al., 2015).

4.2.1 Continuous Structure

The most common structure is continuous to avoid the use of binder in fabricating electronic devices. Polymers are easy to form continuous structures due to the unique molecular structure, flexibility and viscosity. The inorganic components can be distributed in polymer matrix or coated on the surface of polymers

4.2.1.1 Compacted Film

With a low loading or uniform dispersion of inorganic components in polymer matrix, the composite can form compact films just like the bare polymers. But the introduction of inorganic phase will influence the orientation and crystallinity of polymers. From the viewpoint of the relationship between structure and performance, a nanocrystalline structure with high crystallinity is pursued so that relatively higher mechanical and electrical properties can be obtained. For instance, CNTs were incorporated into a thermoplastic elastomer and yielded nanocomposites with CNT contents of 1–5 vol.% (Koerner et al., 2004). In the compact composite (Fig. 4.2A), the CNTs were uniformly dispersed and slightly increased initial polymer crystallinity by serving as heterogeneous nucleation sites. Interestingly, the performance of the photovoltaic device based on poly(3-hexyl thiophene) (P3HT) was strongly dependent on the enhancement of the crystallinity of the active layer. CNT/P3HT composites were found to form a uniform active layer with an interpenetrating network structure, and microcrystals of P3HT were formed on the sidewall of CNTs with enhanced crystallinity, which contributed to pronounced enhancement of photovoltaic performance (Geng and Zeng, 2006). The other inorganic components can also act as nucleating agent for polymers to endow them higher performances.

Aligned CNT/oriented polymer composite films were obtained by coating PEDOT:PSS solution onto CNT sheets. The aligned CNTs induced the orientation of polymer chains. Therefore, this aligned composite film showed remarkable properties, such as higher electrocatalytic activity than the bare aligned CNTs, bare oriented PEDOT:PSS, and a randomly dispersed CNT/PEDOT:PSS composite. The synergetic interaction between aligned CNTs and polymer chains, and the induced orientation of polymers by aligned CNTs contributed to this new phenomenon (Guan et al., 2013). The composite film can also be prepared by cutting highly aligned CNT/polymer composite arrays. The polymers were first incorporated into CNT arrays, and the composite arrays have been sliced into thin films ranging from nanometers to micrometers using an ultramicrotome or thick films ranging from micrometers to millimeters using a microtome. The CNTs could be aligned in a perpendicular or parallel direction in the plane, depending on the slicing direction (Li et al., 2011a). Due to the alignment of CNTs, the composite showed excellent mechanical and electrical properties.

4.2.1.2 Porous

Porous materials have extremely high surface area and free volume, which are favor for the ion or charge adsorption and the wettability of solvent or electrolyte, thus enhancing the performance of electronic devices. Porous polymers can be prepared by using template and direct synthesis (Wu et al., 2012). The use of template is a simple and versatile approach for preparation of porous polymer composites. It involves the casting or molding polymers in the presence of a template, and the porous polymers are obtained by removing the template. The direct synthesis means that the pores are directly generated during the removal of the solvent from polymer solution. Porous polymer composites can be thus prepared by depositing an inorganic layer onto the porous polymer frameworks. They can

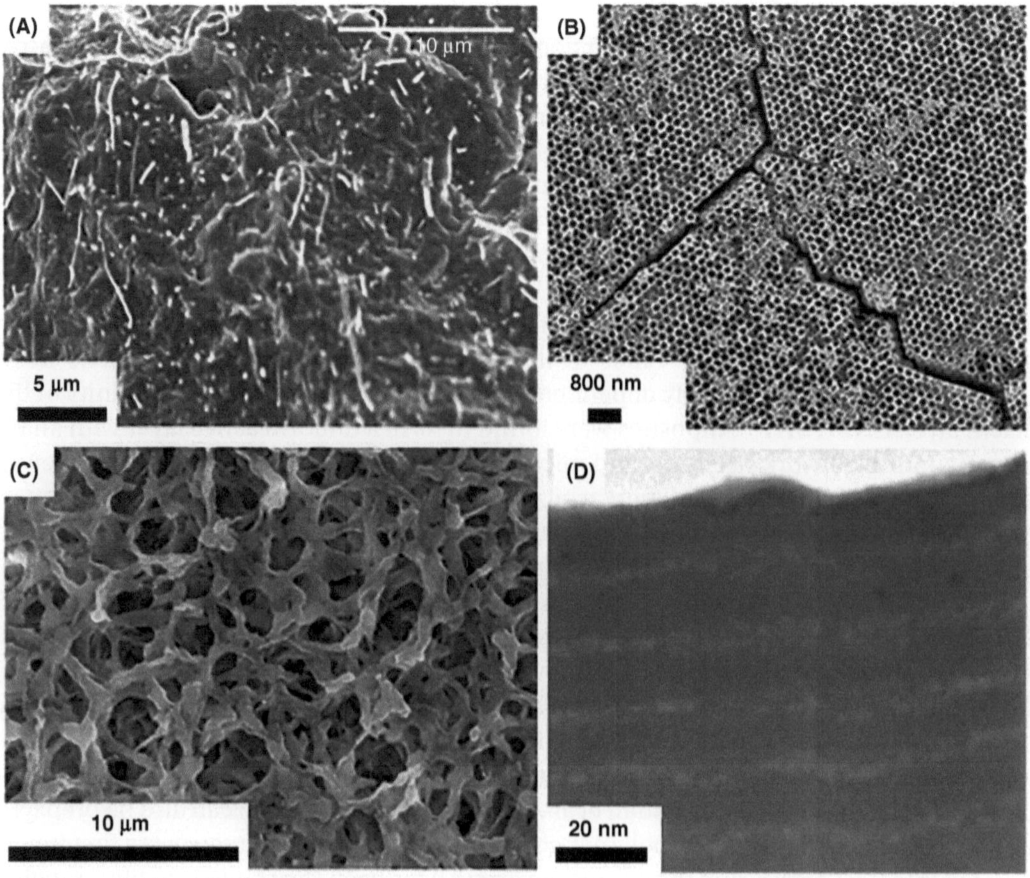

FIGURE 4.2 Continuous structures of polymer nanocomposites. (A) SEM image of thermoplastic elastomer/CNT composite with CNT dispersed in polymer matrix. (B) SEM image of PPy/Au nanocomposite with ordered porous structure. (C) SEM image of P(VDF-HFP)/SiO_2 composite with porous structure. (D) TEM image of PEDOT/graphene composite with a multilayer structure. *(A) Reproduced with permission from reference Koerner, H., Price, G., Pearce, N.A., Alexander, M., Vaia, R.A., 2004. Remotely actuated polymer nanocomposites—stress-recovery of carbon-nanotube-filled thermoplastic elastomers. Nat. Mater. 3, 115–120. Copyright 2004, Macmillan Publishers Ltd. (B) Reproduced with permission from reference Yu, A., Meiser, F., Cassagneau, T., Caruso, F., 2004. Fabrication of polymer-nanoparticle composite inverse opals by a one-step electrochemical co-deposition process. Nano Lett. 4, 177–181. Copyright 2004, American Chemical Society. (C) Reproduced with permission from reference He, X., Shi, Q., Zhou, X., Wan, C., Jiang, C., 2005. In situ composite of nano SiO_2–P(VDF-HFP) porous polymer electrolytes for Li-ion batteries. Electrochim. Acta. 51, 1069–1075. Copyright 2005, Elsevier Ltd. (D) Reproduced with permission from reference Lee, S., Cho, M.S., Lee, H., Nam, J.-D., Lee, Y., 2012. A facile synthetic route for well-defined multilayer films of graphene and PEDOT via an electrochemical method. J. Mater. Chem. 22, 1899–1903. Copyright 2012, Royal Society of Chemistry.*

also be directly prepared by using a polymer/inorganic mixture to replace the pure polymer during the template or direct synthesis. The porosity, size distribution, and even orientation can be controlled during the preparation.

By using the colloidal crystal spheres as templates [typically silica or polystyrene (PS) beads], various types of porous materials with precisely controlled pore sizes and highly ordered three-dimensional porous structures have been prepared. In a typical procedure, a three-dimensional template is first constructed from self-assembled colloid particles. Then, the template is filled with desired materials. Finally, the template is removed by dissolution or calcination, resulting in macroporous arrays that reflect the inverse opal structure. Porous PPy/Au nanoparticles were prepared by electrochemically codepositing Au and PPy into the removable PS sphere templates (Fig. 4.2B) (Yu et al., 2004). The pores were highly ordered in a hexagonal array and interconnected with small channels. The pore size could be controlled by the diameter of template spheres.

Porous polymer composites can also be prepared by mixing the target polymer composite with another removable polymer in solution and evaporating the solvent, followed by the selective dissolution of a polymer. Microporous PVDF-*co*-HFP/$LiAlO_2$ composites were prepared with a preferential polymer dissolution process. The composite solution was mixed with a certain content of PVA, and spread as films on a glass substrate using doctor blade. After dissolving PVA, the porous composite membranes were obtained with a uniform porous structure. The porosity of the polymer film depended on the preferential removal of polymer used. In addition, the micropores in the composite membrane were well interconnected and $LiAlO_2$ ceramic fillers were in the holes of membrane rather than aggregation in the upper surface (Kalyana Sundaram and Subramania, 2007). The porous structured polymer composites can be typically used as separator membrane in lithium ion batteries and as electrodes in electronic devices. The porosity and pore size are determinant in the uptake process of the electrolyte solution which in turn affects the ionic conductivity of the separator. The inorganic particles are incorporated into polymer matrix to improve the thermal, mechanical, and electrochemical properties of the materials.

The incorporation of inorganic components can directly produce the porous structure of polymers without using any template. Poly(vinylidene fluoride-hexafluoropropylene) [P(VDF-HFP)]/SiO_2 composite porous membranes were prepared through direct in situ composite method based on hydrolysis of tetraethoxysilane and phase inversion (Fig. 4.2C) (He et al., 2005). SiO_2 particles were homogeneously dispersed in the polymeric matrix, thus the composite showed similar morphology with the pristine polymer. The incorporation of SiO_2 increased the porosities of membrane compared with pristine polymer membrane, but the increased SiO_2 content would cause the decrease of porosity. Therefore, the content of silica should be controlled in a suitable range for proper porosity and stability to be used as the matrix of porous polymer electrolytes.

4.2.1.3 Layered

Layered composite films can be directly prepared by alternate deposition or vacuum filtration of the mixed dispersions of both components. For instance, PANI/Prussian blue composite films have been fabricated using layer-by-layer assembly of the polycation PANI and a negatively ionized Prussian blue nanoparticle dispersion (DeLongchamp and Hammond, 2004). The resultant organic/inorganic nanocomposites exhibited good smoothness and a classical linear increase in film thickness with assembly steps. In addition, the composite films could be used as a multiply colored electrochromic electrode compared to the bare Prussian blue and PANI with limited color states.

Multilayered PEDOT/graphene composites were also prepared by alternately electrodepositing PEDOT and graphene on metal electrode (Fig. 4.2D) (Lee et al., 2012). Deposition of each layer was repeated in order to obtain the desired number of multilayers. The graphene layer was compact with a thickness less than 10 nm. The multilayered composite film possessed better performance than the conventional composite and bare PEDOT films.

Due to the planar structure, graphene was widely used to make layered composites with polymers. PANI/graphene multilayer films have been prepared via alternate deposition of negatively charged GO and positively charged PANI upon electrostatic interaction, followed by the reduction of their GO components with hydroiodic acid (Sheng et al., 2011). The thicknesses of the PANI and GO monolayers were estimated to be 1.06 and 1.32 nm, respectively (Sarker and Hong, 2012). The morphology of monolayers could be controlled by the concentrations of PANI and GO during assembly. PANI nanofibers and chemically converted graphene were also made into a layered composite film, with PANI nanofibers sandwiched between graphene layers (Wu et al., 2010). The composite film was mechanically stable and had a high flexibility, thus could be bent into large angles or be shaped into various desired structures. Graphene can also form layered structure with graphene sheets as building blocks by vacuum infiltration of graphene suspension. Then the PANI can be electropolymerized onto graphene by using layered graphene film as electrode impregnated in monomer solution (Wang et al., 2009). The composite film maintained the layered structure.

4.2.2 Separated Structure

When the size of a material is reduced to the nanoscale, their physical and chemical properties are dramatically changed. The separated nanostructure of polymer composites is expected to bring important improvements for polymer electronics because the size reduction of materials increases the contact surface area and lowers the interfacial impedance between the electrode and the electrolyte, and decreases the transport pathways for both electrons and ions (Shi et al., 2015). In addition, the mechanical properties for strain accommodation as well as the flexibility will be improved. A variety of nanostructures of polymer composites have been developed including zero-dimensional nanoparticles, one-dimensional nanowires/rods/belts, two-dimensional nanosheets/plates, and three-dimensional porous frameworks/networks.

4.2.2.1 Zero-Dimensional

The zero-dimensional structure mainly corresponds to nanoparticle or nanosphere. Conducting polymers can be coated on the surface of inorganic particles, and inorganic layers can also be coated onto polymer nanoparticles. Due to their low dimensions, the diffusion pathways of charges and ions would be reduced, thus enhance the electronic properties of composites.

Inorganic/polymer core/shell nanoparticles are made of metal, metallic compound, metal oxide, or silica core with a polymer shell. A polymer shell can improve the oxidation stability of the core, serve as a surfactant or stabilizer to prevent the agglomeration of nanoparticles and have good compatibility with other materials. Inorganic nanoparticles coated with conductive polymers, such as PANI, PPy, and polythiophene (PTh), have attracted much interest for use in different applications in electronic devices (Ghosh Chaudhuri and Paria, 2011). For instance, metal/conducting polymer core-shell structured nanocomposites were prepared. Aniline and pyrrole have been adsorbed and in situ polymerized on the surface of silver and gold nanoparticle cores to prepare uniform polymer shells with thickness of tens of nanometers (Fig. 4.3A–B (Xing et al., 2009). These core-shell composite nanoparticles could be stably dispersed in common organic solvents and thus showed promising applications in electronics. Silica core/PPy shell composite particles were also synthesized by chemical polymerization of pyrrole monomer on the surface of the silica sphere (Hao et al., 2003). The steric agent poly(*N*-vinylpyrrolidone) was used as anchor molecule between silica core and monomer during the polymerization. The PPy shell was uniformly coated over the silica surface. Other inorganic moieties, such as Fe, Fe_3O_4, $CaCO_3$, and CdS, have also been coated with polymers, such as conducting polymers, PS, and PMMA. Carbon nanospheres have been made into hollow structure, and aniline was then chemically oxidative polymerized and coated on the external surface to yield composite hollow spheres (Lei et al., 2010).

On the other hand, polymer nanoparticles can be coated with a layer of inorganic moieties to improve the mechanical and electrical properties. These types of particles, in general, have the dual properties of both the inorganic and organic materials. The inorganic material, especially a metal oxide coating on polymers, is beneficial in several respects, such as increased strength of the overall material, resistance to oxidation, thermal and colloidal stability, and abrasion resistance. For instance, PANI/graphene composites with core/shell hollow structures have been fabricated via a solution-based coassembly process (Fig. 4.3C–D). PANI hollow spheres have large specific surface area, providing high electroactive regions and short diffusion lengths for both charge and ion transport. Graphene was coated onto the spheres to offer highly conductive pathways by bridging individual PANI hollow spheres together for better rate and cycling performance of supercapacitors (Fan et al., 2013).

4.2.2.2 One-Dimensional

One-dimensional nanostructured polymer composite materials include nanowires, nanorods, nanotubes, nanobelts, and nanoribbons. Compared to the other three dimensions, the first characteristic of one-dimensional nanostructure is its smaller dimension structure and high aspect ratio, which could efficiently transport electrical carriers along one controllable direction, thus is highly suitable for moving charges in integrated nanoscale systems (Tran et al., 2009). The second characteristic of one-dimensional nanostructure is its device function, which can be exploited as device elements in many kinds of nanodevices. With a rational synthetic design, nanostructures with different diameters/

FIGURE 4.3 Separated structures of polymer nanocomposites. (A–B) TEM images of PANI-coated Au nanoparticles. (C–D) SEM and TEM images of graphene-wrapped PANI hollow spheres, respectively. (E–F) SEM and TEM images of PANI-coated TiO_2 nanorods. (G–H) SEM and TEM images of MnO_2 nanoparticle-enriched PEDOT nanowires. (I–J) TEM images of polypyrrole/Ag coaxial nanocables. (K) SEM image of PPy/RGO composite sheets. (L–M) SEM and TEM images of ordered whisker-like PANI on the surface of mesoporous carbon. *(A–B) Reproduced with permission from reference Xing, S., Tan, L.H., Yang, M., Pan, M., Lv, Y., Tang, Q., Yang, Y., Chen, H., 2009. Highly controlled core/shell structures: tunable conductive polymer shells on gold nanoparticles and nanochains. J. Mater. Chem. 19, 3286–3291. Copyright 2009, Royal Society of Chemistry. (C–D) Reproduced with permission from reference Fan, W., Zhang, C., Tjiu, W.W., Pramoda, K.P., He, C., Liu, T., 2013. Graphene-wrapped polyaniline hollow spheres as novel hybrid electrode materials for supercapacitor applications. ACS Appl. Mater. Interfaces 5, 3382–3391. Copyright 2013, American Chemical Society. (E–F) Reproduced with permission from reference Cai, G., Tu, J., Zhou, D., Zhang, J., Xiong, Q., Zhao, X., Wang, X., Gu, C., 2013. Multicolor electrochromic film based on TiO_2@polyaniline core/shell nanorod array. J. Phys. Chem. C 117, 15967–15975. Copyright 2013, American Chemical Society. (G–H) Reproduced with permission from reference Liu, R., Duay, J., Lee, S.B., 2010. Redox exchange induced MnO_2 nanoparticle enrichment in poly(3,4-ethylenedioxythiophene) nanowires for electrochemical energy storage. ACS Nano 4, 4299–4307. Copyright 2010, American Chemical Society. (I–J) Reproduced with permission from reference Chen, A., Wang, H., Li, X., 2005. One-step process to fabricate Ag-polypyrrole coaxial nanocables. Chem. Commun. 1863–1864. Copyright 2005, Royal Society of Chemistry. (K) Reproduced with permission from reference Yin, J., Chang, R., Shui, Y., Zhao, X., 2013. Preparation and enhanced electro-responsive characteristic of reduced graphene oxide/polypyrrole composite sheet suspensions. Soft Matter 9, 7468–7478. Copyright 2013, Royal Society of Chemistry. (L–M) Reproduced with permission from reference Wang, Y.G., Li, H.Q., Xia, Y.Y., 2006. Ordered whisker like polyaniline grown on the surface of mesoporous carbon and its electrochemical capacitance performance. Adv. Mater. 18, 2619–2623. Copyright 2006, Wiley-VCH.*

widths, thicknesses, and lengths can be fabricated and controlled (Yin and Zheng, 2012). The structure of one-dimensional conducting polymer nanocomposites, the interfacial adhesion between conducting polymers and the second component, and the aspect ratio will all affect the properties of the nanocomposites. A large number of advanced techniques have been developed to fabricate one-dimensional nanostructures with well-controlled morphology and chemical composition. In general, the controlled synthetic strategies could be classified into two main categories: template-based methods and template-free methods, such as, self-assembly and electrospinning.

For the template-based process, one-dimensional polymer composites can be obtained by codeposition of two components in porous templates. AAO and porous silicate are usually employed as templates. The dimension, aspect ratio and orientation of one-dimensional polymer composite structures could be controlled either by adjusting the experimental parameters or by the template itself (Liu et al., 2011c). MnO_2/PEDOT coaxial nanowires have been synthesized by a one-step electrochemical deposition process using a porous alumina template (Liu et al., 2011c). The phase segregation of MnO_2 and PEDOT led to simultaneous growth of MnO_2 core and PEDOT shell resulting in coaxial nanowires. In this composite, the MnO_2 core was utilized for its high energy density, while the PEDOT shell was applied for its high conductivity and its porous and flexible nature. The PEDOT shell facilitated electron transport and ion diffusion into the energy dense MnO_2 core and protected the core from structural collapse and breaking. These combined properties resulted in a synergic composite that had very high specific capacitances at high current densities as opposed to pure MnO_2 nanowires, PEDOT nanowires, and MnO_2 thin films. This method has the advantages of universality and controllable dimensions. However, the synthesis quantity is limited by the size or amount of the template, thus limiting its large-scale fabrication. Moreover, the posttreatment for template removal may destroy the formed nanostructures and increase the processing cost.

Alternatively, preexisting nanostructured inorganic materials, such as nanotube, nanorod, and nanowires, could also be used as templates to guide the growth of one-dimensional polymer nanostructures and finally prepare composites (Lu et al., 2011). In this process, it is important to investigate effective methods to make conducting polymers grow on the surface of inorganic components with one-dimensional nanostructure. Since CNTs are one-dimensional nanostructures, coating a layer of conducting polymers on CNTs represents one of the simplest methods to fabricate one-dimensional conducting polymer/CNT nanocomposites. CNTs are electrically conductive, thus the conducting polymers can be conveniently grown on the CNT surface by electrochemical polymerization. The morphologies and properties of the nanocomposites can be controlled by the electropolymerization conditions, such as the applied potential or current density. One-dimensional PANI/CNT composites have been synthesized by the electrochemical oxidation of aniline in H_2SO_4 on the CNT electrode (Downs et al., 1999). PANI was completely or partially coated on the surface of CNTs. PANI was also formed on the surface of CNTs via covalent bonds after in situ chemical oxidative polymerization (Philip et al., 2005). The as-synthesized PANI/CNT nanocomposites had a uniform core-sheath structure and the PANI formed a thick homogeneous coating on the surface of CNTs. In addition, emulsion polymerization was also used to prepare one-dimensional conducting polymer/CNT nanocomposites with uniform core-sheath structures (Yu et al., 2005). The microemulsion system can be sodium dodecyl-benzenesulfonate surfactant, butanol, hexane, and hydrochloric acid. Monomers of conducting polymers were polymerized in the adsorbed micelles to from conducting polymer particles on the surface of CNTs. Of course, vapor deposition polymerization is also an effective approach for the preparation of one-dimensional core-shell polymer nanocomposites (Lu et al., 2011).

Compared to entangled CNTs, aligned CNT arrays prove to be a promising substrate for the electrodeposition of conducting polymers because conducting polymers can be grown well on the surface of individual CNT, facilitating the fabrication of uniform coaxial CNT/conducting polymer nanocomposites. Aligned CNT arrays have been used to make conducting polymer/CNT coaxial nanowires by electrochemically depositing a layer of conducting polymer onto each CNT (Gao et al., 2000). The coaxial structure allowed the CNT to provide high mechanical stability, efficient thermal and electrical conduction, and form conducting polymer layer. The high surface area increased the charge injection and separation, which would enhance the performance of electronic devices.

One-dimensional metal oxides can be used as templates similarly. The TiO_2/PANI core/shell nanorod array was prepared by electropolymerization of aniline on the surface of TiO_2 nanorod array (Fig. 4.3E–F) (Cai et al., 2013). The core/shell structure and the porous space among the nanorods made the ion diffusion easier and provide larger surface area for charge-transfer reactions. They showed great promise as a potential multicolor electrochromic material. Other metal oxide nanorods and nanotubes have also been coated with conducting polymers, such as MnO_2/PPy nanotubes (Yao et al., 2013) and SnO_2/PANI nanorod arrays (Xu et al., 2014b).

In addition, bare conducting polymer nanowire and nanotube can be first synthesized, and polymer composites could be formed by the post chemically doping of metal or metal oxide to improve the electrochemical performances (Lu et al., 2011). MnO_2 nanoparticle enriched PEDOT nanowires are fabricated by simply soaking the PEDOT nanowires in $KMnO_4$ solution (Liu et al., 2011b). The MnO_2 nanoparticles have uniform sizes and are finely dispersed in the PEDOT nanowire matrix (Fig. 4.3G–H). PANI/Ag composite nanofibers have been prepared by the redox process between PANI nanofibers and the corresponding $AgNO_3$. Similarly, gold, and Pt nanoparticles were also easily produced inside or on the surface of the PANI nanofibers via such an in situ redox reaction to form composite nanofibers. PANI/MnO_2 coaxial nanofibers have also been fabricated by growing MnO_2 on the surface of PANI nanofibers.

Composites with one-dimensional structure can also be fabricated by template-free method. They are formed via an in situ synthetic process. For instance, Ag/PPy nanocables have been prepared through the redox reaction between pyrrole and silver nitrite in aqueous solution in the presence of polyvinylpyrrolidone. During the reaction process, the formation of Ag nanowires and the polymerization of a pyrrole sheath were proceeded simultaneously. The as-synthesized Ag/PPy nanocables displayed a coaxial nanostructure with a dark core and a lighter sheath (Fig. 4.3I–J) (Chen et al., 2005). The diameter of the Ag nanowire core was about 20 nm, while the outer diameter of PPy sheath reached about 50 nm.

Electrospinning is another effective and versatile technique capable of generating one-dimensional nanostructures from a variety of polymers (Lu et al., 2009). Compared to the commercial mechanical spinning process for generating microfibers, electrospinning mainly makes use of the electrostatic repulsions between surface charges to reduce the diameter of a viscoelastic jet or a glassy filament. One of the most important advantages of the electrospinning technique is that it is relatively easy and not expensive to produce large numbers of different kinds of nanofibers. Other advantages of the electrospinning technique are the ability to control the fiber diameters, the high surface to volume ratio, high aspect ratio, and pore size as nonwoven fabrics. In order to insert an inorganic nanocomponent into a polymer fiber for making composite nanofibers, the most straightforward strategy is to disperse the inorganic nanocomponent in a polymer solution, followed by electrospinning (Lu et al., 2009).

The problem first need to be solved is to make the inorganic nanoparticles effectively disperse in polymer nanofibers. CNTs can be well incorporated in electrospun polymer fibers only after they are well dispersed in the electrospinning solution. For example, composite nanofiber sheets of well-aligned polyacrylonitrile (PAN) nanofibers containing CNTs were prepared by electrospinning a CNT-suspended solution of PAN in dimethyl formamide using a moving collector. The CNTs were parallel and oriented along the axes of the nanofibers (Hou et al., 2005). The electrospun polymer/CNT composite fibers showed much improved electrical conductivity, thermal stability, tensile strength, and tensile modulus. CNTs were dispersed in aniline solution, and CNT/PANI thermoelectric composite nanofibers were prepared by a combination of in situ polymerization and electrospinning processes. The obtained composite nanofibers were macroscopically aligned, and all the nanotubes were homogeneously wrapped by PANI and aligned in the direction parallel to the nanofiber axis, forming a bundle-cable-like CNT/PANI structure (Wang et al., 2012).

A gas–solid reaction was further introduced to the electrospinning technique to incorporate semiconductor nanostructures into polymer nanofibers with better dispersion. The production of well-dispersed PbS nanoparticles in polymer fiber matrices has been achieved by this method (Lu et al., 2005). First, metal salt and polymer were codissolved into one solvent to make a homogeneous solution. Then the above solution was electrospun to obtain polymer/metal salt composite nanofibers. The composite nanofibers were finally exposed to H_2S gas at room temperature to synthesize PbS nanoparticles in situ in polymer nanofibers.

4.2.2.3 Two-Dimensional

Two-dimensional nanostructures have two dimensions outside of the nanometric size range, such as nanoplates, nanosheets, and nanodisks. Graphene is a typical two-dimensional film, which is composed of a one-atom-thick planar sheet of sp^2-bonded carbon atoms that are densely packed into a honeycomb crystal lattice. This material exhibits a high electrical conductivity, a high surface area of over 2600 $m^2\ g^{-1}$, an elevated chemical tolerance, and a broad electrochemical window. Therefore, they were used to form two-dimensional nanocomposites with polymers. The graphene not only increases the electrical conductivity of the polymer, but also enhance its mechanical stability. Conducting polymers with various hierarchical structures have been deposited on two-dimensional graphene sheets. For instance, graphene/PPy nanosheet composites have been successfully fabricated based on an in situ chemical oxidation polymerization method. Sheet-like PPy with a thickness of 20–50 nm was grown on the surface of the composites (Xu et al., 2011). Reduced graphene oxide (RGO) was also used as a core to prepare core–shell-structured RGO-supported PPy composite sheets (Yin et al., 2013). The composite sheets possessed free-standing plate-like morphology with a thickness of ~100 nm (Fig. 4.3K). PPy was thoroughly and uniformly covered on RGO sheets and the PPy-coating shell restricted the restacking of RGO sheets.

Metal oxides with two-dimensional structure have also been studied to composite with polymers. V_2O_5/PANI nanocomposite nanosheets have been synthesized by in situ intercalation polymerization of aniline with layered V_2O_5 under hydrothermal conditions. The thickness of the nanosheets was ranged from 10 to 20 nm, and the typical lateral dimensions of the nanosheets varied from hundreds of nanometers to several microns (Pang et al., 2005). The thickness of the nanocomposite nanosheets decreased with the increasing concentration of aniline. MoS_2/PANI composites were also prepared via in situ polymerization of aniline monomer in the presence of two-dimensional MoS_2 suspension. MoS_2 acted as a polymerization substrate in the polymerization system and provided a path for the insertion and extraction of ions within the PANI, and ensured a high reaction rate. As is expected, this highly conductive MoS_2/PANI nanocomposite-based electrode demonstrated a high specific capacitance and a high long-term cycling stability (Huang et al., 2013). $NiCo_2O_4$ nanosheet arrays were grown vertically and polymers can be coated on the surface to form composite nanosheets (Xu et al., 2014a). PPy and PEDOT were studied, and the coating did not deteriorate the ordered structure but increased the thickness of the nanosheets and made the nanosheet surface comparatively rough. In addition, the structures of the two composites were slightly different. For the PEDOT-coated nanosheets, the active surface of $NiCo_2O_4$ electrode material was nearly fully blocked by mushy-like PEDOT nanosheets, while the PPy nanosheets were interconnected with each other but still did not fully cover the entire $NiCo_2O_4$ nanosheet arrays.

4.2.2.4 Three-Dimensional

The three-dimensional continuous nanostructured framework has a large specific surface area and favors rapid electron and ion transport in polymer-based electronics. The previously mentioned polymer composites with three kinds of structures can form three-dimensional porous structures when they were stacked. Besides, polymers can be incorporated into porous silica, carbon, and graphene as the retaining frameworks to form three-dimensional structured composites. The resultant composite materials exhibit surface properties of the polymers and high mechanical strength and high electric conductivity of the frameworks, which will provide new possibilities for advanced applications. The structures, such as the connectivity, pore diameters, and shapes, are mainly decided by the frameworks.

In situ polymerization inside mesoporous silica has been investigated a lot in order to make functional polymer/silica composite materials with well-defined mesoporosity

(Choi et al., 2005). In the polymer/silica composites, the polymers were filled in the pores or selectively coated into a thin layer on silica walls. The location of the polymers may be controlled by varying the structure of the silica framework, the species of polymers, and polymerization condition.

Polymers can also be loaded into porous carbon via different processes to form composites, such as in situ polymerization of monomer in the presence of mesoporous carbon or its precursor and the direct physical mixing method (Li et al., 2009b). For instance, PANI was filled in mesoporous carbon to form composites, and it could also be grown on the surface of mesoporous carbon by different processes (Choi and Ryoo, 2003). For instance, ordered whisker-like PANI has been grown on the surface of mesoporous carbon (Fig. 4.3L–M), (Wang et al., 2006) and a PANI uniform layer may also be coated on a three-dimensional macroporous carbon. The thickness of the PANI layer can be controlled easily through the modulation of the scan rate while keeping the same total synthesis time during electrodeposition. These polymer composites exhibit the same chemical properties of the organic polymers, whereas the stability of the pores against mechanical compression, thermal, and chemical treatments is greatly enhanced.

Recently, three-dimensional porous graphene films have been prepared by CVD, using sacrificial soft and hard templates, self-assembled graphene hydrogel by a hydrothermal method, electrochemical reduction of GO in a GO aqueous dispersion of high concentration, and so on (Han et al., 2014c; Jiang and Fan, 2014; Meng et al., 2013). Such a structure is also easy to composite with other functional polymer materials. Porous composite films with conducting polymers, such as, PANI nanowire arrays on the outer and inner surface of porous graphene were further produced by a dilute polymerization process to combine the advantages of both materials. Three-dimensional porous graphene/PANI composites with pores vertically oriented on current collectors were also obtained by one-step electrodeposition from the homogeneous mixed solution of aniline monomer and GO (Zhou et al., 2014). In this case, the graphene network provided not only a scaffold with large specific surface area for the deposition of PANI, but also highly conductive porous channels for the transfer of electrons and the facile adsorption–desorption of ions in the electrolyte.

4.3 Property

The main purpose of incorporating second phase to polymer matrix is to enhance properties of materials and use them for various applications. The hierarchical structures of polymer composites were also developed for better performances of electronic devices based on them. Here, the mechanical, electrical, electrochemical and electroluminescent, as well as the sensing properties of polymer composites were summarized.

4.3.1 Mechanical Property

It has been shown that dramatic improvements in mechanical properties of polymers can be achieved by incorporation of a few weight percentages of inorganic components.

Generally, the stiffness or Young's modulus can be readily improved by adding inorganic particles due to their much higher stiffness than polymer matrices. However, their strengths depend on the stress transfer between the inorganic particles and the polymer matrix. Incompatibility between phases may lower stress transfer due to poor interfacial adhesion, resulting in a lower composite strength. Therefore, strong interfacial adhesion between the inorganics and polymer matrix is crucial for effective reinforcement. Moreover, the enhancement also depends on the shape, size, content, dispersion, and alignment of inorganic components in polymers.

From the view of interfacial adhesion, it has been reported that functionalized carbon components offer higher interfacial interactions with polymers and better dispersion than their unmodified counterparts, thereby increasing mechanical properties at the same loadings (Sun et al., 2013b). A noncovalent functionalization of either CNTs or graphene sheets makes their surfaces more compatible with polymers while maintaining their structures and properties. The efficiency of load transfer is typically low due to the weak interaction between CNT or graphene and polymer in this case. The covalent attachment of functional groups on their surfaces can largely improve the efficiency of load transfer as the more uniform dispersion of carbon components and stronger interaction between them and polymers are produced. For instance, the modified GO/polymer composite showed an increase of 240% in tensile modulus compared to the only 11% improvement of nonmodified GO.

Particle shape has an obvious effect on these mechanical properties. It was found that fillers of high aspect ratio had high surface areas, thus resulting in more contact area between the filler and the polymer matrix. Therefore, by presuming a better adhesion and bonding, the composites should display much more enhanced strengths for the fiber or sheet-like fillers than the spherical particles (Ahmad et al., 2008). The mechanical properties of epoxy nanocomposites with graphene platelets and CNTs have been compared at low contents for dispersion (Rafiee et al., 2009). The results indicated that graphene platelets significantly out-perform CNTs. For instance, the Young's modulus of the graphene nanocomposite was ~3% higher than that based on single-walled CNTs; compared with the pure epoxy, the tensile strength was enhanced by ~40% after introduction of graphene platelets and only ~14% by incorporation of multiwalled CNTs. The superiority of graphene platelets over CNTs in terms of mechanical properties was attributed to their higher specific surface areas, which increase nanofiller–matrix adhesion/interlocking interactions.

For the particle size, smaller spheres exhibit higher specific surface areas for a given particle load. The tensile strength of polymer composites increased with particle sizes from micro- to nanoscales while with the same volume fraction (Fu et al., 2008). This result indicates that the strength is increased with the increasing specific surface area of the filled particle through a more efficient stress transfer mechanism. In the case of sheet or fiber-shaped particles with one or two dimensional sizes in microscale, the specific surface area is increased with the increasing aspect ratio, thus the larger sheets and longer fibers can more obviously enhance the mechanical property of the polymer composite.

Typically, the strength of polymer composite is increased with the increasing inorganic content at a relatively lower percentage. When the load reaches certain content, the strength will decrease due to the aggregation of inorganic particles. However, for the polymer composites from the robust inorganic networks, such as CNT sponges and porous graphene, still exhibited high performances at relatively high contents.

The alignment of inorganics, especially the fibers, is also a key parameter for high mechanical properties. For instance, after alignment, the CNT/polymer composites showed an increase of several orders of magnitude than composites from the entangled CNTs. The alignment of CNTs in polymer composites could be achieved by using mechanical stretching, flowing, electric field, magnetic field, and with the assistance of liquid crystalline molecules (Xie et al., 2005). However, very low degrees of alignments are obtained, and the alignment of CNTs generally occurs on the surface. Much higher alignments have been realized in polymer composites by the use of highly aligned CNT arrays, sheets, and fibers as templates in recent years. By this template method, the structures including the distances among CNTs and arrangements of polymer molecules could be varied and controlled. Therefore, polymer composite fibers with inorganic nanotubes or nanofibers aligned along a specific direction showed high mechanical properties.

4.3.2 Electrical Property

The electrical conductivity is very important for electrode and active layer in polymer electronics. For the insulated polymer, various conductive fillers, such as metal nanoparticle and nanowire, CNT and graphene, have been added to offer electrical conductivity (Yu et al., 2011c; Han et al., 2012b; Sekitani et al., 2009). In this case, the conduction process can be described by a percolation model. Percolation is a statistical concept that describes the formation of an infinite cluster of connected particles or pathways. Lowering the percolation threshold is a critical issue to achieve light and low-cost conductive composites made of an insulating matrix loaded with conductive particles. Particles with large aspect ratio were reported to have a low threshold than spheres due to a larger excluded volume at the same weight. For instance, the use of CNTs in place of carbon black leads to a lower percolation threshold and higher maximum conductivity in a bulk epoxy matrix (Sandler et al., 2003). Therefore, conductive rod-like fillers, nanofibers, nanowires, and nanotubes are currently widely investigated for making antistatic and electrically conductive materials (Vigolo et al., 2005). Silver nanowires have been also composited with elastic polymers to fabricate flexible electrodes for electronics, such as LED and solar cells. The fillers formed conductive network, thus the composite films showed the areal resistance as low as $10\ \Omega\ sq^{-1}$ (Yu et al., 2011b).

The percolation is usually very low, such as less than 1%. However, the polymers for electronics require high conductivity, thus a much higher content of filler is desired to form highly conductive networks in polymer matrix. Moreover, a good dispersion also helps to achieve percolation at low volume fraction of fillers. The graphene–polymer composites that were prepared via complete exfoliation of graphite and molecular-level dispersion of

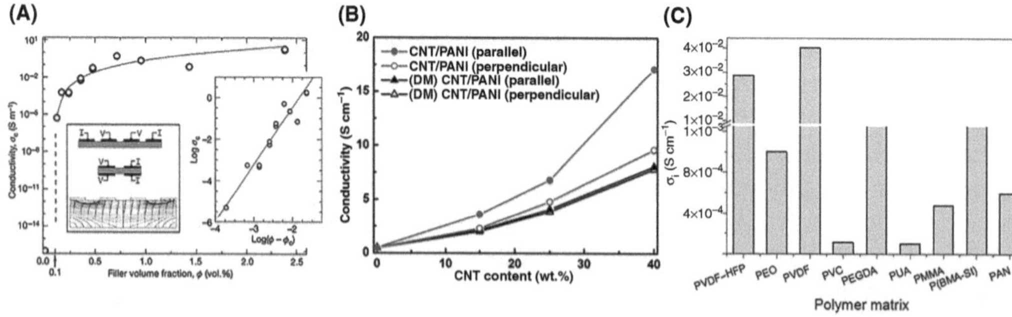

FIGURE 4.4 Electrical properties of polymer composites. (A) The electrical conductivity of the PS/graphene composite as a function of graphene volume fraction. (B) The electrical conductivity of directly mixing *(DM)* CNT/PANI and CNT/PANI nanofibers with different CNT contents in the directions being parallel and perpendicular to the fiber axis. (C) The maximum ionic conductivity of the composite polymer electrolytes with SiO_2 particles for different polymer matrices. *(A) Reproduced with permission from reference Stankovich, S., Dikin, D.A., Dommett, G.H.B., Kohlhaas, K.M., Zimney, E.J., Stach, E.A., Piner, R.D., Nguyen, S.T., Ruoff, R.S., 2006. Graphene-based composite materials. Nature 442, 282–286. Copyright 2006, Macmillan Publishers Ltd. (B) Reproduced with permission from reference Wang, Q., Yao, Q., Chang, J., Chen, L., 2012. Enhanced thermoelectric properties of CNT/PANI composite nanofibers by highly orienting the arrangement of polymer chains. J. Mater. Chem. 22, 17612–17618. Copyright 2012, Royal Society of Chemistry. (C) Reproduced with permission from reference Nunes-Pereira, J., Costa, C.M., Lanceros-Méndez, S., 2015. Polymer composites and blends for battery separators: state of the art, challenges, and future trends. J. Power Sources 281, 378–398. Copyright 2015, Elsevier Ltd.*

individual, chemically modified graphene sheets within polymer hosts exhibited a percolation threshold of 0.1 volume percent for room-temperature electrical conductivity (Fig. 4.4A) (Stankovich et al., 2006). When the content was increased to only 1 volume percent, this composite achieved a conductivity of 0.1 S m^{-1}, which was sufficient for many electrical applications. The polymer composites by infiltrating polymers into the networked conductive fillers showed high conductivity even at very low contents due to connected conductive pathways. The alignment of nanofiber fillers further increased the electrical properties (Sandler et al., 2003). Aligned CNT/epoxy composites showed a conductivity as high as 10^3 S cm^{-1} and could be used as counter electrodes for solar cells (Sun et al., 2013a).

For the intrinsic conductive polymers, the fillers are used to further enhance the electrical properties. Conducting polymers including PTh, PANI, PPy, and PEDOT as well as their derivatives have been added with metal and carbon nanomaterials for high conductivities. The PEDOT:PSS/Ag nanowire composite by laminating Ag nanowires into PEDOT:PSS showed a sheet resistance of 12 Ω sq^{-1}, while the PEDOT:PSS film hold a value greater than 10 MΩ sq^{-1} over large scales (Gaynor et al., 2011). In particular, composites of carbon-based materials and conducting polymers have attracted a great deal of attention. For instance, the pristine poly(phenylene vinylene) (PPV) has a quite low conductivity at 10^{-10} S m^{-1}. After addition with CNTs, the conductivity was dramatically increased to the level of 10^{-2} S m^{-1} (Coleman et al., 1998). The conductivity of the composites is determined significantly by the formation of a CNT conducting pathway that depends on the CNT concentration and dispersibility within the polymer matrix. Alignment of CNTs was also important for high performance (Sun et al., 2013a). The composite nanofibers made from CNT/PANI by in situ polymerization and electrospinning process had higher conductivities than the direct mixing CNT/PANI composites due to the aligned structure in the composite fiber, although both of their conductivities were increased with the increasing CNT content (Fig. 4.4B) (Wang et al., 2012). In addition, the alignment of CNTs made composites exhibit a great anisotropic behavior in the two directions, for example, the conductivity in the parallel direction is higher than that in the perpendicular direction. Metal nanoparticles could be further incorporated for even higher conductivity. A hybrid transparent conductive material composed of Au-decorated CNT and PEDOT:PSS has been successfully synthesized. The sintered Au nanoparticles were connected with the CNT within PEDOT:PSS, forming a three-dimensional network to reduce the contact resistance. As a result, the Au-CNT/PEDOT:PSS composite exhibited a low sheet resistance of 51 Ω sq^{-1} (Cho et al., 2015).

Ion conductivity was important for the separation films or polymer electrolyte membranes in electrochemical energy storage devices, such as, lithium ion battery (Nunes-Pereira et al., 2015). The enhancement of the ionic conductivity has been generally attributed to the decrease of the polymer crystallinity in the presence of the inorganic particles and also to interactions between inorganic particles and the polymers. The addition of SiO_2 within different polymer matrices frequently resulted in the increase of the ionic conductivity (up to 3.5×10^{-2} S cm^{-1} at RT) (Fig. 4.4C) (Nunes-Pereira et al., 2015). It is shown that fluorinated polymers, such as PVDF and P(VDF-HFP), presented the best electrical performance. These polymers have a large complex dielectric constant function which contributes to the increase of the ionic conductivity. The inclusion of TiO_2 into P(VDF-HFP) results in the increase of the ionic conductivity. The ionic conductivity could also be increased due to the polarization of the ferroelectric fillers. $BaTiO_3$/poly(vinylidene fluoride-co-trifluoroethylene) [P(VDF-TrFE)] composites exhibited an enhanced ionic conductivity 3 orders of magnitude (1.04×10^{-4} S cm^{-1}) with 4 wt.% of filler content. The addition of clays to different polymer matrices invariably results in increased ionic conductivity, up to 2.5×10^{-3} S cm^{-1}. A continuous increase of the ionic conductivity up to more than 7 orders of magnitude was reported with montmorillonite contents up to 7 wt.% into a PAN matrix.

For the composite polymer electrolytes, the conductive carbonaceous filler must be below the electrical percolation threshold, due to the need to obtain an electronically insulating material with suitable ionic conductivity. These fillers are also used to improve the thermal stabilization and serve as mechanical reinforcement to improve the electrolyte/electrode compatibility. CNT/P(VDF-TrFE) composites showed higher porosity and electrolyte uptake compared to the pristine polymer. CNT also contributed to increase ionic conductivity (2.6×10^{-6} S cm^{-1}, 0.1 wt.% CNT) and diminished its variations with temperature.

4.3.3 Electrochemical Property

The electrochemical reactions involve charge movements between the electrode and the electrolyte. The electrochemical properties of polymer composites were embodied in electrochemical energy storage devices and DSSCs.

Electrochemical energy storage/release is realized by electron and ion charge/discharge. Batteries and supercapacitors are two kinds of typical electrochemical energy storage devices, both of which store electricity by electrochemical processes. Therefore, the electrochemical properties of electrode materials are essential to the performance of storage devices. Conducting polymers, including PANI, PPy, and PEDOT, are most used as pseudocapacitive charge storage in supercapacitors through electrochemical redox reactions. They show high specific capacitances but are limited by their relatively low electrical conductivities. In addition, the cycle stability is poor due to large volumetric swelling and shrinking during charge/discharge process as a result of ion doping and dedoping. Carbon materials were introduced to composite with conducting polymers to increase ion transport and electronic conduction capabilities and cycle stability of conducting polymers in supercapacitors. For instance, the electrical conductivity was 5.5 S cm^{-1} for graphene/PANI nanofiber composite, which was 20 times higher than that of pure PANI nanofibers (0.5 S cm^{-1}) (Wu et al., 2010). The discharge curve of the supercapacitor with PANI/graphene composite was similar to the bare PANI but with a much lower "IR drop" ("IR" means internal resistance), indicating that the IR of the supercapacitor was decreased by introduction of graphene (Fig. 4.5A). The low IR is of great importance in energy storing devices, because less energy will be wasted to produce unwanted heat during charging/discharging processes. The composite film also showed higher volumetric and gravimetric capacitances than those of polymers and graphene due to the hierarchical structure and synergic effect of both components. The carbonaceous shell-coated PANI and PPy electrodes exhibited exceptionally high capacitance retentions of ~95 and ~85% after 10,000 cycles, which were significantly improved compared to the retention rate of bare PANI (~20%) and PPy (~25%) obtained under the same conditions (Fig. 4.5B–C) (Liu et al., 2014).

In lithium ion battery, the capacity, charging rate, and cycle life are very significant factors for performance. They are associated with volume expansion, surface degradation, and dissolution of active material (Sengodu and Deshmukh, 2015). High-capacity lithium-ion anode materials always lead to problems with volume change and swelling after the electrodes absorb lithium because the structural integrity of the electrode is changed. Flexible conducting polymers can withstand large strain and maintain their conductivity as well as mechanical integrity. Inorganic/polymer nanocomposites possess a high surface area, which can provide high lithium ion flux across the electrode interface, short diffusion pathways for lithium ions and electrons, abundant active sites for lithium storage, and high tolerance for volume change during charge/discharge to enhance the structural stability of the electrodes. Therefore, the lithium ion batteries based on these composites show high capacity and long-term stability. Conductive polymer/Si composites have showed good specific capacity and stable cycling performance of 1400 mAh g^{-1} after over 650 cycles (Liu et al., 2011a). The PEDOT-coated Si nanowires showed an improved cycling stability, and the capacity retention after 100 charge–discharge cycles was increased from 30% to 80% over bare Si nanowires (Yao et al., 2012). These improvements were attributed to the conductive coating layer that maintained the mechanical integrity

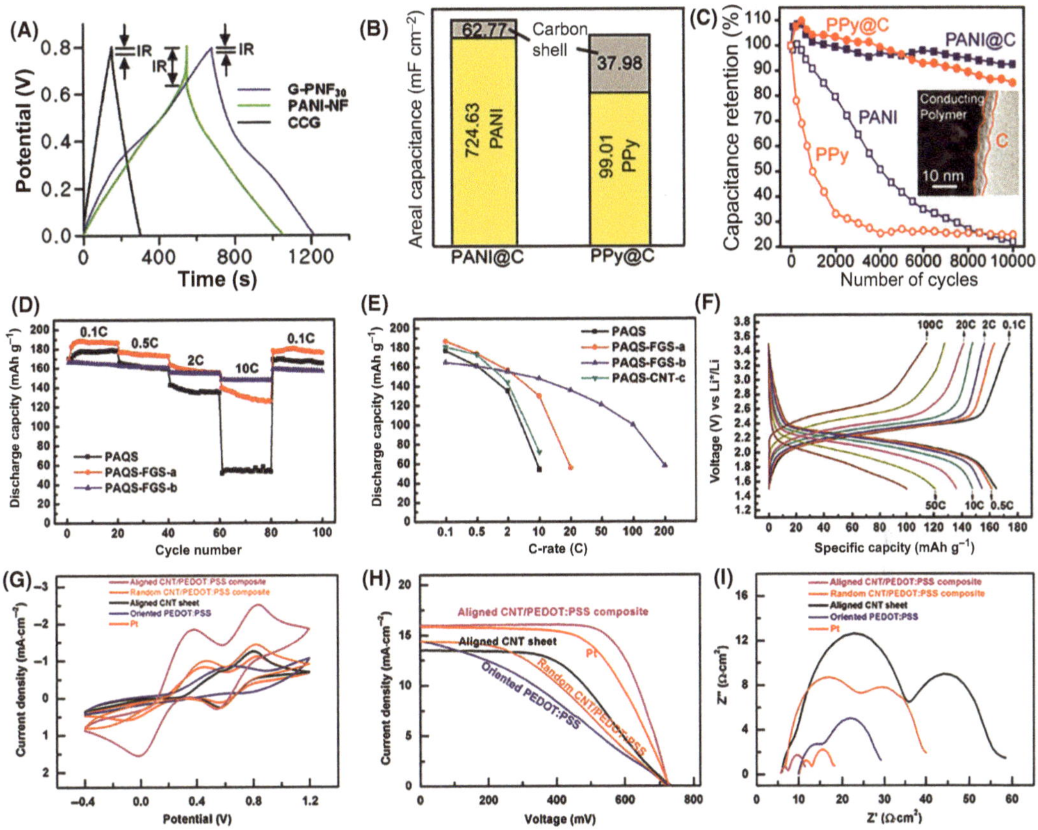

FIGURE 4.5 Electrochemical properties of polymer composites. (A) Galvanostatic charge/discharge curves of the supercapacitors based on graphene/PANI nanofiber composite *(G-PNF$_{30}$)*, PANI nanofiber *(PANI-NF)*, and graphene films (CCG). (B–C) Histogram illustration of capacitance and cycling performance of PANI, carbon-coated PANI, PPy, and carbon-coated PPy electrodes. (D–E) Cycling performance at different C-rates and discharge specific capacity of PAQS and its composites. (F) Voltage profiles of PAQS-FGS at different C-rates. (G–I) Cyclic voltammetry, J–V curves and Nyquist plots of aligned CNT/PEDOT:PSS composites as counter electrodes in DSSCs compared with of platinum, aligned CNT sheet, oriented PEDOT:PSS, and randomly dispersed CNT/PEDOT:PSS, with the polymer weight percentage of 33%. *(A) Reproduced with permission from reference Wu, Q., Xu, Y., Yao, Z., Liu, A., Shi, G., 2010. Supercapacitors based on flexible graphene/polyaniline nanofiber composite films. ACS Nano 4, 1963–1970. Copyright 2010, American Chemical Society. (B–C) Reproduced with permission from reference Liu, T., Finn, L., Yu, M., Wang, H., zhai, T., Lu, X., Tong, Y., Li, Y., 2014. Polyaniline and polypyrrole pseudocapacitor electrodes with excellent cycling stability. Nano Lett. 14, 2522–2527. Copyright 2014, American Chemical Society. (D–F) Reproduced with permission from reference Song, Z., Xu, T., Gordin, M.L., Jiang, Y.-B., Bae, I.-T., Xiao, Q., Zhan, H., Liu, J., Wang, D., 2012. Polymer–graphene nanocomposites as ultrafast-charge and -discharge cathodes for rechargeable lithium batteries. Nano Lett. 12, 2205–2211. Copyright 2012, American Chemical Society. (G–I) Reproduced with permission from reference Guan, G., Yang, Z., Qiu, L., Sun, X., Zhang, Z., Ren, J., Peng, H., 2013. Oriented PEDOT:PSS on aligned carbon nanotubes for efficient dye-sensitized solar cells. J. Mater. Chem. A 1, 13268–13273. Copyright 2013, Royal Society of Chemistry.*

of the cycled Si material, along with preserving electrical connections among Si nanowires that would otherwise have become electrically isolated during volume changes. A core–shell structured PEDOT/V_2O_5 with graphite foam showed ultrafast stable Li-ion storage performance, with higher capacities and improved rate (168 mAh g^{-1} at 60 C) and cycling capabilities (Chao et al., 2014).

The electrochemical redox mechanism of a polymer cathode is based on the reversible redox reaction of the organic functional group, such as quinone, anhydride, and nitroxide radical, accompanied by the association and disassociation of lithium ions or electrolyte anions (Song et al., 2010). A polymer with a stable skeleton and highly electroactive functional group can potentially be a high-power cathode candidate because its redox reaction is intrinsically faster than inorganic intercalation cathodes. However, the intrinsic electronic insulation and low mass density of the polymers result in unsatisfactory rate performance. Therefore, a certain content of conductive fillers, such as carbon and conducting polymers, is required for high utilization of the active polymer materials to improve the rate performance. The electrochemical redox reactions of lithium ions with poly(anthraquinonyl sulfide) (PAQS) and polyimide are based on the quinone and anhydride functional groups, respectively. Their composites with graphene were synthesized through in situ polymerization in the presence of graphene sheets. Compared to the pure polymer, the nanocomposites possess much higher active material utilization ratios and superior ultrafast-charge and -discharge ability (Fig. 4.5D–F) (Song et al., 2012). For example, bare PAQS delivers a discharge capacity of 177 mAh g^{-1} at 0.1 C, corresponding to a utilization ratio of 79% compared to the theoretical capacity of 225 mAh g^{-1}. After adding 6 and 26 wt.% graphene, the reversible capacity of composites were 187 and 165 mAh g^{-1}, corresponding to a polymer utilization ratio of 88 and 95%, respectively. Besides the specific capacity, the rate performance was also greatly improved with the increasing graphene load; the discharge capacity of composites at 10 C remained 90% of that at 0.1 C. It is also worth noting that the composite cathode materials can obtain a high capacity even under ultrafast charging and discharging, for example, a discharge capacity of 100 mAh g^{-1} even at 100 C when the whole discharge process completes in 16 s.

The counter electrodes of DSSC are used to reduce the redox species as a mediator in regenerating the sensitizer (dye) after electron injection. They should have high conductivity for electron transfer and excellent catalytic activity for reduction of an electrolyte (I_3^-/I^-) with high chemical stability and low cost. The counter electrodes are traditionally composed of Pt materials. However, Pt metal is expensive and scarce at nature. The limited supply of Pt cannot meet the increasing demand for its broad applications (Yun et al., 2014). Conductive polymers were used to replace the Pt as counter electrodes due to the good electrical properties. However, they showed lower efficiency than a Pt-based DSSC. Therefore, other materials were incorporated to conducting polymers to provide larger electrochemical active surfaces and increase the electrocatalytic activity and electrical conductivity (Yun et al., 2014). Metals, such as Pt nanoparticles, were introduced into porous PEDOT, and the DSSCs showed similar performance with respect to conventional Pt electrode under the same conditions. Metal oxides, such as TiO_2 nanoparticles were also composited with PEDOT, and the DSSC based on the composite counter electrode exhibited a high conversion efficiency up to 8.49%, which was even higher than that (~7.5%) of a DSSC with a Pt electrode (Maiaugree et al., 2012). It was possibly due to the increase of the polymer active surface area. Furthermore, it was shown that TiO_2 with smaller particle sizes (~25 nm) was better than that with larger particle sizes (~100 nm) for the composite electrode. Other metal oxides/polymer composites were also investigated, such as, ZnO and NiO (Sun et al., 2015).

Carbon nanomaterials were also introduced to conducting polymers to increase the conductivity and surface area of electrodes, leading to a higher electrocatalytic activity and the improvement of power conversion efficiency. The conductivity of the CNT/PEDOT composites is determined significantly by the formation of a CNT conducting pathway that depends on the CNT concentration and dispersibility within the polymer matrix. The alignment of CNTs and orientation of conducting polymers were found to improve the electrocatalytic activity and the power conversion efficiency better than randomly dispersed CNT composites due to higher conductivity (Guan et al., 2013; Allen et al., 2014). When PEDOT:PSS was composited with aligned CNTs, the backbones were induced to orient along the CNT length, with aromatic rings exposed to the surface. As a result, the aligned composite film showed higher catalytic activity, lower contact resistance, and higher power conversion efficiency (8.3%) than randomly dispersed CNT/PEDOT:PSS composite, bare aligned CNTs, oriented PEDOT:PSS and even conventional platinum under the same conditions (Fig. 4.5G–I) (Guan et al., 2013). The functionalization of CNTs would further increase the performance of composite electrode. In functionalized CNT/PEDOT composite film, the PEDOT polymer chains enclosed the CNTs, generating a more porous structure. In contrast, a PEDOT film without CNT possesses a compact structure. As a result, the CNT/PEDOT composite electrode has a larger surface area than the PEDOT electrode, leading to a higher electrocatalytic activity and smaller charge-transfer resistance for the I_3^-/I^- and I_2/I_3^- redox reactions. In addition, the power conversion efficiency of the DSSC based on CNT/PEDOT composite was 13.0% higher than that of the DSSC with bare PEDOT film.

Carbon nanomaterials themselves also have good electrocatalytic activities, thus their composites with insulating polymers could also be used as counter electrodes of DSSCs. For instance, the CNT/epoxy composite films with CNTs being aligned to be perpendicular and parallel to the film surface have been investigated. However, the performances were lower than bare CNTs. They can also be composited with elastic polymers to be used for stretchable electronic devices.

4.3.4 Electroluminescent Property

Electroluminescence is a property of the generation of light by electrical excitation. Conjugated polymers have demonstrated strong light absorption and efficient emission in optoelectronic devices. Among the family of electroluminescent polymers, the conjugated PPV and polyfluorene derivatives have attracted growing interests due to their tunable intrinsic electroluminescence which is dependent on their photoluminescence property, thermal stability, film processability, and chain rearrangement. However, the aggregation phenomenon, which may occur among the polymer units, often results in excimer formation with consequent reduction in the electroluminescence quantum yield of the final device. In addition, most of the undoped polymers often exhibit low charge conductivities, which will require threshold voltage and decrease the electroluminescent property (Bansal et al., 2009). Therefore, the second phase has been combined with conjugated polymers to improve efficiencies. It was demonstrated that the dispersion of inorganic particles into conjugated polymer films led to a modification of the chain organization and the polymer morphology, limiting π–π staking. In addition, as a general feature, inorganic additives led to an increase of the polymer surface roughness, and charge transport may occur between the second phase and the polymer chains due to the close interfacial contacts, which can function as a bridge to improve the bipolar charge injection into the conjugated polymer. These effects play a key role in enhancing the electroluminescence performances and the quantum yield of the polymer matrices in electroluminescence devices.

Inorganic TiO_2, SiO_2, and gold particles were dispersed into conjugated polymer films to modify the chain organization and the polymer morphology. By incorporating gold nanoparticles with diameters of 5–10 nm into poly(9,9′-dioctylfluorene), the electroluminescent devices based on the composite active layers showed higher external quantum efficiencies and enhanced environmental stability than that of the pristine polymer-based device (Park et al., 2004). The gold nanoparticles protected the devices against oxidation by quenching the triplet state of the emissive polymer. In addition, the gold nanoparticles appeared to modify the device interfacial morphology, which not only facilitates electron injection but also blocks hole migration. For instance, the capped gold nanoparticles in poly(9,9′-dioctylfluorene-*alt*-thiophene) had increased the roughness of the composite film 10 times higher than that of the pristine polymer film, resulting in an increased interfacial area between the film and the deposited cathode (Wu et al., 2006). This increased interfacial area, together with the photooxidation-suppressing and hole-blocking characteristics of gold nanoparticles, significantly enhanced the electron injection, lowered the threshold voltage, and increased the electroluminescence by nearly an order of magnitude.

By adding carbon nanomaterials into electroluminescent polymers, the charge transport could be enhanced for high electroluminescent properties. After adding with 0.1 wt.% CNTs, the electroluminescence intensity of poly[2-methoxy-5-2-ethylhexyloxy-1,4-phenylenevinylene] (MEHPPV) has been increased and the threshold voltage was also reduced from 8 V for the pure MEHPPV to 2 V for the composite at forward bias (Xu et al., 2005). It was also reported that a small concentration of CNTs could increase the electrical conductivity of the poly(*m*-phenylenevinylene-*co*-2,5-dioctoxy-*p*-phenylenevinylene) by up to 8 orders of magnitude (Curran et al., 1998). The CNTs in the composite acted as nanometric heat sinks to prevent the buildup of large thermal effects caused either optically (photobleaching) or electrically. Therefore, the stability of this device in air was better than that based on the equivalent pristine polymer, for example, the former with

5 wt.% CNT last approximately 5 times longer than the latter with the turn-on voltage as low as 14 V. In addition, the presence of CNTs diluted the polymer–polymer interaction and thus reduced self-quenching effects. However, the increase in luminescence is observed only until a critical CNT concentration is reached, after which, luminescence started to decrease with any further increase in CNT concentration. CNTs can also improve the color tuning in multiemitting devices due to the controlled hole mobility in the emissive layers. Moreover, the improved electric properties of the hole-conducting layer will also attribute to the performance improvement (Wang et al., 2008). The electroluminescent intensity of the device by using CNT/PEDOT:PSS composite as hole injection layer was significantly increased and the turn-on voltage was decreased compared to the bare polymers.

4.3.5 Sensing

When the second phases are incorporated into responsive polymers, they can not only increase the original properties but also endow them with new properties. Here we focus on the electroactive polymers with actuating and chromic properties.

4.3.5.1 Electroactuation

Electroactive conducting polymers achieve electroactive effect by a reversible counter-ion insertion and expulsion during redox cycling. Oxidation at the anode electrodes and reduction at the cathode electrodes induce a considerable volume change mainly due to the exchange of ions with an electrolyte. The positively charged polymer electrode is oxidized, which causes volume expansion, while the negative electrode of polymer is reduced, which leads to expulsion of anions and volume shrinkage. Therefore, bending actuation is achieved. The electroactive properties of conducting polymers are significantly influenced by the ion diffusion and conductivity of polymer electrodes. A small amount of conductive fillers can significantly improve the electroactivity. The generated force is limited by the breaking strength of the polymer materials. And the inorganic fillers can enhance the mechanical strength of polymers and thus increase the generated forces. In addition, the incorporation of inorganics will also improve the durability of the electroactuation. For instance, the PPy/CNT composite films showed several folds of enhanced strength and modulus than bare PPy film and the electrical conductivity is 50–60% improved compared with bare PPy film (Zheng et al., 2011). Moreover, the incorporation of CNTs also dramatically reduced the creep during cyclic deformation (Fig. 4.6A). Less than 0.2% creep strain occurred in the PPy/CNT composite in 35 min under a load of 6 MPa, while the bare PPy film exhibited more than 2% creep strain in just 15 min under a load of only 4 MPa.

Besides the intrinsic conductive polymers, some deformable polymers, such as shape-memory polymers, are usually activated by heating. After incorporating with conductive fillers, such as carbon nanomaterials, they can be simulated by the electricity through Joule heating (Liu et al., 2009; Hu and Chen, 2010; Koerner et al., 2004). This kind of electrothermally active polymer composites can produce expansion/contraction and bending behaviors upon with the electricity. Moreover, these actuators can work durably

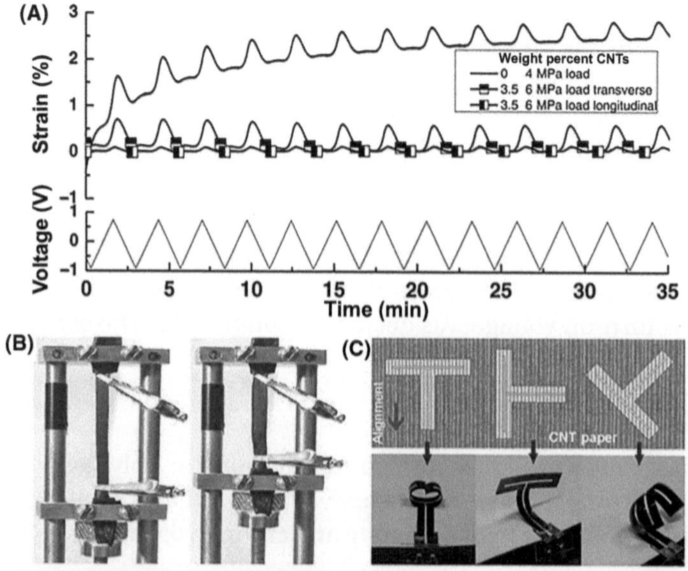

FIGURE 4.6 Enhanced and new electroactuations of polymer composites. (A) Electroactuations of polymer composites. Actuation stroke versus time for both neat PPy and PPy/CNT composites. (B) Electrically stimulated stress recovery of thermoplastic polymer/CNT nanocomposite. (C) Three different ways of processing and actuations of T-shaped polymer/CNT composite films with different CNT alignment: vertically, horizontally, and diagonally. *(A) Reproduced with permission from reference Zheng, W., Razal, J.M., Whitten, P.G., Ovalle-Robles, R., Wallace, G.G., Baughman, R.H., Spinks, G.M., 2011. Artificial muscles based on polypyrrole/carbon nanotube laminates. Adv. Mater. 23, 2966–2970. Copyright 2011, Wiley-VCH. (B) Reproduced with permission from reference Koerner, H., Price, G., Pearce, N.A., Alexander, M., Vaia, R.A., 2004. Remotely actuated polymer nanocomposites—stress-recovery of carbon-nanotube-filled thermoplastic elastomers. Nat. Mater. 3, 115–120. Copyright 2004, Macmillan Publishers Ltd. (C) Reproduced with permission from reference Li, Q., Liu, C., Lin, Y.-H., Liu, L., Jiang, K., Fan, S., 2015. Large-strain, multiform movements from designable electrothermal actuators based on large highly anisotropic carbon nanotube sheets. ACS Nano 9, 409–418. Copyright 2015, American Chemical Society.*

in air without electrolytes compared with ionic actuators and at lower voltages than electronic polymer actuators. CNT is one of the best candidates as electrothermal material due to the high electrical and thermal conductivity, flexibility, and excellent thermal and chemical stability. Tremendous recovery of deformation, in excess of 300%, was found in thermoplastic elastomer/CNT composites. It was demonstrated that the addition of CNTs into thermoplastic elastomer could increase the relative fixity for 25%, and store and subsequently release up to 100% more recovery stress than the neat thermoplastic elastomer under heating (Koerner et al., 2004). The most important thing was that the actuation of the composites can be electrically induced associated with Joule heating of the matrix when the current was passed through the conductive percolative network of the CNTs within the resin (Fig. 4.6B).

Polymer composite actuators with a bilayer structure of different coefficients of thermal expansion (CTE) can generate bending displacement. When one side of the film is introduced with conductive fillers, the actuator can be electrothermally driven. PDMS/CNT composite was reported to be driven by the thermal expansion of the PDMS with CNTs as nanoheaters under the applied voltage (Sellinger et al., 2010). By embedding aligned CNTs within the surface of the PDMS film, bending direction could be controlled in the composite actuator (Fig. 4.6C) (Chen et al., 2011; Li et al., 2015). These actuators also had fast response, large displacement, low driving voltage, and long service life. The displacement performance was further enhanced by adding negative CTE materials into polymers, such as, graphene and its derivatives (Huang et al., 2012; Grigoriadis et al., 2010). For instance, an electromechanical spongy graphene/PDMS bimorph actuator has been designed to show an ultralarge bending displacement output under low voltage stimulation, a high displacement-to-length ratio, and vibration motion at AC voltage, which was much larger and faster than that of the other electromechanical bimorph actuators (Hu et al., 2014). Moreover, in order to increase the displacement of the actuator while decrease the operation voltage, electrothermal bimetallic actuator was developed with the first layer of CNT/polyurethane (PU) composite film and the second layer of CNT/silicone rubber composite film (Zeng et al., 2015). The actuation mechanisms were attributed to not only the mismatch in the CTEs, but also the unique negative temperature coefficient effect of the first composite layer. This method provides a new instructive idea for fabricating a class of electrothermal actuators that is of extensive prospects in smart applications like biomimetic robotics, artificial muscles, switches, microsensors, and so on.

4.3.5.2 Electrochromism

Electrochromic conducting polymers showed high coloration efficiencies at relatively low redox switching potentials as well as flexibility. However, the color transition speed, chemical stability, and mechanical strength were relatively low from the perspective of applications. Therefore, the second phases were introduced to complement these properties. Electrochromic inorganic materials with relatively fast color switching, durability, and long-term stability were incorporated to conducting polymers to show multicolors compared to the bare polymers. For instance, ultrathin WO_3 nanorods were embedded in PANI to fabricate composite films (Zhang et al., 2013). In the composite, WO_3 was colored at a negative potential and PANI was colored at positive potential. The combination of the two materials leads to a dual-electrochromism with multicolors varied from purple to green, light yellow, and finally dark blue (Fig. 4.7A). The modulation of the electrochromism could be enhanced by incorporation of GO (Wei et al., 2013). A transmittance difference in the bleached and colored states in PANI/GO nanocomposite was 60.6%, which was much higher than 37.4% of bare PANI film, when they were applied the same alternative square-wave voltage of 0.8 and −0.2 V (Fig. 4.7B).

The stability of the electrochromic polymers can be improved by the intimate contact with the inorganic component during the redox transformations. The inorganic nanoparticulate matrix, such as mesoporous TiO_2, acts as a skeleton, keeping the polymer intact and avoiding its detachment (Takagi et al., 2012). In addition, significantly faster electrochromic response was obtained in the case of a nanostructured TiO_2/PPy composites compared to the neat PPy layer. This was due to the better charge carrier transport from the network structured freeways for the charge carriers. Also the nanodimension of the

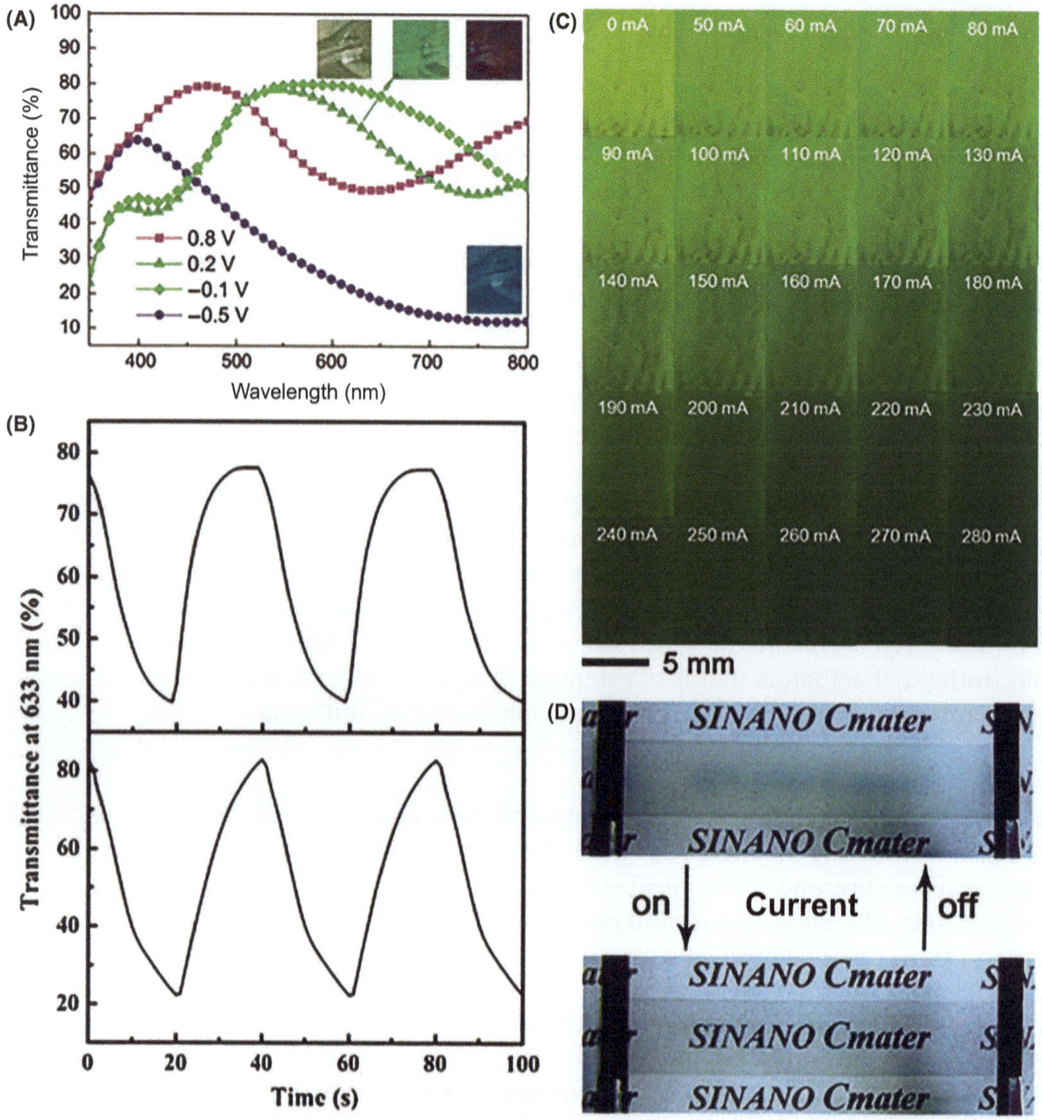

FIGURE 4.7 Enhanced and new electrochromic properties of polymer composites. (A) Visible transmittance spectra of the PANI/WO_3 composite film at different bias potentials and the corresponding photographs. (B) In situ corresponding transmittance of PANI (up) and PANI/GO nanocomposite (down) films at 633 nm under a square-wave voltage of 0.8 and −0.2 V with a pulse width of 20 s. (C) Photographs of PTP/CNT composite film when increasing currents from 0 to 280 mA were applied. The color gradually changes from yellowish green to dark green. (D) The reversible switch of the PU/MWCNT composite film between opaque and transparent with an electric current off and on. *(A) Reproduced with permission from reference Zhang, J., Tu, J.-P., Du, G.-H., Dong, Z.-M., Wu, Y.-S., Chang, L., Xie, D., Cai, G.-F., Wang, X.-L., 2013. Ultra-thin WO_3 nanorod embedded polyaniline composite thin film: synthesis and electrochromic characteristics. Sol. Energy Mater. Sol. Cells 114, 31–37. Copyright 2013, Elsevier Ltd. (B) Reproduced with permission from reference Wei, H., Zhu, J., Wu, S., Wei, S., Guo, Z., 2013. Electrochromic polyaniline/graphite oxide nanocomposites with endured electrochemical energy storage. Polymer 54, 1820–1831. Copyright 2013, Elsevier Ltd. (C) Reproduced with permission from reference Sun, X., Zhang, Z., Lu, X., Guan, G., Li, H., Peng, H., 2013c. Electric current test paper based on conjugated polymers and aligned carbon nanotubes. Angew. Chem. Int. Ed. 52, 7776–7780. Copyright 2013c, Wiley-VCH. (D) Reproduced with permission from reference Meng, F., Zhang, X., Xu, G., Yong, Z., Chen, H., Chen, M., Li, Q., Zhu, Y., 2011. Carbon nanotube composite films with switchable transparency. ACS Appl. Mater. Interfaces 3, 658–661. Copyright 2011, American Chemical Society.*

polymer could avoid limitations from the small exciton diffusion length in conducting polymers. Porous hybrid structures facilitated diffusion/migration of ions and resulted in faster coloration.

Besides enhancement of the electrochromic performance of conducting polymers, the conductive filler can also be composited with other chromic polymers to endow them new electrochromic properties. For instance, PDA can change color under the stimuli of heating, solvent, pH, and biomolecules. However, they cannot change color in response to electricity due to the very low conductivity. CNTs were then introduced into PDA, and the composites showed rapid and reversible color transition upon passing with electric current (Peng et al., 2009). Some thermal chromic polymers can also be composited with conductive fillers to show electrochromic properties due to the electrothermal effect of the fillers. For instance, poly[1-phenyl-2-(*p*-trimethylsilyl) phenylacetylene] (PTP), exhibited a reversible change in fluorescent intensity induced by heat and solvent, which resulted from the molecular perturbation and exciton deconfinement within the PTP backbone. Then electrochromatic composites have been prepared by incorporating aligned CNTs with PTP (Sun et al., 2013c). The resulted composite fibers and films exhibited a rapid change in fluorescent intensity as well as color transition from yellowish green to dark green with electric current ranging from several to hundreds of microamperes (Fig. 4.7C). The color change could repeat for more than a thousand cycles without any fatigue in structure and sensitivity. PU shows a switchable transparency in response to temperature due to the phase transition of the soft segments. Aligned CNTs were introduced into PU film to realize the electrothermal transmittance change and improve the switchability (Meng et al., 2011). The obtained PU/CNT composite film exhibited a rapid transition from opacity to transparence in a few seconds under an electric current, and slowly reversed after the current was cut off (Fig. 4.7D). This method was easy to operate with low energy cost and can be used at a large scale for various other polymers applied in optically switchable windows.

4.4 Summary

Polymers have been investigated for various electronics, but their performances need to be further improved for practical applications. Therefore, the second phase is incorporated to tune their structures and properties. This chapter has summarized the synthesis, structures, and properties of polymer composites which can be used in electronics by incorporating various nanomaterials, such as, metals, carbonaceous materials and inorganic compounds. The polymer composites can be fabricated by diverse methods, including separate synthesis of polymers and inorganics followed by mixing the two components, in situ synthesis of one component in the presence of the other component, and one-pot synthesis of the two components simultaneously. It is essential for polymer composites to achieve the uniform dispersion, proper content, and certain alignment of inorganics in polymer matrix, as well as the effect interfacial interaction between polymers and inorganics. The polymer composites can be designed into various continuous and separated structures with different dimensions, which are determined by the properties of each component and the synthetic process. The incorporation of the second phase can not only enhance the properties of polymers, such as mechanical strength, electrical conductivity, and electrochemical performance, but also generate new properties compared to the bare polymers, such as from insulating to conductive, actuation, and chromism triggered by electricity rather than direct heating. These resulted polymer composites can be employed in a variety of polymer electronics ranging from energy harvesting and storage devices to sensors and actuators based on the electricity.

References

Ahmad, F.N., Jaafar, M., Palaniandy, S., Azizli, K.A.M., 2008. Effect of particle shape of silica mineral on the properties of epoxy composites. Compos. Sci. Technol. 68, 346–353.

Allen, R., Pan, L., Fuller, G.G., Bao, Z., 2014. Using in-situ polymerization of conductive polymers to enhance the electrical properties of solution-processed carbon nanotube films and fibers. ACS Appl. Mater. Interfaces 6, 9966–9974.

Armel, V., Winther-Jensen, O., Kerr, R., Macfarlane, D.R., Winther-Jensen, B., 2012. Designed electrodeposition of nanoparticles inside conducting polymers. J. Mater. Chem. 22, 19767–19773.

Bansal, M., Srivastava, R., Lal, C., Kamalasanan, M.N., Tanwar, L.S., 2009. Carbon nanotube-based organic light emitting diodes. Nanoscale 1, 317–330.

Benson, J., Kovalenko, I., Boukhalfa, S., Lashmore, D., sanghadasa, M., Yushin, G., 2013. Multifunctional CNT-polymer composites for ultra-tough structural supercapacitors and desalination devices. Adv. Mater. 25, 6625–6632.

Berson, S., De Bettignies, R., Bailly, S., Guillerez, S., Jousselme, B., 2007. Elaboration of P3HT/CNT/PCBM composites for organic photovoltaic cells. Adv. Funct. Mater. 17, 3363–3370.

Breimer, M.A., Yevgeny, G., Sy, S., Sadik, O.A., 2001. Incorporation of metal nanoparticles in photopolymerized organic conducting polymers: a mechanistic insight. Nano Lett. 1, 305–308.

Byrne, M.T., Gun'ko, Y.K., 2010. Recent advances in research on carbon nanotube–polymer composites. Adv. Mater. 22, 1672–1688.

Cai, G., Tu, J., Zhou, D., Zhang, J., Xiong, Q., Zhao, X., Wang, X., Gu, C., 2013. Multicolor electrochromic film based on TiO_2@polyaniline core/shell nanorod array. J. Phys. Chem. C 117, 15967–15975.

Chabi, S., Peng, C., Hu, D., Zhu, Y., 2014. Ideal three-dimensional electrode structures for electrochemical energy storage. Adv. Mater. 26, 2440–2445.

Chao, D., Xia, X., Liu, J., Fan, Z., Ng, C.F., Lin, J., Zhang, H., Shen, Z.X., Fan, H.J., 2014. A V2O5/Conductive-polymer core/shell nanobelt array on three-dimensional graphite foam: a high-rate, ultrastable, and freestanding cathode for lithium-ion batteries. Adv. Mater. 26, 5794–5800.

Chen, A., Wang, H., Li, X., 2005. One-step process to fabricate Ag-polypyrrole coaxial nanocables. Chem. Commun. 14, 1863–1864.

Chen, L., Liu, C., Liu, K., Meng, C., Hu, C., Wang, J., Fan, S., 2011. High-performance, low-voltage, and easy-operable bending actuator based on aligned carbon nanotube/polymer composites. ACS Nano 5, 1588–1593.

Cho, E.-C., Li, C.-P., Huang, J.-H., Lee, K.-C., Huang, J.-H., 2015. Three-dimensional conductive nanocomposites based on multiwalled carbon nanotube networks and PEDOT:PSS as a flexible transparent electrode for optoelectronics. ACS Appl. Mater. Interfaces 7, 11668–11676.

Choi, M., Ryoo, R., 2003. Ordered nanoporous polymer-carbon composites. Nat. Mater. 2, 473–476.

Choi, M., Kleitz, F., Liu, D., Lee, H.Y., Ahn, W.-S., Ryoo, R., 2005. Controlled polymerization in mesoporous silica toward the design of organic–inorganic composite nanoporous materials. J. Am. Chem. Soc. 127, 1924–1932.

Coleman, J.N., Curran, S., Dalton, A.B., Davey, A.P., McCarthy, B., Blau, W., Barklie, R.C., 1998. Percolation-dominated conductivity in a conjugated-polymer-carbon-nanotube composite. Phys. Rev. B 58, R7492–R7495.

Curran, S.A., Ajayan, P.M., Blau, W.J., Carroll, D.L., Coleman, J.N., Dalton, A.B., Davey, A.P., Drury, A., Mccarthy, B., Maier, S., Strevens, A., 1998. A Composite from poly(*m*-phenylenevinylene-*co*-2,5-dioctoxy-*p*-phenylenevinylene) and carbon nanotubes: a novel material for molecular optoelectronics. Adv. Mater. 10, 1091–1093.

DeLongchamp, D.M., Hammond, P.T., 2004. Multiple-color electrochromism from layer-by-layer-assembled polyaniline/Prussian blue nanocomposite thin films. Chem. Mater. 16, 4799–4805.

Ding, B., Lu, X., Yuan, C., Yang, S., Han, Y., Zhang, X., Che, Q., 2012. One-step electrochemical composite polymerization of polypyrrole integrated with functionalized graphene/carbon nanotubes nanostructured composite film for electrochemical capacitors. Electrochim. Acta 62, 132–139.

Downs, C., Nugent, J., Ajayan, P.M., Duquette, D.J., Santhanam, K.S.V., 1999. Efficient polymerization of aniline at carbon nanotube electrodes. Adv. Mater. 11, 1028–1031.

Drury, A., Chaure, S., Kröll, M., Nicolosi, V., Chaure, N., Blau, W.J., 2007. Fabrication and characterization of silver/polyaniline composite nanowires in porous anodic alumina. Chem. Mater. 19, 4252–4258.

Fan, W., Zhang, C., Tjiu, W.W., Pramoda, K.P., He, C., Liu, T., 2013. Graphene-wrapped polyaniline hollow spheres as novel hybrid electrode materials for supercapacitor applications. ACS Appl. Mater. Interfaces 5, 3382–3391.

Fu, S.Y., Feng, X.Q., Lauke, B., Mai, Y.W., 2008. Effects of particle size, particle/matrix interface adhesion and particle loading on mechanical properties of particulate–polymer composites. Compos. Part B: Eng. 39, 933–961.

Gao, M., Huang, S., Dai, L., Wallace, G., Gao, R., Wang, Z., 2000. Aligned coaxial nanowires of carbon nanotubes sheathed with conducting polymers. Angew. Chem. Int. Ed. 39, 3664–3667.

Gaynor, W., Burkhard, G.F., Mcgehee, M.D., Peumans, P., 2011. Smooth nanowire/polymer composite transparent electrodes. Adv. Mater. 23, 2905–2910.

Ge, J., Cheng, G., Chen, L., 2011. Transparent and flexible electrodes and supercapacitors using polyaniline/single-walled carbon nanotube composite thin films. Nanoscale 3, 3084–3088.

Geng, J., Zeng, T., 2006. Influence of single-walled carbon nanotubes induced crystallinity enhancement and morphology change on polymer photovoltaic devices. J. Am. Chem. Soc. 128, 16827–16833.

Ghosh Chaudhuri, R., Paria, S., 2011. Core/shell nanoparticles: classes, properties, synthesis mechanisms, characterization, and applications. Chem. Rev. 112, 2373–2433.

Gong, X., Liu, J., Baskaran, S., Voise, R.D., Young, J.S., 2000. Surfactant-assisted processing of carbon nanotube/polymer composites. Chem. Mater. 12, 1049–1052.

Grigoriadis, C., Haase, N., Butt, H.J., Müllen, K., Floudas, G., 2010. Negative thermal expansion in discotic liquid crystals of nanographenes. Adv. Mater. 22, 1403–1406.

Guan, G., Yang, Z., Qiu, L., Sun, X., Zhang, Z., Ren, J., Peng, H., 2013. Oriented PEDOT:PSS on aligned carbon nanotubes for efficient dye-sensitized solar cells. J. Mater. Chem. A 1, 13268–13273.

Gui, X., Wei, J., Wang, K., Cao, A., Zhu, H., Jia, Y., Shu, Q., Wu, D., 2010. Carbon nanotube sponges. Adv. Mater. 22, 617–621.

Han, J., Kim, H., Kim, D.Y., Jo, S.M., Jang, S.-Y., 2010. Water-soluble polyelectrolyte-grafted multiwalled carbon nanotube thin films for efficient counter electrode of dye-sensitized solar cells. ACS Nano 4, 3503–3509.

Han, J., Li, L., Fang, P., Guo, R., 2012a. Ultrathin MnO_2 nanorods on conducting polymer nanofibers as a new class of hierarchical nanostructures for high-performance supercapacitors. J. Phys. Chem. C 116, 15900–15907.

Han, T.-H., Lee, Y., Choi, M.-R., Woo, S.-H., Bae, S.-H., Hong, B.H., Ahn, J.-H., Lee, T.-W., 2012b. Extremely efficient flexible organic light-emitting diodes with modified graphene anode. Nat. Photon. 6, 105–110.

Han, G., Liu, Y., Zhang, L., Kan, E., Zhang, S., Tang, J., Tang, W., 2014a. MnO_2 nanorods intercalating graphene oxide/polyaniline ternary composites for robust high-performance supercapacitors. Sci. Rep. 4, 4824.

Han, J.-H., Zhang, H., Chen, M.-J., Wang, G.-R., Zhang, Z., 2014b. CNT buckypaper/thermoplastic polyurethane composites with enhanced stiffness, strength and toughness. Compos. Sci. Technol. 103, 63–71.

Han, S., Wu, D., Li, S., Zhang, F., Feng, X., 2014c. Porous graphene materials for advanced electrochemical energy storage and conversion devices. Adv. Mater. 26, 849–864.

Hao, L., Zhu, C., Chen, C., Kang, P., Hu, Y., Fan, W., Chen, Z., 2003. Fabrication of silica core–conductive polymer polypyrrole shell composite particles and polypyrrole capsule on monodispersed silica templates. Synth. Met. 139, 391–396.

He, X., Shi, Q., Zhou, X., Wan, C., Jiang, C., 2005. In situ composite of nano SiO_2–P(VDF-HFP) porous polymer electrolytes for Li-ion batteries. Electrochim. Acta 51, 1069–1075.

Hou, H., Ge, J.J., Zeng, J., Li, Q., Reneker, D.H., Greiner, A., Cheng, S.Z.D., 2005. Electrospun polyacrylonitrile nanofibers containing a high concentration of well-aligned multiwall carbon nanotubes. Chem. Mater. 17, 967–973.

Hou, Y., Cheng, Y., Hobson, T., Liu, J., 2010. Design and synthesis of hierarchical MnO_2 nanospheres/carbon nanotubes/conducting polymer ternary composite for high performance electrochemical electrodes. Nano Lett. 10, 2727–2733.

Hu, J., Chen, S., 2010. A review of actively moving polymers in textile applications. J. Mater. Chem. 20, 3346–3355.

Hu, Y., Lan, T., Wu, G., Zhu, Z., Chen, W., 2014. A spongy graphene based bimorph actuator with ultra-large displacement towards biomimetic application. Nanoscale 6, 12703–12709.

Huang, Y., Liang, J., Chen, Y., 2012. The application of graphene based materials for actuators. J. Mater. Chem. 22, 3671–3679.

Huang, K.J., Wang, L., Liu, Y.J., Wang, H.B., Liu, Y.M., Wang, L.L., 2013. Synthesis of polyaniline/2-dimensional graphene analog MoS_2 composites for high-performance supercapacitor. Electrochim. Acta 109, 587–594.

Jiang, L., Fan, Z., 2014. Design of advanced porous graphene materials: from graphene nanomesh to 3D architectures. Nanoscale 6, 1922–1945.

Jiang, H., Ma, J., Li, C., 2012. Polyaniline-MnO_2 coaxial nanofiber with hierarchical structure for high-performance supercapacitors. J. Mater. Chem. 22, 16939–16942.

Kalyana Sundaram, N.T., Subramania, A., 2007. Nano-size $LiAlO_2$ ceramic filler incorporated porous PVDF-*co*-HFP electrolyte for lithium-ion battery applications. Electrochim. Acta 52, 4987–4993.

Koerner, H., Price, G., Pearce, N.A., Alexander, M., Vaia, R.A., 2004. Remotely actuated polymer nanocomposites—stress-recovery of carbon-nanotube-filled thermoplastic elastomers. Nat. Mater. 3, 115–120.

Lee, S., Cho, M.S., Lee, H., Nam, J.-D., Lee, Y., 2012. A facile synthetic route for well-defined multilayer films of graphene and PEDOT via an electrochemical method. J. Mater. Chem. 22, 1899–1903.

Lei, Z., Chen, Z., Zhao, X.S., 2010. Growth of polyaniline on hollow carbon spheres for enhancing electrocapacitance. J. Phys. Chem. C 114, 19867–19874.

Li, C., Bai, H., Shi, G., 2009a. Conducting polymer nanomaterials: electrosynthesis and applications. Chem. Soc. Rev. 38, 2397–2409.

Li, L., Song, H., Zhang, Q., Yao, J., Chen, X., 2009b. Effect of compounding process on the structure and electrochemical properties of ordered mesoporous carbon/polyaniline composites as electrodes for supercapacitors. J. Power Sources 187, 268–274.

Li, L., Yang, Z., Gao, H., Zhang, H., Ren, J., Sun, X., Chen, T., Kia, H.G., Peng, H., 2011a. Vertically aligned and penetrated carbon nanotube/polymer composite film and promising electronic applications. Adv. Mater. 23, 3730–3735.

Li, Q., Liu, J., Zou, J., Chunder, A., Chen, Y., Zhai, L., 2011b. Synthesis and electrochemical performance of multi-walled carbon nanotube/polyaniline/MnO_2 ternary coaxial nanostructures for supercapacitors. J. Power Sources 196, 565–572.

Li, Q., Liu, C., Lin, Y.-H., Liu, L., Jiang, K., Fan, S., 2015. Large-strain, multiform movements from designable electrothermal actuators based on large highly anisotropic carbon nanotube sheets. ACS Nano 9, 409–418.

Liu, R., Lee, S.B., 2008. MnO_2/poly(3,4-ethylenedioxythiophene) coaxial nanowires by one-step coelectrodeposition for electrochemical energy storage. J. Am. Chem. Soc. 130, 2942–2943.

Liu, Y., Lv, H., Lan, X., Leng, J., Du, S., 2009. Review of electro-active shape-memory polymer composite. Compos. Sci. Technol. 69, 2064–2068.

Liu, G., Xun, S., Vukmirovic, N., Song, X., Olalde-Velasco, P., Zheng, H., Battaglia, V.S., Wang, L., Yang, W., 2011a. Polymers with tailored electronic structure for high capacity lithium battery electrodes. Adv. Mater. 23, 4679–4683.

Liu, R., Duay, J., Lee, S.B., 2011b. Electrochemical formation mechanism for the controlled synthesis of heterogeneous MnO_2/poly(3,4-ethylenedioxythiophene) nanowires. ACS Nano 5, 5608–5619.

Liu, R., Duay, J., Lee, S.B., 2011c. Heterogeneous nanostructured electrode materials for electrochemical energy storage. Chem. Commun. 47, 1384–1404.

Liu, T., Finn, L., Yu, M., Wang, H., zhai, T., Lu, X., Tong, Y., Li, Y., 2014. Polyaniline and polypyrrole pseudocapacitor electrodes with excellent cycling stability. Nano Lett. 14, 2522–2527.

Lu, X., zhao, Y., Wang, C., 2005. Fabrication of PbS nanoparticles in polymer-fiber matrices by electrospinning. Adv. Mater. 17, 2485–2488.

Lu, X., Wang, C., Wei, Y., 2009. One-dimensional composite nanomaterials: synthesis by electrospinning and their applications. Small 5, 2349–2370.

Lu, X., Zhang, W., Wang, C., Wen, T.-C., Wei, Y., 2011. One-dimensional conducting polymer nanocomposites: synthesis, properties and applications. Prog. Polym. Sci. 36, 671–712.

Lu, X., Zhang, F., Dou, H., Yuan, C., Yang, S., Hao, L., Shen, L., Zhang, L., Zhang, X., 2012. Preparation and electrochemical capacitance of hierarchical graphene/polypyrrole/carbon nanotube ternary composites. Electrochim. Acta 69, 160–166.

Maiaugree, W., Pimanpang, S., Towannang, M., Saekow, S., Jarernboon, W., Amornkitbamrung, V., 2012. Optimization of TiO_2 nanoparticle mixed PEDOT–PSS counter electrodes for high efficiency dye sensitized solar cell. J. Non-Cryst. Solids 358, 2489–2495.

Mamedov, A.A., Kotov, N.A., Prato, M., Guldi, D.M., Wicksted, J.P., Hirsch, A., 2002. Molecular design of strong single-wall carbon nanotube/polyelectrolyte multilayer composites. Nat. Mater. 1, 190–194.

Meng, C., Liu, C., Chen, L., Hu, C., Fan, S., 2010. Highly flexible and all-solid-state paper like polymer supercapacitors. Nano Lett. 10, 4025–4031.

Meng, F., Zhang, X., Xu, G., Yong, Z., Chen, H., Chen, M., Li, Q., Zhu, Y., 2011. Carbon nanotube composite films with switchable transparency. ACS Appl. Mater. Interfaces 3, 658–661.

Meng, Y., Wang, K., Zhang, Y., Wei, Z., 2013. Hierarchical porous graphene/polyaniline composite film with superior rate performance for flexible supercapacitors. Adv. Mater. 25, 6985–6990.

Moniruzzaman, M., Winey, K.I., 2006. Polymer nanocomposites containing carbon nanotubes. Macromolecules 39, 5194–5205.

Ning, H.M., Hu, N., Kamata, T., Qiu, J.H., Han, X., Zhou, L.M., Chang, C., Liu, Y., Wu, L.K., Qiu, J.H., Ji, H.L., Wang, W.X., Zemba, Y., Atobe, S., Li, Y., Alamusi, Fukunaga, H., 2013. Improved piezoelectric properties of poly(vinylidene fluoride) nanocomposites containing multi-walled carbon nanotubes. Smart Mater. Struct. 22, 065011.

Nunes-Pereira, J., Costa, C.M., Lanceros-Méndez, S, 2015. Polymer composites and blends for battery separators: state of the art, challenges, and future trends. J. Power Sources 281, 378–398.

Pang, S., Li, G., Zhang, Z., 2005. Synthesis of polyaniline-vanadium oxide nanocomposite nanosheets. Macromol. Rapid Commun. 26, 1262–1265.

Park, J.H., Lim, Y.T., Park, O.O., Kim, J.K., Yu, J.-W., Kim, Y.C., 2004. Polymer/gold nanoparticle nanocomposite light-emitting diodes: enhancement of electroluminescence stability and quantum efficiency of blue-light-emitting polymers. Chem. Mater. 16, 688–692.

Park, H.J., Kim, J., Chang, J.Y., Theato, P., 2008. Preparation of transparent conductive multilayered films using active pentafluorophenyl ester modified multiwalled carbon nanotubes. Langmuir 24, 10467–10473.

Peng, C., Zhang, S., Jewell, D., Chen, G.Z., 2008. Carbon nanotube and conducting polymer composites for supercapacitors. Prog. Nat. Sci. 18, 777–788.

Peng, H., Sun, X., Cai, F., Chen, X., Zhu, Y., Liao, G., Chen, D., Li, Q., Lu, Y., Zhu, Y., Jia, Q., 2009. Electrochromatic carbon nanotube/polydiacetylene nanocomposite fibres. Nat. Nanotechnol. 4, 738–741.

Philip, B., Xie, J., Abraham, J.K., Varadan, V.K., 2005. Polyaniline/carbon nanotube composites: starting with phenylamino functionalized carbon nanotubes. Polym Bull. 53, 127–138.

Rafiee, M.A., Rafiee, J., Wang, Z., Song, H., Yu, Z.-Z., Koratkar, N., 2009. Enhanced mechanical properties of nanocomposites at low graphene content. ACS Nano 3, 3884–3890.

Sahoo, N.G., Rana, S., Cho, J.W., Li, L., Chan, S.H., 2010. Polymer nanocomposites based on functionalized carbon nanotubes. Prog. Polym. Sci. 35, 837–867.

Sandler, J.K.W., Kirk, J.E., Kinloch, I.A., Shaffer, M.S.P., Windle, A.H., 2003. Ultra-low electrical percolation threshold in carbon-nanotube-epoxy composites. Polymer 44, 5893–5899.

Sarker, A.K., Hong, J.-D., 2012. Layer-by-layer self-assembled multilayer films composed of graphene/polyaniline bilayers: high-energy electrode materials for supercapacitors. Langmuir 28, 12637–12646.

Sekitani, T., Nakajima, H., Maeda, H., Fukushima, T., Aida, T., Hata, K., Someya, T., 2009. Stretchable active-matrix organic light-emitting diode display using printable elastic conductors. Nat. Mater. 8, 494–499.

Sellinger, A.T., Wang, D.H., Tan, L.-S., Vaia, R.A., 2010. Electrothermal polymer nanocomposite actuators. Adv. Mater. 22, 3430–3435.

Sengodu, P., Deshmukh, A.D., 2015. Conducting polymers and their inorganic composites for advanced Li-ion batteries: a review. RSC Adv. 5, 42109–42130.

Sheng, K., Bai, H., Sun, Y., Li, C., Shi, G., 2011. Layer-by-layer assembly of graphene/polyaniline multilayer films and their application for electrochromic devices. Polymer 52, 5567–5572.

Shi, Y., Peng, L., Ding, Y., Zhao, Y., Yu, G., 2015. Nanostructured conductive polymers for advanced energy storage. Chem. Soc. Rev. 44, 6684–6696.

Song, Z., Zhan, H., Zhou, Y., 2010. Polyimides: promising energy-storage materials. Angew Chem. Int Ed. 49, 8444–8448.

Song, Z., Xu, T., Gordin, M.L., Jiang, Y.-B., Bae, I.-T., Xiao, Q., Zhan, H., Liu, J., Wang, D., 2012. Polymer–graphene nanocomposites as ultrafast-charge and -discharge cathodes for rechargeable lithium batteries. Nano Lett. 12, 2205–2211.

Song, H., Liu, C., Xu, J., Jiang, Q., Shi, H., 2013. Fabrication of a layered nanostructure PEDOT:PSS/SWCNTs composite and its thermoelectric performance. RSC Adv. 3, 22065–22071.

Spitalsky, Z., Tasis, D., Papagelis, K., Galiotis, C., 2010. Carbon nanotube–polymer composites: chemistry, processing, mechanical and electrical properties. Prog. Polym. Sci. 35, 357–401.

Srivastava, S., Kotov, N.A., 2008. Composite layer-by-layer (LBL) assembly with inorganic nanoparticles and nanowires. Acc. Chem. Res. 41, 1831–1841.

Stankovich, S., Dikin, D.A., Dommett, G.H.B., Kohlhaas, K.M., Zimney, E.J., Stach, E.A., Piner, R.D., Nguyen, S.T., Ruoff, R.S., 2006. Graphene-based composite materials. Nature 442, 282–286.

Sun, X., Wang, W., Qiu, L., Guo, W., Yu, Y., Peng, H., 2012. Unusual reversible photomechanical actuation in polymer/nanotube composites. Angew. Chem. Int. Ed. 51, 8520–8524.

Sun, X., Chen, T., Yang, Z., Peng, H., 2013a. The Alignment of carbon nanotubes: an effective route to extend their excellent properties to macroscopic scale. Acc. Chem. Res. 46, 539–549.

Sun, X., Sun, H., Li, H., Peng, H., 2013b. Developing polymer composite materials: carbon nanotubes or graphene? advanced materials 25, 5153–5176.

Sun, X., Zhang, Z., Lu, X., Guan, G., Li, H., Peng, H., 2013c. Electric current test paper based on conjugated polymers and aligned carbon nanotubes. Angew. Chem. Int. Ed. 52, 7776–7780.

Sun, K., Zhang, S., Li, P., Xia, Y., Zhang, X., Du, D., Isikgor, F., Ouyang, J., 2015. Review on application of PEDOTs and PEDOT:PSS in energy conversion and storage devices. J. Mater. Sci.: Mater. Electron. 26, 4438–4462.

Takagi, S., Makuta, S., Veamatahau, A., Otsuka, Y., Tachibana, Y., 2012. Organic/inorganic hybrid electrochromic devices based on photoelectrochemically formed polypyrrole/TiO_2 nanohybrid films. J. Mater. Chem. 22, 22181–22189.

Tran, H.D., Li, D., Kaner, R.B., 2009. One-dimensional conducting polymer nanostructures: bulk synthesis and applications. Adv. Mater. 21, 1487–1499.

Tsakova, V., 2008. How to affect number, size, and location of metal particles deposited in conducting polymer layers. J. Solid State Electrochem. 12, 1421–1434.

Vaisman, L., Wagner, H.D., Marom, G., 2006. The role of surfactants in dispersion of carbon nanotubes. Adv. Coll. Interface Sci. 128–130, 37–46.

Vigolo, B., Coulon, C., Maugey, M., Zakri, C., Poulin, P., 2005. An experimental approach to the percolation of sticky nanotubes. Science 309, 920–923.

Wang, Y.G., Li, H.Q., Xia, Y.Y., 2006. Ordered whisker like polyaniline grown on the surface of mesoporous carbon and its electrochemical capacitance performance. Adv. Mater. 18, 2619–2623.

Wang, G.-F., Tao, X.-M., Wang, R.-X., 2008. Fabrication and characterization of OLEDs using PEDOT:PSS and MWCNT nanocomposites. Compos. Sci. Technol. 68, 2837–2841.

Wang, D.W., Li, F., Zhao, J., Ren, W., Chen, Z.G., Tan, J., Wu, Z.S., Gentle, I., Lu, G.Q., Cheng, H.-M., 2009. Fabrication of graphene/polyaniline composite paper via in situ anodic electropolymerization for high-performance flexible electrode. ACS Nano 3, 1745–1752.

Wang, Q., Yao, Q., Chang, J., Chen, L., 2012. Enhanced thermoelectric properties of CNT/PANI composite nanofibers by highly orienting the arrangement of polymer chains. J. Mater. Chem. 22, 17612–17618.

Wei, H., Zhu, J., Wu, S., Wei, S., Guo, Z., 2013. Electrochromic polyaniline/graphite oxide nanocomposites with endured electrochemical energy storage. Polymer 54, 1820–1831.

Woo, S., Lee, S.-J., Kim, D.-H., Kim, H., Kim, Y., 2014. Conducting polymer/in-situ generated platinum nanoparticle nanocomposite electrodes for low-cost dye-sensitized solar cells. Electrochim. Acta 116, 518–523.

Wu, S.H., Huang, H.M., Chen, K.C., Hu, C.W., Hsu, C.C., Tsiang, R.C.C., 2006. A green polymeric light-emitting diode material: poly(9,9-dioctylfluorene-alt-thiophene) end-capped with gold nanoparticles. Adv. Funct. Mater. 16, 1959–1966.

Wu, Q., Xu, Y., Yao, Z., Liu, A., Shi, G., 2010. Supercapacitors based on flexible graphene/polyaniline nanofiber composite films. ACS Nano 4, 1963–1970.

Wu, D., Xu, F., Sun, B., Fu, R., He, H., Matyjaszewski, K., 2012. Design and preparation of porous polymers. Chem. Rev. 112, 3959–4015.

Xia, X., Hao, Q., Lei, W., Wang, W., Sun, D., Wang, X., 2012. Nanostructured ternary composites of graphene/Fe_2O_3/polyaniline for high-performance supercapacitors. J. Mater, Chem. 22, 16844–16850.

Xie, X.L., Mai, Y.W., Zhou, X.P., 2005. Dispersion and alignment of carbon nanotubes in polymer matrix: a review. Mater. Sci. Eng.: R: Rep. 49, 89–112.

Xing, S., Tan, L.H., Yang, M., Pan, M., Lv, Y., Tang, Q., Yang, Y., Chen, H., 2009. Highly controlled core/shell structures: tunable conductive polymer shells on gold nanoparticles and nanochains. J. Mater. Chem. 19, 3286–3291.

Xu, Z., Wu, Y., Hu, B., Ivanov, I.N., Geohegan, D.B., 2005. Carbon nanotube effects on electroluminescence and photovoltaic response in conjugated polymers. Appl. Phys. Lett. 87, 263118.

Xu, C., Sun, J., Gao, L., 2011. Synthesis of novel hierarchical graphene/polypyrrole nanosheet composites and their superior electrochemical performance. J. Mater. Chem. 21, 11253–11258.

Xu, K., Huang, X., Liu, Q., Zou, R., Li, W., Liu, X., Li, S., Yang, J., Hu, J., 2014a. Understanding the effect of polypyrrole and poly(3,4-ethylenedioxythiophene) on enhancing the supercapacitor performance of $NiCo_2O_4$ electrodes. J. Mater. Chem. A 2, 16731–16739.

Xu, W., Zhao, K., Niu, C., Zhang, L., Cai, Z., Han, C., He, L., Shen, T., Yan, M., Qu, L., Mai, L., 2014b. Heterogeneous branched core–shell SnO_2–PANI nanorod arrays with mechanical integrity and three dimensional electron transport for lithium batteries. Nano Energy 8, 196–204.

Yan, Y., Cheng, Q., Pavlinek, V., Saha, P., Li, C., 2012. Fabrication of polyaniline/mesoporous carbon/ MnO_2 ternary nanocomposites and their enhanced electrochemical performance for supercapacitors. Electrochim. Acta 71, 27–32.

Yang, X., Lu, Y., 2005. Preparation of polypyrrole-coated silver nanoparticles by one-step UV-induced polymerization. Mater. Lett. 59, 2484–2487.

Yao, Y., Liu, N., Mcdowell, M.T., Pasta, M., Cui, Y., 2012. Improving the cycling stability of silicon nanowire anodes with conducting polymer coatings. Energy Environ. Sci. 5, 7927–7930.

Yao, W., Zhou, H., Lu, Y., 2013. Synthesis and property of novel MnO_2@polypyrrole coaxial nanotubes as electrode material for supercapacitors. J. Power Sources 241, 359–366.

Yin, Z., Zheng, Q., 2012. Controlled synthesis and energy applications of one-dimensional conducting polymer nanostructures: an overview. Adv. Energy Mater. 2, 179–218.

Yin, J., Chang, R., Shui, Y., Zhao, X., 2013. Preparation and enhanced electro-responsive characteristic of reduced graphene oxide/polypyrrole composite sheet suspensions. Soft Matter 9, 7468–7478.

Yu, A., Meiser, F., Cassagneau, T., Caruso, F., 2004. Fabrication of polymer-nanoparticle composite inverse opals by a one-step electrochemical co-deposition process. Nano Lett. 4, 177–181.

Yu, Y., Ouyang, C., Gao, Y., Si, Z., Chen, W., Wang, Z., Xue, G., 2005. Synthesis and characterization of carbon nanotube/polypyrrole core–shell nanocomposites via in situ inverse microemulsion. J. Polym. Sci. Part A: Polym. Chem. 43, 6105–6115.

Yu, G., Hu, L., Liu, N., Wang, H., Vosgueritchian, M., Yang, Y., Cui, Y., Bao, Z., 2011a. Enhancing the supercapacitor performance of graphene/MnO2 nanostructured electrodes by conductive wrapping. Nano Lett. 11, 4438–4442.

Yu, Z., Li, L., Zhang, Q., Hu, W., Pei, Q., 2011b. Silver nanowire-polymer composite electrodes for efficient polymer solar cells. Adv. Mater. 23, 4453–4457.

Yu, Z., Niu, X., Liu, Z., Pei, Q., 2011c. Intrinsically stretchable polymer light-emitting devices using carbon nanotube-polymer composite electrodes. Adv. Mater. 23, 3989–3994.

Yun, S., Hagfeldt, A., Ma, T., 2014. Pt-free counter electrode for dye-sensitized solar cells with high efficiency. Adv. Mater. 26, 6210–6237.

Zeng, Z., Jin, H., Zhang, L., Zhang, H., Chen, Z., Gao, F., Zhang, Z., 2015. Low-voltage and high-performance electrothermal actuator based on multi-walled carbon nanotube/polymer composites. Carbon 84, 327–334.

Zhang, J., Shu, D., Zhang, T., Chen, H., Zhao, H., Wang, Y., Sun, Z., Tang, S., Fang, X., Cao, X., 2012. Capacitive properties of PANI/MnO_2 synthesized via simultaneous-oxidation route. J. Alloys Compd. 532, 1–9.

Zhang, J., Tu, J.-P., Du, G.-H., Dong, Z.-M., Wu, Y.-S., Chang, L., Xie, D., Cai, G.-F., Wang, X.-L., 2013. Ultra-thin WO3 nanorod embedded polyaniline composite thin film: synthesis and electrochromic characteristics. Sol. Energy Mater. Sol. Cells 114, 31–37.

Zheng, W., Razal, J.M., Whitten, P.G., Ovalle-Robles, R., Wallace, G.G., Baughman, R.H., Spinks, G.M., 2011. Artificial muscles based on polypyrrole/carbon nanotube laminates. Adv. Mater. 23, 2966–2970.

Zhou, Q., Li, Y., Huang, L., Li, C., Shi, G., 2014. Three-dimensional porous graphene/polyaniline composites for high-rate electrochemical capacitors. J. Mater. Chem. A 2, 17489–17494.

Zhu, C., Zhai, J., Wen, D., Dong, S., 2012. Graphene oxide/polypyrrole nanocomposites: one-step electrochemical doping, coating and synergistic effect for energy storage. J. Mater. Chem. 22, 6300–6306.

5

Energy Harvesting Based on Polymer

5.1 Introduction

Energy harvesting is of vital importance for human society sustainability (Cook et al., 2010; Dillon, 2010). The most popular process to obtain energy is fossil fuel by converting chemical energy to mechanical energy and then electrical energy. However, this traditional source will be exhausted someday. And there are ever increasing requirement of electricity in expanding fields. With the increasing amount of knowledge of science and discovery of new phenomenon, various energy conversion devices have been designed for harvesting ambient energy, including light, heat, and mechanical vibration (Zeng et al., 2014). Polymer material, a light-weight, strong, structure- and property-tunable, and solution-processable material, is particularly promising for fabricating large-scale flexible and low-cost energy harvesting devices (Coakley and McGehee, 2004). This chapter will focus on energy harvesting devices based on photovoltaic (Li et al., 2012), thermoelectric (Zhang et al., 2014a), piezoelectric (Ramadan et al., 2014), and triboelectric (Wang et al., 2015b) phenomena, and the related polymer materials will be highlighted in these applications. We start from the principles of energy harvesting devices and then describe the requirements on the used materials. The application of polymers in these devices and the recent advancement in designing new structures are highlighted. The polymer structure and morphology that affect the performances of these devices will be discussed further. Finally, the challenges and prospects will be summarized.

5.2 Photovoltaic Device

The photovoltaic effect is the creation of voltage or electric current in a special structure upon exposure to light and could be applied to convert photon energy into electricity (Tang, 1986). Based on this photovoltaic effect, devices with various structures and materials are designed for solar energy conversion application, including silicon-based solar cell (Savin et al., 2015), thin film solar cell (Major et al., 2014), and third-generation solution-processed solar cell (Yan and Saunders, 2014). Here we mainly focus on the third-generation solution processed solar cell that applying polymer as functional components, including polymer solar cell, dye sensitized solar cell, and perovskite solar cell. Here the conducting polymer functions as photoactive layer, hole transporting layer and electrocatalytic counter electrodes in these three devices.

5.2.1 Working Mechanism

5.2.1.1 Polymer Solar Cell

Polymer solar cell, also named plastic solar cell, using conjugated polymers as light absorber, electron donor, acceptor, and/or hole transporting material, has been investigated for 2 decades (Li et al., 2012). The recorded highest energy conversion efficiency currently is 10% (Dun et al., 2015). In order to further improve the performance, new polymers with various molecular structures and their applications in photovoltaic devices are under intensive investigations.

At the very beginning of polymer solar cell, the structure is similar to conventional silicon-based solar cell with a planar junction. People expect this device functions as P–N junction solar cell with simply coating of p-type and n-type organic semiconductor materials. However, due to the short excitons diffusion length, the length during which excitons return to ground state, the distance between the generated excitons and junction interface should be short enough for generating free charge carriers. On the other hand, the light absorber should be thick enough to harvest as much as incident light. To balance this dilemma, bulk heterojunction in which donor (p-type) and acceptor (n-type) form interpenetrated phases has been successfully developed to achieve high performances. In 1995, poly(2-methoxy, 5-(2′-ethylhexyloxy)-1,4-phenylene vinylene) (MEH-PPV) was blended with C_{60} and its derivatives to realize the first polymer solar cell with a high power conversion efficiency of 2.9% under 20 mW cm^{-2} illumination (Yu et al., 1995).

Fig. 5.1 illustrates the structure and basic working principle of the today's common polymer solar cell. A typical polymer solar cell contains a donor/acceptor bulk-heterojunction light-harvesting layer, sandwiched between electron and hole extraction layer then anode and cathode. The energy diagram is displayed in Fig. 5.1B (Li et al., 2012). A photovoltaic process generally contains the generation of excitons and production of free carriers from incident photons. First, conjugated polymer absorbs incident light and generates excitons, which are bonded electron–hole pairs. In polymer solar cell, these excited hole–electron pairs or excitons will dissociate into free carriers in donor/acceptor interface and then migrate into charge extraction and transporting materials, which is driven by the build-in electric field. Afterward, these free charge carriers, electrons and holes, will be collected in anode and cathode, respectively. Finally these free charge carriers will transfer to external circuit to fulfill a conversion from light to electricity (Zhu et al., 2009; Collavini et al., 2015). Here the polymer functions as photoactive layer for light absorption and charge generation and transport.

Power conversion efficiency is used to evaluate the photovoltaic performances of solar cells, which is defined by dividing the maximum output power with incident power. The parameters of polymer solar cell performance including open-circuit voltage (V_{OC}), short-circuit current density (J_{SC}) are originated from the intrinsic properties of the photoactive polymer. For detailed information of the relationship between polymer properties and solar cell performance will be discussed in the following sections.

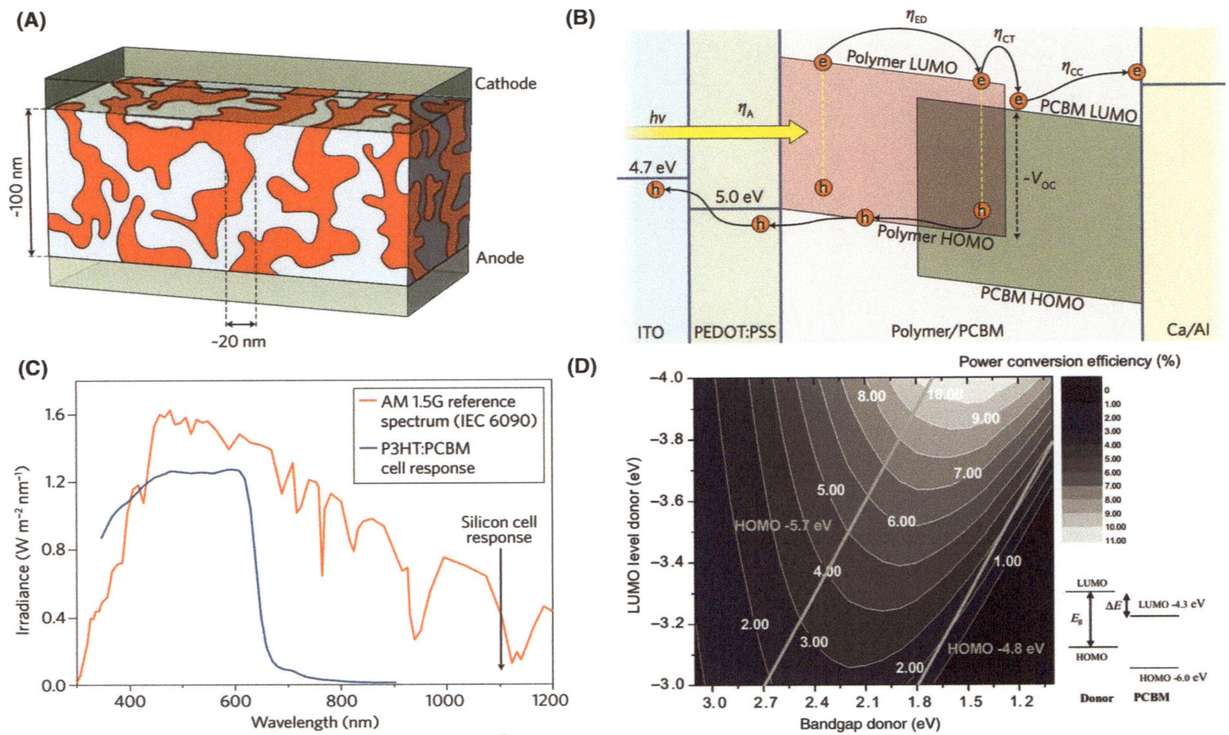

FIGURE 5.1 (A) Typical structure of polymer solar cell. (B) Energy diagram level and charge transport in a polymer solar cell. (C) AM1.5 reference spectrum and typical absorbing curve in polymer solar cell. (D) Contour plot showing the theoretic power conversion efficiency *(contour lines and colors)* versus the bandgap and the LUMO level of the donor polymer (PCBM as acceptor). *(C) Reproduced with permission from reference Li, G., Zhu, R., Yang, Y., 2012. Polymer solar cells. Nat. Photon. 6, 153–161. Copyright 2012, Macmillan Publishers Ltd. (D) Reproduced with permission from reference Scharber, M.C.C., Mühlbacher, D., Koppe, M., Denk, P., Waldauf, C., Heeger, A.J., Brabec, C.J., 2006. Design rules for donors in bulk-heterojunction solar cells—towards 10% energy-conversion efficiency. Adv. Mater. 18, 789–794. Copyright 2006, Wiley-VCH*

5.2.1.2 Dye-Sensitized Solar Cell

Dye-sensitized solar cell (DSSC) is another kind of solution-processed solar cells applying functional polymer component as electrode or hole transporting material. Basically, the structure of DSSC contains three parts, namely dye-sensitized photoanode, counter electrode, and redox electrolyte, which looks like a sandwich. The dye-sensitized photoanode is responsible for light absorbing and charge injection, while counter electrode is for redox pair reduction, and redox electrolyte or hole transporting material is for dye reduction. The mechanism of DSSC is depicted later. First, light travel through transparent electrode and is absorbed by dye to reach the excited state. Then the excited electrons would be injected into the conduction band of semiconductor metal oxide and transfer to external circuit. The oxidized dye would be reduced by redox pair in the electrolyte, which will be reduced by counter electrode with external circuit electrons to complete a circle. There are two grand challenges facing DSSC, as high cost platinum counter electrode and corrosive liquid electrolyte. Here the high performance counter electrode can be fabricated with conducting polymer and using a polyelectrolyte can eliminate the use of liquid electrolyte that might leak out (Saranya et al., 2015).

5.2.1.3 Perovskite Solar Cell

The perovskite solar cell is derived from DSSC and developed into a new family of solar cells with excellent power conversion efficiency. The efficiency increased from 3.8% to a certificated 20.2% in less than 5 years and show promising application in a near future (Yang et al., 2015b). Due to the intrinsic properties of this perovskite material, the perovskite solar cell develops into a simpler structure compared with its origination of DSSC counterpart. Typically, in a high performance and stable perovskite solar cell, a transparent conducting electrode is used as a substrate, followed with subsequence deposition of compact blocking layer (TiO_2, ZnO), perovskite light harvesting layer, hole transporting layer, and back contact electrode. With the illumination of sunlight, photons are absorbed by perovskite light harvesting materials. Due to a small binding energy between electron–hole pair, free charge carriers are generated instead of excitons (Collavini et al., 2015). The electrons and holes would transfer to electron transporting layer and hole transporting layer, and collected in anode and cathode, respectively. In a complete operation circle, a hole-transporting layer is required for efficient hole extraction and transporting. The representative polymer materials functions as hole transporting layer will also be discussed.

5.2.2 Polymer as Light Harvesting Material

Power conversion efficiency is used to evaluate the photovoltaic performances of solar cells, which is defined by dividing the maximum output power with incident power (Smestad et al., 2008).

$$\eta = \frac{P_{max}}{P_{in}} = \frac{J_{SC} \times V_{OC} \times FF}{P_{in}} \tag{5.1}$$

Here J_{SC} is short-circuit current density; V_{OC} is open-circuit voltage; *FF*, fill factor, is defined by the following equation, which is indicative of the similarity to rectangular of the shape of *J–V* curve.

$$FF = \frac{J_{P_{max}} \times V_{P_{max}}}{J_{SC} \times V_{OC}} \tag{5.2}$$

It can be concluded that to improve the photovoltaic performances, short-circuit current density, open-circuit voltage, and fill factor should be improved. Generally, the short-circuit current density directly relates to the bandgap of light harvesting materials. When a solar cell is illuminated, photons with energy higher than the bandgap can be harvested then generating excitons. The light harvesting process is evaluated by external quantum efficiency (EQE), which is the ratio between the collected photo-generated charges and the number of incident photons. Short-circuit current density could be calculated by equation:

$$J_{SC} = e\int_{E_g}^{\infty} EQE_{(E)} \times n\text{AM}1.5(E)dE \tag{5.3}$$

Here E_g is the bandgap of light absorber; $nAM1.5(E)$ is number of photons with different energy in the standard incident spectrum. Smaller bandgap is promising for more light absorption thus higher J_{SC}.

Another factor affecting the performance of solar cell is open-circuit voltage which is also determined by the bandgap of photoactive materials. In polymer solar cells, practically, the open-circuit voltage is dictated by the energy difference between the lowest unoccupied molecular orbital (LUMO) of acceptor and the highest occupied molecular orbital (HOMO) of donor (Brabec et al., 2001). The practical open-circuit voltage is empirically described in the following equation.

$$V_{OC} = \frac{1}{e} \times \left(\left|\text{HOMO}_{\text{Donor}}\right| - \left|\text{LUMO}_{\text{Acceptor}}\right|\right) - 0.3V \tag{5.4}$$

Besides, the LUMO of acceptor should locate beneath the LUMO of donor, because electrons are to transfer from the donor to the acceptor. Designing photoactive materials should strike a balance between the narrowed bandgap to benefit light harvesting efficiency and the enlarged HOMO/LUMO energy difference to maintain a high open-circuit voltage.

The third factor that influences the performance of solar cell is the fill factor that relates to the microscale morphology of active layer, charge extraction layer, and posttreatment of deposition process (Guo et al., 2013). The fill factor will also influence the short-circuit current density and open-circuit voltage that are relative to internal series and parallel resistance in a device.

Innovations in materials science and technology have provided promising strategies to realize a high photovoltaic performance. Besides the intrinsic properties of photoactive polymer materials, morphology is also critical in bulk heterojunction polymer solar cells.

Building highly efficient polymer solar cells, therefore, require an insightful understanding of three key issues: materials design, morphology control, and interface engineering.

5.2.2.1 Donor Material

Typically, the synthesized conjugated polymer is p-type semiconductor and used as donor material combined with [6,6]-phenyl-C61-butyric acid methyl ester (PCBM) as acceptor. Fig. 5.2A shows the chemical structures of some representative conducting polymer materials. In this polymer solar cell, a donor material is used to absorb light and generate

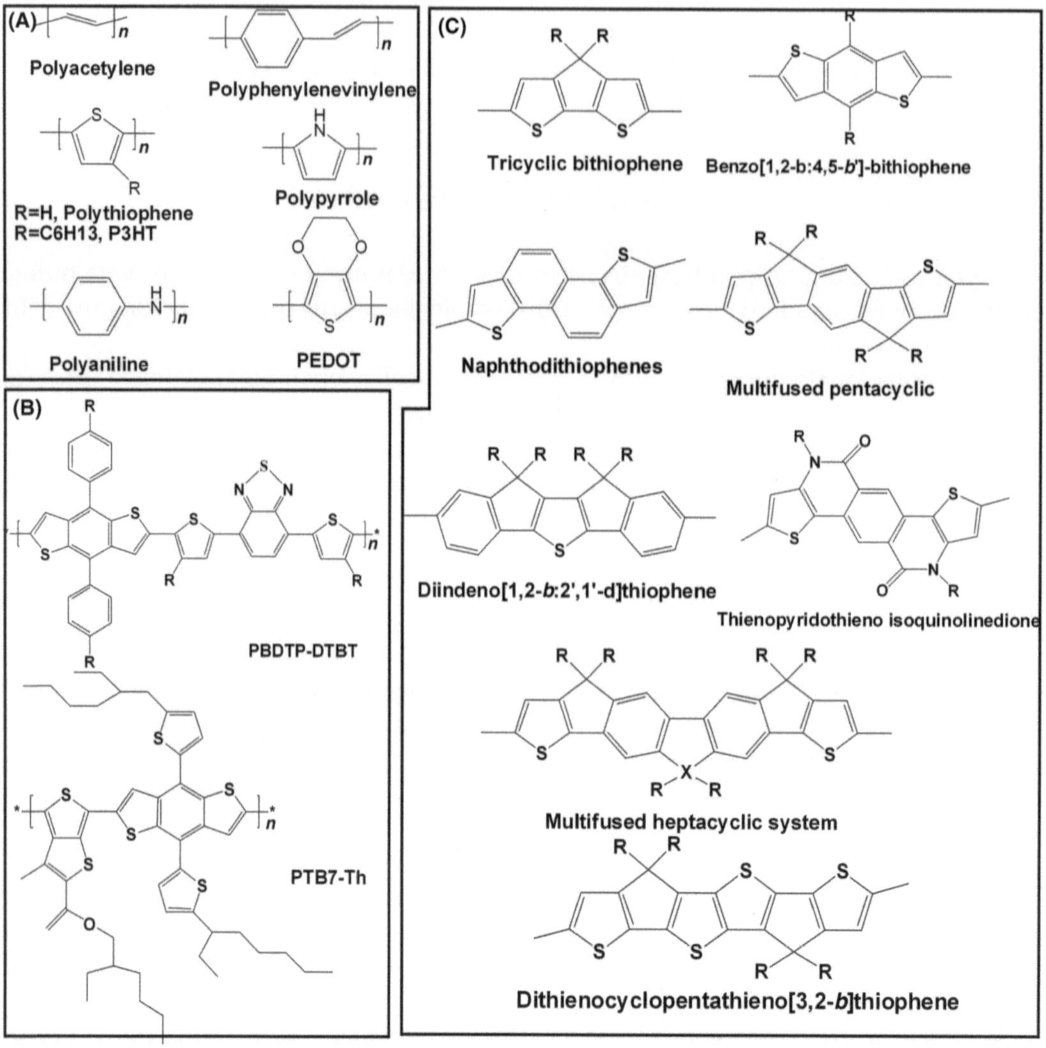

FIGURE 5.2 (A) Structures of conducting polymers. (B) Representative high performance donor materials in polymer solar cell. (C) Representative fused aromatic ring building blocks for high performance donor materials.

excitons. As shown in Fig. 5.1C–D, developing new polymer donor material provides opportunity to realize high-performance solar cells with power conversion efficiency higher than 10% (Scharber et al., 2006; Li, 2012). In the early days of polymer solar cells, one of the most popular polymers was MEH-PPV (Yu et al., 1995). With further optimization of acceptor materials, PPV-based polymer solar cells obtained an efficiency over 3% (Wienk et al., 2003). But the low hole mobility and large bandgap limit the performance of MEH-PPV based devices.

Then comes the most successful conjugated polymers, polythiophene (PTh) and its derivatives, especially poly(3-hexylthiphene) (P3HT). This material has provided a prototype for investigation of structure–performance relation and for morphology optimization (Li et al., 2005). A highest efficiency of 5% has been obtained applying P3HT as donor and PCBM as acceptor. People learn from this material that its large bandgap and mismatched energy diagram level with acceptor material is not suitable for high-performance solar cells (Scharber et al., 2006). Also a bright future has been pointed out for the structure design of high performance polymer and posttreatment for morphology control.

For the design of high performance donor materials, several aspects should be considered. First, a narrow bandgap contributes to a broad absorption of incident light. Second, a suitable energy diagram level, especially a relative low lying HOMO is beneficial to a high open circuit voltage. Third, the high planarity of polymer can afford higher hole mobility. Finally, a high solubility permits an easy processing, which is related to molecular weight and side chain groups. However, a high planarity which can enhance π–π stacking and increase hole mobility is detrimental to solubility. Hence, these aspects should be comprehensively balanced (Zhang et al., 2015b).

The innovations in materials science boost the development of polymer solar cell. Considering the four aspects mentioned previously, three methods have been developed accordingly (Li, 2012), namely side chain engineering, donor–acceptor (D–A) copolymerization and large fused aromatic ring structure. Some of the representative high performance conjugated polymers are displayed in Fig. 5.2B.

First method is introducing alkyl side chains or other electron donating or withdrawing groups into conjugated polymers (Nishinaga et al., 2015; Li et al., 2015a; Verstappen et al., 2015; Zhang and Li, 2014). The properties of conjugated polymers would be changed with the introduction of side chains. For example, the solubility can be significantly enhanced with alkyl side chains. The morphology of heterojunction blend is also affected by the side chain. And the energy diagram level can be tuned by substituting electron donating or withdrawing groups.

Alkyl was introduced into insoluble PTh as side chain, producing the soluble and most successful P3HT polymer (Li et al., 2005). However, this alkyl side chain also decreases the crystallinity of polymer in the heterojunction blend, which is negative for high carrier mobility. It has been reported that with lower side chain density, the performance can be greatly improved. The latest report shows that the performance and stability can be both enhanced by reducing the side chain density on fluorinated poly(4-(2′-ethylhexyl)-4-octyl-4H-cyclopenta[2,1-*b*:3,4-*b*′]-dithiophene-2,6-diyl-*alt*-6,7-difluoro-2,3-bis

[5′-(2″-ethylhexyl)-thiophene-2′-yl]quinoxaline-5,8-diyl) (PCPDTQx) D–A copolymers (Verstappen et al., 2015). Besides carrier mobility, the energy level can also be tuned with lower side chain density. In another case, using flexible side chain can both enhance the solubility and performance, and also significantly influence the morphology of heterojunction blend (Li et al., 2015a). When electron withdrawing groups like fluorine are used for side chain, it is available to lower the HOMO of donor and achieve a high open circuit voltage (Zhang and Li, 2014).

The second method is to design alternative copolymer with alternating donor and acceptor unit (D–A structure) to systematically tune the bandgap and energy level (Li, 2012). The most straightforward way is to polymerize a donor monomer and an acceptor monomer through a transition-metal catalyzed cross-coupling reaction. The polymerization method has been discussed in Chapter 2. Electron-rich (donor)–electron-deficient (acceptor) (D–A) copolymerization is able to broaden the light absorption and tune the energy levels of conjugated polymers. In 2003, the first reported D–A copolymer was synthesized from donor units of fluorene, acceptor units of benzothiadiazole (BT), and thiophene π-bridges (Svensson et al., 2003). Currently, the most popular donor units are dithienosilole (DTS), benzodithiophene (BDT), indacenodithiophene (IDT). As shown in Fig. 5.2B, an certified efficiency of 9.94% was obtained applying poly [[2,6′-4,8-di(5-ethylhexylthienyl)benzo[1,2-*b*;3,3-*b*] dithiophene] [3-fluoro-2[(2-ethylhexyl)carbonyl] thieno[3,4-*b*]thiophenediyl]] (PTB7-Th) as photoactive material (Dun et al., 2015).

The third strategy is applying large fused aromatic ring as building block in the main chain to obtain high performance solar cells (Wu et al., 2015). The most used building blocks for high performance polymers include benzene, thiophene, furan, selenophene, pyrrole, silole, germole, and cyclopentadienyl moieties. As shown in Fig. 5.2C, a large fused aromatic ring unit in backbone is beneficial for large planarity and enhancing stacking properties, which is capable of promoting intermolecular π–π interactions and enhancing charge carrier mobility. In recent years there are many well-defined new large aromatic fused conjugated polymers which can serve as the superb donor monomers to prepare a range of donor–acceptor copolymers and are reviewed recently (Wu et al., 2015). However, it should be noted that such an aromatic fused high planarity copolymer is not highly soluble and the side-chain engineering plays a critical role as mentioned before. The length, density, and location of side chain on a polymer backbone will dramatically influence the solubility, crystallinity, and molecular packing which alter the charge transport and morphology of composites, resulting in variations in device performance. The combination of these three methods could design and synthesize new family high performance polymers for photovoltaic devices.

5.2.2.2 Acceptor Material

In a heterojunction blend, another important component is acceptor material for electron injection and transport. Most used acceptor materials are fullerene derivatives such as [60]PCBM, [70]PCBM, etc. In early studies, [60]PCBM was replaced with [70]PCBM to enhance the light absorption because of the more asymmetric structure of [70]PCBM

(Wienk et al., 2003). To enhance the performance of solar cells, researchers need to strengthen absorption in the visible spectrum, up shift the LUMO and increase the electron mobility in acceptor materials (Li, 2012). However, the fullerene derivatives barely absorb light in the visible spectrum due to a low absorption coefficient.

Recently a new alternative to traditional fullerene derivatives—n-type conjugated polymers—has been developed as acceptor materials, which produces an all-polymer solar cell (Facchetti, 2013). Unlike fullerene derivative acceptors, the n-type conjugated polymers can function as both light absorbers and electron acceptors due to their high light absorption coefficient. Moreover, the energy level of the n-type polymers is tunable through molecular engineering, thereby being compatible to p-type donors to optimize the photovoltaic performance. The all-polymer solar cell has flexibility in tuning the chemical properties and energy levels of its polymer components as well as stability in adapting various circumstances.

Fig. 5.3 displays the typical developed n-type polymer as acceptor materials in polymer solar cells (Facchetti, 2013). Normally there are several kind of polymer backbone structures for n-type polymers, with the first reported cyanated polyphenylenevinylenes in 1995 (Halls et al., 1995). However, this first type acceptor has little potential for high performance polymer solar cell and fullerene came into charge since then (Yu et al., 1995). For high performance acceptor polymer materials, high electron affinity and mobility are required. Nowadays the most promising n-type conjugated polymers are the ones based on the use of strong electron-withdrawing naphthalene di-imides (NDIs) backbone (Suraru and Würthner, 2014). These NDI-based conjugated polymers exhibit low-lying LUMO levels and high electron mobility. The properties of acceptor polymer can be easily tuned with suitable backbone unit and designed D–A structure. A series of naphthalene di-imide-selenophene copolymer acceptor was synthesized and the energy conversion efficiency has been reported at around 3.3%, which is comparable with PCBM counterpart (Earmme et al., 2013).

One of the challenges for applying all-polymer active blend is the miscibility and processability of two polymer mixture, which greatly influence the phase separation of donor/acceptor and performance of this active layer. The poor performance of polymer/polymer BHJ solar cells reported so far is mainly attributed to the undesirable phase separation of the blends, such as, large phase-separated domain sizes, inhomogeneous internal phase composition, and reduced crystallinity (Hwang et al., 2015). To enhance the performance of all polymer heterojunction, various strategies have been developed. The factors that affect the intrinsic properties of n-type polymer, processability and following heterojunction blend including the molecular weight, side chain engineering, film deposition, and posttreatment process are covered later.

Various new acceptor polymers are designed and synthesized via side chain engineering. This process can not only tune the intrinsic HOMO and LUMO energy levels of polymer but also modify the morphology of heterojunction blend. With different lengths of alkyl side chains, the performance varies significantly. With a short alkyl side chain, the copolymer was proven to be effective for promoting their crystalline orientation with

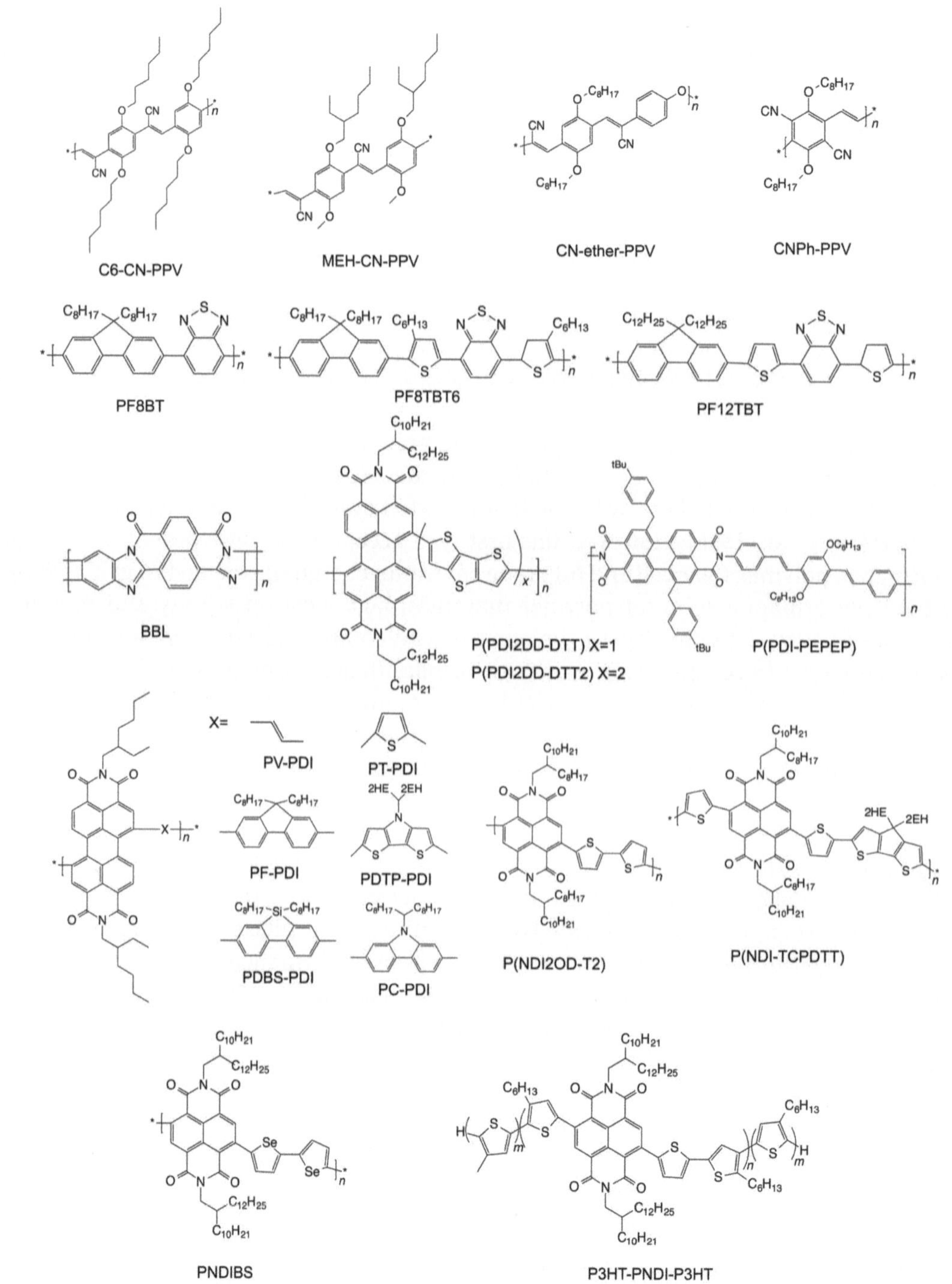

FIGURE 5.3 Structures of typical n-type conducting polymers used as acceptors in polymer solar cells. *Reproduced with permission from reference Facchetti, A., 2013. Polymer donor–polymer acceptor (all-polymer) solar cells. Mater. Today 16, 123–132. Copyright 2013, Elsevier Ltd.*

face-on geometry and, at the same time, producing the optimal blend morphology with finely separated phase domains. The properties like light absorbing and bandgap are not changed but the blend morphology is varied significantly. Those two results (crystalline and morphology) cooperated with each other to induce well-balanced hole and electron mobility ratio (μ_h/μ_e) in heterojunction blend and generate highly efficient all-polymer PSCs with power conversion efficiency approaching 6% (Lee et al., 2015).

With other substitutes group instead of alkyl side chains, the energy levels of acceptor polymer can be modified to match that of donor materials. Fluoro-substituted n-type conjugated polymers were introduced for all-polymer solar cells with the highest reported energy conversion efficiency of 6.71% at the state of art (Jung et al., 2015).

The other factor that influence the processability and performance of all polymer heterojunction blend is the molecular weight. The molecular weight of a conjugated polymer is one of the key factors determining the electrical and mechanical properties as well as its solubility in organic solvents and miscibility with other polymers (Kang et al., 2015). High molecular weight promoted a strong "face-on" geometry in the blend film, suppressed the formation of an excessively large crystalline domain, and facilitated molecularly intermixed phases with donor polymers. Therefore, the optimized all-polymer solar cells based on high molecular weight acceptor polymer blend showed substantially higher hole and electron mobility than those containing low molecular weight acceptor polymer blend.

5.2.2.3 Morphology Control

The morphology, typically means the phase separation and crystalline of conjugated polymer in heterojunction blend. The morphology is dependent on the intrinsic properties of polymer, including molecule weight, solubility, and planarity. However, combining with suitable deposition process and posttreatment, a satisfactory morphology can be prepared. Among this morphology control technology, thermal annealing and solvent annealing are currently the most popular two methods.

Besides this post treatment to obtain good morphology, nanostructure of active polymer material in heterojunction blend is preferable for high performance, especially one-dimensional nanostructure (Sun et al., 2015a). This aligned one-dimensional nanostructure might provide direct charge transport path in heterojunction blend. For example, an effective route to enhance the photovoltaic performance of bulk heterojunction device lies in the use of nanofiber networks that increase the charge separation and transport. To better understand the enhancement of polymer solar cells with active nanofiber network layers, Sum et al. investigated the charge generation and recombination dynamics in P3HT nanofiber using transient absorption spectroscopy. The spectra showed a more efficient charge generation in the nanofiber network compared with the control bulk sample (Kurniawan et al., 2012). To increase the charge transport and decrease charge localization, external electric field treatment of the active layer was conducted to align the randomly dispersed polymer chains to be perpendicular to the electrode. A 22.7% improvement in power conversion efficiency and 37.5% improvement in charge mobility had been achieved after the treatment (Solanki et al., 2014).

To obtain the morphology information, including phase separation and crystalline, we can now use microscopic techniques, atomic force microscopy, transmission electron microscopy, electron tomography, variable-angle spectroscopic ellipsometry, X-ray photoemission spectroscopy, and grazing-incidence X-ray diffraction. The detailed information of this characterization methods can be found from the specific reference (Li et al., 2012; Huang et al., 2014).

5.2.3 Polymer as Charge Transport Material

5.2.3.1 Counter Electrode

As depicted in Section 5.2.1.2, counter electrode is an important component in dye sensitized solar cells. The electrons transfer from external circuit to the counter electrode, which is a catalytic material to transfer electrons into electrolyte to regenerate the redox couple. To maintain a sufficiently high concentration of reducing agent in electrolyte, a high electrocatalytic property of counter electrode is required. Platinum is the most investigated material and exhibits the highest performance till now due to its excellent electroccatalytic property and high conductivity (Sun et al., 2015a). However, platinum is highly cost compared with other alternatives.

Various kinds of conducting polymers can work as counter electrode in DSSC. Poly(3,4-ethylenedioxythiophene) (PEDOT), polypyrrole (PPy), PANI are most investigated polymer materials. With the development of nanomaterial, the nanostructure has been introduced into previously mentioned polymers. These polymers are readily deposited by either chemical or electrochemical polymerization process. And they are readily modified into aligned nanostructure that might enhance charge transport and surface area simultaneously (Sun et al., 2015a).

For instance, a soft-template method was used to synthesize nanofibers of PEDOT with diameters of 10–50 nm. A high electrical conductivity up to 83 S cm^{-1} was obtained. Use the nanofibers as counter electrode produced a power conversion efficiency of 9.2%, much higher than its bulk PEDOT counterpart (6.8%) and comparable to that of the platinum electrode (8.6%) (Lee et al., 2012).

Furthermore, by introduction of a second component, high performance polymer composite can be easily obtained, such as, carbon nanotube additive. Guan et al. developed an aligned carbon nanotube–oriented poly(3,4-ethylenedioxythiophene):poly(styrene sulfonate) (PEDOT:PSS) composites which show an unexpectedly high catalytic activity. When used as counter electrode in dye sensitized solar cell, a high power conversion efficiency of 8.3% is obtained (Guan et al., 2013). The aligned carbon nanotube could enhance both the conductivity and charge transfer properties of this composite.

5.2.3.2 Hole Transporting Material

Polymer solar cell: Typically the p-type conducting polymer can be applied as hole transporting material, and the mostly used is PEDOT:PSS in polymer solar cell. This material has been studied thoroughly and fabricated into one-dimensional nanostructured network to increase the surface area and enhance the hole extraction and transporting properties (Nasybulin et al., 2011).

DSSC: The most studied DSSCs are based on liquid electrolyte that might leak out. The grand challenge of the organic solvents employed in traditional DSSCs is their high volatility, which strongly limits their durability. To eliminate this drawback, various solid-state electrolyte or p-type conductor are developed for all solid-state DSSCs. Among these materials conducting polymers are promising alternatives for all-solid-state DSSCs. There are many relative research works focusing on this area. However, the main problem lies in a poor infiltration of polymer material into mesoporous nanocrystal metal oxide anode (Schmidt-Mende and Grätzel, 2006). There might be large volume of void inside the solid-state DSSCs that decreases the interface area for efficient charge transfer.

There are two kinds of polymer material that used in quasi-solid/solid state DSSCs. For quasi-solid electrolytes, polyionic liquids have been proposed as solvent and redox couple as solute. They appear in molten salts and present many promising properties, such as, high chemical and thermal stability and high ionic conductivity. Their main drawback is related to its high viscosity, which makes the ions diffusion rather slow. As the transport of I^{3-} ions to the counter electrode in an ionic liquid matrix represents a rate-limiting step in DSSC (Bella, 2015), the performance of quasi-solid electrolytes based solar cell is unsatisfied.

The other kind of polymers as hole transporting materials in DSSCs are p-type conducting polymers, such as, PEDOT, polythiophene (PTh), and PPy. The use of polymer as hole transporting material has been reported for solid-state DSSCs, while the performances was relatively low (Nogueira et al., 2004). Polymer electrolytes offer a high long-term stability for the DSSC. They also provide the solar cells to be lightweight and flexible. However, the resulting devices typically display lower energy conversion efficiencies compared to the liquid electrolyte due to lower photocurrent values caused by the limited ionic conductivity within a polymeric matrix. Also it is critical to further enhance the penetration of polymer electrolyte into the nanostructured photoanode.

Perovskite solar cell: A newly developed material, perovskite, is then investigated for fabricating high performance solid-state solar cells. The perovskite materials satisfy the requirement of high light absorbance and carrier mobility and long diffusion length. In these devices the use of mesoporous anode that cause poor infiltration of polymer is eliminated. Here polymer functions as efficient hole extraction and transport material combined with light-harvesting perovskite layer. These polymer includes PEDOT:PSS, P3HT, poly-[2,1,3-benzothiadiazole-4,7-diyl[4,4-bis(2-ethylhexyl)-4H-cyclopenta[2,1-b:3,4-b′]dithiophene-2,6-diyl]] (PCPDTBT), (poly-[[9-(1-octylnonyl)-9H-carbazole-2,7-diyl]-2,5-thiophenediyl-2,1,3-benzothiadiazole-4,7-diyl-2,5-thiophenediyl]) (PCDTBT), poly-triarylamine (PTAA) and related derivatives (Malinkiewicz et al., 2013; Heo et al., 2013). Normally all the high performance p-type polymer could be used to fabricate high performance perovskite solar cells. Furthermore, these polymer materials are promising for high flexible and robust perovskite solar cell.

In summary, with the increasing interests on the next-generation solar cells that can be continuously fabricated through solution processes, a lot of efforts have been made to develop highly efficient devices based on polymers. A variety of new polymers are designed and synthesized for desired functions, including electron donor, electron acceptor, hole transport, and electrocatalysis. However, the stability of these solar cells currently remains a great challenge from a viewpoint of practical applications (Jorgensen et al., 2012). This requires much more work to push this field forward.

5.3 Thermoelectric Generator

Thermoelectric generator represents another approach to harvest energy and provide electricity. A variety of materials are used in this kind of devices, mainly inorganic semiconductors. However, such inorganic materials are fragile and heavyweight, and a high temperature is required for annealing. As previously mentioned, polymers are made through solution processes with low cost and high efficiency; they are also lightweight and can be fabricated at a large scale through printing technology. Therefore, the development of polymers for thermoelectric generators attracts increasing interests aiming at practical applications (Taroni et al., 2014).

5.3.1 Working Mechanism

There are three kinds of effects based on the thermoelectric phenomenon, namely Seebeck effect, Peltier effect, and Thomson effect. The first thermoelectric generator (Fig. 5.4) was invented from the Seebeck effect in 1820s. Typically in a junction composed of two different materials, the charge carrier like electron or hole would transfer at the interface due to different carrier concentrations. The diffusion of carrier causes a diffusing potential at the junction, which hinders the further diffusion through a drift current. Finally, a balance is achieved between the diffusion current and drift current, forming an electric field in this junction. The intensity of the field is dependent on temperature, so it can be used to measure temperatures. Vice versa when connecting two materials into a circuit with two junctions and generating a temperature gradient in the two junctions, there will be a current through this circuit, which is called Seebeck effect. The current direction depends on the intrinsic properties of materials that are used to construct the junction and temperature gradient direction. Conversely, when a current comes through these two junctions, a temperature difference in the two junctions has been created, which is called Peltier effect. For a thermoelectric device based on the Peltier effect, whether the heterojunction would absorb or emit heat depends on the current direction and the used materials. Normally the semiconductor heterojunction causes a high temperature difference which is beneficial for thermoelectric conversion (Culebras et al., 2014).

One might make use of the Seebeck effect to fabricate a thermoelectric generator to transfer thermal energy to electricity. For instance, the abundance of solar thermal energy and the widespread waste heat from automobile make thermoelectric generators attractive as low-cost sources in harvesting energy. Fig. 5.4A shows a unit of flexible thermoelectric generator that based on a polymer p−n junction (Chen et al., 2015c). To measure the performance of thermoelectric generator, the output power (P_{out}) should be divided by input

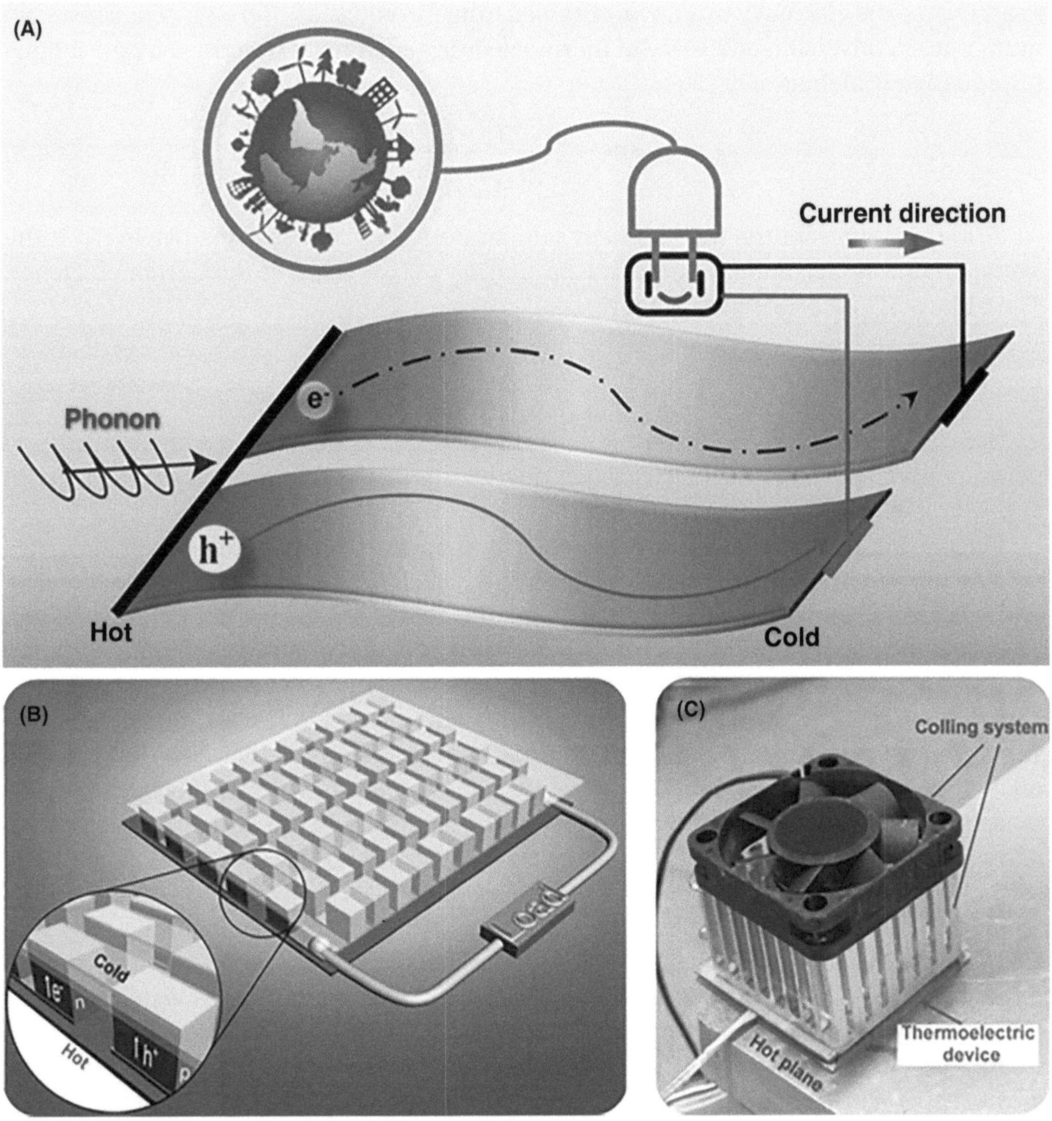

FIGURE 5.4 (A) Typical structure of a flexible thermoelectric device module. (B) Illustration of single module connected in series and parallel. (C) Photograph of a practical thermoelectric device. *(A) Reproduced with permission from reference Chen, Y., Zhao, Y., Liang, Z., 2015c. Solution processed organic thermoelectrics: towards flexible thermoelectric modules. Energy Environ. Sci. 8, 401–422. Copyright 2015, Royal Society of Chemistry. (B, C) Reproduced with permission from reference Sun, Y., Sheng, P., Di, C., Jiao, F., Xu, W., Qiu, D., Zhu, D., 2012. Organic thermoelectric materials and devices based on p- and n-type poly(metal 1,1,2,2-ethenetetrathiolate)s. Adv. Mater. 24, 932–937. Copyright 2012, Wiley-VCH.*

heat (Q_{in}), so the efficiency (η) can be obtained from the equation of $\eta = P_{out}/Q_{in}$. The maximum power conversion efficiency of thermoelectric generator is described by the following equation (Culebras et al., 2014):

$$\eta_{max} = \Phi_C \frac{\sqrt{1+ZT_{av}}-1}{\sqrt{1+ZT_{av}}+T_h/T_c} \equiv \Phi_C \gamma \quad (5.5)$$

T_h and T_c represent the high and low temperature, respectively. Meanwhile, T_{av} is the average value of high and low temperature, and $\Phi_C = (T_h - T_c)/T_h$ is the Carnot efficiency. The value of ZT is called figure of merit:

$$ZT = \frac{\alpha^2 \sigma}{\kappa} T \quad (5.6)$$

Here α is the Seebeck coefficient, which is the potential difference as a function of temperature difference; σ is the electrical conductivity, and κ is the thermal conductivity of applied material.

The maximum efficiency directly depends on the figure of merit (ZT) of specific material and the structure of thermoelectric generator, which determines the Carnot efficiency and γ. When the temperature difference of hot and cold is set, the Carnot efficiency is set. Then it is the figure of merit of specific material that is most important. To increase the performance of thermoelectric generator, one should focus on the Seebeck coefficient, electrical conductivity, and thermal conductivity of fabricated materials. Actually one advantage of using polymer materials in thermoelectric generator is their low thermal conductivity. So, the design and modification of polymer materials are mainly based on the increase of electrical conductivity through various strategies. In this case the performance of a thermoelectric polymer material can also be evaluated by a parameter called power factor (PF) as $PF = \alpha^2 \sigma$ (Zhang et al., 2014a; Shi et al., 2015).

Polymer-based lightweight thermoelectric generator can be fabricated through low-cost solution process and shows flexibility, and thermoelectric generators do not have moving parts so it is quiet. However, as the output voltage of a single thermoelectric generator is relatively low, thermoelectric generator normally consist of an array of thermoelectric generator modulus in series/parallel connection, which is illustrated in Fig. 5.4B. A practical large area in series connection thermoelectric generator is shown in Fig. 5.4C (Sun et al., 2012). This in series connection structure might increase the complexity of fabrication process.

Here we will discuss the fabrication of high performance thermoelectric generator based on polymer materials and their composites. The design of molecule structure, modification of component, and microstructure of polymer material will be thoroughly discussed on the effect of thermoelectric performance.

5.3.2 Conducting Polymers for Thermoelectric Generator

To increase the performance of thermoelectric generator, one should consider to increase electrical conductivity and Seebeck coefficient simultaneously while leave the thermal conductivity unchanged. And the electrical conductivity σ is related to carrier charge q (C), carrier concentration n (cm−3), and carrier mobility μ (cm^2 V−1 s−1) as $\sigma = nq\mu$. Consequently, high carrier mobility and an efficient doping process are prerequisites for high performances in both p-type and n-type thermoelectric materials. On the other hand, the Seebeck coefficient might be vary in a reverse way compared with electrical conductivity. A higher electrical conductivity might be detrimental to Seebeck coefficient. A balance should be met between these two factors (Zhang et al., 2014a; Culebras et al., 2014).

5.3.2.1 P-Type Conducting Polymers

Like polymers that function as donor part in polymer solar cell, these materials can also be used in thermoelectric generator as p-type conducting component. The structure–property relationships that learn from polymer solar cells may contribute to the design of high thermoelectric performance polymer materials. And various design of molecular structure and modification with doping are achieved for high performance thermoelectric materials. The conducting polymers that investigated most in thermoelectricity including PANI, polyalkyl thiophenes (PTh and its derivatives), PEDOT, polyacetylene, PPy and poly(2,7-carbazolyenevinylene) as shown in Fig. 5.2A (Wei et al., 2015). The structure–property relationship, the structure design and doping effect will be summarized in this section.

5.3.2.1.1 MOLECULE DESIGN AND ENGINEERING

Abundant monomer structures are promising for polymerization or copolymerization for the design of high thermoelectric performance polymer. Unlike polymer solar cell with a heterojunction blend, the miscibility with other material should not be worried unless in the case of nanocomposite material. The structure of the polymer, such as side chain length and backbone structure, might affect the electrical conductivity and crystalline. Nowadays, side-chain engineering has been recognized as an effective method to tune the electrical properties of semiconducting conjugated polymers. Mai et al. investigated the side chain effect on the electrical conductivity, morphology, and thermoelectric properties of a series of self-doped narrow-bandgap conjugated polymers based on a π-conjugated cyclopenta-[2,1-*b*;3,4-*b*′]-dithiophene-*alt*-4,7-(2,1,3-benzothiadiazole) backbones (Mai et al., 2014). Similarly, shortening the side chain of conjugated polymer leads to a higher crystallinity and a more edge-on molecular orientation, which provide useful clues to modulate electrical conductivity.

In other case, a more general polymer, PTh, was used to investigate the side chain effect (Endrődi et al., 2015). These polymers were recrystallized to a well-ordered lamellar one-dimensional self-assembled nanofiber. With the change of the length of the alkyl side chain, the macroscopic morphology structure and intrinsic properties of assembly are varied. For the shortest alkyl side chain, the highest electrical conductivity was obtained due to a more densely packed structure. This side chain can not only affect the morphology of the assembly but also the electronic properties of this condensed material.

5.3.2.1.2 DOPING AND DEDOPING

However, even though the properties of the polymer can be engineered with elaborately design, the performance is still far from requirement. It has been well recognized in inorganic thermoelectric materials that the best performance is found in highly doped semiconductors (Kim et al., 2013). The doping could significantly enhance the electrical conductivity of inorganic thermoelectric materials. However a balance between electrical conductivity and Seebeck coefficient should be reached to obtain a maximum thermoelectric efficiency (maximum power factor $PF = \alpha^2\sigma$ or figure of merit ZT), considering the dilemma in inorganic materials that the electrical conductivity and the Seebeck coefficient are changed in a reverse direction with the changing of doping level (Mengistie et al., 2015). In the case of conducting polymer, the doping level of material also influences the Seebeck coefficient and electrical conductivity thus the thermoelectric properties. The dilemma between electrical conductivity and Seebeck coefficient in polymer materials is not significant but still need to be taken into consideration. In the past years, many works dealing with the change of the doping level in conductive polymers to enhance the thermoelectric properties have been investigated. The doping can be introduced within the polymerization process or posttreatment with secondary dopant.

Normally the doping comes from the synthesis process, which introduce counter ions. For instance, the conductive polymers, like PEDOT, P3HT, and PANI synthesized from oxidative chemical polymerization. These polymers are p-type semiconductors, which exhibit an electron deficiency in the backbone. The positive charges in the backbone are neutralized by the dopant agent from oxidative chemical polymerization (Culebras et al., 2014). The doping level is related to the positive charges in the backbone. In a higher doping level, a high electrical conductivity is observed.

The types of doping can be tuned by using different kinds of oxidative salt catalysts to polymerize the reaction, which absorbs different counter ions. And the doping level can be controlled with the concentration of molar ratio between the monomer and acid or with different sizes of counter ions. Mai et al. investigate the effect of counter ions with different sizes (Mai et al., 2014). The ions used include Na^+, K^+, and tetrabutylammonium. They found that with a smaller size counter ion the polymer exhibited a higher doping level. The smallest counter ion (Na^+) endowed thin films with the highest electrical conductivity while a comparable Seebeck coefficient, thereby leading to the highest power factor of 0.44 m K^2.

Besides oxidative chemical polymerization, the electrochemical synthesis is another method for the preparation of conductive polymers, such as, PPy, PEDOT, and PANI, sometimes with uniform nanoscale morphology. Compared with oxidative chemical polymerization, this method provides an easier way to tune the doping level by varying the synthesis voltage. For example, PEDOT and PPy films can be prepared by an electrochemical method in a three-electrode cell. With different voltages for electrochemical polymerization the films with different oxidation levels, namely bipolaron, polaron, and neutral states, can be obtained (Culebras et al., 2015). The different electrochemical polymerization conditions produced a variation of one order of magnitude in the electrical conductivity and the Seebeck coefficient. Actually these changes are related to the doping level of the polymer film and they can be accurately controlled by the applied voltage. This is a much easier way to operate in order to obtain required performance with optimized performance (Lu et al., 2013).

Other than the polymerization process that resulted in different dopant ions or doping levels, the thermoelectric performance can be further optimized by posttreatments when sometimes a further doping or dedoping is required. This appears to be an efficient way to enhance the performance. With a posttreatment of H_2SO_4, a copolymer poly(3,4-ethylenedioxythiophene)–tosylate–polyethylene glycol–polypropylene glycol–polyethylene glycol (PEDOT–Tos–PPP) film showed a significantly enhanced electrical conductivity and slightly reduced Seebeck coefficient compared to those without treatment. This gives a higher thermoelectric power factor thus a higher figure of merit value of around 0.024 (Wang et al., 2014a).

The figures of merit values of conductive polymers were established to be much lower than inorganic semiconductors until 2013, until Pipe et al. reported the optimization of ZT in PEDOT:PSS by reducing dopant volume using ethylene glycol as a secondary doping agent (Kim et al., 2013). When a PEDOT:PSS film was subjected to ethylene glycol with different time, the ethylene glycol can be mixed into the conductive polymer film and exchanged with the first doping agent. Owing to the selective removal of counter ions in PSS under such treatment, the film thickness was reduced by half, the doping level was decreased, and the thermal conductivity was also suppressed. After doping by ethylene glycol, a figure of merit of $ZT \approx 0.42$ had been reported. Subsequent work with the other secondary doping agents also verified the aforementioned conclusion, and it provided an efficient way to enhance the performance of PEDOT-based thermoelectric generator (Mengistie et al., 2015).

5.3.2.1.3 MORPHOLOGY CONTROL

As we have mentioned in polymer solar cell, the alignment and crystalline are two key factors which determine the performance of the polymer film. The polymer conformation and morphology may play an important role in its physical properties, especially in the electrical conductivity. Due to a high binding energy the electron–hole pair appears as strongly bonded excitons, which greatly affects the transport properties in conducting polymers (Li et al., 2012; Ihnatsenka et al., 2015). Typically the microscopic morphologies like crystallinity are correlated with such properties as bulk carrier mobility. Therefore microscopic morphologies are expected to remarkably impact thermoelectric property of polymeric materials. The same characterization technologies are applied in the measurement of such properties (Li et al., 2012).

The benefit on controlling polymer morphologies has been reported in the process of electrochemical polymerization, which is readily operated to obtained nanoscale structure or dense bulk material in combination with template or nontemplate process. This electrochemical polymerization also can readily tune the doping level, which is beneficial for high performance thermoelectric material (Sun et al., 2015a). PTh and PPy films were respectively prepared through electrochemical polymerization. By varying polymerization conditions including current density, temperature, concentration of monomer and electrolyte, the best thermoelectric property was produced from the smoothest and densest morphology of polymer film (Hiraishi et al., 2009). In addition, the crystalline is much better in a smoother and denser film which is beneficial for the electrical conductivity.

Sometimes the morphology change comes from a postdoping treatment. The effect of morphology is also confirmed where secondary doping treatment is widely adopted to increase the electrical conductivity due to a reduced doping volume and a higher figure of merit of 0.46 (Kim et al., 2013; Mengistie et al., 2015; Wang et al., 2014a). Such remarkable increase in electrical conductivity is ascribed to the improvement in mobility in a more condense and crystallized polymer materials. (Zhang et al., 2014a). Interestingly, Seebeck coefficients are less sensitive to this postdoping treatment than electrical conductivity, so a continuously increase of figure of merit with the increasing electrical conductivity can be obtained through this process.

Among the nanoscale morphology, one-dimensional nanostructure, especially in an aligned format, provide great potential for high-performance conducting polymer film (Sun et al., 2015a). For example, the figure of merit value of silver-doped poly(3-hexylthiophene) with nanofiber network structure is 0.0026, 6 times higher than its bulk counterpart (Endrődi et al., 2014). It is the enhancement of carrier mobility that leads to an increase of electrical conductivity of this nanofiber network. Combined with a heavily doped condition, a silver nanofiber doped film can produce a high value of figure of merit. In another case, it has been reported that the power factor of an aligned PEDOT nanoarray is three times higher than that of the PEDOT deposited directly as a bulk material, owing to improved crystallinity and carrier mobility in a nanoarray (Taggart et al., 2011). Overall, material morphology is a critical factor that affect the carrier mobility thus electrical conductivity of conducting polymer. Similar to doping, this control of molecular organization offers another way to improve thermoelectric property in consideration to electrical conductivity. Characterizations related to crystallinity, density, and mobility should be considered carefully for performance optimization.

In summary, much effort is required to illustrate the effect of molecular structures of polymers, counter ions (doping level), and morphology on the electrical conductivity and thermoelectric properties (Zhang et al., 2014a). Furthermore, considering the dilemma between Seebeck coefficient and electrical conductivity, for example, Seebeck coefficient would be decreased when electrical conductivity is increased, a trade-off between Seebeck coefficient and electrical conductivity should be met for high performance. By now, several research results have demonstrated that it is possible to enhance Seebeck coefficient and electrical conductivity simultaneously, by means of tuning transport pathways of charge carriers (Ihnatsenka et al., 2015). An example on tuning carrier transport was reported for a blend of P3HT and poly(3-hexylthiothiophene) (P3HTT) (Sun et al., 2010). There is an upward shift of HOMO level by 0.15 eV in P3HTT due to the sulfur atom. In a blend with 80% weight concentration of P3HT and suitable additive, electrical conductivity and Seebeck coefficient are increased simultaneously. The second one is a new structure with a conducting polymer film sandwiched between two electrodes (Anno et al., 2015). In this structure the thermal conduction is inhibited but the electrical conductivity unchanged. The insightful understanding of the relation between structure, composition, and properties would be a great help to enhance the thermoelectric performance.

5.3.2.2 N-Type Conducting Polymers

Practical thermoelectric generators require both electron and hole conducting materials with high power factors as shown in Fig. 5.4. As the conventional synthesized polymers are normally p-type conductors, the thermoelectric performance of p-type polymer materials is rapidly advancing, as mentioned previosuly. Traditionally used n-type materials are vapor-doped fullerenes (Menke et al., 2012) and powder-processed organometallic poly(Ni 1,1,2,2-ethenetetrathiolate) derivatives (Sun et al., 2012), which have exhibited the highest thermoelectric performance with power factors reaching 30 and 66 $\mu W\ m^{-1}\ K^{-2}$, respectively. However, these materials cannot be made by solution processing, which severely restricts their practical applications. And there are few examples of high performance n-type polymer thermoelectric materials.

Recently, several solution-processed n-type thermoelectric polymer materials were reported (Schlitz et al., 2014; Russ et al., 2014). Many efforts are made to improve the properties using strategies that applied in p-type conducting polymers, like doping and side chain engineering. Pei et al. developed a series of n-type polymers to investigate the relation between the chemical structures of n-type polymers and thermoelectric properties (Shi et al., 2015). They develop a series of high performance n-type polymers, benzodifurandione-based poly(*p*-phenylene vinylene) (BDPPV), chloridized benzodifurandione-based poly(*p*-phenylene vinylene) (ClBDPPV), and fluorinated benzodifurandione-based poly(*p*-phenylene vinylene) (FBDPPV), with varied LUMO levels and electron mobility caused by the halogen atoms. A n-type dopant, (4-(1,3-dimethyl-2,3-dihydro-1Hbenzoimidazol-2-yl)phenyl) dimethylamine (N-DMBI), was chosen for performance enhancement. They found that introduction of halogen atoms to the polymer backbones has a dramatic influence on not only the electron mobility but also the doping level, both of which are critical to the electrical conductivity. With fluorine substitute and optimized doping level, the polymer FBDPPV shows a highest electron mobility, leading to electrical conductivity of 14 $S\ cm^{-1}$ and power factors up to 28 $\mu W\ m^{-1}\ K^{-2}$. However, the material performances are much lower than those obtained for p-type thermoelectric materials, such as PEDOT, which requires more efforts to develop comparable n-type thermoelectric materials.

5.3.2.3 Ionic Seebeck Effect in Conducting Polymers

During the development of thermoelectric generator, ionic thermoelectric effects have been observed in inorganic ionic materials where ions are the only charge carriers (Anno et al., 2015). Recently, researches focusing on conducting polymers also reveal an ionic thermoelectric effect in addition to the known electronic thermoelectric effect. Their Seebeck coefficient is as large as ~200 $\mu V\ K^{-1}$. This finding discloses a new possible approach to improve the thermoelectric properties of conducting polymers by combining various types of charge carriers (Wang et al., 2015a).

The role of ions in the thermoelectric responses of different PEDOT derivatives was investigated (Wang et al., 2015a). Unexpected significant increases in the thermo-induced voltage was observed at high humidity levels, which is identified as an ionic Seebeck effect. The total thermo-power appears as the sum of the ionic Seebeck coefficient and the electronic Seebeck coefficient. Importantly, the ionic Seebeck effect in polymeric mixed conductors PEDOT:PSS leads to an enhancement of maximum power factor by 2–4 orders of magnitudes, which becomes even higher than pure electronic conductors. This effect might be taken into consideration for further enhancement or new application of thermoelectric generator.

5.3.3 Coordination Polymers for Thermoelectric Generator

Other than conducting polymers applied in thermoelectric generator, there is another kind of polymers. Coordination polymers are a kind of polymers with metal ions acting as connectors and ligands as linkers (Zhang et al., 2014a). The first use of coordination polymer in thermoelectric phenomenon can be dated back to 1979. At that time, the Seebeck coefficient was around 1.2 mV K^{-1}. The positive value indicates a p-type conduction, which is usually the case in the coordination polymer.

After many years of development, coordination polymers was discovered that is promising thermoelectric materials for high performance devices in 2012 (Sun et al., 2012). A series of coordination polymers have been synthesized (Fig. 5.5). More detail information of these coordination polymers for thermoelectric generator is summarized in a recent review (Zhang et al., 2014a). An all-polymer thermoelectric generator shown in Fig. 5.4B–C was fabricated based on the as-synthesized polymers. This kind of polymers has expanded the selection of polymer materials in thermoelectric applications.

5.3.4 Polymer Composite for Thermoelectric Generator

As mentioned previously, the efficiency of thermoelectric generator is determined by the value of figure of merit. The value of figure of merit should be as high as possible. If the value of figure of merit is close to a positive infinite, then the value of γ should be 1, which

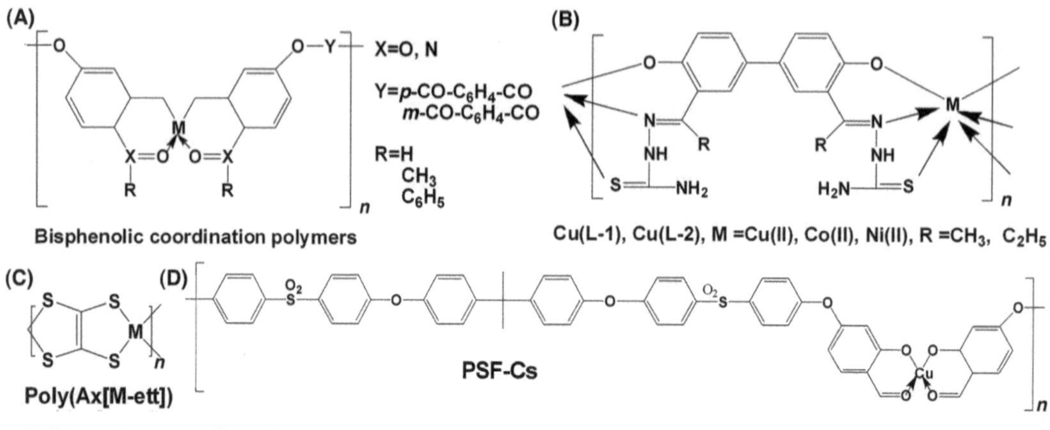

FIGURE 5.5 Structures of coordination polymers used in thermoelectric generators. *Reproduced with permission from reference Zhang, Q., Sun, Y., Xu, W., Zhu, D., 2014. Organic thermoelectric materials: emerging green energy materials converting heat to electricity directly and efficiently. Adv. Mater. 26, 6829–6851. Copyright 2014, Wiley-VCH.*

makes the efficiency of thermoelectric generator the same as Carnot efficiency (Culebras et al., 2014). However, the figure of merit is lower than 5 at the state of the art, with the highest value of 0.42 for polymer materials (Kim et al., 2013). Even though a variety of strategies has been used to enhance the thermoelectric properties including molecular structure design, side chain engineering, doping, and dedoping, as previously mentioned, the polymer-based materials are still far from practical application. The low Seebeck coefficient and low electrical conductivity still cannot meet the requirement.

If conducting additive are introduced, polymers may be used as matrices and frameworks to prepare conducing composites for thermoelectric generators. After incorporated with inorganic additives particularly at the nanoscale, the performances of polymer materials can be enhanced significantly (McGrail et al., 2015; Culebras et al., 2014). In this way the polymer could be conducting polymer or normal plastic, thus extends the applicable polymer materials.

Among the used inorganic moieties, carbon nanomaterials are the most explored ones, including carbon nanotube, graphene, and graphene oxide. These carbon nanomaterials possess excellent electrical conductivity and mechanical properties and promising for the enhancement of nanocomposite. However, a great challenge remains, that is to obtain uniform distribution of carbon nanomaterial into polymer matrices. Various strategies have been developed for distribution like in situ polymerization and modification of the nanocarbon materials (Sun et al., 2013).

For instance, a composites of CNT and poly(vinyl acetate) was fabricated from aqueous solutions (Yu et al., 2008). At a 20 wt.% content of CNTs the nanocomposite exhibited a high electrical conductivity (48 $S \cdot cm^{-1}$), thermal conductivity (0.34 $W\,m^{-1} \cdot K^{-1}$), and a thermoelectric figure of merit (6×10^{-3}). In another nanocomposite, PEDOT:PSS doped with DMSO, in a SWCNTs dispersion, a *ZT* values around 0.02 is obtained. The nanocomposite of SWCNT/PEDOT:PSS with 35 wt.% of SWCNT increased the conductivity to 400 $S \cdot cm^{-1}$ while Seebeck coefficient was maintained to be unchanged (Kim et al., 2010). To further enhance the dispersion of CNT in the nanocomposite, poly(vinyl acetate) was added. The nanocomposite with poly(vinyl acetate) gives high electrical conductivities at the order of 10^3 $S \cdot cm^{-1}$ (Yu et al., 2011). The enhancement of electrical conductivity is promising for high performance thermoelectric generator.

However, the content of used CNT is relatively high. Even though the electrical conductivity increases, the thermal conductivity might increase as well. This is due to a poor dispersion of CNTs inside the polymer matrices. Normally there are three kinds of methods to obtain CNT composite, namely melting of matrices, solution process, and in situ polymerization (Sun et al., 2013). For dispersion enhancement, in situ polymerization is better than the other two processes. A template-directed in situ polymerization was developed to prepare nanocomposites from PEDOT:PSS and MWCNT with greatly enhanced thermoelectric property (Zhang et al., 2015c). The results reveal that 3,4-ethylenedioxythiophene was successfully polymerized, enwrapping the surfaces of dispersed MWCNTs as templates with the aid of PSS. The thermoelectric measurements show that the PEDOT:PSS/MWCNT nanocomposite exhibited much enhanced electrical conductivities, Seebeck coefficients, and power factors compared with bare PEDOT:PSS. The enhancement of thermoelectric properties may be explained by the formation of a nanostructure of PEDOT:PSS as well as the introduced CNTs.

Graphene sheets and nanoribbons are another promising additive for high performance nanocomposite. Simultaneous increase in electrical conductivity and Seebeck coefficient in graphene sheet/PANI nanocomposite have been obtained (Du et al., 2012). Similar to the CNT/polymer composite, in situ polymerization also represents a solid method to achieve high performance composites. Recently, three different in situ chemical oxidative polymerization routes were investigated to make thermoelectric nanocomposites (Xu et al., 2015). The three routes, for example, spin-coating and subsequent liquid layer polymerization, spin-coating followed by vapor phase polymerization, and in situ polymerization and then posttreatment by immersion in ethylene glycol, have been developed to achieve reduced graphene oxide (RGO)/PEDOT nanocomposites. As demonstrated by scanning electron microscopic and energy-dispersive X-ray spectroscopic techniques, PEDOT has been successfully coated on the surface of the RGO nanosheet. Importantly, all nanocomposites display a greatly enhanced thermoelectric performance in a comparison to a pure PEDOT.

Except CNT and graphene, several pioneering works focus on the influence of inorganic structure on the thermoelectric efficiency of polymer matrices, especially the newly developed two-dimensional graphene-like materials, such as, TiS_2 (Wan et al., 2015). Several groups enhanced the thermoelectric performance by combining a polymer matrix with layered bismuth telluride (Bi_2Te_3) (Dun et al., 2015). By mixing layered Bi_2Se_3 nanoplates with polyvinylidene fluoride (PVDF), a flexible thermoelectric film was prepared with Seebeck coefficient of $-80\ \mu V\ K^{-1}$ and electrical conductivity of $5100\ S\ m^{-1}$, resulting in a power factor approaching $30\ \mu W\ m^{-1}\ K^{-2}$.

In conclusion, polymer materials are promising for flexible and printable thermoelectric devices. A low processing temperature expands the alterative substrate, such as, paper or other polymers (Wei et al., 2014a; Anno et al., 2015). The properties of polymer materials can be easily tuned by molecular structure, doping level, and composite morphology. However, the figure of merit is much lower than inorganic counterparts. It is the low electrical conductivity that limits the performance, and the increasing efforts should be made to solve the problem in the future.

5.4 Piezoelectric Transducer

Besides solar energy and thermal energy, there are abundant sources for mechanical energy in ambient environment, which is potential for electricity generation. The most popular application is the electromagnetic induction used to generate electricity from mechanical energy. However, a high mechanical power input is required in the aforementioned process. It remains difficult for the low-power mechanical energy harvesting, such as movements in human body. To this end, a new family of piezoelectric devices has been designed for the generation of electricity from the mechanical vibrations. The piezoelectric transducer or generator had been developed many years ago, for example, the first PVDF film to convert water wave energy to electricity learned from eels (Taylor et al., 2001). The piezoelectric generator attracted increasing attentions since 2006 when ZnO nanowire were used to make generator (Wang and Song, 2006).

Piezoelectric generator can be used to scavenge ambient vibration energy into useful electricity to drive miniaturized devices that need a small portion of energy for a long running period. However, in the state of the art, there are only a small part of generators on polymer materials due to poor piezoelectric coefficients compared with inorganic materials (Crossley et al., 2014). However, the unique advantages of flexible, robust, lightweight, and low-cost make the polymers attractive in the development of piezoelectric generators.

5.4.1 Working Mechanism

Piezoelectric effect was discovered in 1880 and is a phenomenon that produce electrical charges across material boundaries with application of a mechanical stress. This effect is obtained from specific materials, named piezoelectric materials and the relationship between induced electrical field and the mechanical stress is linear and reversible (Ramadan et al., 2014).

As shown in Fig. 5.6A a typical structure of piezoelectric generator is illustrated (Cha et al., 2011). A piezoelectric material is sandwiched between two electrodes. After application of an external force, the piezoelectric material might generate a strain to induce a voltage (Crossley et al., 2014). Typically, there are two different electric field directions with a mechanical stress and describe by d_{31} and d_{31} coefficient, respectively (Fig. 5.6B) (Anton and Sodano, 2007; Bowen et al., 2014). Stress and strain indicate the normal stress and strains or shear stress, and strain. The piezoelectric coefficient d is related to the input stress and induced charge collection direction (Cook-Chennault et al., 2008). In the case of d_{31}, the generated electric field is perpendicular to the applied stress. While d_{33} means the generated electric field is in the same direction to the applied stress. One should note that the generated electric field is in a pulse format and reverse the direction with compress and release. (Crossley et al., 2014).

The piezoelectric effect is originated from the polarization of materials and mainly divided into two types. For the first type, similar to inorganic crystal materials, the semicrystalline polymer materials whose positive and negative centers aligned in a specific way can cause piezoelectric effect with input stress and strain (Ramadan et al., 2014). Another type is electret materials that keep alignment of polarity to the direction of polarized electric field and may preserve the poling effect after removing the electric field (Yang et al., 2014). After treatment by high electric field or plasma, a highly polarized polymer material can be obtained.

The efficiency of a generator is usually defined as the electrical energy recovered against the input strain energy applied to the generator during a mechanical cycle (Shu and Lien, 2006). The electrical energy recovered is measured by calculating the output power $V_R{}^*I_R$ (where V, I, and R are instantaneous voltage, current, and external resistance,

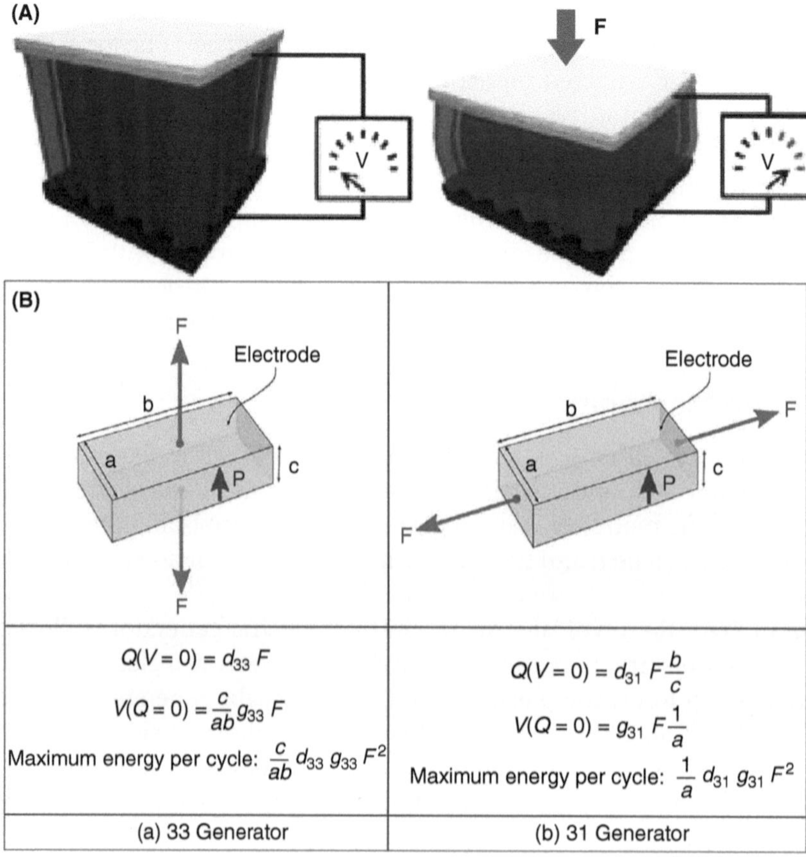

FIGURE 5.6 Working principle of piezoelectric nanogenerator based on polymer. (A) A typical piezoelectric nanogenerator with active polymer sandwiched between two electrodes. (B) Two typical modes in piezoelectric nanogenerator. *(A) Reproduced with permission from reference Cha, S., Kim, S.M., Kim, H., Ku, J., Sohn, J.I., Park, Y.J., Song, B.G., Jung, M.H., Lee, E.K., Choi, B.L., Park, J.J., Wang, Z.L., Kim, J.M., Kim, K., 2011. Porous PVDF as effective sonic wave driven nanogenerators. Nano Lett. 11, 5142–5147. Copyright 2011, American Chemical Society. (B) Reproduced with permission from reference Bowen, C.R., Kim, H.A., Weaver, P.M., Dunn, S., 2014. Piezoelectric and ferroelectric materials and structures for energy harvesting applications. Energy Environ. Sci. 7, 25–44. Copyright 2014, Royal Society of Chemistry.*

respectively). The input strain energy applied to the sample can be estimated as $0.5S^2Yv$, where v is the volume of generator and S can be determined experimentally. An efficiency up to 30% has been discovered for the ZnO nanowires in generator (Wang and Song, 2006; Zhu et al., 2010), while for polymer-based piezoelectric generator the efficiency is relatively low right now.

Although this field attracts a great deal of interest from academy and industry, more efforts should be made for insight understanding of energy generating and harvesting process. The detailed mechanism and related factors that determine the performance of piezoelectric generator are still need further investigation.

5.4.2 Materials

There are mainly two kinds of piezoelectric polymer materials as mentioned before. First, the polymer materials intrinsically have the piezoelectric effect. This kind of polymer materials mainly are PVDF and its copolymer of trifluoroethylene (PVDF-TrFE) (Furukawa, 1989), nylon-11 (Newman et al., 1980), and polyuria (Hattori et al., 1996). However, most polymer-based piezoelectric generators are fabricated from PVDF and its copolymers. The other polymer materials might endow the generator with thermo-resisting properties, while it has not been verified yet.

Besides the previously mentioned piezoelectric material, the second one lies in electret that can preserve the induced charges for a long time. It can not only be used in the fabrication of piezoelectric generators but also triboelectric generators based on electrostatic induction effect. The electret and related devices are discussed further. Here we mainly focus on PVDF and its copolymers in polymer based-piezoelectric transducer/generator.

The piezoelectric effect stems from hydrogen and fluorine atoms in the VDF, which are positioned perpendicularly to the polymer backbone. Fig. 5.7A shows a typical molecular structure of PVDF with different crystalline phases (Chang et al., 2012). The piezoelectric performance of PVDF is dependent on the nature of the crystalline phase (Crossley et al., 2014). Typically, PVDF has three crystalline phases, namely α, β, and γ, and it is the α-phase that typically forms in most situations. While it is polar β-phase that shows the strongest piezoelectric behavior so this material needs to be electrically poled using an electric field with the order of 100 MV m^{-1} or mechanically stretched. A higher β-phase crystalline can lead to a higher piezoelectric coefficient d_{33}. Note that the copolymer of P(VDF-TrFE) [$(CH_2\text{-}CF_2)_n\text{-}(CHF\text{-}CF_2)_m$] crystallizes more easily into the β-phase due to steric factors (Furukawa, 1989). So, the most applied material in piezoelectric generator is P(VDF-TrFE).

5.4.3 Poling Process

The poling treatment uses a high electric field to reorient polarization alignment within the polymer material. And this electric field must be higher to drive domain wall motion. There are typical two systems for poling process namely electrode poling and corona poling (Ramadan et al., 2014; Crossley et al., 2014).

For the electrode poling metal electrodes are required to be first deposited on both sides of the polymer film. And a high voltage (5–100 MV m^{-1}) is applied between the two electrodes. To preserve the polymer under such strong electric field, the poling process must be done in a vacuum chamber (Park et al., 2004).

Another method is corona poling which requires only one electrode to be covered on the polymer film. A conductive needle and a grid are used for corona poling (Park et al., 2004). These two component are on top of the piezoelectric polymer with a high electric field applied to the needle and a low electric field to the grid. The tip of the needle is used to get ionized gas under high voltage and the ionized gas is accelerated toward the piezoelectric polymer. A hot plate is applied for better control of the poling process (Bauer, 1996).

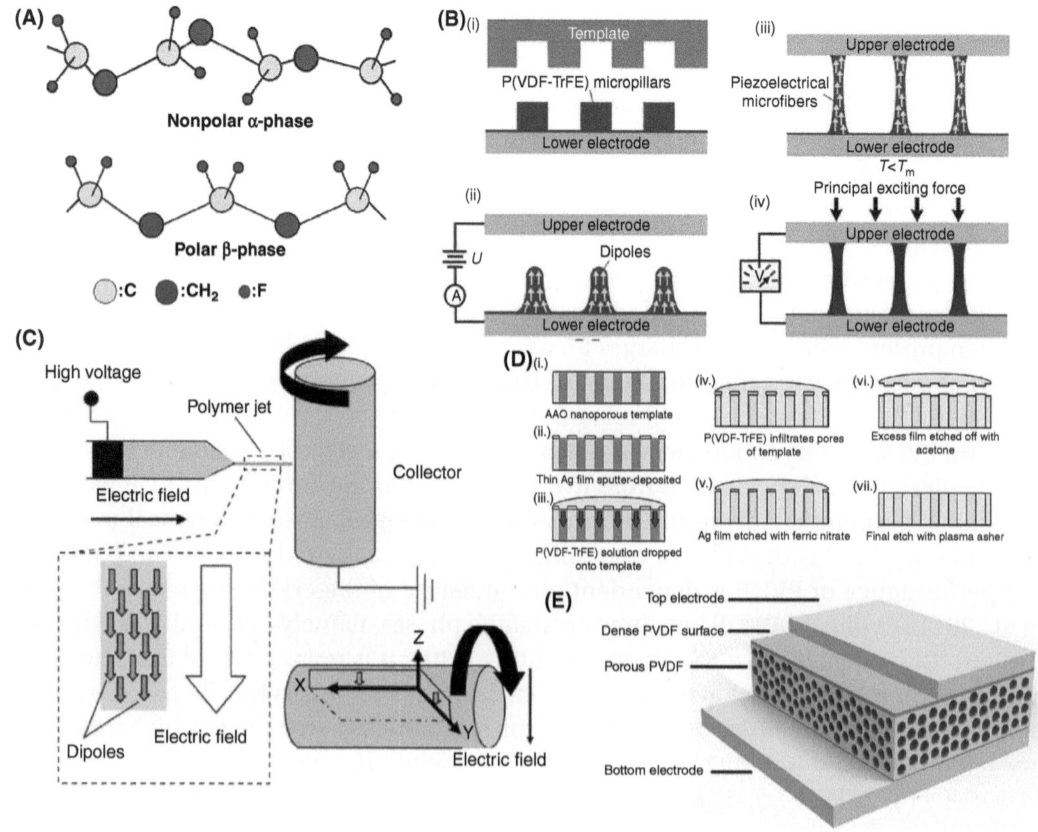

FIGURE 5.7 (A) Piezoelectric polymer materials (PVDF and its copolymers). (B) Self-poled and self-connected vertically integrated fibers of piezoelectric PVDF structure. (C) Electrospinning process for self-poling nanofiber. (D) Self-poled piezoelectric polymer nanowires fabricated by confined template method. (E) Piezoelectric polymer materials with a nanoporous structure. *(A) Reproduced with permission from reference Chang, J., Dommer, M., Chang, C., Lin, L. 2012. Piezoelectric nanofibers for energy scavenging applications. Nano Energy 1, 356–371. Copyright 2012. Elsevier Ltd. (B) Reproduced with permission from reference Chen, X., Tian, H., Li, X., Shao, J., Ding, Y., An, N., Zhou, Y., 2015b. A high performance P(VDF-TrFE) nanogenerator with self-connected and vertically integrated fibers by patterned EHD pulling. Nanoscale 7, 11536–11544. Copyright 2015, Royal Society of Chemistry. (C) Reproduced with permission from reference Mandal, D., Yoon, S., Kim, K.J., 2011. Origin of piezoelectricity in an electrospun poly(vinylidene fluoride-trifluoroethylene) nanofiber web-based nanogenerator and nano-pressure sensor. Macromol. Rapid Commun. 32, 831–837. Copyright 2011, WILEY-VCH. (D) Reproduced with permission from referenceWhiter, R.A., Narayan, V., Kar-Narayan, S., 2014. A scalable nanogenerator based on self-poled piezoelectric polymer nanowires with high energy conversion efficiency. Adv. Energy Mater. 4, 1400519. Copyright 2014, WILEY-VCH. (E) Reproduced with permission from reference Chen, D., Zhang, J.X.J., 2015. Microporous polyvinylidene fluoride film with dense surface enables efficient piezoelectric conversion. Appl. Phys. Lett. 106, 193901. Copyright 2015, American Institute of Physics.*

5.4.4 Morphology and Performance

The typical structure of piezoelectric generator includes a PVDF thin film sandwiched between two electrodes. The polymer materials can be easily deposited through a solution process like spin coating or casting. Wu et al. fabricated a flexible piezoelectric generator based on poly(vinylidenefluoride-*co*-trifluoroethylene) (PVDF-TrFE) thin film (Pi et al., 2014). The thin film was prepared via spin-coating technique and functioned as an active layer. After thermal poling under electric field ranged from 0.5 to 0.8 MV cm^{-1}, the crystalline of β-phase increased significantly. The as-fabricated generator exhibited the open-circuit voltage up to 7 V and short-circuit current of 58 nA with current density of 0.56 μA cm^{-2}.

To enhance the alignment of molecular chain, the fabrication process is modified with a patterned electrohydrodynamic pulling technology (Fig. 5.7B) (Chen et al., 2015b). A new self-connected, piezoelectric microfiber array of P(VDF-TrFE) was vertically integrated between two electrodes. An array of micropillar P(VDF-TrFE) was first deposited onto one electrode by casting and hot embossing process. The microfiber was prepared by applying a high voltage across an electrode pair sandwiching an air gap and an array of shallow micropillars. During the electrohydrodynamic pulling process, the pulling force tends to pull the micropillars upward, generating a microfiber array finally in robust contact with the upper electrode (Fig. 5.7B). With a highly alignment of molecular chain orientation, a well-bridged electrode pair for efficient charge collection for maximized conversion is fabricated. The as-fabricated piezoelectric generator exhibited an enhanced output voltage up to 4.0 V and a current of 2.6 μA, therefore the piezoelectric voltage was enhanced to 5.4 times of the bulk film.

However, the thin film fabricated by casting or spin coating required large amount of solution and postpoling process which increases both the cost and complexity. With the advancement in fabrication methods, PVDF nanowires can be prepared by an electrospinning process (Fig. 5.7C). The ferroelectric β-phase is promoted by a combination of the large electric fields applied during the electrospinning process and the mechanical stretching (Mandal et al., 2011). For only the electrospinning process at ambient temperature without post poling process, the overall crystallinity in the aligned array can reach 48%, which is comparable to the previous best result produced in thin films by poling and stretching (Persano et al., 2013). Chang et al., (2010) showed a high energy conversion efficiency up to 21.8% based on the PVDF nanofibers.

While the previously mentioned processes require complex instruments and high voltages up to 20 kV with limited scale, an alternative fabrication process is realized through a template wetting. Dissolved or molten PVDF is soaked into a nanoporous template (Fig. 5.7D) (Bhavanasi et al., 2014; Whiter et al., 2014). During the formation of nanowires, the PVDF is subjected to substantial stresses which result in preferential formation of a ferroelectric β-phase. Suitable templates include anodized aluminum oxide that is available commercially with a range of sizes, pore diameters, and pore densities. Furthermore, it is possible to free the nanowire from the template via a selective etch in phosphoric acid

(Whiter et al., 2014). The confined geometry of the template promotes the production of an oriented β crystalline phase required for superior piezoelectric energy harvesting, without the electrical poling. This room-temperature fabrication process is attractive and can be implemented at a large scale (Liew et al., 2015).

Along with the aligned nanowire structure, a mesoporous structure has also been designed to enhance the device performance (Fig. 5.7E). The increased piezoelectric efficiency is due to the optimized stress–strain response (Mao et al., 2014). Compared with the previous piezoelectric generator, the mesoporous PVDF thin film generator possesses several unique merits that are critical for harvesting ambient mechanical energy and developing practical self-powering electronic systems. The method is simple and effective for a large-scale fabrication of mesoporous piezoelectric PVDF thin films and does not require high mechanical strain or high electric field for the formation of β-phase PVDF. Furthermore, the generator takes advantage of the device's own weight to modulate its displacement and amplifies its electrical output. This unique operation principle simplifies the design in generator. It also provides an ideal solution for directly integrating multiple generators into one system to boost the output. The controllable porosity also enables adjustable mechanical property of the PVDF film. At high porosity, the PVDF film is flexible and soft like sponge. Although the high-porosity film would dissipate more mechanical energy, a good portion of it can be converted into electrical energy due to the large film displacement resulting in a high electric output under small surface oscillation. In addition, the mechanical properties can be controlled by dispersing microscale pores in a polymer matrix with a dense top layer (Fig. 5.7E) (Chen and Zhang, 2015).

Annealing temperature is another factor that might influence the crystalline structure and thus the piezoelectric response property. It had been discovered that small grains with weak piezo responses were observed for generator films annealed above the Curie temperature. For instance, acicular grains were obtained and the piezo response increased when the annealing temperature approached the melting point. Annealing above the melting point decreased the piezo response, and the morphology was changed dramatically into plate-like structures (Lau et al., 2013).

Combining the fabrication process, morphology and posttreatment, the performance of PVDF and its copolymer-based piezoelectric generator can be further enhanced for various applications, like wearable device or self-powering remote sensor nodes (Huang et al., 2015; Dewan et al., 2014). Incorporation of high performance inorganic piezoelectric materials, such as, PZT and $BeTiO_3$ to form nanocomposite with polymers, a variety of high performance generators may be produced (Nunes-Pereira et al., 2013; Shin et al., 2014).

In summary, researchers have made a great effort to enhance the performance of polymer-based piezoelectric generator, through the optimization of fabrication process, material morphology and posttreatment. Although a lot of efforts have been made to understand the piezoelectric effect for ambient mechanical energy scavenging, the efficiency and output power are still too low to power daily electronic devices directly, and the rectifier and capacitor are generally required for energy storage. The output energy is first stored in a capacitor that can be discharged to power the external electronic devices. Recently, a new kind of devices based on the triboelectric phenomenon has been developed to output a voltage up to 1000 V. This kind of generator might be a promising alternative for piezoelectric-based generator for mechanical energy harvesting. And this triboelectric device will be discussed in the following section.

5.5 Triboelectric Generator

Since the first report of energy harvesting device based on the triboelectric phenomenon in 2012, increasing interests are attracted to develop high performance triboelectric generator (Zhu et al., 2012; Fan et al., 2012a; Wang et al., 2012). Triboelectric effect is that charge transfer between any two materials and observed in the very beginning of human history. After contact with each other, the two materials become electrically charged in opposite signals (Williams, 2012). This effect has been usually considered as a negative effect and is avoided in many areas. But if considering from another point of the charges on the triboelectric material, the charged surface will drive the motion of electrons, thus inducing the formation of electric currents when forms a circuit.

5.5.1 Working Mechanism

Triboelectric generators combine two effects, for example, contact triboelectrification and electrostatic induction. For the triboelectric generator, the ability to trigger triboelectrification and to preserve the generated charge on a dielectric surface should be considered. Although the mechanism of how charges transfer between two materials and the amount of charges transferred remained unknown yet, the output voltage, current, and charges that are induced in a triboelectric generator can be measured (Niu et al., 2013b).

Fig. 5.8 schematically shows the structure of triboelectric generator and generation of charged surface (Zhu et al., 2012). Typically, there are two materials with different charge preserve properties like Kapton and polymethyl methacrylate (PMMA). These dielectric materials are attached to conductive electrodes like copper, silver, or aluminum. Compared with PMMA, Kapton is more attractive to the electron and negatively charged after contact with PMMA (Wang et al., 2012; Zhu et al., 2012). After being positively and negatively charged of PMMA and Kapton respectively, the charges can be preserved for a long time. When the two charged surfaces are departed from each other, the net charges in each surface would induce the movement of carriers in attached conductive electrodes, which is known as electrostatic induction. Under this case, the electrons will move from Kapton to PMMA if there is a load in external circuit or a positive potential from PMMA to Kapton will form if an open circuit exists. After the departure of two electrodes the distribution of charges in electrodes then reaches a new equilibrium to counteract the net charges in polymer film. If the two charged surfaces contact again, the surface charge would counteract first, then the charges collected in each electrode would transport to the other side into a neutral condition, so electrons transfer from PMMA to Kapton. Since the surface charge can be preserved for a long time, this process can be repeated again

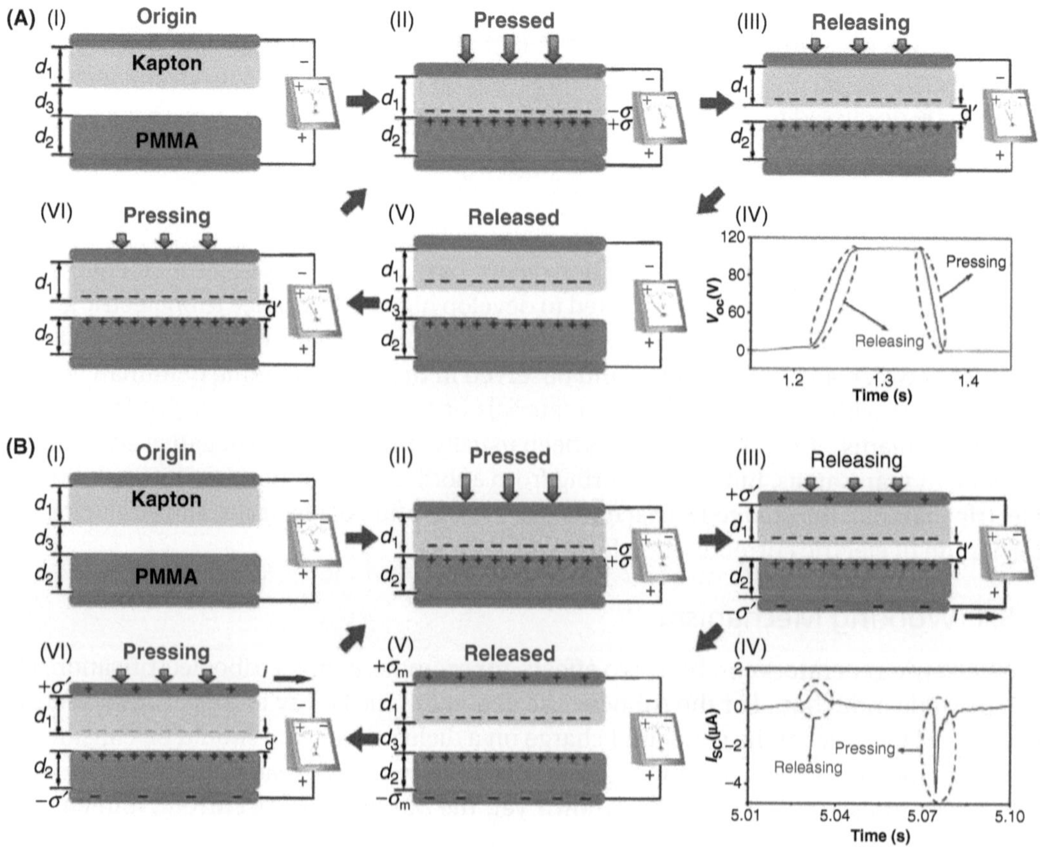

FIGURE 5.8 Working principle and structure of typical triboelectric nanogenerator. Structure design and application of triboelectric nanogenerators. *Reproduced with permission from reference Zhu, G., Pan, C., Guo, W., Chen, C.Y., Zhou, Y., Yu, R., Wang, Z.L., 2012. Triboelectric-generator-driven pulse electrodeposition for micropatterning. Nano Lett. 12, 4960–4965. Copyright 2012, American Chemical Society.*

and again. Driven by the external mechanical motions, the relative position of the triboelectrically charged surface will be changed periodically, which leads to the periodic variation of the induced potential difference between the two electrodes. The applied mechanical energy is converted to alternative electric current. For practical applications in daily electronic devices, a rectifier is used to transfer alternative current to direct current (Wang, 2013).

Based on the contact triboelectrification and electrostatic induction effect, a variety of generator structures have been designed for specific applications. Typically, the structures can be divided into four categories (Fig. 5.9) (Wang et al., 2015b; Wang, 2014), for example, contact mode structure (Fan et al., 2012a, Niu et al., 2013b), sliding mode structure (Niu et al., 2013a), single electrode mode structure (Li et al., 2015c) and freestanding triboelectric-layer mode structure (Niu et al., 2015). The theoretical models for these four modes are developed to guide the design and fabrication of high output triboelectric

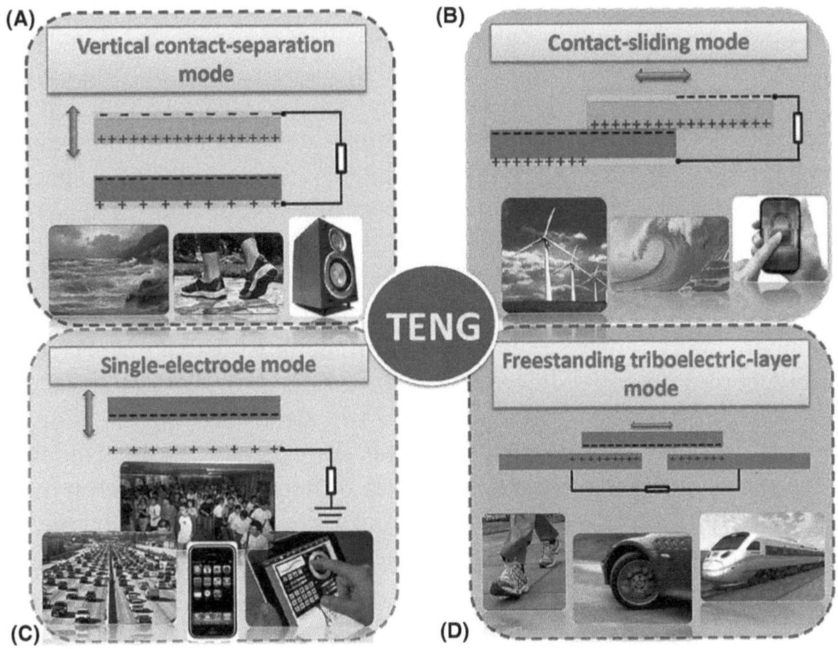

FIGURE 5.9 Structure design and application of triboelectric nanogenerators. *Reproduced with permission from referenceWang, Z.L., 2014. Triboelectric nanogenerators as new energy technology and self-powered sensors—principles, problems and perspectives. Farad. Discuss. 176, 447–458. Copyright 2014, Royal Society of Chemistry.*

generator. Besides the structure, the material selection, morphology, and thickness also influence the performance of generator.

The contact-separation mode is the first discovered mode of triboelectric generator with the simplest structure (Figs. 5.8 and 5.9A) (Fan et al., 2012a). In this mode, two films with different triboelectric property are assembled into a stacked configuration. In this mode, the electricity generation comes from the repeated movement of this two films from contacting to separating as depicted in Fig. 5.8. To achieve the contact and separation movement of two films and keep separated when they stand still, several structures, such as arch-shaped electrode structure (Zhong et al., 2013) and spring-supported structure (Kim et al., 2015a) have been explored. This triboelectric generator can be used for harnessing mechanical energy from mechanical vibrations.

The lateral sliding mode is similar to the contact-separation mode (Fig. 5.9B) (Guo et al., 2015; Xie et al., 2014). The electricity is also generated through the repeated contact and separation between the two polymer surfaces. However, the separation of two surface-charged materials take place in a shear movement rather than vertical movement. This shearing between the two surfaces first creates triboelectric charges on the two surfaces. When two surfaces are separated with each other after shearing motion, electron movement in attached metal electrodes will be induced. Through the periodical sliding, an alternate current output is generated (Niu et al., 2013a).

The previously two modes require two electrodes to be attached onto the two moving triboelectric layers and connecting to external circuit. It is known that a moving object is usually charged when it contacts with air or other objects. Thus moving objects can be used as the triboelectric layer to induce the movement of electrons in a still metal electrode. And single-electrode triboelectric generator is developed with only one electrode attached with the moving triboelectric layer (Fig. 5.9C) (Niu et al., 2015). A special application is to use this structure to harvesting the energy of rain drops (Lin et al., 2014).

Freestanding triboelectric-layer mode make use of single electrode mode by placing a charged polymer film in between two metal electrodes (Fig. 5.9D) (Wang et al., 2014c; Wang et al., 2015c). The polymer film is required to be very thin and flexible. The movement of the flexible polymer film brings it alternatively approaching the two metal electrodes, and a periodically reversed potential difference will be induced. Unlike single electrode mode where a reference electrode is required, there are two electrodes that are connected to an external circuit. The two electrodes can sit on the same plane with the triboelectric layer slide between them, or the triboelectric layer can be sandwiched between the two electrodes (Wang et al., 2014c). This design is potential for harvesting wind energy by flowing the freestanding charged polymer film (Yang et al., 2013; Wang et al., 2015c).

5.5.2 Materials

Theoretical models have been developed for contact separation mode, lateral sliding mode, single electrode mode, and freestanding triboelectric-layer mode (Niu et al., 2013a,b; Niu et al., 2015). These theory can serve as an important guidance for design of the high-performance triboelectric generator (Niu et al., 2013b). The most important theoretical equation for representing the real-time power generation of a triboelectric generator is related to the relationship among three parameters: the voltage (V) between the two electrodes, the amount of transferred charge (Q) in between (divided by time is current), and the separation distance (x) between the two triboelectric charged layers, which can be named as V–Q–x relationship. In an external circuit with different load resistances, the output voltage and current depend on the external load. Based on the analytical V–Q–x equation, dynamic output performance may be calculated with arbitrary external load resistance.

From the derived calculations for the four structures (Niu et al., 2013a,b; Niu et al., 2015) one can conclude that the transferred charge, current, and voltage are directly proportional to the triboelectric charged surface density (σ). The charge surface density σ affects the magnitude of I and V. The σ is mainly determined by the triboelectrification properties of the two materials and the roughness of the contact surfaces (Niu et al., 2013b).

5.5.2.1 Triboelectric Series

With a rational selection of materials, the charge surface density could be enhanced for higher output performance. The first triboelectric series on the ability of a material to gain or lose electrons was published by John Carl Wilcke. Such a series for some conventional materials was summarized previously (Davies, 1969). A material listed at the bottom of the series, upon touching a material near the top of the series, will attain a more negative charge by gaining electrons (Niu et al., 2013a). The larger the distance is between two materials at the series, the more the charge is transferred and the higher the charge density is produced. Typically, the used dielectric materials are Teflon (PTFE), Kapton (polyimide) and polydimethylsiloxane, which exhibit the most negative polarity in the series. For the application in biomedical field, biocompatibility should be taken into consideration. For example, ethyl cellulose and polylactic acid were used in triboelectric generators that have the potential of harvesting energy from biological activities to drive biomedical devices possibly in vivo (Sun et al., 2015b). However, the understanding of its fundamental mechanisms for charge transfer is rather poor, the proposed mechanisms remain doubts, and how to effectively control the density of the triboelectric surface charge through material or surface modification requires further investigation.

5.5.2.2 Morphology Control

Another factor that may enhance the performance of triboelectric generator is the surface area of triboelectrification materials, especially the dielectric layer. It is proposed that a higher surface area can preserve more charges. Generally, the voltage and current are enhanced by increasing the roughness of friction surface. Then comes to the morphology control of the triboelectric-layer surface. There are four methods to modify the surface morphology, for example, template method (soft lithography) (Fan et al., 2012b), self-assembling (Jeong et al., 2014), reactive ion etching (Yang et al., 2015a; Kim et al., 2015b), and electrospinning (Huang et al., 2015; Zhang et al., 2015a) (Fig. 5.10A–D).

The surfaces can be modified by template method which also named soft lithography technology with the creation of pyramid-, square-, or hemisphere-based micro- or nanopatterns, which are effective for enhancing the contact area and possibly the triboelectrification property (Fan et al., 2012b). It is relatively easy to synthesize nanostructured inorganic materials like anodization to growth aluminum oxide nanotube array or morphology-controlled silicon wafer. After the fabrication of nanostructured templates, a polymer solution is poured into the nanostructured surface, followed by curing and peeling off (Fig. 5.10A). A smooth surface of triboelectric layer produces an open circuit voltage of around 3 V. However, after the formation of nanostructured surface, the open-circuit voltage can be enhanced to around 20 V also with increased short-circuit current (Fan et al., 2012a,b).

The ion etching is a simple process to prepare aligned nanowire array on polymer surface (Kim et al., 2015b). This technology also benefits the enhancement of triboelectric device from the view point of surface area (Huang et al., 2015). A facile and robust route to nanoscale tunable triboelectric energy harvesters is realized by the formation of highly functional and controllable nanostructures via self-assembly of block copolymer. Similarly, a surface with nanodots, nanogrates, or nanomeshes significantly enhances the voltage, current, and output power compared with a smooth surface. The maximal output power can be up to 122.8 mW m^{-2} (Jeong et al., 2014).

However, the created bumpy structure on the surface may increase the friction force, which may possibly reduce the energy conversion efficiency of the triboelectric generator

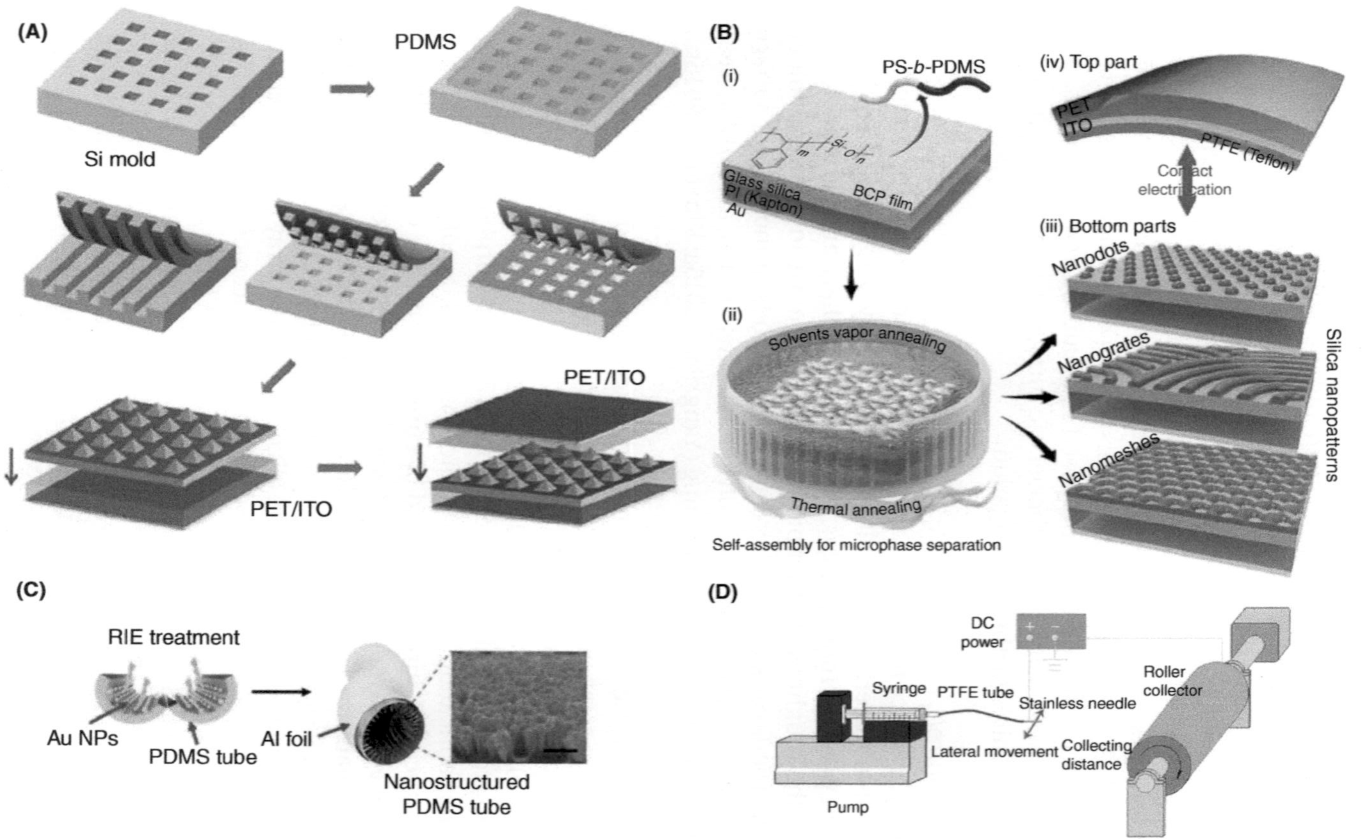

FIGURE 5.10 Morphology control in triboelectric nanogenerator using different methods. (A) Template. (B) Self-assembling methods. (C) Reactive ions etching *(RIE)* process. (D) Electrospinning. *(A) Reproduced with permission from reference Fan, F.R., Lin, L., Zhu, G., Wu, W., Zhang, R., Wang, Z.L., 2012b. Transparent triboelectric nanogenerators and self-powered pressure sensors based on micropatterned plastic films. Nano Lett., 12, 3109–3114. Copyright 2012, American Chemical Society. (B) Reproduced with permission from referenceJeong, C.K., Baek, K.M., Niu, S., Nam, T.W., Hur, Y.H., Park, D.Y., Hwang, G.T., Byun, M., Wang, Z.L., Jung, Y.S., Lee, K.J., 2014. Topographically-designed triboelectric nanogenerator via block copolymer self-assembly. Nano Lett. 14, 7031–7038. Copyright 2014, American Chemical Society. (C) Reproduced with permission from reference Kim, K.N., Chun, J., Kim, J.W., Lee, K.Y., Park, J.U., Kim, S.W., Wang, Z.L., Baik, J.M., 2015b. Highly stretchable 2D fabrics for wearable triboelectric nanogenerator under harsh environments. ACS Nano 9, 6394–6400. Copyright 2015, American Chemical Society. (D) Reproduced with permission from reference Huang, T., Wang, C., Yu, H., Wang, H., Zhang, Q., Zhu, M., 2015. Human walking-driven wearable all-fiber triboelectric nanogenerator containing electrospun polyvinylidene fluoride piezoelectric nanofibers. Nano Energy 14, 226–235. Copyright 2015, Elsevier Ltd.*

and decrease the stability and robustness of this generator (Li et al., 2015b). Therefore, an optimization to balance the surface charge density and friction force has to be met to maximize the conversion efficiency. To overcome the mentioned dilemma, a new structure of sponge-like triboelectric layer is developed to increase the inner surface area by forming holes inside the bulk materials (Lee et al., 2014). The increase in the inner surface increase the surface charge density, while a smooth surface could decrease the friction force.

5.5.2.3 Surface Charged Process

The third factor that might increase the surface charge density is to apply injection of ions (Zhang et al., 2014b; Wei et al., 2014b; Wang et al., 2014b). As mentioned before, the static surface charges are created by the triboelectrification process between the two different surfaces, in which the driving force is possibly related to the difference of the surface chemical potential. However, the maximum surface charge density is still unclear. Because a key process in the generator is electrostatic induction due to the presence of static surface charges, the performance enhancement can be obtained by the other methods in addition to the surface triboelectrification. Due to the intrinsic properties of dielectric materials of preserving charges (known as electret), a process of injecting single-polarity charged particles/ions onto the surface of an electret is introduced for enhancing the output of generator. Through adjusting the injection cycles, the surface charge density can be well controlled (Zhang et al., 2014b). However, the surface charge density is limited by the breakdown electric field of the air. This ion-injection method serves as an effective approach for gigantic elevation of the output power by as high as 25 times, which was proven to be stable over 5 months and 400,000 continuous operation cycles. The power output density reached the maximum value of $\sim$315 W m^{-2} at the resistance of 300 MΩ. Therefore, the ion injection is a very effective method for the improvement of power output. This power output is much higher than commercial silicon solar cells with power conversion efficiency around 20% under AM1.5 spectrum.

5.5.3 Application

Similar to the piezoelectric generator, a triboelectric generator is used to convert mechanical energy of a periodic physical process to an electric signal through the triboelectrification and electrostatic induction. There is a variety of sources, such as the movement of wind, water wave, periodical movement like tire rotation and people walking. A lot of efforts are made for the application of triboelectric generator to transfer environment mechanical energy to electricity (Niu et al., 2013a; Wang et al., 2015b; Zhu et al., 2015).

For instance, 70% of the earth's surface is covered with water and there is full of wave energy as environmentally benign forms of sustainable energy. A network design made of triboelectric generators for large-scale harvesting of kinetic water energy was reported (Chen et al., 2015a). Each generator network consists of four single generator units. These four units construct a closed railing with a ball inside. Along with the movement of water wave and making use of the collision of the rolling ball caused contact and separation of triboelectric polymer and metal electrode in generator unit, the generator network converts the slow, random, and high-force all-directional oscillatory motions into electricity. On the basis of the measured output of a single triboelectric generator, the triboelectric generators network is expected to give an average power output of 1.15 MW from 1 km^2 surface area (Chen et al., 2015a).

Wind energy is also could be transferred into electricity with special designed triboelectric generator. By utilizing the wind-induced resonance vibration of a fluorinated ethylene-propylene film between two aluminum foils, a wind energy based triboelectric generator is developed (Yang et al., 2013). With the periodical vibration of thin, light weight fluorinated ethylene-propylene film, a contact and separation mode is induced in between the two aluminum foils thus electricity generated. The output electricity is related to the wind velocity, which endows this generator dual functions as a sustainable power source by harvesting wind energy and as a self-powered wind velocity sensor system for wind speed and direction detection.

Human walking and body movement also generate abundant mechanical energy. This movement is appropriate source for triboelectric generator. Integrated into shoes is promising for harvesting human walking energy. Integrated into wearable textiles is promising for body movement energy collected. A patterned textile-based wearable triboelectric generator was developed with a high output voltage and current of about 120 V and 65 μA, respectively (Seung et al., 2015). The author successfully demonstrated the operation of light-emitting diodes, a liquid crystal display, and a keyless vehicle entry system only with the output power of textile-based wearable triboelectric generator without any help of external power sources.

Apart from energy harvesting application, this triboelectric generator could be applied in self-powered devices that required mechanical pressing as input. These devices include keyboard, touch screen, and remote control (Zhong et al., 2013). One intriguing application is a self-powered burglar alarm using transparent electrode materials and covered on precious paintings (Zhong et al., 2015). The system is based on an electrostatic induction mechanism with no extra power system. The self-powered generator is composed of two transparent components, which are the carbon nanotubes/transparent-paper and polyethylene/carbon nanotubes/transparent-paper. With a touching to this system an electric signal would be induced due to a contact and separation mode of the two electrodes.

5.6 Perspective

There is a huge amount energy in our environment. And the nature has its own laws of the transformation between the solar, thermal and vibration energy, and electricity, the most universal energy form used in daily life. Currently, only a part of the photovoltaic effect, thermoelectric effect, piezoelectric effect, and triboelectric effect has been investigated, and more efforts are required to carefully understand these effects to further optimize the device structure and enhance the device performance.

For the application, the active material plays a key role in the effectiveness of these devices. Polymer materials have been widely synthesized through solution or melting processes with low costs, and they can be designed to be lightweight and deformable. Therefore, the polymer materials are particularly promising for flexible and wearable electronics, the promising next-generation technology. To this end, more attentions should be paid to further enhance the device efficiency by modifying their structures. In particular, it is necessary to increase the stability of the polymer-based devices as polymer materials are well known to be less stable compared with their inorganic counterparts.

So far the harvesting energy devices typically appear in a planar structure, and they have been made to be thin to meet the flexible requirement in the modern electronics. However, although these thin-film devices can be bent into curved structures, it is difficult for them to be twisted three-dimensionally. On the other hand, a twisting deformation is needed for a variety of flexible electronic products. To this end, a new family of one-dimensional fiber-shaped harvesting energy devices is proposed and attracts increasing interests in recent years. These harvesting energy fibers may be twisted into various forms without obvious damages in structure. In addition, they can be further woven into flexible electronic textiles and are promising for wearable applications as well as many new emerging fields. However, the energy conversion efficiencies of such fiber-shaped harvesting energy devices are low, and it is also challenging to produce them at an application scale. The low electrical conductivity of the used fiber electrode is demonstrated to be the main limitation on the earlier directions. A lot of efforts should be made to develop high-performance conducting fibers and fabrication methods of the fiber device in the future.

References

Anno, H., Nishinaka, T., Hokazono, M., Oshima, N., Toshima, N., 2015. Thermoelectric power-generation characteristics of PEDOT:PSS thin-film devices with different thicknesses on polyimide substrates. J. Electron. Mater. 44, 2105–2112.

Anton, S.R., Sodano, H.A., 2007. A review of power harvesting using piezoelectric materials (2003–2006). Smart Mater. Struct. 16, R1–R21.

Bauer, S., 1996. Poled polymers for sensors and photonic applications. J. Appl. Phys. 80, 5531–5558.

Bella, F., 2015. Polymer electrolytes and perovskites: lights and shadows in photovoltaic devices. Electrochim. Acta 175, 151–161.

Bhavanasi, V., Kusuma, D.Y., Lee, P.S., 2014. Polarization orientation, piezoelectricity, and energy harvesting performance of ferroelectric PVDF-TrFE nanotubes synthesized by nanoconfinement. Adv. Energy Mater. 4, 1400723.

Bowen, C.R., Kim, H.A., Weaver, P.M., Dunn, S., 2014. Piezoelectric and ferroelectric materials and structures for energy harvesting applications. Energy Environ. Sci. 7, 25–44.

Brabec, C.J., Cravino, A., Meissner, D., Sariciftci, N.S., Fromherz, T., Rispens, M.T., SAnchez, L., Hummelen, J.C., 2001. Origin of the open circuit voltage of plastic solar cells. Adv. Funct. Mater. 11, 374–380.

Cha, S., Kim, S.M., Kim, H., Ku, J., Sohn, J.I., Park, Y.J., Song, B.G., Jung, M.H., Lee, E.K., Choi, B.L., Park, J.J., Wang, Z.L., Kim, J.M., Kim, K., 2011. Porous PVDF as effective sonic wave driven nanogenerators. Nano Lett. 11, 5142–5147.

Chang, C., Tran, V.H., Wang, J., Fuh, Y.K., Lin, L., 2010. Direct-write piezoelectric polymeric nanogenerator with high energy conversion efficiency. Nano Lett. 10, 726–731.

Chang, J., Dommer, M., Chang, C., Lin, L., 2012. Piezoelectric nanofibers for energy scavenging applications. Nano Energy 1, 356–371.

Chen, D., Zhang, J.X.J., 2015. Microporous polyvinylidene fluoride film with dense surface enables efficient piezoelectric conversion. Appl. Phys. Lett. 106, 193901.

Chen, J., Yang, J., Li, Z., Fan, X., Zi, Y., Jing, Q., Guo, H., Wen, Z., Pradel, K.C., Niu, S., Wang, Z.L., 2015a. Networks of triboelectric nanogenerators for harvesting water wave energy: a potential approach toward blue energy. ACS Nano 9, 3324–3331.

Chen, X., Tian, H., Li, X., Shao, J., Ding, Y., An, N., Zhou, Y., 2015b. A high performance P(VDF-TrFE) nanogenerator with self-connected and vertically integrated fibers by patterned EHD pulling. Nanoscale 7, 11536–11544.

Chen, Y., Zhao, Y., Liang, Z., 2015c. Solution processed organic thermoelectrics: towards flexible thermoelectric modules. Energy Environ. Sci. 8, 401–422.

Coakley, K.M., McGehee, M.D., 2004. Conjugated polymer photovoltaic cells. Chem. Mater. 16, 4533–4542.

Collavini, S., Volker, S.F., Delgado, J.L., 2015. Understanding the outstanding power conversion efficiency of perovskite-based solar cells. Angew. Chem. Int. Ed. 54, 9757–9759.

Cook, T.R., Dogutan, D.K., Reece, S.Y., Surendranath, Y., Teets, T.S., Nocera, D.G., 2010. Solar energy supply and storage for the legacy and nonlegacy worlds. Chem. Rev. 110, 6474–6502.

Cook-Chennault, K.A., Thambi, N., Sastry, A.M., 2008. Powering MEMS portable devices—a review of non-regenerative and regenerative power supply systems with special emphasis on piezoelectric energy harvesting systems. Smart Mater. Struct. 17, 043001.

Crossley, S., Whiter, R.A., Kar-Narayan, S., 2014. Polymer-based nanopiezoelectric generators for energy harvesting applications. Mater. Sci. Technol. 30, 1613–1624.

Culebras, M., GóMez, C.M. S Cantarero, A., 2014. Review on polymers for thermoelectric applications. Materials 7, 6701–6732.

Culebras, M., Uriol, B., Gomez, C.M., Cantarero, A., 2015. Controlling the thermoelectric properties of polymers: application to PEDOT and polypyrrole. Phys. Chem. Chem. Phys. 17, 15140–15145.

Davies, D.K., 1969. Charge generation on dielectric surfaces. J. Phys. D-Appl. Phys. 2, 1533–1537.

Dewan, A., Ay, S.U., Karim, M.N., Beyenal, H., 2014. Alternative power sources for remote sensors: a review. J. Power Sources 245, 129–143.

Dillon, A.C., 2010. Carbon nanotubes for photoconversion and electrical energy storage. Chem. Rev. 110, 6856–6872.

Du, Y., Shen, S.Z., Yang, W., Donelson, R., Cai, K., Casey, P.S., 2012. Simultaneous increase in conductivity and Seebeck coefficient in a polyaniline/graphene nanosheets thermoelectric nanocomposite. Synth. Met. 161, 2688–2692.

Dun, C., Hewitt, C.A., Huang, H., Xu, J., Montgomery, D.S., Nie, W., Jiang, Q., Carroll, D.L., 2015. Layered Bi_2Se_3 nanoplate/polyvinylidene fluoride composite based n-type thermoelectric fabrics. ACS Appl. Mater. Interfaces 7, 7054–7059.

Earmme, T., Hwang, Y.-J., Murari, N.M., Subramaniyan, S., Jenekhe, S.A., 2013. All-polymer solar cells with 3.3% efficiency based on naphthalene diimide-selenophene copolymer acceptor. J. Am. Chem. Soc. 135, 14960–14963.

Endrődi, B., Mellár, J.S., Gingl, Z., Visy, C., Janáky, C., 2014. Reasons behind the improved thermoelectric properties of poly(3-hexylthiophene) nanofiber networks. RSC Adv. 4, 55328–55333.

Endrődi, B., Mellár, J., Gingl, Z., Visy, C., Janáky, C., 2015. Molecular and supramolecular parameters dictating the thermoelectric performance of conducting polymers: a case study using poly(3-alkylthiophene)s. J. Phys. Chem. C 119, 8472–8479.

Facchetti, A., 2013. Polymer donor–polymer acceptor (all-polymer) solar cells. Mater. Today 16, 123–132.

Fan, F.-R., Tian, Z.-Q., Lin Wang, Z., 2012a. Flexible triboelectric generator. Nano Energy 1, 328–334.

Fan, F.R., Lin, L., Zhu, G., Wu, W., Zhang, R., Wang, Z.L., 2012b. Transparent triboelectric nanogenerators and self-powered pressure sensors based on micropatterned plastic films. Nano Lett. 12, 3109–3114.

Furukawa, T., 1989. Ferroelectric properties of vinylidene fluoride copolymers. Phase Transit. 18, 143–211.

Guan, G., Yang, Z., Qiu, L., Sun, X., Zhang, Z., Ren, J., Peng, H., 2013. Oriented PEDOT:PSS on aligned carbon nanotubes for efficient dye-sensitized solar cells. J. Mater. Chem. A 1, 13268–13273.

Guo, X., Zhou, N., Lou, S.J., Smith, J., Tice, D.B., Hennek, J.W., Ortiz, R.P., Navarrete, J.T.L., Li, S., Strzalka, J., Chen, L.X., Chang, R.P.H., Facchetti, A., Marks, T.J., 2013. Polymer solar cells with enhanced fill factors. Nat. Photon. 7, 825–833.

Guo, H., Leng, Q., He, X., Wang, M., Chen, J., Hu, C., Xi, Y., 2015. A Triboelectric generator based on checker-like interdigital electrodes with a sandwiched PET Thin film for harvesting sliding energy in all directions. Adv. Energy Mater. 5, 1400790.

Halls, J.J.M., Walsh, C.A., Greenham, N.C., Marseglia, E.A., Friend, R.H., Moratti, S.C., Holmes, A.B., 1995. Efficient photodiodes from interpenetrating polymer networks. Nature 376, 498–500.

Hattori, T., Takahashi, Y., Iijima, M., Fukada, E., 1996. Piezoelectric and ferroelectric properties of polyurea-5 thin films prepared by vapor deposition polymerization. J. Appl. Phys. 79, 1713–1721.

Heo, J.H., Im, S.H., Noh, J.H., Mandal, T.N., Lim, C.-S., Chang, J.A., Lee, Y.H., Kim, H.-J., Sarkar, A., Nazeeruddin, M.K., Grätzel, M., Seok, S.I., 2013. Efficient inorganic–organic hybrid heterojunction solar cells containing perovskite compound and polymeric hole conductors. Nat. Photon. 7, 486–491.

Hiraishi, K., Masuhara, A., Nakanishi, H., Oikawa, H., Shinohara, Y., 2009. Evaluation of thermoelectric properties of polythiophene films synthesized by electrolytic polymerization. Jpn. J. Appl. Phys. 48, 071501.

Huang, Y., Kramer, E.J., Heeger, A.J., Bazan, G.C., 2014. Bulk heterojunction solar cells: morphology and performance relationships. Chem. Rev. 114, 7006–7043.

Huang, T., Wang, C., Yu, H., Wang, H., Zhang, Q., Zhu, M., 2015. Human walking-driven wearable all-fiber triboelectric nanogenerator containing electrospun polyvinylidene fluoride piezoelectric nanofibers. Nano Energy 14, 226–235.

Hwang, Y.J., Earmme, T., Courtright, B.A., Eberle, F.N., Jenekhe, S.A., 2015. n-type semiconducting naphthalene diimide-perylene diimide copolymers: controlling crystallinity, blend morphology, and compatibility toward high-performance all-polymer solar cells. J. Am. Chem. Soc. 137, 4424–4434.

Ihnatsenka, S., Crispin, X., Zozoulenko, I.V., 2015. Understanding hopping transport and thermoelectric properties of conducting polymers. Phys. Rev. B 92, 035201.

Jeong, C.K., Baek, K.M., Niu, S., Nam, T.W., Hur, Y.H., Park, D.Y., Hwang, G.T., Byun, M., Wang, Z.L., Jung, Y.S., Lee, K.J., 2014. Topographically-designed triboelectric nanogenerator via block copolymer self-assembly. Nano Lett. 14, 7031–7038.

Jorgensen, M., Norrman, K., Gevorgyan, S.A., Tromholt, T., Andreasen, B., Krebs, F.C., 2012. Stability of polymer solar cells. Adv. Mater. 24, 580–612.

Jung, J.W., Jo, J.W., Chueh, C.-C., Liu, F, Jo, W.H., Russell, T.P., Jen, A.K., 2015. Fluoro-substituted n-type conjugated polymers for additive-free all-polymer bulk heterojunction solar cells with high power conversion efficiency of 6.71%. Adv. Mater. 27, 3310–3317.

Kang, H., Uddin, M.A., Lee, C., Kim, K.-H., Nguyen, T.L., Lee, W., Li, Y., Wang, C., Woo, H.Y., Kim, B.J., 2015. Determining the role of polymer molecular weight for high performance all-polymer solar cells: its effect on polymer aggregation and phase separation. J. Am. Chem. Soc. 137, 2359–2365.

Kim, D., Kim, Y., Choi, K., Grunlan, J.C., Yu, C., 2010. Improved thermoelectric behavior of nanotube-filled polymer composites with poly(3,4-ethylenedioxythiophene) poly(styrenesulfonate). ACS Nano 4, 513–523.

Kim, G.H., Shao, L., Zhang, K., Pipe, K.P., 2013. Engineered doping of organic semiconductors for enhanced thermoelectric efficiency. Nat. Mater. 12, 719–723.

Kim, D., Jeon, S.-B., Kim, J.Y., Seol, M.-L., Kim, S.O., Choi, Y.-K., 2015a. High-performance nanopattern triboelectric generator by block copolymer lithography. Nano Energy 12, 331–338.

Kim, K.N., Chun, J., Kim, J.W., Lee, K.Y., Park, J.U., Kim, S.W., Wang, Z.L., Baik, J.M., 2015b. Highly stretchable 2D fabrics for wearable triboelectric nanogenerator under harsh environments. ACS Nano 9, 6394–6400.

Kurniawan, M., Salim, T., Tai, K.F., Sun, S., Sie, E.J., Wu, X., Yeow, E.K.L., Huan, C.H.A., Lam, Y.M., Sum, T.C., 2012. Carrier dynamics in polymer nanofiber: fullerene solar cells. J. Phys. Chem. C 116, 18015–18022.

Lau, K., Liu, Y., Chen, H., Withers, R.L., 2013. Effect of annealing temperature on the morphology and piezoresponse characterisation of poly(vinylidene fluoride-trifluoroethylene) films via scanning probe microscopy. Adv. Cond. Matter Phys. 2013, 1–5.

Lee, T.H., Do, K., Lee, Y.W., Jeon, S.S., Kim, C., Ko, J., Im, S.S., 2012. High-performance dye-sensitized solar cells based on PEDOT nanofibers as an efficient catalytic counter electrode. J. Mater. Chem. 22, 21624–21629.

Lee, K.Y., Chun, J., Lee, J.H., Kim, K.N., Kang, N.R., Kim, J.Y., Kim, M.H., Shin, K.S., Gupta, M.K., Baik, J.M., Kim, S.W., 2014. Hydrophobic sponge structure-based triboelectric nanogenerator. Adv. Mater. 26, 5037–5042.

Lee, C., Kang, H., Lee, W., Kim, T., Kim, K.-H., Woo, H.Y., Wang, C., Kim, B.J., 2015. High-performance all-polymer solar cells via side-chain engineering of the polymer acceptor the importance of the polymer packing structure and the nanoscale blend morphology. Adv. Mater. 27, 2466–2471.

Li, Y., 2012. Molecular design of photovoltaic materials for polymer solar cells: toward suitable electronic energy levels and broad absorption. Acc. Chem. Res. 45, 723–733.

Li, G., Shrotriya, V., Huang, J., Yao, Y., Moriarty, T., Emery, K., Yang, Y., 2005. High-efficiency solution processable polymer photovoltaic cells by self-organization of polymer blends. Nat. Mater. 4, 864–868.

Li, G., Zhu, R., Yang, Y., 2012. Polymer solar cells. Nat. Photon. 6, 153–161.

Li, G., Zhao, B., Kang, C., Lu, Z., Li, C., Dong, H., Hu, W., Wu, H., Bo, Z., 2015a. Side chain influence on the morphology and photovoltaic performance of 5-fluoro-6-alkyloxybenzothiadiazole and benzodithiophene based conjugated polymers. ACS Appl. Mater. Interfaces 7, 10710–10717.

Li, S., Wang, S., Zi, Y., Wen, Z., Lin, L., Zhang, G., Wang, Z.L., 2015b. Largely improving the robustness and lifetime of triboelectric nanogenerators through automatic transition between contact and noncontact working states. ACS Nano 9, 7479–7487.

Li, Y., Cheng, G., Lin, Z.-H., Yang, J., Lin, L., Wang, Z.L., 2015c. Single-electrode-based rotationary triboelectric nanogenerator and its applications as self-powered contact area and eccentric angle sensors. Nano Energy 11, 323–332.

Liew, W.H., Mirshekarloo, M.S., Chen, S., Yao, K., Tay, F.E., 2015. Nanoconfinement induced crystal orientation and large piezoelectric coefficient in vertically aligned P(VDF-TrFE) nanotube array. Sci. Rep. 5, 9790.

Lin, Z.H., Cheng, G., Lee, S., Pradel, K.C., Wang, Z.L., 2014. Harvesting water drop energy by a sequential contact-electrification and electrostatic-induction process. Adv. Mater. 26, 4690–4696.

Lu, B., Chen, S., Xu, J., Zhao, G., 2013. Thermoelectric performances of different types of polyselenophene and its copolymers with 3-methylthiophene via electropolymerization. Synth. Met. 183, 8–15.

Mai, C.-K., Schlitz, R.A., Su, G.M., Spitzer, D., Wang, X., Fronk, S.L., Cahill, D.G., CHABINYC, M.L., BAZAN, G.C., 2014. Side-chain effects on the conductivity, morphology, and thermoelectric properties of self-doped narrow-band-gap conjugated polyelectrolytes. J. Am. Chem. Soc. 136, 13478–13481.

Major, J.D., Treharne, R.E., Phillips, L.J., Durose, K., 2014. A low-cost non-toxic post-growth activation step for CdTe solar cells. Nature 511, 334–337.

Malinkiewicz, O., Yella, A., Lee, Y.H., Espallargas, G.M., Graetzel, M., Nazeeruddin, M.K., Bolink, H.J., 2013. Perovskite solar cells employing organic charge-transport layers. Nat. Photon. 8, 128–132.

Mandal, D., Yoon, S., Kim, K.J., 2011. Origin of piezoelectricity in an electrospun poly(vinylidene fluoride-trifluoroethylene) nanofiber web-based nanogenerator and nano-pressure sensor. Macromol. Rapid Commun. 32, 831–837.

Mao, Y., Zhao, P., Mcconohy, G., Yang, H., Tong, Y., Wang, X., 2014. Sponge-like piezoelectric polymer films for scalable and integratable nanogenerators and self-powered electronic systems. Adv. Energy Mater. 4, 1301624.

McGrail, B.T., Sehirlioglu, A., Pentzer, E., 2015. Polymer composites for thermoelectric applications. Angew. Chem. Int. Ed. 54, 1710–1723.

Mengistie, D.A., Chen, C.H., Boopathi, K.M., Pranoto, F.W., Li, L.J., Chu, C.W., 2015. Enhanced thermoelectric performance of PEDOT:PSS flexible bulky papers by treatment with secondary dopants. ACS Appl. Mater. Interfaces 7, 94–100.

Menke, T., Ray, D., Meiss, J., Leo, K., Riede, M., 2012. In-situ conductivity and Seebeck measurements of highly efficient n-dopants in fullerene C60. Appl. Phys. Lett. 100, 093304.

Nasybulin, E., Wei, S., Cox, M., Kymissis, I., Levon, K., 2011. Morphological and spectroscopic studies of electrochemically deposited poly(3,4-ethylenedioxythiophene) (PEDOT) hole extraction layer for organic photovoltaic device (OPVd) fabrication. J. Phys. Chem. C 115, 4307–4314.

Newman, B.A., Chen, P., Pae, K.D., Scheinbeim, J.I., 1980. Piezoelectricity in nylon 11. J. Appl. Phys. 51, 5161–5164.

Nishinaga, S., Mori, H., Nishihara, Y., 2015. Phenanthrodithiophene–isoindigo copolymers: effect of side chains on their molecular order and solar cell performance. Macromolecules 48, 2875–2885.

Niu, S., Liu, Y., Wang, S., Lin, L., Zhou, Y.S., Hu, Y., Wang, Z.L., 2013a. Theory of sliding-mode triboelectric nanogenerators. Adv. Mater. 25, 6184–6193.

Niu, S., Wang, S., Lin, L., Liu, Y., Zhou, Y.S., Hu, Y., Wang, Z.L., 2013b. Theoretical study of contact-mode triboelectric nanogenerators as an effective power source. Energy Environ. Sci. 6, 3576–3583.

Niu, S., Liu, Y., Chen, X., Wang, S., Zhou, Y.S., Lin, L., Xie, Y., Wang, Z.L., 2015. Theory of freestanding triboelectric-layer-based nanogenerators. Nano Energy 12, 760–774.

Nogueira, A.F., Longo, C., De Paoli, M.A., 2004. Polymers in dye sensitized solar cells: overview and perspectives. Coord. Chem. Rev. 248, 1455–1468.

Nunes-Pereira, J., Sencadas, V., Correia, V., Rocha, J.G., Lanceros-Méndez, S., 2013. Energy harvesting performance of piezoelectric electrospun polymer fibers and polymer/ceramic composites. Sensors Actuat. A: Phys. 196, 55–62.

Park, C., Ounaies, Z., Wise, K.E., Harrison, J.S., 2004. In situ poling and imidization of amorphous piezoelectric polyimides. Polymer 45, 5417–5425.

Persano, L., Dagdeviren, C., Su, Y., Zhang, Y., Girardo, S., Pisignano, D., Huang, Y., Rogers, J.A., 2013. High performance piezoelectric devices based on aligned arrays of nanofibers of poly(vinylidenefluoride-*co*-trifluoroethylene). Nat. Commun. 4, 1633.

Pi, Z., Zhang, J., Wen, C., Zhang, Z.-B., Wu, D., 2014. Flexible piezoelectric nanogenerator made of poly(vinylidenefluoride-*co*-trifluoroethylene) (PVDF-TrFE) thin film. Nano Energy 7, 33–41.

Ramadan, K.S., Sameoto, D., Evoy, S., 2014. A review of piezoelectric polymers as functional materials for electromechanical transducers. Smart Mater. Struct. 23, 033001.

Russ, B., Robb, M.J., Brunetti, F.G., Miller, P.L., Perry, E.E., Patel, S.N., Ho, V., Chang, W.B., Urban, J.J., Chabinyc, M.L., Hawker, C.J., Segalman, R.A., 2014. Power factor enhancement in solution-processed organic n-type thermoelectrics through molecular design. Adv. Mater. 26, 3473–3477.

Saranya, K., Rameez, M., Subramania, A., 2015. Developments in conducting polymer based counter electrodes for dye-sensitized solar cells–an overview. Eur. Polym. J. 66, 207–227.

Savin, H., Repo, P., Von Gastrow, G., Ortega, P., Calle, E., Garin, M., Alcubilla, R., 2015. Black silicon solar cells with interdigitated back-contacts achieve 22.1% efficiency. Nat. Nanotechnol. 10, 624–628.

Scharber, M.C.C., Mühlbacher, D., Koppe, M., Denk, P., Waldauf, C., Heeger, A.J., Brabec, C.J., 2006. Design rules for donors in bulk-heterojunction solar cells—towards 10% energy-conversion efficiency. Adv. Mater. 18, 789–794.

Schlitz, R.A., Brunetti, F.G., Glaudell, A.M., Miller, P.L., Brady, M.A., Takacs, C.J., Hawker, C.J., Chabinyc, M.L., 2014. Solubility-limited extrinsic n-type doping of a high electron mobility polymer for thermoelectric applications. Adv. Mater. 26, 2825–2830.

Schmidt-Mende, L., GräTzel, M., 2006. TiO_2 pore-filling and its effect on the efficiency of solid-state dye-sensitized solar cells. Thin Solid Films 500, 296–301.

Seung, W., Gupta, M.K., Lee, K.Y., Shin, K.S., Lee, J.H., Kim, T.Y., Kim, S., Lin, J., Kim, J.H., Kim, S.W., 2015. Nanopatterned textile-based wearable triboelectric nanogenerator. ACS Nano 9, 3501–3509.

Shi, K., Zhang, F., Di, C.A., Yan, T.W., Zou, Y., Zhou, X., Zhu, D., Wang, J.Y., Pei, J., 2015. Toward high performance n-type thermoelectric materials by rational modification of BDPPV backbones. J. Am. Chem. Soc. 137, 6979–6982.

Shin, S.H., Kim, Y.H., Jung, J.Y., Lee, M.H., Nah, J., 2014. Solvent-assisted optimal $BaTiO_3$ nanoparticles-polymer composite cluster formation for high performance piezoelectric nanogenerators. Nanotechnology 25, 485401.

Shu, Y.C., Lien, I.C., 2006. Analysis of power output for piezoelectric energy harvesting systems. Smart Mater. Struct. 15, 1499–1512.

Smestad, G.P., Krebs, F.C., Lampert, C.M., Granqvist, C.G., Chopra, K.L., Mathew, X., Takakura, H., 2008. Reporting solar cell efficiencies in solar energy materials and solar cells. Sol. Energy Mater. Sol. Cells 92, 371–373.

Solanki, A., Wu, B., Salim, T., Yeow, E.K.L., Lam, Y.M., Sum, T.C., 2014. Performance improvements in polymer nanofiber/fullerene solar cells with external electric field treatment. J. Phys. Chem. C 118, 11285–11291.

Sun, J., Yeh, M.L., Jung, B.J., Zhang, B., Feser, J., Majumdar, A., Katz, H.E., 2010. Simultaneous increase in Seebeck coefficient and conductivity in a doped poly(alkylthiophene) blend with defined density of states. Macromolecules 43, 2897–2903.

Sun, Y., Sheng, P., Di, C., Jiao, F., Xu, W., Qiu, D., Zhu, D., 2012. Organic thermoelectric materials and devices based on p- and n-type poly(metal 1,1,2,2-ethenetetrathiolate)s. Adv. Mater. 24, 932–937.

Sun, X., Sun, H., Li, H., Peng, H., 2013. Developing polymer composite materials: carbon nanotubes or graphene? Adv. Mater. 25, 5153–5176.

Sun, H., Deng, J., Qiu, L., Fang, X., Peng, H., 2015a. Recent progress in solar cells based on one-dimensional nanomaterials. Energy Environ. Sci. 8, 1139–1159.

Sun, J., Li, W., Liu, G., Li, W., Chen, M., 2015b. Triboelectric nanogenerator based on biocompatible polymer materials. J. Phys. Chem. C 119, 9061–9068.

Suraru, S.-L., Würthner, F., 2014. Strategies for the synthesis of functional naphthalene diimides. Angew. Chem. Int. Ed. 53, 7428–7448.

Svensson, M., Zhang, F.L., Veenstra, S.C., Verhees, W.J.H., Hummelen, J.C., Kroon, J.M., Inganas, O., Andersson, M.R., 2003. High-performance polymer solar cells of an alternating polyfluorene copolymer and a fullerene derivative. Adv. Mater. 15, 988–991.

Taggart, D.K., Yang, Y., Kung, S.C., Mcintire, T.M., Penner, R.M., 2011. Enhanced thermoelectric metrics in ultra-long electrodeposited PEDOT nanowires. Nano Lett. 11, 125–131.

Tang, C.W., 1986. Two-layer organic photovoltaic cell. Appl. Phys. Lett. 48, 183–185.

Taroni, P.J., Hoces, I., Stingelin, N., Heeney, M., Bilotti, E., 2014. Thermoelectric materials: a brief historical survey from metal junctions and inorganic semiconductors to organic polymers. Israel J. Chem. 54, 534–552.

Taylor, G.W., Burns, J.R., Kammann, S.M., Powers, W.B., Welsh, T.R., 2001. The energy harvesting eel: a small subsurface ocean/river power generator. IEEE J. Oceanic Eng. 26, 539–547.

Verstappen, P., Kesters, J., D'Olieslaeger, L., Drijkoningen, J., Cardinaletti, I., Vangerven, T., Bruijnaers, B.J., Willems, R.E.M., D'Haen, J., Manca, J.V., Lutsen, L., Vanderzande, D.J.M., Maes, W., 2015. Simultaneous enhancement of solar cell efficiency and stability by reducing the side chain density on fluorinated PCPDTQx copolymers. Macromolecules 48, 3873–3882.

Wan, C., Gu, X., Dang, F., Itoh, T., Wang, Y., Sasaki, H., Kondo, M., Koga, K., Yabuki, K., Snyder, G.J., Yang, R., Koumoto, K., 2015. Flexible n-type thermoelectric materials by organic intercalation of layered transition metal dichalcogenide TiS_2. Nat. Mater. 14, 622–627.

Wang, Z.L., 2013. Triboelectric nanogenerators as new energy technology for self-powered systems and as active mechanical and chemical sensors. ACS Nano 7, 9533–9557.

Wang, Z.L., 2014. Triboelectric nanogenerators as new energy technology and self-powered sensors—principles, problems and perspectives. Farad. Discuss. 176, 447–458.

Wang, Z.L., Song, J., 2006. Piezoelectric nanogenerators based on zinc oxide nanowire arrays. Science 312, 242–246.

Wang, S., Lin, L., Wang, Z.L., 2012. Nanoscale triboelectric-effect-enabled energy conversion for sustainably powering portable electronics. Nano Lett. 12, 6339–6346.

Wang, J., Cai, K., Shen, S., 2014a. Enhanced thermoelectric properties of poly(3,4-ethylenedioxythiophene) thin films treated with H_2SO_4. Org. Electron. 15, 3087–3095.

Wang, S., Xie, Y., Niu, S., Lin, L., Liu, C., Zhou, Y.S., Wang, Z.L., 2014b. Maximum surface charge density for triboelectric nanogenerators achieved by ionized-air injection: methodology and theoretical understanding. Adv. Mater. 26, 6720–6728.

Wang, S., Xie, Y., Niu, S., Lin, L., Wang, Z.L., 2014c. Freestanding triboelectric-layer-based nanogenerators for harvesting energy from a moving object or human motion in contact and non-contact modes. Adv. Mater. 26, 2818–2824.

Wang, H., Ail, U., Gabrielsson, R., Berggren, M., Crispin, X., 2015a. Ionic Seebeck effect in conducting polymers. Adv. Energy Mater. 5, 1500044.

Wang, S., Lin, L., Wang, Z.L., 2015b. Triboelectric nanogenerators as self-powered active sensors. Nano Energy 11, 436–462.

Wang, X., Wang, S., Yang, Y., Wang, Z.L., 2015c. Hybridized electromagnetic-triboelectric nanogenerator for scavenging air-flow energy to sustainably power temperature sensors. ACS Nano 9, 4553–4562.

Wei, Q., Mukaida, M., Kirihara, K., Naitoh, Y., Ishida, T., 2014a. Polymer thermoelectric modules screen-printed on paper. RSC Adv. 4, 28802–28806.

Wei, X.Y., Zhu, G., Wang, Z.L., 2014b. Surface-charge engineering for high-performance triboelectric nanogenerator based on identical electrification materials. Nano Energy 10, 83–89.

Wei, Q., Mukaida, M., Kirihara, K., Naitoh, Y., Ishida, T., 2015. Recent progress on PEDOT-based thermoelectric materials. Materials 8, 732–750.

Whiter, R.A., Narayan, V., Kar-Narayan, S., 2014. A scalable nanogenerator based on self-poled piezoelectric polymer nanowires with high energy conversion efficiency. Adv. Energy Mater. 4, 1400519.

Wienk, M.M., Kroon, J.M., Verhees, W.J.H., Knol, J., Hummelen, J.C., Van Hal, P.A., Janssen, R.A.J., 2003. Efficient methano[70]fullerene/MDMO-PPV bulk heterojunction photovoltaic cells. Angew. Chem. Int. Ed. 42, 3371–3375.

Williams, M.W., 2012. Triboelectric charging of insulating polymers—some new perspectives. AIP Adv. 2, 010701.

Wu, J.S., Cheng, S.W., Cheng, Y.J., Hsu, C.S., 2015. Donor–acceptor conjugated polymers based on multifused ladder-type arenes for organic solar cells. Chem. Soc. Rev. 44, 1113–1154.

Xie, Y., Wang, S., Niu, S., Lin, L., Jing, Q., Yang, J., Wu, Z., Wang, Z.L., 2014. Grating-structured freestanding triboelectric-layer nanogenerator for harvesting mechanical energy at 85% total conversion efficiency. Adv. Mater. 26, 6599–6607.

Xu, K., Chen, G., Qiu, D., 2015. In situ chemical oxidative polymerization preparation of poly(3,4-ethylenedioxythiophene)/graphene nanocomposites with enhanced thermoelectric performance. Chem. Asian J. 10, 1225–1231.

Yan, J., Saunders, B.R., 2014. Third-generation solar cells: a review and comparison of polymer: fullerene, hybrid polymer and perovskite solar cells. RSC Adv. 4, 43286–43314.

Yang, Y., Zhu, G., Zhang, H.L., Chen, J., Zhong, X.D., Lin, Z.H., Su, Y.J., Bai, P., Wen, X.N., Wang, Z.L., 2013. Triboelectric nanogenerator for harvesting wind energy and as self-powered wind vector sensor system. ACS Nano 7, 9461–9468.

Yang, Y., Zhang, H., Zhong, X., Yi, F., Yu, R., Zhang, Y., Wang, Z.L., 2014. Electret film-enhanced triboelectric nanogenerator matrix for self-powered instantaneous tactile imaging. ACS Appl. Mater. Interfaces 6, 3680–3688.

Yang, P.K., Lin, L., Yi, F., Li, X., Pradel, K.C., Zi, Y., Wu, C.I., He, J.H., Zhang, Y., Wang, Z.L., 2015a. A flexible, stretchable and shape-adaptive approach for versatile energy conversion and self-powered biomedical monitoring. Adv. Mater. 27, 3817–3824.

Yang, W.S., Noh, J.H., Jeon, N.J., Kim, Y.C., Ryu, S., Seo, J., Seok, S.I., 2015b. High-performance photovoltaic perovskite layers fabricated through intramolecular exchange. Science 348, 1234–1237.

Yu, G., Gao, J., Hummelen, J.C., Wudl, F., Heeger, A.J., 1995. Polymer photovoltaic cells enhanced efficiencies via a network of internal donor acceptor heterojunctions. Science 270, 1789–1791.

Yu, C., Kim, Y.S., Kim, D., Grunlan, J.C., 2008. Thermoelectric behavior of segregated-network polymer nanocomposites. Nano Lett. 8, 4428–4432.

Yu, C., Choi, K., Yin, L., Grunlan, J.C., 2011. Light-weight flexible carbon nanotube based organic composites with large thermoelectric power factors. ACS Nano 5, 7885–7892.

Zeng, W., Shu, L., Li, Q., Chen, S., Wang, F., Tao, X.M., 2014. Fiber-based wearable electronics: a review of materials, fabrication, devices, and applications. Adv. Mater. 26, 5310–5336.

Zhang, Z.-G., Li, Y., 2014. Side-chain engineering of high-efficiency conjugated polymer photovoltaic materials. Sci. China Chem. 58, 192–209.

Zhang, Q., Sun, Y., Xu, W., Zhu, D., 2014a. Organic thermoelectric materials: emerging green energy materials converting heat to electricity directly and efficiently. Adv. Mater. 26, 6829–6851.

Zhang, X.-S., Han, M.-D., Wang, R.-X., Meng, B., Zhu, F.-Y., Sun, X.-M., Hu, W., Wang, W., Li, Z.-H., Zhang, H.-X., 2014b. High-performance triboelectric nanogenerator with enhanced energy density based on single-step fluorocarbon plasma treatment. Nano Energy 4, 123–131.

Zhang, F., Li, B., Zheng, J., Xu, C., 2015a. Facile fabrication of micro-nano structured triboelectric nanogenerator with high electric output. Nanoscale Res. Lett. 10, 298.

Zhang, H., Ye, L., Hou, J., 2015b. Molecular design strategies for voltage modulation in highly efficient polymer solar cells. Polym. Int. 64, 957–962.

Zhang, Z., Chen, G., Wang, H., Li, X., 2015c. Template-directed in situ polymerization preparation of nanocomposites of PEDOT:PSS-coated multi-walled carbon nanotubes with enhanced thermoelectric property. Chem. Asian J. 10, 149–153.

Zhong, J., Zhong, Q., Fan, F., Zhang, Y., Wang, S., Hu, B., Wang, Z.L., Zhou, J., 2013. Finger typing driven triboelectric nanogenerator and its use for instantaneously lighting up LEDs. Nano Energy 2, 491–497.

Zhong, J., Zhu, H., Zhong, Q., Dai, J., Li, W., Jang, S.-H., Yao, Y., Henderson, D., Hu, Q., Hu, L., Zhou, J., 2015. Self-powered human-interactive transparent nanopaper systems. ACS Nano 9, 7399–7406.

Zhu, X.Y., Yang, Q., Muntwiler, M., 2009. Charge-transfer excitons at organic semiconductor surfaces and interfaces. Acc. Chem. Res. 42, 1779–1787.

Zhu, G., Yang, R., Wang, S., Wang, Z.L., 2010. Flexible high-output nanogenerator based on lateral ZnO nanowire array. Nano Lett. 10, 3151–3155.

Zhu, G., Pan, C., Guo, W., Chen, C.Y., Zhou, Y., Yu, R., Wang, Z.L., 2012. Triboelectric-generator-driven pulse electrodeposition for micropatterning. Nano Lett. 12, 4960–4965.

Zhu, G., Peng, B., Chen, J., Jing, Q., Lin Wang, Z., 2015. Triboelectric nanogenerators as a new energy technology: from fundamentals, devices, to applications. Nano Energy 14, 126–138.

6

Energy Storage Devices Based on Polymers

6.1 Introduction

The evergrowing global demand of energy together with the depletion of fossil fuels makes it critical to develop sustainable and renewable energy resources. Developing relevant energy storage systems, for example, supercapacitors and lithium-ion batteries (LIBs), is essential for utilizing sustainable and renewable energy resources. Especially, the growing demand of portable and flexible cutting-edge electronics in modern society greatly stimulates the development of inexpensive, flexible, lightweight, and sustainable energy storage systems with high-power and energy densities. Among energy storage systems, supercapacitors have drawn considerable attentions in recent years due to their merits of high power density (10 kW kg^{-1}), superior rate capability, rapid charging/discharging rate, long cycle life (>10,0000 cycles), etc. So the supercapacitor can bridge the gap between batteries and traditional capacitors in terms of both storing energy and power bursts. For example, they can deliver an energy density several orders of magnitude higher than that of the dielectric capacitors and also store and deliver a large amount of charges within seconds, which allows them to provide higher powers than batteries.

In recent years, tremendous research effort has been aimed at increasing the energy density of supercapacitors without sacrificing their high power capability so that they reach the levels achieved in batteries. For this purpose, it is critical to develop high performance electrode materials. Carbon nanomaterials, transition metal oxides/hydroxides, and conducting polymers (CPs) are three commonly used electrode materials for supercapacitors. In particular, CPs as one of the pseudocapacitive materials, have the advantages of high-redox active capacitance, high conductivity, high-intrinsic flexibility, low cost and ease of fabrication. Thus, several approaches have been adopted to employ either the bare CPs alone or composites of CPs with carbon nanomaterials and metal oxides as the electrode materials to maximize the electrochemical performance of the devices. Although the energy densities of commercially available supercapacitors (less than 10 Wh kg^{-1}) typically exceed the conventional dielectric capacitors, they are still significantly inferior to LIBs (180 Wh kg^{-1}). Currently, supercapacitors are commonly used along with LIBs to provide the additional power required in a variety of electronic devices.

Commercialized in 1990, LIBs become a ubiquitous energy supply in electronics. An LIB basically comprises two electrodes, that is, cathode (positive electrode) and anode (negative electrode), separated by an ionic conductive electrolyte, providing a voltage derived from the potential difference between the two electrodes. An electrode is commonly a composite including active materials, binders, and conductive additives, which is casted onto a current collector. Moreover, a separator absorbing electrolyte is placed between the two electrodes to avoid a short circuit in practical applications. Therefore, an LIB may contain three major categories of materials, that is, metal-based, inorganic nonmetal-based, and organic materials. Among them, polymer plays an important role in the construction of LIBs.

In this chapter, we focus on the recent progress of polymers as well as their composites for the application in supercapacitors and LIBs. The main requirements of polymers for supercapacitors or LIBs are carefully investigated. The relationships between the polymer structure and electrochemical performance including specific capacitance, rate capability, and cycle stability are further highlighted for the guidance on the design of the next-generation of supercapacitors and LIBs based on polymers.

6.2 Supercapacitor

Typically, a supercapacitor is composed of two electrodes dipped in an electrolyte solution with a suitable separator. It is generally accepted that the energy storage mechanism of supercapacitors can be classified into electrical double layer capacitors (EDLCs) and pseudocapacitors (Fig. 6.1A) (Jost et al., 2014). In EDLCs, the charge storage is based on a reversible ion adsorption from an electrolyte onto electrodes with high specific surface areas to form a double layer structure. The capacitance comes from the pure electrostatic

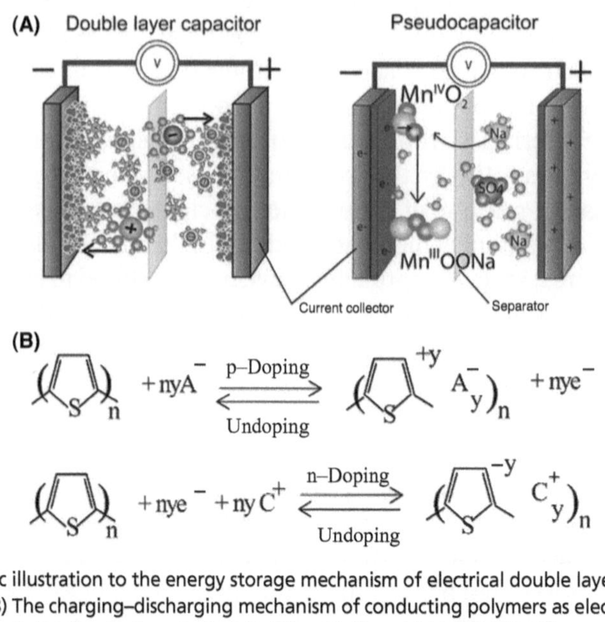

FIGURE 6.1 (A) Schematic illustration to the energy storage mechanism of electrical double layer capacitors and pseudocapacitors. (B) The charging–discharging mechanism of conducting polymers as electrode materials. *(A) Reproduced with permission from reference Jost, K., Dion, G., Gogotsi, Y., 2014. Textile energy storage in perspective. J. Mater. Chem. A 2, 10776–10787. Copyright 2014, Royal Society of Chemistry.*

charge accumulated at the electrode/electrolyte interface. Carbon-based materials including activated carbon, ordered mesoporous carbon, carbide-derived carbon, carbon nanotube, and graphene have been widely explored as electrode materials for EDLCs. For pseudocapacitors, the capacitance is mainly derived from reversible redox reactions between the electrolyte and electroactive species on the electrode surface. The most commonly known pseudocapacitive materials include transition metal oxide/hydroxide, such as RuO_2, MnO_2, CoO/Co_3O_4, $Co(OH)_2$, $NiO/Ni(OH)_2$, and Fe_2O_3; and CPs like polypyrrole (PPy), polyaniline (PANI), poly[3,4-ethylenedioxythiophene] (PEDOT) and their derivatives.

ELDCs can deliver high power density coupled with high charge–discharge cycling stability, but they suffer from low energy density owing to the relatively low capacitance of carbon-based materials. Differently, pseudocapacitive materials have been commonly employed to enhance the capacitances due to their fast surface redox reactions. Although metal oxide/hydroxide shows an ultrahigh theoretical specific capacitance, its poor electrical conductivity results in a lower power density. CPs becomes attractive candidate materials for pseudocapacitors due to their combined high electrical conductivity, high environmental stability, low cost, easy fabrication, and flexibility. The ability of CPs to store charges arises from a doping process. After doping, the CPs becomes electronically conductive. A schematic for the p- or n-doping and dedoping process of CPs is shown in Fig. 6.1B, in which polythiophene is reversibly oxidized and reduced with the insertion and deinsertion of counter-ions (A^- and C^+) from electrolytes (Mastragostino et al., 2001).

6.2.1 Polymer-Based Electrode

6.2.1.1 Conducting Polymer

A variety of CPs has been investigated as electrode materials for pseudocapacitors. In most cases, they are synthesized either chemically or electrochemically by oxidation of the relevant monomer in solution (Li et al., 2009; Yuan et al., 2013). According to the charge storage mechanism of CPs, the rapid and reversible redox reactions not only occur in the active materials but also at the electrode–electrolyte interface. Therefore, increasing interests are attracted to design various nanostructured CPs. The high surface area and nanoporous structure make CPs more accessible to the electrolyte for redox reactions. Meanwhile, reducing the size of bulk CPs can also greatly minimize the transport length for both electrons and ions, which is beneficial for maintaining high capacitance at a high scan rate or current density (Zhang et al., 2013b; Yu et al., 2015). Therefore, the controllable synthesis of CPs with different morphologies is highly desired. It was found that the preparation of one-dimensional (1D) ordered nanostructure, particularly nanowire, represents an effective strategy for high-performance electrode materials in supercapacitor (Wang et al., 2010, 2014a). For instance, a perpendicularly aligned PANI nanowire arrays showed a high specific capacitance of 950 F g^{-1} at 1 A g^{-1} and retained as high as 780 F g^{-1} even at a high current density of 40 A g^{-1} (Fig. 6.2) (Wang et al., 2014a). Similarly, PPy nanowire arrays also displayed a specific capacitance of 566 F g^{-1} at 1.10 A g^{-1} in comparison with

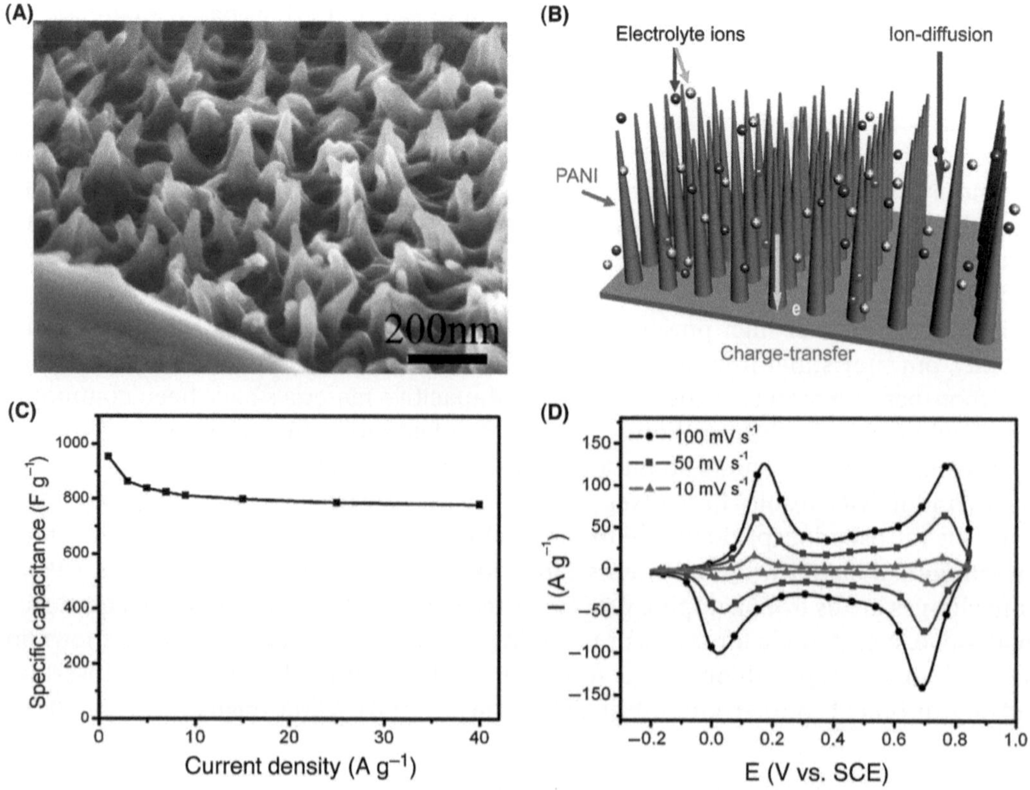

FIGURE 6.2 (A) Cross-sectional scanning electron microscopy *(SEM)* image of PANI nanowire array. (B) Schematic illustration to the working mechanism of aligned conducting polymer nanowire array. (C) Dependence of specific capacitance on the current density. (D) CV curves at increasing scan rates. *Reproduced with permission from reference Wang, K., Wu, H., Meng, Y., Wei, Z., 2014a. Conducting polymer nanowire arrays for high performance supercapacitors. Small 10, 14–31. Copyright 2014, Wiley-VCH.*

414 F g^{-1} for disordered PPy nanowire networks and 378 F g^{-1} for PPy film at the same current density (Huang et al., 2010). The reasons for high specific capacitance and high rate capability can be attributed to the 1D nanostructure of PPy, which allows for an efficient charge transport and provides a reduced ion-transport length. Meanwhile, the space among oriented nanowires can accommodate the large volume change without incurring fracture during charge–discharge process.

However, there remain some drawbacks for the application of CPs as electrode materials in supercapacitor. For instance, the electrodes usually suffer from a relatively low electrical conductivity and poor cycle stability during long-term charge–discharge processes. The poor cyclic stability is ascribed to the inferior mechanical stability. During charging and discharging process, the repeated inserting/deinserting of ions gives rise to large volumetric changes. To date, several approaches for enhancing the mechanical stability and electrical conductivity of the CPs have been explored. One of the commonly used strategies is designing various composites by combining CPs with other carbon materials, metal, and metal oxides/hydroxides. These nanocomposites combine the excellent properties of individual components and may even display synergistic effects from them.

6.2.1.2 Carbon/Conducting Polymer Composite

The poor cycle stability and relative poor electrical conductivity of bare CPs have partially restricted their practical applications in electrode materials. On the other hand, carbon materials share the merits of low cost, high chemical and thermal stability, and high electrical and mechanical properties (Zhang and Zhao, 2009; Yan et al., 2014). Therefore, a combination of CPs with carbon materials at the molecular scale seems to be an effective strategy to improve the electrical conductivity and mechanical property of CPs.

6.2.1.2.1 CARBON PARTICLES

Porous carbon materials in different forms (Zhang and Zhao, 2009; Pech et al., 2010; Zhu et al., 2011; Zhai et al., 2011; Ma et al., 2013; Kajdos et al., 2010; Chmiola et al., 2010) (e.g., activated carbon, mesoporous carbon, templated carbon, carbide-derived carbon, porous carbon sphere, and carbon onion) possess merits of large specific surface area, high porosity, and high electrical conductivity, making them promising candidates for electrode materials with high energy and power densities. Compositing CPs with porous carbon materials is thus widely explored. It has been demonstrated that the significant synergistic effects between CPs and porous carbon materials could lead to improved electrochemical properties superior to those of individual components (Yan et al., 2011, 2013; Wang et al., 2013c; Liu et al., 2014a; Zhang et al., 2010a; Nyholm et al., 2011; Yu et al., 2013). For example, perpendicularly aligned PANI nanowhiskers were successfully synthesized on the outer surface of ordered mesoporous carbon (CMK-3) by a chemical oxidative polymerization process, and the resulting PANI/CMK-3 composite exhibited a high specific capacitance of 470 F g^{-1} and high capacitance retention of 90.4% after 1000 cycles (Yan et al., 2011). These remarkable electrochemical properties originate mainly from the aligned pseudocapacitive PANI nanowhiskers which could create electrochemical accessibility for electrolyte ions and reduce their transport distances. In addition, the bimodal pore distribution of mesoporous carbon substrate and unique hierarchical nanostructures also endow the composite with a high specific surface area of 470 m^2 g^{-1}, which also contributes to the superior electrochemical performances. Recently, monolithic coral-like porous carbon with a higher specific surface area of 1125 m^2 g^{-1} was used as skeleton to support various CPs like PPy, PANI, and polythiophene (Wang et al., 2013c). The composite materials with 19 wt.% of PPy yielded a high specific capacitance of 1488 F g^{-1} in an aqueous H_2SO_4 electrolyte, and the maximum energy and power densities reached 50 Wh kg^{-1} and 12 kW kg^{-1}, respectively. To alleviate the structural instability of CPs derived from the repeated volumetric swelling and shrinking during charge–discharge process, a thin layer of carbonaceous shell was deposited onto polymer surface by hydrothermal carbonization of glucose (Liu et al., 2014a). For instance, the carbonaceous shell-coated PANI electrode exhibited exceptionally high capacitance retention of ~95% after 10,000 cycles, in contrast to ~20% retention in the bare PANI (Fig. 6.3A and B).

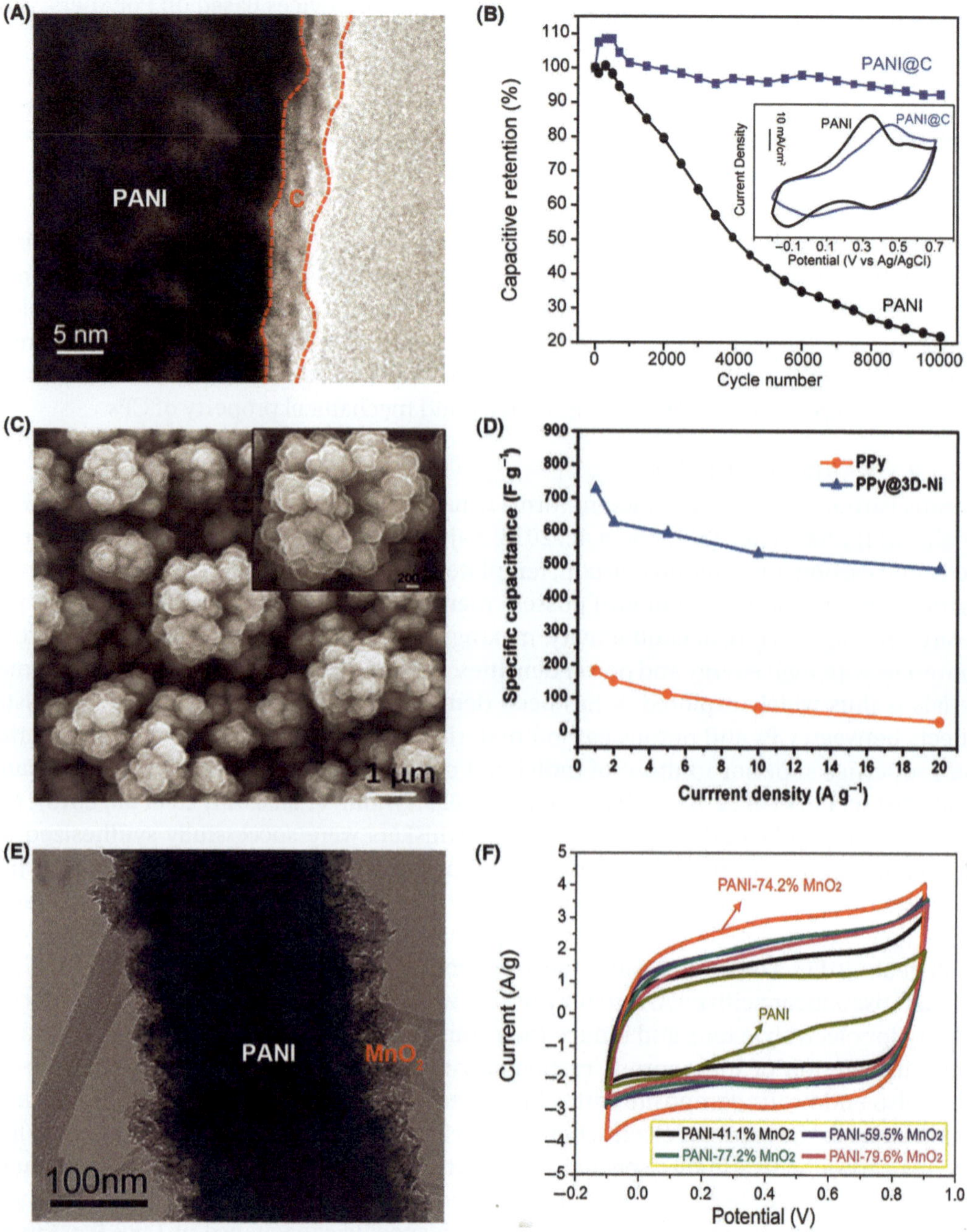

FIGURE 6.3 (A) Transmission electron microscopy *(TEM)* image of PANI/C composite. (B) comparison on the cyclic performance of PANI and PANI/C composite electrodes. (C) SEM image of the PPy shell/3D-Ni core composite. (D) Dependence of the specific capacitance of the composite on the current density. (E) TEM image of PANI/MnO_2 coaxial nanofibers. (F) CV curves of the PANI/MnO_2 composites with different MnO_2 contents. *(A,B) Reproduced with permission from reference Liu, T., Finn, L., Yu, M., Wang, H., Zhai, T., Lu, X., Tong, Y., Li, Y., 2014a. Polyaniline and polypyrrole pseudocapacitor electrodes with excellent cycling stability. Nano Lett. 14, 2522–2527. Copyright 2014, American Chemical Society. (C,D) Reproduced with permission from reference Chen, G.F., Su, Y.Z., Kuang, P.Y., Liu, Z.Q., Chen, D.Y., Wu, X., Li, N., Qiao, S.Z., 2015a. Polypyrrole shell@ 3D-Ni metal core structured electrodes for high-performance supercapacitors. Chem. Eur. J. 21, 4614–4621. Copyright 2015, Wiley-VCH. (E,F) Reproduced with permission from reference Jiang, H., Ma, J., Li, C., 2012. Polyaniline–MnO2 coaxial nanofiber with hierarchical structure for high-performance supercapacitors. J. Mater. Chem. 22, 16939. Copyright 2012, Royal Society of Chemistry.*

6.2.1.2.2 1D CARBON MATERIALS

1D nanostructures (such as nanowires, nanotubes, and nanorods) are emerging as a representative family of new energy materials for supercapacitors because they can not only provide an efficient pathway to transport charges but also reduce the electronic and ionic transport lengths (Yan et al., 2014; Yu et al., 2015). Compared with porous carbon, carbon nanotubes (CNTs) show a high aspect ratio with high electrical conductivity and high mechanical, chemical, and thermal stability. They have been thus considered as promising candidates for high-power electrode materials. In particular, their high mechanical resilience and open tubular network make them a good substrate for depositing active materials. In recent years, binary nanocomposites based on CNTs and CPs has been explored extensively (Hu et al., 2012; Fan et al., 2012; Benson et al., 2013; Zhong et al., 2013; de Souza et al., 2014). CPs in different forms of nanoparticle, nanowire/rod, and nanosheet are uniformly grown on the surfaces of the CNTs through electrostatic forces and π–π interactions. Especially, well-ordered 1D CP nanostructures have been studied as ideal candidates for further increasing the capacitive performance of nanocomposites owing to their large specific surface areas and optimized ion diffusion pathways. For example, a hierarchical nanocomposite with PANI nanorods perpendicularly aligned on the surfaces of CNTs showed a higher specific capacitance of 568 F g^{-1} at a current density of 10 A g^{-1} and better cycling stability than individual components (Fan et al., 2012). The enhanced electrochemical performances are derived from the synergistic effect of two components as well as the fact that the hierarchical structure not only increases the specific surface area of the nanocomposite but also minimizes the charge transport distance. In another case, a pulsed electrodeposition technique was employed to deposit PANI onto the unwoven CNT fabric, and the resulting PANI/CNT composites demonstrated a high stability without significant degradation for over 30,000 cycles. In contrast, a bare PANI rapidly dropped to 30% in specific capacitance after 1000 cycles (Benson et al., 2013). The robust electrically interconnected CNT networks and interactions between the electrodeposited PANI and CNT are likely responsible for the high cyclic stability.

In recent years, aligned CNT materials synthesized by chemical vapor deposition have aroused increasing interests for supercapacitors due to their intriguing advantages over these randomly entangled CNTs (Lima et al., 2011; Zhang et al., 2014). Their relatively regular pore structures and conductive channels and higher effective specific surface areas result in rapid ion and electron transports and enhance charge storage/deliver properties, which is highly desirable for high-rate applications. In particular, the CNT yarn can be spun or twisted from a continuous CNT sheet that was dry-drawn from an aligned CNT array, and it has been extensively studied for wearable energy storage devices (Lima et al., 2011). The flexible CNT fibers are also ideal current collectors for depositing various CPs (Cai et al., 2013; Chen et al., 2014, 2015b). For example, ordered PANI nanowire arrays were incorporated into the CNT yarn by an in situ electrochemical polymerization to improve the performance of the supercapacitor (Wang et al., 2013a). The as-prepared PANI/CNT composite yarn produced a specific area capacitance of 38 mF cm^{-2}, 16 times higher than that of pure CNT yarn (ca. 2.3 mF cm^{-2}). Compared with the porous carbon materials, a main limitation for CNTs is the relatively low specific surface area (generally less than 500 $m^2\ g^{-1}$). To this end, several attempts including chemical or plasma activation are made to introduce more defects and increase the specific surface area.

Apart from CNTs, commercially available carbon nanofibers are specifically selected as the current collector due to their high electrical conductivity and chemical stability, light weight, large porosity, and flexibility (Chen et al., 2011). In most case, carbon cloth woven from carbon nanofibers is an attractive candidate for flexible electrodes. It was reported that PANI nanowires deposited on carbon cloth can deliver a high gravimetric capacitance of 1079 F g^{-1} and an exceptionally high specific area capacitance of 1.8 F cm^{-2} (Horng et al., 2010). The outstanding electrochemical performances make the composite electrodes preferable for flexible supercapacitors. Hierarchical PEDOT nanowires directly grown on carbon cloth have also been employed as electrodes (Hsu et al., 2013). A wide operated potential window of 1.6 V in neutral Na_2SO_4 electrolyte with energy density of 182.1 Wh kg^{-1} and specific power of 13.1 kW kg^{-1} had been achieved. Moreover, the capacitive performance of PEDOT nanowire/carbon cloth was higher than that of PEDOT film/carbon cloth electrode no matter in acidic or neutral electrolyte.

6.2.1.2.3 2D CARBON MATERIALS

As an intriguing two-dimensional (2D) carbon material, graphene exhibits several attractive properties, such as high theoretical specific surface area (ca. 2600 $m^2\ g^{-1}$), extraordinarily high electrical conductivity, and high flexibility. Therefore, graphene is widely explored as an excellent support skeleton for growing various CPs upon being discovered (Yan et al., 2010; Wang et al., 2014b; Xu et al., 2011; Liu et al., 2014b; Mini et al., 2011). To further enhance the electrochemical performance of the composite electrode, considerable efforts are devoted to control the morphology of CPs grown on graphene nanosheets. Nanostructured CPs are expected to provide higher electroactive regions and shorten ion diffusion compared to their bulk materials, so graphene nanosheets served as highly conductive substrates to deposit PANI nanoparticles with diameters of ca. 2 nm (Yan et al., 2010). The resulting composite delivered a maximum specific capacitance of 1046 F g^{-1}, compared to 115 F g^{-1} for the bare PANI. The PANI nanoparticles in turn also effectively prevented the agglomeration of graphene sheets and thus significantly enhance the electrode/electrolyte interface area to reduce the diffusion and migration length of ions in electrolyte during charge–discharge processes, which gave rise to remarkably enhanced electrochemical performances. For instance, it displayed an energy density of 39 Wh kg^{-1} at a power density of 70 kW kg^{-1}.

It was reported that the synthesis of mesoporous PANI film on ultrathin graphene nanosheets is favorable for high-performance supercapacitors (Wang et al., 2014b). Due to the mesoporous structure of the composite electrode, it dramatically improved the specific capacitance, rate capability, and cyclic stability in comparison to the bare PANI. The composite displayed specific capacitance of 749 F g^{-1} at 0.5 A g^{-1} in 1 M H_2SO_4 electrolyte and retained by 73% at 5.0 A g^{-1}, much higher than 315 F g^{-1} at 0.5 A g^{-1} and 39% retention at 5.0 A g^{-1} for the pristine PANI. In addition, the composite can maintain 88% of its initial capacitance compared to 45% for the pristine PANI after 1000 cycles. The much improved long-term cycle stability and rate capability was mainly attributed to the formed mesoporous structure that effectively solves the serious swelling and shrinking problem of PANI during rapid charge–discharge process.

The synthesis of hierarchical graphene/PPy nanosheet composite offered an electrochemical capacitance of 318.6 F g^{-1}, much higher than 130 F g^{-1} of pure PPy nanoparticles. These hierarchical composite further enables a high rate capability and excellent cycling stability with 95% retention after 1000 cycles (Xu et al., 2011). However, due to the van der Waals interaction among neighboring graphene sheets, the agglomeration or restacking of graphene is usually unavoidable. This phenomenon will reduce the specific surface area and hinder the rapid diffusion of ions in electrolyte to the surface of active material, consequently decreasing the performance of electrode. To avoid this drawback, graphene oxide (GO) was often used to replace graphene as the substrate (Zhang et al., 2013a; Wang et al., 2013b; Xu et al., 2010). Hierarchical nanocomposites with the PANI nanowires array on 2D GO nanosheets were successfully constructed for a higher electrochemical capacitance of 555 F g^{-1} and better cycling stability than that of a randomly interconnected PANI nanowires, mainly due to the synergistic effect between PANI and GO (Xu et al., 2010).

6.2.1.2.4 3D CARBON MATERIALS

Both CNT and graphene exhibit large specific surface areas and high electrical conductivity at nature, so the assembly of them into three-dimensional (3D) porous structures like carbon aerogel, sponge, foam, and framework is expected to combine their advantages for fabricating high-performance supercapacitors. The 3D porous nanostructures with large specific surface areas and well-defined micro/mesopores facilitate the sufficient contact between electrolyte and active material and shorten the ion transport pathway, giving rise to fast reaction kinetics that is crucial for achieving high power density. As far as the electrical conductivity is concerned, the active CPs with a high electrical and ionic conductivity will lower the polarization and accelerate the reaction kinetics.

3D porous carbon materials are becoming a new family of lightweight substrates instead of commercially used nickel foam or aluminum foil in recent years, which are particularly beneficial for applications in portable electronic devices. It is demonstrated that an electrode with 3D porous structure and feasible charge transport can sustain a longer cyclic stability. Several recent reports have demonstrated that utilizing 3D conductive carbon materials as skeletons to deposit CPs is a highly efficient method to construct high-performance electrodes for supercapacitors (Patil et al., 2015; Kulkarni et al., 2014; Li et al., 2013; Zhang et al., 2010b). For example, a composite electrode prepared by chemically depositing PANI nanofibers on 3D graphene framework showed a higher specific capacitance of 1024 F g^{-1} in 1 M H_2SO_4, good cycling stability with 87% retention after 5000 cycles, and improved rate capability as compared to the composite based on stainless steel substrate. The PANI nanofibers provided a large specific surface area while the 3D porous graphene backbone offered efficient conducting pathways (Kulkarni et al., 2014). In another case, deposition PANI on 3D-ordered macroporous carbon achieves a high specific capacitance of 1490 F g^{-1} and good cycling stability (Zhang et al., 2010b). The 3D-ordered pore structure of the carbon favors the diffusion of the ions in electrolyte. Meanwhile, a thin and porous layer of PANI deposited on the macroporous carbon wall can greatly shorten the diffusion length, hence leading to an enhanced energy storage capacity and a good rate capability.

6.2.1.3 Metal/Conducting Polymer Composite

A lot of attention has been paid to minimize the contact resistance between the active material and the current collector. Due to high electrical conductivities of metals, they are frequently used as matrices for CPs. The resultant composite electrodes generally show much enhanced electrochemical performances, such as high energy density, excellent rate performance, and good cycling stability in comparison to the bare CPs (Zhang et al., 2015; Huang et al., 2015; Chen et al., 2015a). Among them, gold (Au), nickel (Ni) foam, copper (Cu), aluminum (Al) foil, and stainless steel represent the most used candidates. For instance, a novel PPy shell/3D-Ni core composite was developed by applying 3D Ni films as current collector to enhance the electrochemical performance of conventional PPy film (Fig. 6.3C) (Chen et al., 2015a). The obtained composite electrode exhibited a higher specific capacitance of 726 F g^{-1} and better rate capability at increasing current densities from 1 to 20 A g^{-1} than those of the pristine PPy (Fig. 6.3D). The highly conductive 3D metal substrates can not only provide a short diffusion pathway for fast ion transport but also benefits efficient electron transfer. Supercapacitors derived from the 3D PANI/nanoporous Au composite showed an excellent rate capability with ultrahigh volumetric capacitance of 1500 F cm^{-3} at 1 A cm^{-3} and 866 F cm^{-3} even at a current density of 80 A cm^{-3} as well as a high energy density of 78 mWh cm^{-3}, which is about four orders of magnitude higher than those of electrolytic capacitors and thin film lithium battery with the similar power density of 190 W cm^{-3} (Lang et al., 2012). The outstanding electrochemical performances are mainly derived from the unique bicontinuous nanoporous skeleton in which pseudocapacitive PANI shells are incorporated into pore channels, rendering fast ion diffusion and electron transfer and accordingly ensuring the sufficient redox reaction of PANI during the charge–discharge processes.

6.2.1.4 Metal Compound/Conducting Polymer Composite

Compared with carbonaceous materials for EDLCs, transition metal oxide/hydroxide can provide much higher specific capacitance based on fast and reversible redox reactions. Unfortunately, the low power density and relatively poor cycling stability have partially restricted their practical applications. The former originates from poor electrical conductivity of the metal compound to confine fast electron transport, while the latter can be attributed to the fragile structure of electrode materials which is easily damaged by swelling and shrinkage during the charge–discharge process. On the contrary, CPs has a relatively high electrical conductivity. Therefore, it can be expected that the synergistic effect between metal oxide/hydroxide and CPs may lead to enhanced electrochemical properties superior to those of each individual components.

In such a composite, the metal oxide/hydroxide mainly contributed to high specific capacitance depending on redox properties as well as structural morphologies. The CPs is crucial for high mechanical stability and high electrical conductivities to facilitate fast electron transport within the composites for high charge–discharge rate. It has been demonstrated that the morphology, surface area, and crystallinity of metal oxide/hydroxide significantly influence their electrochemical properties (Yan et al., 2014; Wei et al., 2011). Among various oxides, RuO_2 has been recognized as the most promising pseudocapacitive material because of its ultrahigh theoretical specific capacitance, high chemical stability, wide potential window, and long cycle life. Unfortunately, the use of RuO_2 is limited by the relatively high cost, low abundance, and environmental harmfulness. To this end, more studies focus on replacing RuO_2 with the other inexpensive transition metal oxides/hydroxides such as MnO_2, Co_3O_4/$Co(OH)_2$, NiO/$Ni(OH)_2$, TiO_2, and Fe_2O_3.

Recently, 1D pseudocapacitive nanomaterials have been intensively investigated as electrodes because the short diffusion pathways for charges can result in high charge–discharge rates (Wang et al., 2015b). Therefore, the combination of metal oxide/hydroxide and conducting polymer into a core/shell coaxial nanostructure has attracted greater attentions. Typically, the synthetic approach for the core/shell coaxial nanowires/rods requires two steps: the preparation of 1D metal oxide/hydroxide backbones as core and subsequently coating a layer of CPs as shell by chemical or electrochemical methods. Several kinds of coaxial metal compound/conducting polymer composites like MnO_2/PEDOT, MnO_2/PPy, and CoO/PPy were successfully prepared and exhibited both high electrochemical and mechanical properties (Zhou et al., 2013; Xie et al., 2011; Yao et al., 2013; Liu and Lee, 2008). For instance, a PPy-incorporated CoO nanowire array exhibited an ultrahigh specific capacitance of 2223 F g^{-1} (this hybrid has a theoretical value of 2467 F g^{-1}), good rate capability, and excellent cycling stability with 99.8% capacitance retention after 2000 cycles (Zhou et al., 2013). Here the CoO core mainly provided high energy storage capacity, while the highly conductive, porous, and flexible PPy shell not only facilitated the electron transport and ion diffusion into the CoO core but also protected it from collapsing and breaking during charge–discharge process. Moreover, the coaxial nanowire structures also led to the short ion diffusion pathway in the composites. A 1.8 V asymmetric supercapacitor was further prepared by using PPy/CoO composite as positive electrode and activated carbon as the negative electrode. It achieved high energy density of ~43.5 Wh kg^{-1}, high power density of ~5.5 kW kg^{-1} at 11.8 Wh kg^{-1} and good cyclic stability for ~20000 cycles.

Unfortunately, the conducting polymer shell, to some extent, will prevent the full contact of the metal oxide core with electrolyte, and hence sacrifice the energy density. To enhance the utilization of pseudocapacitive metal oxide/hydroxide materials, they were commonly grown on the surface of conducting polymer nanofibers to produce PPy/RuO_2, PEDOT/RuO_2, PANI/MnO_2, and PANI/TiO_2 (Gui et al., 2014; Jiang et al., 2012; Han et al., 2012; Mai et al., 2014; Liu et al., 2010b). For example, hierarchical structured PANI/MnO_2 coaxial nanofibers were prepared by simply soaking PANI nanofibers in a $KMnO_4$ solution (Fig. 6.3E) (Jiang et al., 2012). The obtained composite electrode with 74.2% MnO_2 content exhibited a high specific capacitance of 268 F g^{-1} in 1 M Na_2SO_4 electrolyte, while the pure pristine revealed only 109 F g^{-1} (Fig. 6.3F). However, the composite electrode suffered from inferior cyclic stability, for example, ~75.5% retention after 2000 cycles. Recently, the synthesis of more complex ternary "core–shell–shell" nanostructures like Co_3O_4/PPy/MnO_2 coaxial nanowires (Han et al., 2014) and layered V_2O_5/PEDOT/MnO_2 structure (Guo et al., 2015) attracted increasing interests. Such ternary architectures combine the advantages of three different pseudocapacitive materials for improving electrochemical performances. The intermediate conducting polymer layers play two roles by enhancing the electrical conductivity of the composites and serving as a reactive template to grow metal oxides.

6.2.1.5 Ternary Conducting Polymer-Based Composite

6.2.1.5.1 CARBON/METAL OXIDE/CONDUCTING POLYMER

Apart from binary composites, the construction of ternary composites that combine the advantages of each component (carbon material, metal oxide/hydroxide, and conducting polymer) have been explored intensively to fulfill the supercapacitor application in recent years (Xia et al., 2012; Yoon and Kim, 2013; Tang et al., 2014; Grover et al., 2014; Wang et al., 2012b). Through taking full advantages of the synergistic effects from the ternary composite, each component in the composite may play their potential role to a maximal extent. This strategy provides an important technique for future design of next-generation supercapacitors for industrial and consumer applications. Recently, it was demonstrated that the performance of graphene/MnO_2 textile electrodes can be substantially enhanced by wrapping with a thin layer of conductive PEDOT:PSS through a simple dip-dry process (Fig. 6.4A) (Yu et al., 2011a). The PEDOT:PSS coating further lowered the internal resistance of the graphene/MnO_2, which was verified by electrochemical impedance spectra (Fig. 6.4B). The specific capacitance of the ternary composite electrodes was greatly increased by 45% to a value of 380 F g^{-1} at 0.5 mA cm^{-2} as compared to the graphene/MnO_2 composite electrodes and the advantages can extend to higher current density (Fig. 6.4C). In such a ternary composite, graphene offered high specific surface area for the deposition of hierarchical porous MnO_2 nanospheres and also enhanced the electrical conductivity and mechanical stability of the composite. Highly porous MnO_2 nanospheres provided numerous active sites contributing large pseudocapacitances. The conductive PEDOT:PSS polymer layers served as both conductive additives and binder materials. It can potentially offer better interconnectivity within MnO_2 nanoparticles and meanwhile participate in the charge storage process through the redox reaction-based pseudocapacitance. More importantly, the polymer layer prevented collapsing and breaking of MnO_2 and dissolution of metal ions during repeated charge–discharge cycling. All the aforementioned factors in turn contribute to the excellent electrochemical performances for the ternary composite.

The preliminary success has greatly encouraged the application of "conductive wrapping" method to construct ternary nanocomposite film by mixing the preformed CNT/MnO_2 (Hou et al., 2010) or CNT/amorphous-$Ni(OH)_2$ (Jiang et al., 2015) with PEDOT-PSS. Here PEDOT-PSS acted as both an effective dispersant for CNT/MnO_2 and a binder to improve the cohesion of CNT/MnO_2 nanoparticles and adhesion to the substrate. Such a

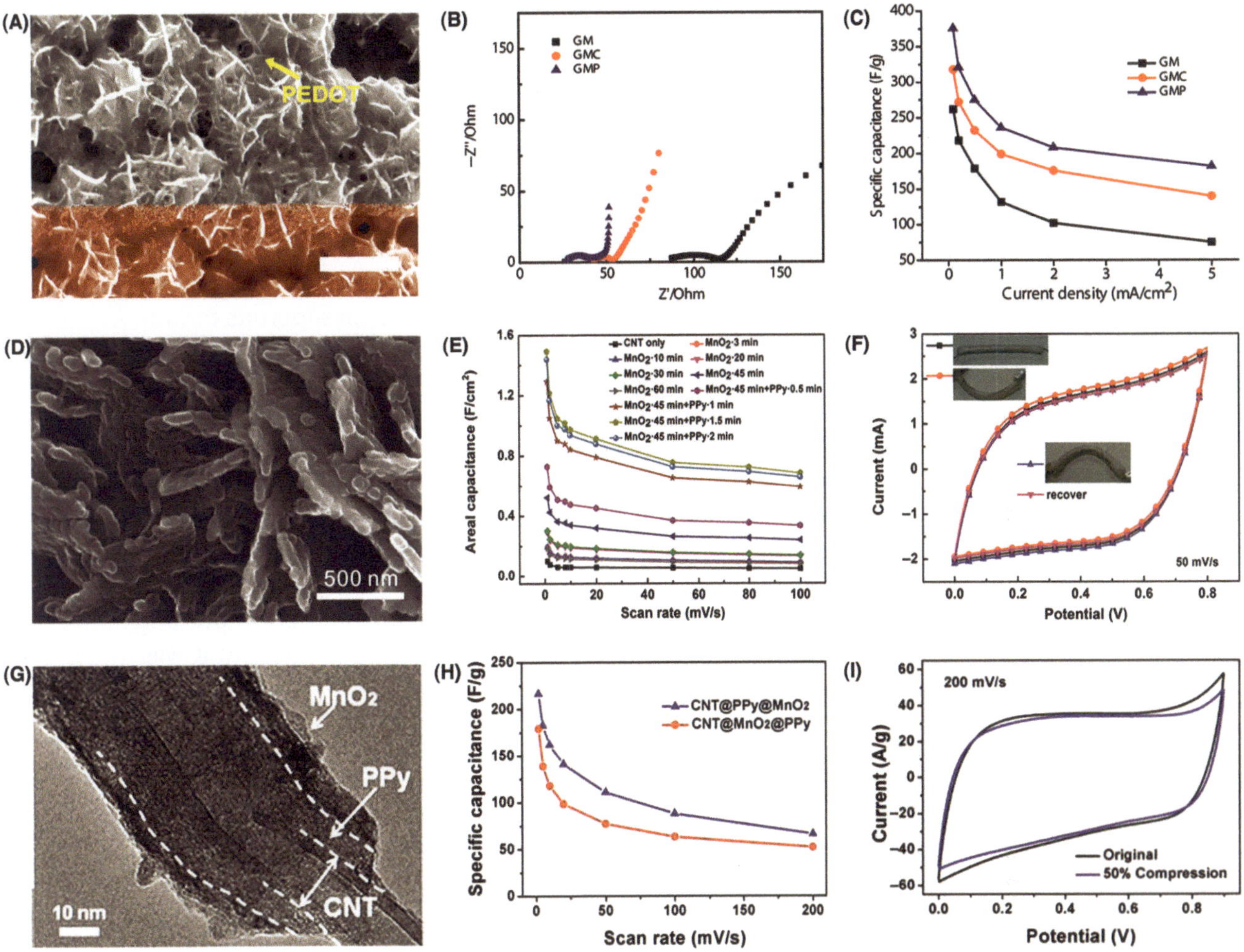

FIGURE 6.4 (A) SEM image of the PEDOT:PSS wrapped graphene/MnO_2 electrode, (B) electrochemical impedance spectra of graphene/MnO_2 *(GM)*, graphene/MnO_2/CNT *(GMC)* and graphene/MnO_2/CNTs/PEDOT:PSS *(GMP)* textile electrodes, and (C) specific capacitances for three different electrode materials at various current densities. (D) SEM image of the PPy coated CNT/MnO_2 electrode. (E) Dependence of specific areal capacitance on the scan rates of different electrode systems, and (F) CVs of the cable-type supercapacitor at different bending states. (G) TEM image of the core-double-shell structure. (H) Comparison of specific capacitances of the two sponges with different sequences at different scan rates. (I) CV curves of a CNT@PPy@MnO_2 sponge before and after compressing (strain of 50%) at 200 mV/s. *(A–C) Reproduced with permission from reference Yu, G., Hu, L., Liu, N., Wang, H., Vosgueritchian, M., Yang, Y., Cui, Y., Bao, Z., 2011a. Enhancing the supercapacitor performance of graphene/MnO_2 nanostructured electrodes by conductive wrapping. Nano Lett. 11, 4438–4442. Copyright 2011, American Chemical Society. (D–F) Reproduced with permission from reference Liu, N., Ma, W., Tao, J., Zhang, X., Su, J., Li, L., Yang, C., Gao, Y., Golberg, D., Bando, Y., 2013. Cable-type supercapacitors of three-dimensional cotton thread based multi-grade nanostructures for wearable energy storage. Adv. Mater. 25, 4925–4931. Copyright 2013, Wiley-VCH. (G–I) Reproduced with permission from reference Li, P., Yang, Y., Shi, E., Shen, Q., Shang, Y., Wu, S., Wei, J., Wang, K., Zhu, H., Yuan, Q., Cao, A., Wu, D., 2014. Core-double-shell, carbon nanotube@polypyrrole@MnO(2) sponge as freestanding, compressible supercapacitor electrode. ACS Appl. Mater. Interfaces 6, 5228–5234. Copyright 2014, American Chemical Society.*

3D conductive wrapping approach can be generalized for designing the next-generation high-performance energy storage devices. To form effective ternary composite electrode, it is very favorable to precisely control the thickness of polymer coating layers. In this regard, the electrochemical deposition method seems more applicable as compared to the dip-coating method (Xia et al., 2015; Liu et al., 2013; Yuan et al., 2008). Recently, flexible cable-type supercapacitors based on 3D CNT-cotton thread/MnO_2/PPy composite electrodes were developed via a facile three-step process (Fig. 6.4D), in which single-walled CNT ink was first coated on cotton threads, and then MnO_2 nanomaterials and PPy film were simply grown on the CNT-coated cotton threads via in situ electrochemical deposition process (Liu et al., 2013). The conductive CNT backbone, active mesoporous flower-like MnO_2 nanoplates and PPy conductive wrapping layer synergistically contribute to the high areal specific capacitance (Fig. 6.4E and F). In some cases, the external conducting polymer in the ternary composite also serves as a physical barrier to prevent the intermediate metal oxide/hydroxide from reductive-dissolution process when being exposed to acidic electrolytes. For instance, the protective PANI coating layer on MnO_2/CNTs nanocomposite makes the ternary composite electrode work more effectively in the mixed Na_2SO_4–H_2SO_4 acidic electrolytes to achieve much better energy storage and electrochemical stability (Yuan et al., 2008).

Design and fabrication of structurally optimized electrode materials are very important for stable pseudocapacitors with excellent energy storage performance. Recently, it was demonstrated that a thin layer of RuO_2 grown by atomic layer deposition onto carbon-cloth-supported PANI nanowire array can significantly enhance its pseudocapacitive stability and performance (Xia et al., 2015). The ternary composite exhibited very high specific capacitance of 710 F g^{-1} and high power density of 42.2 kW kg^{-1} at an energy density of 10 Wh kg^{-1}, as well as remarkable capacitance retention of ~88% after 10,000 cycles even at high current density of 20 A g^{-1}. In such a ternary composite, the middle polymer layers are expected to improve the interaction between the outer metal oxide layer and the inner conductive carbon, reduce the interface resistance and improve the charge transfer. A strong adhesion of metal oxide nanomaterials to the carbon skeleton is important for making stable electrodes. In contrast, it was found that the performance of the 3D composite electrode in the sequence of CNT sponge, PPy, and MnO_2 was better than the CNT, MnO_2, and PPy sequence (Figs. 6.4G–I) (Li et al., 2014). Here the PPy layer served as an intermediate layer for anchoring MnO_2 nanosheets, which otherwise could easily detach from the smooth CNT surface. In addition, the rough MnO_2 shell also lead to higher specific surface area and increased the percentage of pores in the hybrid sponge, which are favorable for the redox reactions.

6.2.1.5.2 CARBON/METAL-ORGANIC FRAMEWORKS/CONDUCTING POLYMER

Metal-organic frameworks (MOFs), as a new class of porous materials, are well-known for high specific surface areas and high porosity that have potential applications in gas storage and separation, catalysis, sensor, and drug delivery (Furukawa et al., 2010). In recent years, the exploration of MOFs as electrode materials for supercapacitors has also been reported because their exceptional porosities and high specific surface areas could make ions in the electrolytes easily accessible to MOFs electrode by diffusion effect. Typically, MOF-based supercapacitors can be classified into two types: (1) MOFs serving as novel templates for preparing porous metal oxide or carbon; (2) pristine MOFs being directly used as a new type of electrode materials to store energy on their internal surfaces through EDLCs mechanism or exploiting the pseudocapacitive redox reaction of metal centers (Liu et al., 2010a; Yang et al., 2014b). Unfortunately, it remains challenging to fabricate supercapacitors based on pure MOFs due to their poor electrical conductivity, incompatibility with the electrolyte and steric hindrance to the insertion of ions arising from limited pore sizes. Although 23 MOFs nanocrystals (with a variety of organic functionalities and metal ions, different pore sizes and shapes) have been prepared for electrodes, only a zirconium MOF exhibited a moderate areal specific capacitance of 5.1 mF cm^{-2} and volumetric specific capacitance of 0.64 F cm^{-3} (Choi et al., 2014).

In another case, a Co-based MOF also exhibited a reasonable specific capacitance of approximately 207 F g^{-1} in LiOH electrolyte (Lee et al., 2012). The incorporation of CPs with MOFs was proved to be a feasible strategy for achieving high-performance supercapacitors. Very recently, a flexible MOF-based superapacitor consisting of cobalt-based MOFs (ZIF-67) that was coated on a carbon cloth with electrochemically polymerized PANI was successfully fabricated (Wang et al., 2015a). Electrochemical studies showed that the PANI-ZIF-67-carbon cloth exhibited an ultrahigh areal specific capacitance of 2146 mF cm^{-2} at 10 mV s^{-1}, while only 1.47 and 727 mF cm^{-2} were calculated for the ZIF-67-carbon cloth and PANI-carbon cloth at the same scan rate, respectively. These results clearly demonstrate that a synergistic effect occurs upon interweaving ZIF-67 aggregate and PANI. The hybrid structure combines the EDLC capacitance of the internal surface area of MOF and the pseudocapcitance of PANI.

6.2.2 Polymer-Based Electrolyte

Apart from the electrode materials, the performance of a supercapacitor is also strongly affected by the electrolyte employed. As we know, both energy and power densities of the device are proportional to the square of operating voltage, which is determined by the decomposition voltage of electrolytes. In general, the liquid electrolytes used in supercapacitors are classified into three types: aqueous electrolytes, organic electrolytes, and ionic liquids (ILs). The commonly used aqueous electrolytes include KOH, H_2SO_4 and Na_2SO_4. Aqueous electrolytes share advantages of low cost, nonflammability, low viscosity, high safety, high ionic conductivity, and easiness of fabrication and usage. However, the main disadvantages of aqueous electrolytes lie in a low voltage window around 1.0 V due to the thermodynamic decomposition of water at 1.23 V and low overpotential for hydrogen evolution. Organic electrolytes can provide voltage windows of 2.5–3 V to provide much higher energy densities to be 6–14 times of the aqueous electrolyte. However, the organic electrolytes suffer from some drawbacks like safety concerns due to the flammability and toxicity, low conductivity, high cost, and hygroscopicity. Due to an even wider electrochemical voltage window up to 4.2 V, nontoxicity, nonflammability, and higher conductivity compared to the organic electrolyte, ILs have attracted increasing interests as promising electrolytes, but the relatively high cost has limited their applications.

Conventional supercapacitors based on aqueous solutions, organic solutions, and ILs as electrolytes are generally not suitable for a variety of applications, such as portable and flexible electronics because of the packaging difficulty. To this end, flexible all-solid-state supercapacitors are emerging as a new class of energy storage devices due to many important advantages, such as light weight, flexibility, and high safety compared to those based on liquid electrolytes (Wang et al., 2014d). A flexible supercapacitor consists of flexible electrode, solid-state electrolyte, separator, and packaging material. The solid-state electrolyte can not only serve as the ionic conducting media but also as electrode separator. It should share the following properties (Gao and Lian, 2014): (1) high ionic conductivity and low electrical conductivity at room temperature; (2) good mechanical strength; (3) good formability; (4) stable, and wide potential window; and (5) high electrochemical stability. Among various solid-state electrolytes, polymer gel electrolytes have been mostly studied as they display relatively high ionic conductivities (Lu et al., 2014; Zhong et al., 2015a). For example, gel polymer electrolytes exhibit ionic conductivities of 10^{-4} to 10^{-3} S cm^{-1} under ambient conditions and 10^{-8} to 10^{-7} S cm^{-1} at a dry state.

Polymer gel electrolytes typically consist of a polymeric framework as host material that gives high mechanical integrity; an organic/aqueous solvent provide the medium for ionic conduction; the electrolytic salt offers an ionic conduction. Therefore, the selection of host polymer, solvent and salt is the key to prepare high-performance polymer gel electrolytes. The most commonly used host polymers include poly(vinyl alcohol) (PVA), poly(acrylic acid) (PAA), potassium polyacrylate (PAAK), poly(ethylene oxide) (PEO), poly(vinylpyrrolidone) (PVP), poly(methylmethacrylate) (PMMA), polyacrylonitrile (PAN), carboxymethyl cellulose (CMC), chitosan, poly(vinylidene fluoride) (PVDF), polytetrafluoroethylene (PTFE), poly(vinylidene fluoride-*co*-hexafluoropropylene) (PVdF-*co*-HFP) and blend of poly(ethylene glycol), and poly(acrylonitrile) (PAN-b-PEG-b-PAN). Organic solvents are used as plasticizers and include ethylene carbonate (EC), propylene carbonate (PC), diethyl carbonate (DEC), dimethyl carbonate (DMC), dimethyl formamide (DMF), dimethyl sulphoxide (DMSO), ethyl methyl carbonate (EMC), and tetrahydrofuran (THF). The electrolytic salt is generally required to have large anions and low dissociation energy so that it can easily dissociate to provide free/mobile ions. According to the electrolytic salt, there are four categories of gel polymer electrolytes, that is, based on proton, alkaline, conducting salts, and IL.

6.2.2.1 Proton-conducting gel polymer electrolyte

Protons (H^+) have a higher mobility than Li^+ ions, which makes them promising as charge carriers for ultrafast charging–discharging supercapacitors. Proton-conducting polymer gel electrolytes are typically prepared by adding an acidic solution to an aqueous polymer solution under stirring. It works only when the mixing with aqueous acids does not result in chemical degradation of the polymer. H_2SO_4 and H_3PO_4 are the most commonly used proton donors. The commonly used polymers include PVA, PEO, PAAM, PVP, CMC, and chitosan. Among them, PVA is most explored because of its good film-forming capability, high water solubility, and adhesive characteristics, which simplifies the fabrication. As one of the first proton-conducting polymer gel electrolytes, H_3PO_4/PVA was described by Polak and coworkers in 1987. It displayed an ionic conductivity ranging from 10^{-4} to 10^{-2} S cm^{-1} in a gel state and 10^{-6} to 10^{-4} S cm^{-1} in the anhydrous state. Currently, proton-conducting polymer gel electrolytes have been extensively investigated to fabricate flexible all-solid-state energy storage devices. For example, an ultrathin all-solid-state supercapacitor was made by using CNT/PANI nanocomposite as electrode and H_2SO_4/PVA gel as electrolyte, which offers high and stable electrochemical performances under deforming (Meng et al., 2010).

Recently, it was demonstrated that the addition of redox-active components (such as hydroquinone, m-phenylenediamine, p-benzenediol, KI, Na_2MO_4, and $VOSO_4$) into the proton-conducting polymer electrolyte can contribute additional pseudocapacitance to the overall capacitance of supercapacitors. For example, PVA-H_2SO_4 polymer gel electrolytes containing redox additives of hydroquinone (HQ) and methylene blue (MB) were prepared to work on positive and negative electrodes, respectively. Due to the reversible Faradaic reactions related to HQ and MB in the corresponding gel polymer electrolyte, this quasi-solid-state supercapacitor showed high specific capacitance of 563.7 F g^{-1} and energy density of 18.7 Wh kg^{-1}, which is almost four times higher than that of a supercapacitor without redox additives (Zhong et al., 2015b).

6.2.2.1.1 CONDUCTING SALTS-BASED POLYMER GEL ELECTROLYTE

Commonly used conducting salts include Na^+ (e.g., Na_2SO_4, NaCl, and $NaNO_3$), Li^+ [e.g., $LiClO_4$, LiCl, Li_2SO_4, $LiBF_4$, $LiCF_3SO_3$, and $LiN(CF_3SO_2)_2$], and K^+ (e.g., KCl, KNO_3, and K_2SO_4,), etc. Conducting salts-based polymer gel electrolytes are typically prepared by mixing the polymers (e.g., PVP, PVA, CMC, PEO, and PAAK) with neutral salts in an aqueous solution or organic solvents like EC, PC, DMF, DMSO, etc. The formation of polymer gel electrolytes was demonstrated to enhance the electrochemical stability of electrode materials for supercapacitors. For instance, the vanadium oxide nanowire electrode-based pseudocapacitor device with PVA/LiCl gel electrolyte can exhibit higher specific capacitance, substantially enhanced structural stability, and cycling performance than that in the aqueous LiCl electrolyte (Wang et al., 2012a) (Fig. 6.5). The capacitance retention in the gel electrolyte achieves 85% over 5000 cycles while only 10% of the initial capacitance was maintained in LiCl aqueous solution. Here the gel electrolyte with polymeric flexibility is critical for enhancing the mechanical stability of vanadium oxide electrodes by serving as an elastic coating. Among various neutral salts, Li^+-based polymer gel electrolytes are typically used for high-performance lithium ion batteries or supercapacitors because of the wide electrochemical voltage window of Li^+ electrolytes. By using PAN-b-PEG-b-PAN copolymer as a host, DMF as a plasticizer and $LiClO_4$ as an electrolytic salt, the obtained polymer gel electrolyte exhibited an ionic conductivity of 6.9×10^{-3} S cm^{-1} with large potential window of 2.1 V (Huang et al., 2012). As for metal oxides-based pseudocapacitive

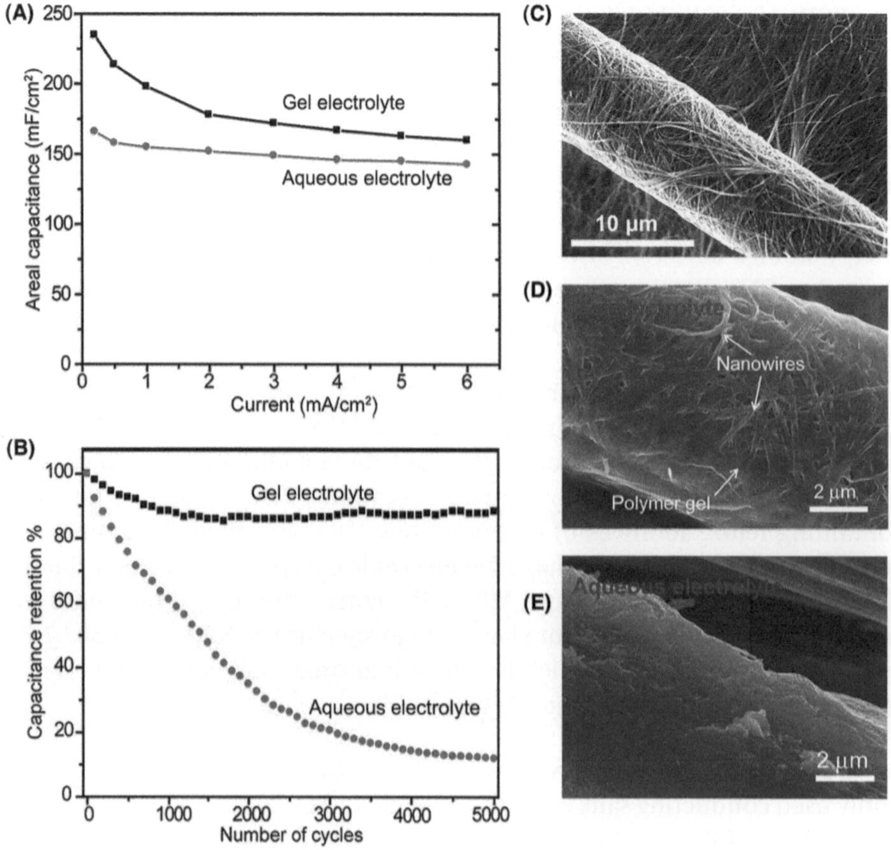

FIGURE 6.5 (A) Comparison on the specific areal capacitances of vanadium oxide in LiCl/PVA gel and LiCl aqueous electrolytes measured at different current densities. (B) Comparison of cycling performance in LiCl aqueous electrolyte and LiCl/PVA gel electrolyte. (C) SEM image of the as-prepared vanadium oxide nanowire networks. (D) and (E) SEM images of vanadium oxide nanowire electrode after testing for 5000 cycles in the gel electrolyte and 3000 cycles in the LiCl aqueous electrolyte, respectively. *Reproduced with permission from reference Wang, G., Lu, X., Ling, Y., Zhai, T., Wang, H., TONG, Y., Li, Y., 2012a. LiCl/PVA gel electrolyte stabilizes vanadium oxide nanowire electrodes for pseudocapacitors. ACS Nano 6, 10296–10302. Copyright 2012, American Chemical Society.*

electrode materials, especially the MnO_2 and V_2O_5, a neutral Na_2SO_4 was the most often used electrolyte since they are instable in an acidic medium. By applying the polymer PVP/Na_2SO_4 gel as electrolyte, a fiber-shaped asymmetric solid-state microsupercapacitors based on MnO_2-coated rGO/SWCNT fiber as positive electrode and N-doped rGO/SWCNT as negative electrode was fabricated (Yu et al., 2014). It delivered a high voltage window of 1.8 V, high volumetric energy density of 5 mWh cm^{-3}, as well as good cycling stability with 87% capacitance retention over 10,000 cycles.

6.2.2.1.2 ALKALINE POLYMER GEL ELECTROLYTES

Alkaline polymer gel electrolytes are commonly used in all-solid-state alkaline rechargeable batteries and supercapacitors. The typically used electrode materials for supercapacitors in alkaline polymer gel electrolytes include transition metal oxides/hydroxides [e.g., NiO/ $Ni(OH)_2$, Co_3O_4/$Co(OH)_2$, MnO_2, $NiCo_2O_4$], nitrides (e.g., TiN, VN), and sulfides (e.g., cobalt sulfide). Alkaline polymer gel electrolytes based on KOH (like PEO/KOH/H_2O and PVA/KOH/H_2O) have been widely used for solid-state supercapacitors because of its high ionic conductivity in comparison with NaOH and LiOH. The ionic conductivity of PEO/KOH/H_2O gel electrolyte reached the level of 10^{-3} S cm^{-1}, (Lewandowski et al., 2001) while an alkaline PVA/KOH/H_2O gel electrolyte gives even higher of 10^{-2} S cm^{-1} (Yang et al., 2005). The ionic conductivity was further enhanced to 0.53 S cm^{-1} at room temperature by incorporation of PVP into PVA/KOH/H_2O to form interpenetrating PVP chains in the highly cross-linked PVA network (Qiao et al., 2010). Similarly, the addition of redox-active components into alkaline polymer gel electrolytes, such as PVA–KOH–KI (Yu et al., 2011b) and PVA–KOH–PPD (Ma et al., 2014) had been also designed to further enhance the supercapacitor performance.

6.2.2.1.3 POLYMERIC IONIC LIQUIDS

Although ILs is studied for nonflammability, nonvolatility, high thermal, and chemical stability, wide electrochemical voltage window and high conductivity, they still suffer from problems of leakage, nonportability and impossibility of miniaturization due to the liquid nature. To solve the problem of fluidity in ILs, they are immobilized into various polymer matrices like PVdF-*co*-HFP, PTFE, PEO, PVA, PMMA, CMC, and chitosan to form solid or quasi-solid-state electrolytes (Ye et al., 2013). The resulting IL-based gel electrolytes exhibit promising advantages, such as high safety, good stability, improved operating temperature range, and flexibility. For example, a flexible bicomponent IL electrolyte film based on 1-ethyl-3-methylimidazolium tetrafluoroborate and 1-ethyl-3-methylimidazolium bistrifluoromethylsulfonylimide in PVdF-*co*-HFP matrix displayed a wide electrochemical window of 4 V, high ionic conductivity up to 3.5 mS cm^{-1} at room temperature and high thermal stability up to 350°C (Lu et al., 2008).

6.3 Lithium-Ion Batteries Based on Polymers

Commercialized from 1990, LIB has become a prevalent power supply for a large variety of electronic devices. An LIB basically comprises two electrodes, that is, cathode (positive electrode) and anode (negative electrode) separated by an ionic conductive electrolyte, providing a voltage derived from the potential differences between the two electrodes (Fig. 6.6) (Lee et al., 2014). The capacity is mainly derived from reversible electrochemical redox reactions between the two electrodes. As shown in Fig. 6.6, during charge process, lithium ions are removed from the cathode, transferred to the electrolyte, and intercalated into the anode. Meanwhile, the electrons flow from the cathode to anode in the external circuit. The process is reversed during discharging. An electrode, which is commonly a composite including active materials, binders, and conductive additives, is typically casted onto a current collector. A separator that absorbs electrolyte is generally placed between the two electrodes to avoid a short circuit. Polymers play an important role in

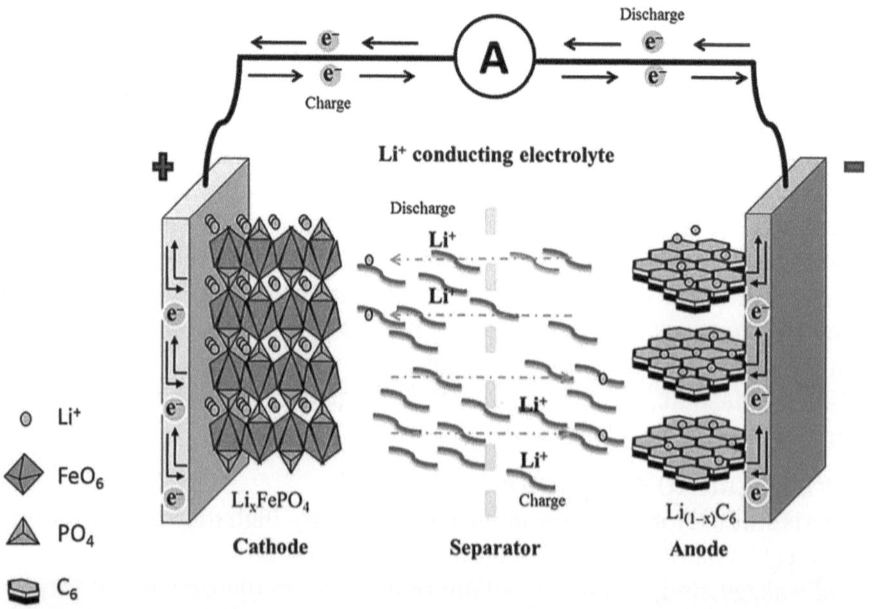

FIGURE 6.6 Schematic illustration to the energy storage mechanism of a typical LIB. *Reproduced with permission from reference Lee, H., Yanilmaz, M., Toprakci, O., Fu, K., Zhang, X., 2014. A review of recent developments in membrane separators for rechargeable lithium-ion batteries. Energy Environ. Sci. 7, 3857–3886. Copyright 2014, Royal Society of Chemistry.*

the construction of LIBs. The contributions of polymers will be discussed according to the following order, that is, as active materials in electrodes, as separators and finally as electrolytes.

6.3.1 Polymers as Active Materials in Electrode

The performance of an LIB, including voltage, capacity, longevity, and charge–discharge rate, fundamentally depends on active materials in cathode and anode. Inorganic materials currently prevail at both academy and industry. On one hand, $LiCoO_2$, $Li_4Ti_5O_{12}$, $LiMn_2O_4$, and $LiFePO_4$ are widely used as cathode materials. On the other hand, graphite, Si and Sn are widely explored for anode materials. Although these LIBs have achieved great success compared to other rechargeable batteries such as lead-acid, Ni-Cd and Ni-MH batteries, the further development of LIBs faces a bottleneck at the moment. First, the traditional inorganic materials in cathode are restricted by low theoretical specific capacity and poor structure stability. Second, the large-scale production of transition metal-based cathode materials is hindered by the limited resource and possible environmental pollution. Compared with inorganic materials, polymer electrode can offer much higher theoretical capacities. Besides, since many organic redox mechanisms have faster reaction kinetics, polymer electrodes show much higher rate performance and power densities than inorganic electrodes. Therefore, a lot of efforts are made to investigate polymeric materials to replace inorganic electrodes (Tarascon and Armand, 2001).

Table 6.1 The Structure and Redox Mechanism of Various Types of Polymer Electrodes

Structure	Redox mechanism	Examples	Advantages	Disadvantages
Conducting polymer	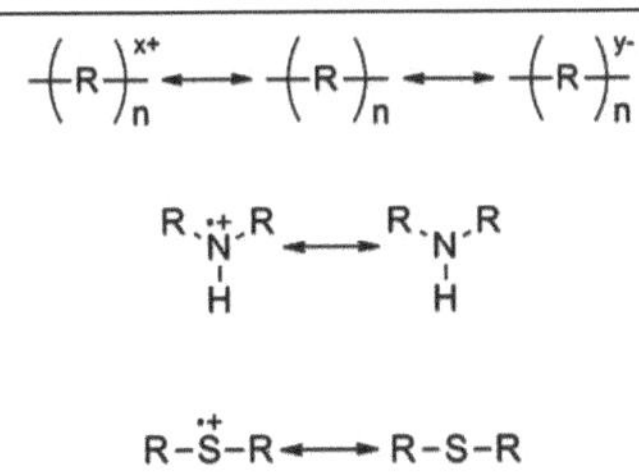	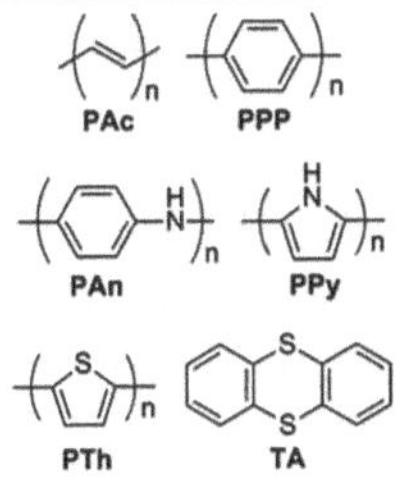	High conductivity Fast kinetics	Low energy density Low stability
Organodisulfide	R-S-S-R ⟷ R-S˙ + ˙S-R	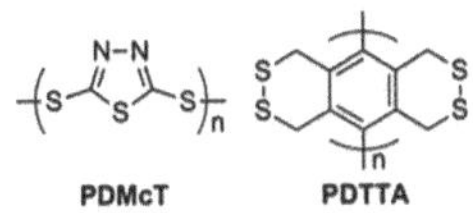	High capacity	Slow kinetics Low stability Low conductivity
Nitroxyl radical	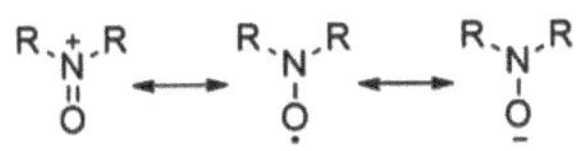	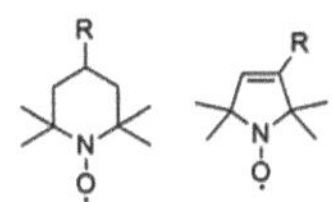	Fast kinetics	Low energy density Low conductivity
Conjugated carbonyl		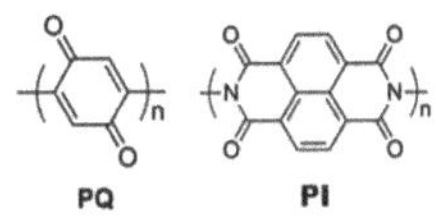	High capacity Fast kinetics	Low conductivity

The use of organic or polymeric electrode materials can be traced back to 1969. Among them, conducting polymers, organodisulfides, nitroxyl radical polymers, and conjugated carbonyl compounds have been mostly explored. The typical structure, redox mechanism, for example, advantage and disadvantage of each type of polymer material are summarized at Table 6.1. An active material in cathode or anode of the rechargeable battery is required to undergo a reversible electrochemical redox reaction. For inorganic materials, the redox reaction is related to the valence change of the transitional metal or elemental substance; for organic materials, the redox reaction is based on the charge state change of the electroactive organic group or moiety (Song and Zhou, 2013).

6.3.1.1 Conducting Polymers

Among various conducting polymers, there are five main types of polymers including polyacetylene (PAc), polyparaphenylene (PPP), polyaniline (PAn), polypyrrole (PPy), and polythiophene (PTh) that have been widely studied as electrode materials in the past decades. Due to the low doping levels of PAc, PPP, PPy, and PTh, their theoretical capacities are lower than 150 mAh g^{-1} and the practical capacities are even much lower (Zhu et al., 2013). Another critical problem of conducting polymers lies in their poor cycling stability, particularly in the case that the cutoff voltage is increased for a higher specific capacity. The irreversible overoxidation reaction at high voltage causes more structural units inactive during charge and discharge cycling. Besides, the serious self-discharge phenomenon also hinders the practical applications of conducting polymers in rechargeable LIBs.

6.3.1.2 Organodisulfides

The S–S bond in organodisulfides can be reversibly broken and rebuilt. Besides, a much higher capacity can be achieved due to the two-electron redox reaction compared with the conducting polymers. Therefore, organodisulfides were recognized as promising organic materials for rechargeable LIBs and there have been numerous studies on organodisulfide materials since 1988. However, this kind of polymers is limited by the problem of dissolution and slow kinetics. The typical example is PDMcT, and the structure is shown at Table 6.1. PDMcT owns a high theoretical capacity of 362 mAh g^{-1}, but the cycling performance is poor, which is derived from the serious dissolution of discharge product (Oyama et al., 1995). In addition, the low reaction kinetics of S–S bond is another critical problem. In order to overcome these problems, considerable efforts have been made to optimize the polymer structure and incorporating electrocatalytic additives. However, it is difficult to resolve the intrinsic drawbacks, and the electrochemical performances of organodisulfides are still far from an acceptable level for practical applications. Therefore, this research direction becomes dying down in recent years.

6.3.1.3 Nitroxyl Radical Polymers

As shown at Table 6.1, nitroxyl radical polymers are a bipolar organic material, and in most cases, they are used as p-type polymers in order to obtain high discharge voltage and stable cycling performance. Due to one-electron reaction and the large molecular weight, the theoretical specific capacity of nitroxyl radical polymers (100 ~ 150 mAh g^{-1}) is much lower than organodisulfides, and it is unlikely to be further improved by structural optimization. Poly(2,2,6,6-tetramethylpiperidine-1-oxyl-4-yl methacrylate), namely PTMA, is the most investigated material among various nitroxyl radical polymers. The charge–discharge plateau occurs at 3.5 V. The specific capacity is 110 mAh g^{-1} and can be well maintained in 100 cycles due to the stable structure (Nakahara et al., 2007a). Besides, the rate performance of PTMA are also excellent due to the fast reaction kinetics of the radical. For 2,2,6,6-tetramethylpiperidine-1-oxyl based LIB, a high current rate of 50 C can be also achieved, which arises from the fast electron exchange reaction (Oyaizu and Nishide, 2009). The fast reaction kinetics is the greatest advantage of nitroxyl radical polymers in the use for LIBs.

6.3.1.4 Conjugated Carbonyl Polymers

Conjugated carbonyl polymers are the most promising type of organic materials in LIB because of their high theoretical capacity, fast reaction kinetics, and structural diversity. The electrochemical redox mechanism can be generalized as an enolization reaction and a reverse reaction of the carbonyl group, and the conjugated structure is pivotal to promote the reaction. We take polyimide (PI) as an example, to show the typical redox reaction. During the reduction process, PI first transforms into the radical anion (PI^{-}) as transition state, and then transforms into dianion (PI^{2-}), along with the incorporation of lithium ions. This process is reversed in the oxidation process. Although PI has four carbonyl groups, only two of them can be utilized. If lowering the cutoff voltage to further reduce PI, the structure may be destroyed due to the serious charge repulsion interaction. Therefore, a two-electron reaction is usually adopted for rechargeable LIBs to obtain high reversibility and cycling stability.

In the past decades, a large number of conjugated carbonyl materials have been investigated as electrode materials for rechargeable LIBs. Overall, there are three types: small organic molecules, organic polymers, and organic salts. Herein, we focus on organic polymers. Polymeric quinone (PQ) were first synthesized from poly(1,4-dimethoxybenzene) and studied as a cathode for LIBs. The structure of PQ is shown at Table 6.1. It has a relatively high discharge voltage (2.5 – 3.0 V vs. Li^+/Li) and high theoretical capacity (505 mAh g^{-1}). However, the real discharge capacity was only about 120 mAh g^{-1}, and the cycling performance was poor, too, possibly due to the immature synthesis. After that, many other polymer electrode materials based on quinone and dianhydride have been investigated, which demonstrated higher electrochemical performances. For example, PI can obtain an average discharge voltage of 2.47 V and a practical capacity of 222 mAh g^{-1}. Besides, the capacity can be well maintained by 183 mAh g^{-1} after 100 charge and discharge cycles (Song et al., 2010).

Constructing polymer electrode materials with a stable, inactive skeleton and highly electroactive functional group is a good strategy to avoid the unwanted dissolution in the electrolyte and achieve fast kinetic properties. A small molar mass of the monomer unit is an important factor to achieve high specific capacity. In addition, the low electrical conductivity of the polymer electrode materials is also a drawback, which results in enlarged electrochemical polarization and slow electron transport in the electrode. In situ polymerization of active polymer material on conducting substrate such as carbon nanotube, graphene may represent an alternative approach to optimize the performance. If the hybrid electrode is well prepared, conjugated carbonyl based polymers can achieve high rate performances due to the fast reaction kinetics.

6.3.1.5 Polymer/Carbon Hybrid Electrodes

Electroactive polymer materials represent one of the most promising candidates for the next generation LIB electrodes due to their unique advantages, such as good design-versatility, high capacity, and good flexibility. However, they are limited by some intrinsic drawbacks in the practical applications. On one hand, except conducting polymers, most polymer electrode materials are plagued by their poor electrical conductivity, which results in slow electron transport during electrode reaction processes. Besides, rapid capacity decay during cycling is another drawback since some discharge products of polymer electrodes are highly soluble in organic electrolytes (Sengodu and Deshmukh, 2015). Consequently, more and more efforts have been devoted to overcoming this problem and improving their electrochemical performance. Among various strategies, coupling the polymer with conductive carbon, such as carbon fiber (CF), carbon nanotube (CNT), graphene, and porous carbon, to form composite electrodes offers the possibility to simultaneously solve these two main problems. In such elegant design systems, polymer component mainly contributes to the charge storage and conductive carbon suppresses the dissolution and facilitates the electron transfer of the polymer materials, which could promote the practical applications of polymer electrode materials for LIBs.

6.3.1.5.1 CARBON FIBER BASED HYBRID ELECTRODES

CFs have attracted increasing attentions due to their large surface areas and combined high strength, electrical and thermal conductivity, and chemical resistance. The combination of CFs with polymer electrode materials not only contributes to enhance the electronic conductivity of insulated polymer, but also increases the contact area between the electrode and electrolyte. As a result, the cyclability and rate capability can be remarkably improved. For instance, a homogeneous hybrid electrode with poly(2, 2, 6, 6-tetramethylpoperidinyloxy-4-ylmethacrylate) (PTMA) and conductive vapor-grown CF by a doctor blade method was designed (Nakahara et al., 2007a; Nakahara et al., 2007b). It almost realized a full utilization of PTMA, yielding a discharge capacity of 110 mAh g^{-1}, approaching to the theoretical capacity of PTMA (111 mAh g^{-1}) (Nakahara et al., 2007a). The hybrid electrode also exhibited high cycle stability, for example, 89% of the initial capacity after 1000 cycles. Moreover, the power capability was also very impressive as the LIB can be charged and discharged in 1 min (Nakahara et al., 2007b). It is noted that these polymer/CF hybrid electrodes also exhibited high mechanical strength and flexibility, which renders them as promising candidates for flexible devices.

6.3.1.5.2 CARBON NANOTUBE BASED HYBRID ELECTRODES

CNT is a candidate material for use in LIBs due to its unprecedented physical and chemical properties, such as high electrical conductivity, good chemical stability, and high mechanical strength. Attracted by these intriguing properties, a lot of efforts have been devoted to integrating CNTs into polymer materials to form hybrid electrodes. The prospect of incorporating CNTs into hybrid electrodes as a conductive additive presents an effective strategy to increase capacity and rate capability of the LIBs. A hybrid electrode of polyaniline and CNTs was obtained through the in situ chemical polymerization of aniline in CNT solution. The resulting LIB displayed a maximum discharge capacity of 86 mAh g^{-1} at the 80th cycle and showed an attractive discharge capacity of 65 mAh g^{-1} at 2 C rate, which is much higher than that gained by the composite of polyaniline and super-p carbon (Sivakkumar and Kim, 2007).

Although the typical solution fabrication is easy to operate with relatively high efficiency, the random aggregation of CNTs in polymer matrix greatly decreases the properties of the hybrid electrodes (Huang et al., 2011). To this end, aligned CNTs have been incorporated polymer materials for better electrodes. For instance, an aligned CNT/poly(3,4-ethylenedioxythiophene) hybrid electrode was synthesized to fabricate LIBs (Chen et al., 2007). They exhibited a highly stable discharge capacity of 265 mAh g^{-1} after 50 cycles, which is remarkably higher than the value of 173 mAh g^{-1} obtained from the CNT paper under the same conditions. This is attributed to the high accessible surface area and high electrical conductivity of the aligned CNTs, given to the fact that the CNTs in a CNT paper aggregate to form networks that decrease the electroactive surface area and electrical conductivity of the electrode.

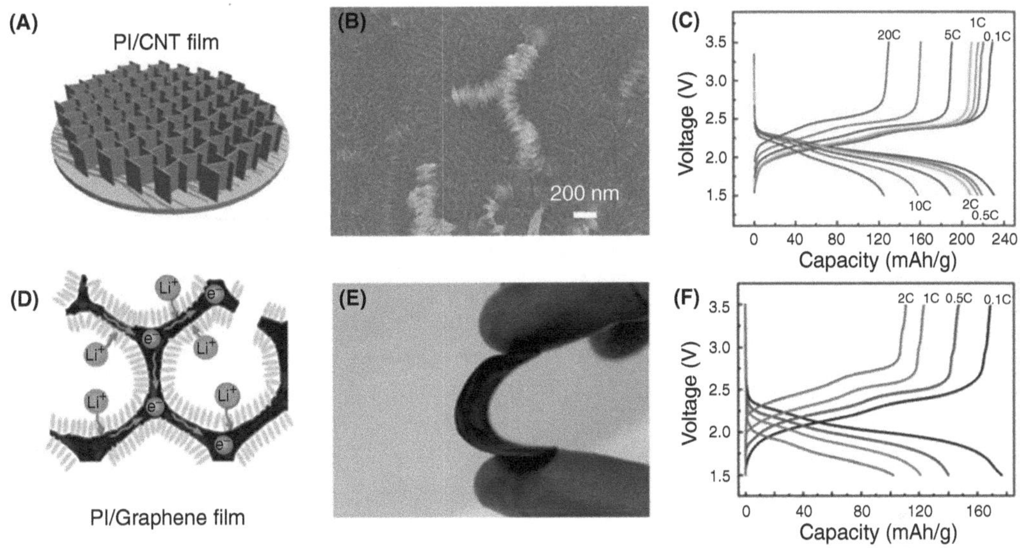

FIGURE 6.7 (A) Schematic of the PI/CNT film electrode. (B) SEM image of the PI/CNT film electrode. (C) Galvanostatic charging/discharging curves of PI/CNT film electrode. (D) Schematic of the PI/graphene film electrode. (E) High bendable property of the composite film. (F) Galvanostatic charging/discharging curves of PI/graphene film electrode. *(A–C) Reproduced with permission from reference Wu, H., Shevlin, S.A., Meng, Q., Guo, W., Meng, Y., Lu, K., Wei, Z., Guo, Z., 2014. Flexible and binder-free organic cathode for high-performance lithium-ion batteries. Adv. Mater. 26, 3338–3343. Copyright 2014, Wiley-VCH. (D–F) Reproduced with permission from reference Meng, Y., Wu, H., Zhang, Y., Wei, Z., 2014. A flexible electrode based on a three-dimensional graphene network-supported polyimide for lithium-ion batteries. J. Mater. Chem. A 2, 10842–10846. Copyright 2014, Royal Society of Chemistry.*

Recently, flexible LIBs are extensively studied for modern electronics (Ren et al., 2013). To this end, the composite of polymer with CNT film is an efficient strategy to construct flexible electrodes, which are indispensable in flexible LIBs. Wei and coworkers prepared a binder-free flexible electrode material using the CNT film as current collector and in situ polymerized polyimide (PI) as active material (Fig. 6.7A) (Wu et al., 2014). The flexible and porous structures of the CNT film guarantee good contact between the electrode and the current collector, which is important for the stability of flexible devices. Moreover, the highly conductive CNT film endows the hybrid electrode with high electrical conductivity and electrons can therefore rapidly transport from the electrode to external circuit. In addition, vertically aligned PI nanoflakes are formed in the processes of in situ polymerized (Fig. 6.7C), which benefit the electrolyte infiltration and realize high utilization of active materials. As a result, the PI/CNT hybrid electrode showed good rate capacity with the capacity of 226 mAh g^{-1} at 0.1 C and 120 mAh g^{-1} at 20 C (Fig. 6.7B). In view of its universality, this simple method can be extended to other polymer materials and is also promising for the application for other flexible electrodes.

6.3.1.5.3 GRAPHENE BASED HYBRID ELECTRODES

Graphene has shown unprecedented physical and chemical properties since it was purified in 2004 by Andre Geim and Konstantin Novoselov (Novoselov et al., 2004). These excellent properties include remarkable electrical conductivity, huge specific surface area in theory, high stability, and high mechanical strength. Therefore, graphene has been reckoned to be a kind of great material in polymer electrode for LIBs. A typical example is introducing graphene into PTMA through dispersing–depositing process (Guo et al., 2012). This PTMA/graphene composite had a good performance in rate capability, delivering a reversible capacity of 226 mAh g^{-1} at 0.1 C and ~80 mAh g^{-1} at 200 C. Moreover, its stable capacity was also surprising, which could reach 100 mAh g^{-1} for 20,000 cycles. However, the dispersing–depositing process resulted in low content of active material (10 wt.%), which made the composite less favorable.

To solve the problem and fabricate polymer/graphene hybrid with high loading of active materials, one effective strategy is in situ polymerization. A poly(anthraquinonyl sulfide) (PAQS)/graphene was demonstrated for the success of this method (Song et al., 2012). Due to a high dispersion capability of graphene, the electrical conductivity of the nanocomposite was increased by more than 6 times compared to pure polymers. As a result of high electrical conductivity, the PAQS-based cathode enabled ultrafast discharging and charging, which led to a capacity of about 100 mAh g^{-1} within only 16 s. There is an increasing expectation that this route can be used in other polymer materials.

Similar to CNTs, graphene can also be used to fabricate flexible and freestanding electrode. A polypyrrol/reduced graphene oxide (PPy/r-GO) hybrid film was prepared to show a high rate capability and good cycling stability when it was used as a cathode material in the flexible LIB (Yang et al., 2012). A 3D graphene was further developed as a conductive skeleton and PI as an active material (Fig. 6.7D and E) (Meng et al., 2014). The 3D graphene skeleton and porous structure facilitated the electron and ion transport. The resulting LIB exhibited a specific capacitance of 175 mAh g^{-1} at 0.1 C as well as high rate performance (Fig. 6.7F). In addition, the flexible PI/graphene hybrid film also has high potential as a cathode for flexible LIBs.

6.3.1.5.4 POROUSCARBON BASED HYBRID ELECTRODES

Porous carbon materials contain interconnected nanochannels which can be used in enveloping redox-active electrode materials. These materials help to improve the performance of polymer electrodes mainly in two ways. On one hand, the nanochannels can accommodate polymer, facilitate the electrolyte impregnation, and prevent the dissolution process. On the other hand, the carbon matrix structure enables the rapid electron flow. However, the performance of the organic-carbon hybrid is generally influenced by the diameter of the pore. Large pores can ensure the filling of molecules, but the prevention of dissolution cannot be ensured. For instance, an ordered nanocomposite of PTMA and mesocellular carbon foam (MCF) was made to show high discharge capacity of 113 mAh g^{-1}. Although the PTMA/MCF showed a stable cycle stability at 25°C, the capacity decay of the hybrid electrode was remarkable (80% after 50 cycles) when the temperature increased to 50°C (Kim et al., 2012). To improve the cycle stability of polymer/porous carbon, surface coating is a good strategy. The PTMA/MCF hybrid electrode was modified with a polyethylene glycol (PEG) coating. The resulting nanocomposite electrode exhibited an improvement in cycling performance (70% capacity retention after 50 cycles) at 50°C with a high capacity of 111 mAh g^{-1} (Kim et al., 2012).

6.3.2 Polymers as Separators

Separator plays an important role in LIB and is placed between cathode and anode to prevent the short circuit between two electrodes. It also helps to isolate the electronic flow as well as enable free ionic transport. The properties of the separator have a strong impact on the cycling stability, reliability, and safety of the LIB. It is critical to improve the safety and reliability without sacrificing the performance and increasing the cost. To choose an appropriate separator, several factors should be considered. The factors include chemical and electrochemical stability, wettability, pore size, thickness, porosity, permeability, mechanical property, puncture strength, and thermal shrinkage. Separators for LIBs can be classified into three major categories: microporous polymeric membrane, modified membrane, and nonwoven mat. Their advantages and disadvantages are illustrated at Table 6.2.

6.3.2.1 Microporous Polymeric Membrane

Microporous polymeric membrane separators are characterized by pore sizes in the micrometer scale. Microporous polymeric membrane separators are mainly made of polyethylene (PE), polypropylene (PP), and the combinations of them (PE/PP and PP/PE/PP) because of their high chemical and mechanical stabilities. According to the number of layers, they can be classified into monolayer and multilayer polymeric microporous membranes.

The dry and wet processes are two main manufacturing methods to prepare microporous polymeric membranes. Both methods are conducted through an extruder and a stretching process to increase the porosity and improve the tensile strength. Generally, separators made by dry process exhibit distinct slit-pore and straight microstructures, whereas those made by wet process show interconnected spherical or elliptical pores. Both methods use cheap polyolefin materials, so the microporous polymeric membranes are not expensive.

In a typical dry process, the melted polyolefin films are extruded and thermally annealed immediately below the polymer melting temperature to induce crystallization and to increase the size and amount of crystallites. The crystallites have a row lamellae. During the extrusion process, the crystallites are highly aligned. During the annealing process, the crystalline structure of the polymer film is significantly improved which will be beneficial to the formation of porous structure in the following stretching step. The uniaxial stretching process is conducted at a low temperature to form ordered micropores, which is caused by fracturing of the tearing apart of the lamellar crystallites. In industry, both uniaxial and biaxial stretching are applied but the uniaxial stretching method obtains better slit-like pore structures. By using this method, a higher tensile strength is achieved, which is desired for handling the separators during assembling an LIB. A typical dry process is simple without the use of solvents, but it only applies to polymers that can form semicrystalline structures.

Table 6.2 Summary of Membranes Utilized in LIB and Their Comparison Based on Fabrication Technology

Category	Material	Substrate	Membrane Thickness	Liquid Electrolyte	Electrolyte Uptake	Ionic Conductivity	Performance	References
Monolayer polymeric membrane	PP, PE		25 μm	$LiPF_6$-EC/DMC	15 ± 2.7%	0.4×10^{-4} S × cm^{-1}	Thin thickness, high porosity, and large pore size	Ryou et al. (2012)
Monolayer polymeric membrane	PVDF		43 ± 2 μm	$LiPF_6$-EC/DMC		2.1(±0.12) × 10^{-3} S cm^{-1}	High rate performance and good cycling ability	Djian et al. (2009)
Multilayer membrane	PAN	PET	~40 μm	$LiPF_6$-EC/DMC/EMC	50–67%	5.37×10^{-4} S cm^{-1}	High specific capacity and good cycling performance	Huai et al. (2010)
Multilayer membrane	PMMA	PE	~25 μm	$LiPF_6$-EC/DEC	200%	3.1×10^{-4} to 5.2×10^{-4} S cm^{-1}	Better discharge C-rate performance	Park et al. (2011)
Modified membrane via coating	PDA	PE	25 μm	$LiPF_6$-EC/DMC/PC	112 ± 3.1%	3×10^{-4} S cm^{-1}	Better C-rate performance and high power density	Ryou et al. (2012)
Modified membrane via deposition	Al_2O_3	PP	25 μm	$LiPF_6$-PC	30%	0.4×10^{-4} S × cm^{-1}	Thin thickness and better wettability	Jung et al. (2012)
Modified membrane via plasma modification	AN	PE	23 μm	$LiPF_6$-DEC/EMC/EC		1.4×10^{-3} S cm^{-1}	Good wettability, stable columbic efficiency, and better cycle performance	Kim et al. (2009)
Electrospinning	PVDF		30 μm	$LiPF_6$-DEC/EMC/EC	320–350%	1.0×10^{-3} S cm^{-1}	Stable charge–discharge and good cycle performance	Costa et al. (2012)

PP, Polypropylene; PE, polyethylene; PVDF, poly(vinylidene fluoride); PAN, polyacrylonitrile; PMMA, poly(methylmethacrylate); AN, acrylonitrile; PET, poly(ethylene terephthalate).

A typical wet process involves five steps: (1) adding plasticizers or low molecular weight substances (paraffin oil and mineral oil) to the polymer, (2) raising the temperature to melt the mixture, (3) extruding or blowing the mixture into a gel-like film, (4) stretching the film biaxially to enlarge the pore size and increase the porosity, and (5) extracting the plasticizer from the film with a volatile solvent to form submicron-sized pores. Compared with the dry process, this wet process can be applied in a wider range of polymers. The use of plasticizers significantly reduces the viscosity which is helpful for the polymer processing. However, the plasticizer extraction in this wet process adds the cost of the microporous polymeric membrane production.

6.3.2.1.1 MONOLAYER POLYMERIC MEMBRANES

Static and dynamic thermomechanical analysis of the commercial polymer separators (PP and PE) were made to understand their properties (Love, 2011). Anisotropic separators manufactured from a dry process showed limited tensile performance in the transverse direction. The separators prepared from a wet process displayed a uniform and biaxial structure and nearly showed identical mechanical strength on both directions. It was also found that small losses in mechanical integrity were observed after the separators were exposed to various electrolytes.

Due to the intrinsic characteristics of the polyolefin microporous monolayer membranes, such as low wettability and poor electrolyte retention as well as limited thermal stability, some other polymers, such as polyvinylidene fluoride (PVDF) (Djian et al., 2009; Saunier et al., 2004) and its copolymers (Jeschke et al., 2014; Costa et al., 2013), polyacrylonitrile (PAN) (Jung et al., 2005; Appetecchi et al., 1997) and poly(methyl methacrylate) (PMMA) (Lee et al., 1996) have also been used for the preparation of microporous membranes. PVDF membranes have great wettability due to the good affinity of PVDF to electrolyte solutions. PVDF and its copolymers have advantages over the usual polyolefin separators due to their high polarity and high dielectric constant, which can assist the ionization of lithium salts in the electrolyte. Another strategy for enhancing the ionic conductivity, mechanical and thermal properties is the preparation of microporous monolayer membranes using polymer blends. Microporous PAN membranes exhibit good thermal stability, desirable electrolyte retention and excellent electrochemical stability, which can also act as a polymer matrix to help conserve the electrolyte. PMMA is another candidate for making monolayer polymeric membranes because of its affinity to the electrolyte.

Due to the low preparation cost, monolayer polymeric membranes have been widely used as separators for LIBs. However, limited by the relatively single function of monolayer polymeric membranes with relatively poor puncture strength and thermal stability, monolayer polymeric membranes may not be able to meet many application demands.

6.3.2.1.2 MICROPOROUS MULTILAYER MEMBRANES

Monolayer polymeric membranes cannot satisfy all of the optimal characteristics that are related to the LIB safety and performance. Therefore, microporous multilayer separators combining different polymers with different functions have been fabricated. A variety of processes including lamination, slot-die, dip coating, gravure, and casting can be used to make the multilayer membranes.

Commonly used separators, such as Celgard 2325 are composed of a PP/PE/PP trilayer structure that provides both high puncture strength and thermal stability. In this structure, the low-melting PE layer can act as a thermal fuse, and the higher-melting PP layer provides a high mechanical strength. At the same time, this method has effectively prevented the external short circuit and accidental overcharge and improved the safety of the LIB. Other substrates, such as nonwoven substrates [polyethylene and poly(ethylene terephthalate)] have also been used to produce microporous multilayer membranes. For the preparation process, the polymer is first dissolved in a solvent to prepare a coating solution. The resultant coating solution is uniformly coated onto the nonwoven mats. Nonwoven PE substrates are the most widely used to make microporous multilayer membranes. The nonwoven PE mat provides good ionic conductivity, high mechanical strength, and great electrochemical performance for the LIB. Nonwoven poly(ethylene terephthalate) (PET) substrates have also been coated with PVDF to make microporous multilayer membranes. The obtained PET multilayer membranes show high wettability and ionic conductivity, good mechanical and electrochemical stability. For instance, a nonwoven PET substrate was coated with PVDF to achieve a combined high conductivity, wettability, and mechanical and electrochemical stability.

To prepare multilayer membranes, another irradiation method to prepare cross-linked microporous multilayer membranes with enhanced thermal stability has been developed. It is realized by two steps. First, the polymer-blended layers, such as poly(ethylene glycol) diacrylate/poly(ethylene glycol) methyl ether acrylate are coated onto polyolefin microporous membranes. Second, the resultant membranes are irradiated to form chemically cross-linked membranes. They exhibited higher thermal and electrochemical properties compared to conventional separators. With the increase of irradiation dose, the thermal stability of the resultant membranes increases accordingly. By using the microporous multilayer membranes, the advantages of each component layers are well combined.

6.3.2.2 Modified Microporous Polyolefin Membranes

Polyolefin separators, especially PE and PP membranes, reveal the advantages of low cost, proper mechanical strength, controllable pore structure, high electrochemical stability, and good thermal shutdown properties. They have been thus mostly used for LIBs over decades. However, pristine polyolefin membranes lack sufficient compatibility with the electrode and carbonate electrolyte resulting from their intrinsically hydrophobic surface, low surface energy and wettability, and unsatisfactory electrolyte retention and ion conductivity. In a typical LIB, good contact between electrode and separator is of great significance because a poor interface may cause an uneven current distribution and increase the interface resistance and eventually lead to the formation of lithium dendrites. This will impair the electrochemical performance of the LIB, particularly, lowering the power capability and cycle life. A number of efforts have been made to alter and improve surface hydrophobicity, mainly focusing on surface modification and doping extra functional components into the polyolefin matrices in order to optimize the overall performance of the LIB. The mostly explored method is based on the surface modification with polymer by either physical coating or chemical grafting.

6.3.2.2.1 PHYSICAL COATING PROCESS

Traditionally, for improving the interfacial adhesion, a surface modification is performed by a solution-based spraying or dipping method with the liquid electrolyte serving as the glue component. There exists a subsequent heating step to facilitate the formation of polymer gel electrolyte. But for some occasions, the dense polymer layer may block the penetration of the electrolyte and afterward undermine the LIB performance. Therefore, several attempts have been conducted to alter the surface character without sacrificing the porous structure.

Dip-coating and phase inversion methods had been proposed as promising solution processes to maintain the porous structure. The original PE membrane was initially immersed into dimethylfluoramine (DMF) solution containing acrylonitrile (AN)–methyl methacrylate (MMA) copolymer (Jeong and Kim, 2004). Afterward, the coated membrane was soaked into water solution, enabling efficient solvent exchange and thus ensuring the formation of microporous structure. The LIB fabricated from the AN–MMA modified membrane showed high capacity retention and good rate capability.

Another possible solution lies in coating ultrathin films with robust adhesion ability. A polydopamine-coated PE separator (Ryou et al., 2011; Ryou et al., 2012) was prepared through a simple dipping process. Due to the ultrathin polydopamine film, the original advantages of the PE membrane, such as the pore size have been well maintained. Meanwhile, the uptake amount of liquid electrolyte increased from $15 \pm 2.7\%$ to $112 \pm 3.1\%$, and the ionic conductivity increased from 0.04×10^{-3} S $\times$ cm^{-1} to 0.3×10^{-3} S $\times$ cm^{-1} after coating. Furthermore, attributing to the versatile adhesion capability and inherent hydrophilic nature of polydopamine, the formation of lithium dendrites has been suppressed. Thus, the modified membrane exhibit greatly enhanced power and cycle performance. With the assistance of dopamine, the other polymer-coated separator, such as PMMA (Shi et al., 2013) and PEG (Fang et al., 2013) coated propylene separator have been also fabricated. They had shared a relatively high ion conductivity.

Besides, it has been reported that some inorganic nanoparticles such as silicon oxide (SiO_2), aluminum oxide (Al_2O_3), and titanium oxide (TiO_2) can significantly enhance the physical and chemical properties of the membrane (Locquet et al., 1998; Croce et al., 1999). Thus in order to improve the thermal stability and wettability of the polyolefin membrane, inorganic nanoparticles were also deposited onto the surfaces with hydrophilic polymer as a binder. For example, a trilayer composite membrane was prepared by coating SiO_2 nanoparticle particles onto PE membranes (Jeong and Lee, 2011) with PVDF-*co*-HFP as a binder. The coating solutions were prepared by mixing inorganic nanoparticles and PVDF-*co*-HFP binder in acetone with a filler-to-binder ratio of 9:1 (by weight). Then the microporous membrane was dipped into the coating solution for several times. The resulting membranes allowed higher porosity and better ionic conductivity.

Besides, the fabricated composite membranes demonstrated good thermal stability and capacity retention. However, the binding method also faced several drawbacks, such as the increase of thickness and decrease of porosity. Fortunately, based on the controllable thickness of the deposition layer, the earlier problems may be lowered in some systems. For instance, a thin layer of Al_2O_3 (Jung et al., 2012) was coated on the surface of porous polymer microframework only with an extra thickness of less than 10 nm and with the pores well maintained.

6.3.2.2.2 CHEMICAL GRAFTING PROCESS

Physical coating processes previously mentioned are usually simple and low cost, and they can also maintain the original mechanical strength of the polyolefin separator. However, the weak interfacial interaction (usually wan der Waals attraction) can only make the membrane hydrophilic temporally since the physical coating layers are subjected to delaminate from the membrane when immersing in the electrolyte and through the repeated cycling and storage process.

Chemical grafting polymerization represents another efficient method to modify the structure and property of the microporous membrane, which can generate a strong coat on the surface through covalent bonds so as to change the surface polarity permanently. For instance, a porous separator was immersed into a mixed solution of photoinitiator and monomer, followed by polymerization under UV irradiation (Senyarich and Viaud, 2000). In this case, the electrolyte retention capacity and wettability depends on the degree of the polymerization and the type of the used monomer.

However, traditional polymerizations always require a preengineered separator surface with active groups prior to grafting, which may involve complex even sometimes violet, multistep treatments, such as ultraviolet or heat. Inevitably, these treatments may introduce chemical residue (e.g., catalyst and initiator) and cause irreversible shrinkage and aging of the separators, which adversely degrade the electrochemical reaction and decrease the durability of the membrane.

Alternately, energetic particle radiation-assisted methods (e.g., plasma, gamma irradiation, and electron beam irradiation) enabled a rapid formation of active sites to initialize the all-solid-state polymerization process and avoid the introduction of chemical impurities. A PE membrane had been coated with acrylonitrile (AN) (Kim et al., 2009) via a plasma technology. The ion conductivity was improved from 0.4 to 1.4 mS/cm, and the average peeling force was increased to 22.6 N/m or by up to 18%. The resulting LIB also exhibited rather a stable cycle performance.

Electron beam irradiation is another commonly used method to modify the surface hydrophlicity of microporous polyolefin membranes. In a typical process, a microporous PP membrane was grafted with acrylic acid and diethylene glycol-dimethacrylate (DEGDM) (Nasef and Hegazy, 2004) monomers using electron beam irradiation to create a hydrophilic surface. The modified membranes showed enhanced electrolyte uptake, ionic conductivity, and cycle life as compared with unmodified PP membranes. Glycidyl methacrylate (GMA) and methyl methacrylate (MMA) (Kim and Lim, 2010) have also been grafted onto the surface of PE membranes through the same method. The grafted monomers on the surface of PE membranes increased the electrolyte uptake and retention, leading to improved electrochemical performance.

Gamma ray irradiation (Shi et al., 2013) is another energetic resource which can benefit the chemical grafting process. The effect of gamma irradiation on morphological, thermal, and electrochemical properties of polyethylene membranes was investigated. The cross-linked polyethylene membrane revealed better thermal shrinkage resistance and the melt integrity temperature was increased when the higher dose of irradiation was used. As expected, a graphite/$LiCoO_2$ based LIB using the modified membrane demonstrated a higher discharge capacity than those using unmodified membranes, particularly at high rates.

The previous methods can generate a lot of active sites (carbon radicals) on backbones of organic materials to initiate the cross-linking reactions between targeting molecules and separators, which also significantly simplify the preparation compared with the previous dry and wet process. However, the high-energy particles arised from electrons, ions, and radicals may lead to the degradation of the polymer backbone, the damaging of targeting molecules, and the formation of undesirable chemical functionalities simultaneously.

6.3.2.3 Nonwoven Mat as Separator

Nonwoven mat as a traditional separator material has certain advantages, such as low processing cost, high porosity, and light weight. In addition, mats with different organic and inorganic fibers can be conveniently fabricated to provide high heat resistances and good mechanical properties. However, nonwoven mat separators typically have large pore sizes with the consideration of preventing dendrites from growing through the pores. Therefore, they have been rarely considered for LIBs in the past decade. Fortunately, some new emerging technologies make the fabrication of nanofiber nonwoven mats available for satisfying LIBs in the near future.

A typical nonwoven mat can usually be produced through a dry laid process, a wet-laid process, a melt-blown process, or an electrospinning process. The air-laid process is a dry process where individualized staple fibers are air conveyed onto a moving belt to form a mat. A latex binder is used to incorporate the mechanical integrity. The wet-laid process is usually conducted on a papermaking machine. Fibers are first suspended in water. Paper machines are used to separate water to form a uniform web, which is then bonded and dried to achieve mechanical integrity. Melt-blown mats are formed by extruding molten polymers through a die and then attenuating the resulting filaments with a high-velocity air to fine diameters. Melt-blown is a relatively low-cost process to fabricate thermally stable thin membranes that are suitable for the separator in LIB. The preparation of thin but strong separators made of polyester, polyamide, and polymethylpentene and its copolymers through this process has been widely studied.

Electro-spinning/electrospraying is another attractive method for preparing nonwoven mats, especially those composed of nanofibers. The electrospinning/electrospraying process is conducted by applying a strong electric field to a polymer solution and collecting a fine-charged jet from a small orifice on a grounded collector. The mat morphology is affected by the solvent evaporation rate, the polymer concentration, the distance from the nozzle tip to the target, the accelerating voltage, etc. This process is applicable for submicron or nanometer-sized fibers. Thus, electrospinning method can produce much thinner nonwoven mats. Due to the challenges of designing high-throughput electrojetting process, no commercial separators are currently available. Porous membranes composed of submicron fibers made of polyacrylonitrile (PAN) (Cho et al., 2007), and PVDF (Lee et al., 2006) have been also investigated. The as-prepared PVDF membrane demonstrated extremely high electrolyte uptake such as 300%. Compared with the conventional porous separators, PAN nonwoven mats showed better cycling performance and rate capability at room temperature. Some modifications are also conducted on the nonwoven mat to broaden its applications with better performances. For one instance, the electrospun PVDF nanofibers (Choi et al., 2004) with ethylene plasma provided the web surface with a low melting PE layer, in which polymerized PE layer acted as a shutter by melting at elevated temperatures, thus enhancing the safety of the resulting LIB. For another instance, a ceramic separator was modified by coating an ultrathin PET nonwoven support layer with oxides including alumina, zirconia, and silica to enhance its mechanical strength and thermal stability (Zhang, 2007; Cho et al., 2008).

Electro-spun nonwoven mats reveal many advantages as LIB separators, including small pore size, high porosity, large surface area, interconnected open structure, and high permeability. However, the relatively low mechanical and thermal properties have greatly limited their practical applications. In order to build more robust separators, more adapting material should be explored. Furthermore, considering the disadvantages of electrospinning, such as low safety, slow production, and high cost, some new production techniques should be further explored aiming at large-scale applications.

6.3.3 Polymers as Electrolytes

The desired electrolyte properties are determined by the working mechanism as well as the structure and type of the energy storage devices. The primary function of an electrolyte in an LIB is to efficiently transport Li^+ ions between two electrodes but impede electrons. An electrolyte typically contains ion-conducting components that are composed of functional or structural components, such as additives and structural enhancements. Based on the states of matter at room temperature, there are basically three types of electrolytes used for LIBs, that is, liquid, solid, and gel electrolytes. The promising merits of liquid electrolytes are high ionic conductivity and the ability to form good/stable contact with electrodes. With the development of flexible and wearable electronics, solid, or gel electrolytes are preferred as they are safer and adaptable to the design of flexible and multifunctional LIB. Many polar polymers, such as PEO, PVDF, PVDF-HFP, PMMA, and PAN have been employed as the polymer host (solid component) for gel or solid electrolytes owing to their good affinity with liquid components.

6.3.3.1 Poly(ethylene oxide)-Based Electrolyte System

Among various electrolytes, PEO-based electrolyte systems are extensively studied since the ionic conductivity of PEO/alkali metal salt complexes was discovered by Wrights in 1973. PEO is a typical semicrystalline polymer with poor electrical conductivity, but it can form various complexes with lithium salts because of the coupling effects between Li^+ and the oxygen atoms on the PEO chains, which improve the dissolution of lithium salts (Agapov and Sokolov, 2011; Chandrasekaran and Selladurai, 2001). However, PEO-based gel electrolytes offer very low ionic conductivity in the range of 10^{-8}–10^{-4} S cm^{-1} at the temperatures between 40 and 100°C, which cannot meet the demand of practical applications. Many previous studies indicate that the amorphous phase with activated chain segments can assist the ion transport. Therefore, much attention has been devoted to increase the percentage of the amorphous phase of a PEO electrolyte. The methods including addition of nanoparticles (such as Al_2O_3, SiO_2, TiO_2, and CuO) or plasticizers and blending with other polymers, (Ji et al., 2011; Tang et al., 2012; Johan et al., 2011) have been employed to enhance ionic conductivity. Copolymers with conductive blocks (usually PEO block) are also promising candidates for high-performance gel electrolytes. The ionically conducting PEO block provides a pathway for ion transport, and other blocks, such as polyethylene (PE) or polystyrene (PS) blocks, form 3D connected frameworks for good mechanical properties. More importantly, the two blocks can form unique patterns with special pathways for ion conduction. (Young and Epps, 2012)

6.3.3.2 Poly(methyl methacrylate)-Based Electrolyte

With carbonyl groups existing in MMA unit, the strong interactions of PMMA with oxygen atoms in carbonate plasticizers endow the gel electrolyte with a good compatibility and high transport for lithium ions. A main limitation of PMMA electrolyte is the poor mechanical strength. Many recent works have undertaken to prepare the PMMA-based gel electrolytes with high mechanical strength as well as high ion conductivity and chemical stability. Polymer matrix embedded with nanofillers like montmorillonite organoclays, mesoporous silica, and aluminum oxide has been shown to be an effective technique to improve the mechanical strength, electrochemical and interfacial stability with electrodes. Specially, the reinforcing capabilities of nanofibers are superior to those of nanoparticles because of their high aspect ratios with good flexibility (Endo et al., 2001; Paul and Robeson, 2008). Carbon nanofibers are famous for their extraordinary mechanical, chemical, and thermal properties. The addition of carbon fibers into the PMMA-based $LiClO_4$-(PC + DEC) gel electrolyte was demonstrated to improve the thermal stability, electrical conductivity, and mechanical strength (Sil et al., 2015). The composite gel electrolyte yields a high strength of 5.28 MPa at 2 wt.% carbon fiber.

6.3.3.3 Polyacrylonitrile-Based Electrolyte

PAN-based gel electrolytes have been widely investigated as the plasticizer and salt components are molecularly dispersed to form homogenous, hybrid electrolyte films. The PAN in the gel electrolyte is deemed to mainly act as an encapsulating matrix material. PAN-based gel electrolytes have several remarkable properties including high electrical conductivities (10^{-3} S cm^{-1}) at room temperature, wide electrochemical voltage window (>4.5 V), high thermal stability, flame retardancy, and good film-processing ability, which are more suitable for the use in flexible LIBs. Due to the absence of O atoms in the PAN matrix, the PAN-$LiClO_4$ based gel electrolytes were found to have a high lithium ion transference number of 0.5, and a even higher value of 0.7 has also been achieved in LiTFSI and LiTFSM salts with a better dissociation (Appetecchi and Scrosati, 1998, Feuillade and Perche, 1975).

To further improve the ionic conductivity and mechanical properties of the PAN-based gel electrolyte, a lot of efforts have been made to reduce the viscosity and increase the amorphous phase. The PAN/clay nanocomposite gel electrolyte exhibited a high ionic conductivity of 1.4×10^{-2} S cm^{-1}, superior film-processing ability, dimensional stability, and plasticizer-maintaining ability in comparison to a bare PAN gel electrolyte (Hwang and Liu, 2002). The exfoliation and intercalation of organophilic clay layers makes the polymer chain of PAN to be amorphous, and the dispersion of inorganic nanoparticles in polymer electrolytes gives many free Li^+, which are all helpful for conductivity. Copolymerization with other polymers is commonly used to reduce the crystallinity of the polymer host and enhance the mechanical stability of the gel electrolyte. With the incorporation of poly(vinyl acetate) and poly(methyl methacrylate), the blending gel electrolytes shared a wide electrochemical voltage window (>6.5 V), high Li^+ transference number of 0.6 and high ionic conductivity of 3.5×10^{-3} S cm^{-1} at room temperature (Wang et al., 2014c). Based on the robust contact of the gel electrolyte with two electrodes in full LIB, a long-term cycling performance at high charge–discharge rates, such as 93% capacity retention for 100 cycles and 63% for 1000 cycles at 17°C, was obtained.

6.3.3.4 Poly(vinylidene fluoride)-Based Electrolyte

PVDF-based gel electrolytes display attractive electrochemical stability and flame resistance because of the electron-withdrawing functional group (–C–F) in PVDF. Meanwhile, the intrinsic high dielectric constant ($e = 8.4$) is helpful for greater dissolution of lithium salts, and then providing high concentration of charge carriers. Two main drawbacks for the application of PVDF in gel electrolyte include: (1) C–F bond can be easily reacted with Li electrode, so PVDF-based gel electrolytes can only be used in LIB with graphite anode and (2) the crystallization of PVDF impedes the ionic conductivity. Modification through grafting, copolymerization, cross-linking and blending has been widely explored to reduce the crystallinity of the PVDF. PVDF-HFP-based gel electrolytes are the commonly used ones for LIB as the addition of HFP in the system can reduce the crystallinity of PVDF and also improve the electrolyte-absorption ability. The PVDF-HFP copolymer is semicrystalline with the HFP content in the range of 12–19%. Further increasing the amount of HFP leads to a completely amorphous elastomer. Influences of lithium salts ($LiCF_3SO_3$, $LiBF_4$, and $LiClO_4$) on the ionic conductivity, thermal stability, and compatibility of PVDF-HFP-based gel electrolytes have been studied. The highest electrical conductivity occurred at the $LiBF_4$-based gel electrolyte, which might be ascribed by the greater dissolution of $LiBF_4$ in comparison to the others (Stephan et al., 2005). The addition of mesoporous silica to the

PVDF-HFP-based gel electrolyte was demonstrated to enhance the discharge capacities of $LiNi_{0.5}Co_{0.2}Mn_{0.3}O_2$ polymer batteries form 159.5–198.6 mAh g^{-1} with a high coulomb efficiency of ca. 99% (Yang et al., 2014a). The mesoporous structure and high specific surface area of mesoporous silica are speculated to trap a large amount of liquid electrolyte to enhance the ionic conductivity.

6.4 Summary

In summary, the recent progress in the development of high performance electrode materials and electrolytes based on polymers and their composites have been discussed for supercapacitors and LIBs. The formation of composite materials is summarized as an effective strategy toward high performance electrodes for the electrochemical energy storage devices. The rational design and morphology control to optimize the polymer materials represent another strategy in developing promising supercapacitors and LIBs. In particular, the two strategies are simultaneously taken into consideration in the advancement of electrode materials. For one instance, the combination of electrochemical active polymers and highly conductive carbon materials in a 1D coaxial nanostructure display remarkable electronic, electrochemical, and mechanical properties for efficient energy storage because this configuration can provide an efficient pathway to transport charges and also reduce the transport length. For another instance, the synthesis of 3D hierarchical nanocomposites with large specific surface area and high porosity proves effective in enhancing the storage capability of the previous devices. In addition, the synergistic effect from multiple components is expected to overcome the limitation of pseudocapacitor materials (e.g., low conductivity and poor cyclic stability) and generate further improved specific capacitance, rate capability, and cycling stability at high electric currents.

The design of appropriate gel electrolytes with a wide and stable operation voltage window is necessary for high-performance supercapacitors and LIBs, particularly, flexible and all-solid-state energy storage devices. Although considerable achievements have been made in synthesizing various nanostructured composites based on polymers over the past several years, there remain some challenges that need to be addressed, such as complicated synthesis procedures, low load of active materials, low specific capacitance, low rate capability, and poor cycling stability. Therefore, increasing interests are attracted to develop large-scale synthetic methods with low cost and environmental friend. In addition, the effective control on the composition, structure, and morphology of the gel electrolytes is also important to enhance their electrochemical properties aiming at supercapacitors and LIBs with high and stable performances.

References

Agapov, A.L., Sokolov, A.P., 2011. Decoupling ionic conductivity from structural relaxation: a way to solid polymer electrolytes? Macromolecules 44, 4410–4414.

Appetecchi, G., Scrosati, B., 1998. A lithium ion polymer battery. Electrochim. Acta 43, 1105–1107.

Appetecchi, G., Croce, F., Scrosati, B., 1997. High-performance electrolyte membranes for plastic lithium batteries. J. Power Sources 66, 77–82.

Benson, J., Kovalenko, I., Boukhalfa, S., Lashmore, D., Sanghadasa, M., Yushin, G., 2013. Multifunctional CNT-polymer composites for ultra-tough structural supercapacitors and desalination devices. Adv. Mater. 25, 6625–6632.

Cai, Z., Li, L., Ren, J., Qiu, L., Lin, H., Peng, H., 2013. Flexible, weavable and efficient microsupercapacitor wires based on polyaniline composite fibers incorporated with aligned carbon nanotubes. J. Mater. Chem. A 1, 258–261.

Chandrasekaran, R., Selladurai, S., 2001. Study of a new polymer electrolyte poly(ethylene oxide): NaClO3 with several plasticizers for battery application. Polym. Int. 50, 89–94.

Chen, J., Liu, Y., Minett, A.I., Lynam, C., Wang, J., Wallace, G.G., 2007. Flexible, aligned carbon nanotube/conducting polymer electrodes for a lithium-ion battery. Chem. Mater. 19, 3595–3597.

Chen, Y.-C., Hsu, Y.-K., Lin, Y.-G., Lin, Y.-K., Horng, Y.-Y., Chen, L.-C., Chen, K.-H., 2011. Highly flexible supercapacitors with manganese oxide nanosheet/carbon cloth electrode. Electrochim. Acta 56, 7124–7130.

Chen, X., Lin, H., Deng, J., Zhang, Y., Sun, X., Chen, P., Fang, X., Zhang, Z., Guan, G., Peng, H., 2014. Electrochromic fiber-shaped supercapacitors. Adv. Mater. 26, 8126–8132.

Chen, G.F., Su, Y.Z., Kuang, P.Y., Liu, Z.Q., Chen, D.Y., Wu, X., Li, N., Qiao, S.Z., 2015a. Polypyrrole shell@3D-Ni metal core structured electrodes for high-performance supercapacitors. Chem. Eur. J. 21, 4614–4621.

Chen, T., Hao, R., Peng, H., Dai, L., 2015b. High-performance, stretchable, wire-shaped supercapacitors. Angew. Chem. Int. Ed. 54, 618–622.

Chmiola, J., Largeot, C., Taberna, P.-L., Simon, P., Gogotsi, Y., 2010. Monolithic carbide-derived carbon films for micro-supercapacitors. Science 328, 480–483.

Cho, T., Sakai, T., Tanase, S., Kimura, K., Kondo, Y., Tarao, T., Tanaka, M., 2007. Electrochemical performances of polyacrylonitrile nanofiber-based nonwoven separator for lithium-ion battery. Electrochem. Solid-State Lett. 10, A159–A162.

Cho, T.-H., Tanaka, M., Onishi, H., Kondo, Y., Nakamura, T., Yamazaki, H., Tanase, S., Sakai, T., 2008. Silica-composite nonwoven separators for lithium-ion battery: development and characterization. J. Electrochem. Soc. 155, A699–A703.

Choi, S.-S., Lee, Y.S., Joo, C.W., Lee, S.G., Park, J.K., Han, K.-S., 2004. Electrospun PVDF nanofiber web as polymer electrolyte or separator. Electrochim. Acta 50, 339–343.

Choi, K.M., Jeong, H.M., Park, J.H., Zhang, Y.-B., Kang, J.K., Yaghi, O.M., 2014. Supercapacitors of Nanocrystalline Metal–Organic Frameworks. ACS Nano 8, 7451–7457.

Costa, C.M., Rodrigues, L., Sencadas, V., Silva, M., Rocha, J.G., Lanceros-Mndez, S., 2012. Effect of degree of porosity on the properties of poly(vinylidene fluoride–trifluorethylene) for Li-ion battery separators. J. Membr. Sci. 407, 193–201.

Costa, C.M., Sencadas, V., Rocha, J.G., Silva, M.M., Lanceros-Mndez, S., 2013. Evaluation of the main processing parameters influencing the performance of poly(vinylidene fluoride–trifluoroethylene) lithium-ion battery separators. J. Solid State Electrochem. 17, 861–870.

Croce, F., Curini, R., Martinelli, A., Persi, L., Ronci, F., Scrosati, B., Caminiti, R., 1999. Physical and chemical properties of nanocomposite polymer electrolytes. J. Phys. Chem. B 103, 10632–10638.

De Souza, V.H.R., Oliveira, M.M., Zarbin, A.J.G., 2014. Thin and flexible all-solid supercapacitor prepared from novel single wall carbon nanotubes/polyaniline thin films obtained in liquid–liquid interfaces. J. Power Sources 260, 34–42.

Djian, D., Alloin, F., Martinet, S., Lignier, H., 2009. Macroporous poly(vinylidene fluoride) membrane as a separator for lithium-ion batteries with high charge rate capacity. J. Power Sources 187, 575–580.

Endo, M., Kim, Y., Hayashi, T., Nishimura, K., Matusita, T., Miyashita, K., Dresselhaus, M., 2001. Vapor-grown carbon fibers (VGCFs): basic properties and their battery applications. Carbon 39, 1287–1297.

Fan, H., Wang, H., Zhao, N., Zhang, X., Xu, J., 2012. Hierarchical nanocomposite of polyanilinenanorods grown on the surface of carbon nanotubes for high-performance supercapacitor electrode. J. Mater. Chem. 22, 2774–2780.

Fang, L.-F., Shi, J.-L., Zhu, B.-K., Zhu, L.-P., 2013. Facile introduction of polyether chains onto polypropylene separators and its application in lithium ion batteries. J. Membr. Sci. 448, 143–150.

Feuillade, G., Perche, P., 1975. Ion-conductive macromolecular gels and membranes for solid lithium cells. J. Appl. Electrochem. 5, 63–69.

Furukawa, H., Ko, N., Go, Y.B., Aratani, N., Choi, S.B., Choi, E., Yazaydin, A.Ö., Snurr, R.Q., O'keeffe, M., Kim, J., 2010. Ultrahigh porosity in metal-organic frameworks. Science 329, 424–428.

Gao, H., Lian, K., 2014. Proton-conducting polymer electrolytes and their applications in solid supercapacitors: a review. RSC Adv. 4, 33091–33113.

Grover, S., Shekhar, S., Sharma, R.K., Singh, G., 2014. Multiwalled carbon nanotube supported polypyrrole manganese oxide composite supercapacitor electrode: Role of manganese oxide dispersion in performance evolution. Electrochim. Acta 116, 137–145.

Gui, Z., Duay, J., Hu, J., Lee, S.B., 2014. Redox-exchange induced heterogeneous RuO2-conductive polymer nanowires. Phys. Chem. Chem. Phys. 16, 12332–12340.

Guo, W., Yin, Y.-X., Xin, S., Guo, Y.-G., Wan, L.-J., 2012. Superior radical polymer cathode material with a two-electron process redox reaction promoted by graphene. Energy Environ. Sci. 5, 5221–5225.

Guo, C.X., Yilmaz, G., Chen, S., Chen, S., Lu, X., 2015. Hierarchical nanocomposite composed of layered V2O5/PEDOT/MnO2 nanosheets for high-performance asymmetric supercapacitors. Nano Energy 12, 76–87.

Han, J., Li, L., Fang, P., Guo, R., 2012. Ultrathin MnO2Nanorods on Conducting Polymer Nanofibers as a New Class of Hierarchical Nanostructures for High-Performance Supercapacitors. J. Phys. Chem. C 116, 15900–15907.

Han, L., Tang, P., Zhang, L., 2014. Hierarchical Co3O4@PPy@MnO2 core–shell–shell nanowire arrays for enhanced electrochemical energy storage. Nano Energy 7, 42–51.

Horng, Y.-Y., Lu, Y.-C., Hsu, Y.-K., Chen, C.-C., Chen, L.-C., Chen, K.-H., 2010. Flexible supercapacitor based on polyaniline nanowires/carbon cloth with both high gravimetric and area-normalized capacitance. J. Power Sources 195, 4418–4422.

Hou, Y., Cheng, Y., Hobson, T., Liu, J., 2010. Design and synthesis of hierarchical MnO2 nanospheres/carbon nanotubes/conducting polymer ternary composite for high performance electrochemical electrodes. Nano Lett. 10, 2727–2733.

Hsu, Y.-K., Chen, Y.-C., Lin, Y.-G., Chen, L.-C., Chen, K.-H., 2013. Direct-growth of poly(3,4-ethylenedioxythiophene) nanowires/carbon cloth as hierarchical supercapacitor electrode in neutral aqueous solution. J. Power Sources 242, 718–724.

Hu, Y., Zhao, Y., Li, Y., Li, H., Shao, H., Qu, L., 2012. Defective super-long carbon nanotubes and polypyrrole composite for high-performance supercapacitor electrodes. Electrochim. Acta 66, 279–286.

Huai, Y., Gao, J., Deng, Z., Suo, J., 2010. Preparation and characterization of a special structural poly(acrylonitrile)-based microporous membrane for lithium-ion batteries. Ionics 16, 603–611.

Huang, J., Wang, K., Wei, Z., 2010. Conducting polymer nanowire arrays with enhanced electrochemical performance. J. Mater. Chem. 20, 1117–1121.

Huang, S., Li, L., Yang, Z., Zhang, L., Saiyin, H., Chen, T., Peng, H., 2011. A new and general fabrication of an aligned carbon nanotube/polymer film for electrode applications. Adv. Mater. 23, 4707–4710.

Huang, C.W., Wu, C.A., Hou, S.S., Kuo, P.L., Hsieh, C.T., Teng, H., 2012. Gel electrolyte derived from poly(ethylene glycol) blending poly(acrylonitrile) applicable to roll-to-roll assembly of electric double layer capacitors. Adv. Funct. Mater. 22, 4677–4685.

Huang, Y., Tao, J., Meng, W., Zhu, M., Huang, Y., Fu, Y., Gao, Y., Zhi, C., 2015. Super-high rate stretchable polypyrrole-based supercapacitors with excellent cycling stability. Nano Energy 11, 518–525.

Hwang, J., Liu, H., 2002. Influence of organophilic clay on the morphology, plasticizer-maintaining ability, dimensional stability, and electrochemical properties of gel polyacrylonitrile (PAN) nanocomposite electrolytes. Macromolecules 35, 7314–7319.

Jeong, Y.-B., Kim, D.-W., 2004. Effect of thickness of coating layer on polymer-coated separator on cycling performance of lithium-ion polymer cells. J. Power Sources 128, 256–262.

Jeong, H.-S., Lee, S.-Y., 2011. Closely packed SiO 2 nanoparticles/poly(vinylidene fluoride-hexafluoropropylene) layers-coated polyethylene separators for lithium-ion batteries. J. Power Sources 196, 6716–6722.

Jeschke, S., Mutke, M., Jiang, Z., Alt, B., Wiemh Fer, H.D., 2014. Study of carbamate-modified disiloxane in porous PVDF-HFP membranes: new electrolytes/separators for lithium-ion batteries. Chemphyschem 15, 1761–1771.

Ji, J., Keen, J., Zhong, W.-H., 2011. Simultaneous improvement in ionic conductivity and mechanical properties of multi-functional block-copolymer modified solid polymer electrolytes for lithium ion batteries. J. Power Sources 196, 10163–10168.

Jiang, H., Ma, J., Li, C., 2012. Polyaniline–MnO2 coaxial nanofiber with hierarchical structure for high-performance supercapacitors. J. Mater. Chem. 22, 16939.

Jiang, W., Yu, D., Zhang, Q., Goh, K., Wei, L., Yong, Y., Jiang, R., Wei, J., Chen, Y., 2015. Ternary hybrids of amorphous nickel hydroxide-carbon nanotube-conducting polymer for supercapacitors with high energy density, excellent rate capability, and long cycle life. Adv. Funct. Mater. 25, 1063–1073.

Johan, M.R., Shy, O.H., Ibrahim, S., Yassin, S.M.M., Hui, T.Y., 2011. Effects of Al_2O_3 nanofiller and EC plasticizer on the ionic conductivity enhancement of solid PEO–$LiCF_3SO_3$ solid polymer electrolyte. Solid State Ionics 196, 41–47.

Jost, K., Dion, G., Gogotsi, Y., 2014. Textile energy storage in perspective. J. Mater. Chem. A 2, 10776–10787.

Jung, B., Yoon, J.K., Kim, B., Rhee, H.-W., 2005. Effect of crystallization and annealing on polyacrylonitrile membranes for ultrafiltration. J. Membr. Sci. 246, 67–76.

Jung, Y.S., Cavanagh, A.S., Gedvilas, L., Widjonarko, N.E., Scott, I.D., Lee, S.H., Kim, G.H., George, S.M., Dillon, A.C., 2012. Improved functionality of lithium-ion batteries enabled by atomic layer deposition on the porous microstructure of polymer separators and coating electrodes. Adv. Energy Mater. 2, 1022–1027.

Kajdos, A., Kvit, A., Jones, F., Jagiello, J., Yushin, G., 2010. Tailoring the pore alignment for rapid ion transport in microporous carbons. J. Am. Chem. Soc. 132, 3252–3253.

Kim, J.Y., Lim, D.Y., 2010. Surface-modified membrane as a separator for lithium-ion polymer battery. Energies 3, 866–885.

Kim, J.Y., Lee, Y., Lim, D.Y., 2009. Plasma-modified polyethylene membrane as a separator for lithium-ion polymer battery. Electrochim. Acta 54, 3714–3719.

Kim, Y., Jo, C., Lee, J., Lee, C.W., Yoon, S., 2012. An ordered nanocomposite of organic radical polymer and mesocellular carbon foam as cathode material in lithium ion batteries. J. Mater. Chem. 22, 1453–1458.

Kulkarni, S.B., Patil, U.M., Shackery, I., Sohn, J.S., Lee, S., Park, B., Jun, S., 2014. High-performance supercapacitor electrode based on a polyaniline nanofibers/3D graphene framework as an efficient charge transporter. J. Mater. Chem. A 2, 4989.

Lang, X., Zhang, L., Fujita, T., Ding, Y., Chen, M., 2012. Three-dimensional bicontinuous nanoporous Au/polyaniline hybrid films for high-performance electrochemical supercapacitors. J. Power Sources 197, 325–329.

Lee, M.-H., Kim, H.J., Kim, E., Rhee, S.B., Moon, M.J., 1996. Effect of phase separation on ionic conductivity of poly(methyl methacrylate)-based solid polymer electrolyte. Solid State Ionics 85, 91–98.

Lee, S.W., Choi, S.W., Jo, S.M., Chin, B.D., Kim, D.Y., Lee, K.Y., 2006. Electrochemical properties and cycle performance of electrospun poly(vinylidene fluoride)-based fibrous membrane electrolytes for Li-ion polymer battery. J. Power Sources 163, 41–46.

Lee, D.Y., Yoon, S.J., Shrestha, N.K., Lee, S.-H., Ahn, H., Han, S.-H., 2012. Unusual energy storage and charge retention in Co-based metal–organic-frameworks. Micropor. Mesopor. Mater. 153, 163–165.

Lee, H., Yanilmaz, M., Toprakci, O., Fu, K., Zhang, X., 2014. A review of recent developments in membrane separators for rechargeable lithium-ion batteries. Energy Environ. Sci. 7, 3857–3886.

Lewandowski, A., Zajder, M., Frąckowiak, E., Beguin, F., 2001. Supercapacitor based on activated carbon and polyethylene oxide–KOH–H_2O polymer electrolyte. Electrochim. Acta 46, 2777–2780.

Li, C., Bai, H., Shi, G., 2009. Conducting polymer nanomaterials: electrosynthesis and applications. Chem. Soc. Rev. 38, 2397–2409.

Li, P., Shi, E., Yang, Y., Shang, Y., Peng, Q., Wu, S., Wei, J., Wang, K., Zhu, H., Yuan, Q., Cao, A., Wu, D., 2013. Carbon nanotube-polypyrrole core-shell sponge and its application as highly compressible supercapacitor electrode. Nano Res. 7, 209–218.

Li, P., Yang, Y., Shi, E., Shen, Q., Shang, Y., Wu, S., Wei, J., Wang, K., Zhu, H., Yuan, Q., Cao, A., Wu, D., 2014. Core-double-shell, carbon nanotube@polypyrrole@MnO(2) sponge as freestanding, compressible supercapacitor electrode. ACS Appl. Mater. Interfaces 6, 5228–5234.

Lima, M.D., Fang, S., Lepro, X., Lewis, C., Ovalle-Robles, R., Carretero-Gonzalez, J., Castillo-Martinez, E., Kozlov, M.E., Oh, J., Rawat, N., Haines, C.S., Haque, M.H., Aare, V., Stoughton, S., Zakhidov, A.A., Baughman, R.H., 2011. Biscrolling nanotube sheets and functional guests into yarns. Science 331, 51–55.

Liu, R., Lee, S.B., 2008. MnO2/poly(3, 4-ethylenedioxythiophene) coaxial nanowires by one-step coelectrodeposition for electrochemical energy storage. J. Am. Chem. Soc. 130, 2942–2943.

Liu, B., Shioyama, H., Jiang, H., Zhang, X., Xu, Q., 2010a. Metal–organic framework (MOF) as a template for syntheses of nanoporous carbons as electrode materials for supercapacitor. Carbon 48, 456–463.

Liu, R., Duay, J., Lane, T., Lee, S.B., 2010b. Synthesis and characterization of RuO 2/poly(3, 4-ethylenedioxythiophene) composite nanotubes for supercapacitors. Phys. Chem. Chem. Phys. 12, 4309–4316.

Liu, N., Ma, W., Tao, J., Zhang, X., Su, J., Li, L., Yang, C., Gao, Y., Golberg, D., Bando, Y., 2013. Cable-type supercapacitors of three-dimensional cotton thread based multi-grade nanostructures for wearable energy storage. Adv. Mater. 25, 4925–4931.

Liu, T., Finn, L., Yu, M., Wang, H., Zhai, T., Lu, X., Tong, Y., Li, Y., 2014a. Polyaniline and polypyrrole pseudocapacitor electrodes with excellent cycling stability. Nano Lett. 14, 2522–2527.

Liu, Y., Ma, Y., Guang, S., Xu, H., Su, X., 2014b. Facile fabrication of three-dimensional highly ordered structural polyaniline–graphene bulk hybrid materials for high performance supercapacitor electrodes. J. Mater. Chem. A 2, 813–823.

Locquet, J.-P., Perret, J., Fompeyrine, J., M Chler, E., Seo, J.W., Van Tendeloo, G., 1998. Doubling the critical temperature of La1. 9Sr0. 1CuO4 using epitaxial strain. Nature 394, 453–456.

Love, C.T., 2011. Thermomechanical analysis and durability of commercial micro-porous polymer Li-ion battery separators. J. Power Sources 196, 2905–2912.

Lu, W., Henry, K., Turchi, C., Pellegrino, J., 2008. Incorporating ionic liquid electrolytes into polymer gels for solid-state ultracapacitors. J. Electrochem. Soc. 155, A361–A367.

Lu, X., Yu, M., Wang, G., Tong, Y., Li, Y., 2014. Flexible solid-state supercapacitors: design, fabrication and applications. Energy Environ. Sci. 7, 2160–2181.

Ma, T.-Y., Liu, L., Yuan, Z.-Y., 2013. Direct synthesis of ordered mesoporous carbons. Chem. Soc. Rev. 42, 3977–4003.

Ma, G., Feng, E., Sun, K., Peng, H., Li, J., Lei, Z., 2014. A novel and high-effective redox-mediated gel polymer electrolyte for supercapacitor. Electrochim. Acta 135, 461–466.

Mai, L., Tian, X., Xu, X., Chang, L., Xu, L., 2014. Nanowire electrodes for electrochemical energy storage devices. Chem. Rev. 114, 11828–11862.

Mastragostino, M., Arbizzani, C., Soavi, F., 2001. Polymer-based supercapacitors. J. Power Sources 97, 812–815.

Meng, C., Liu, C., Chen, L., Hu, C., Fan, S., 2010. Highly flexible and all-solid-state paperlike polymer supercapacitors. Nano Lett. 10, 4025–4031.

Meng, Y., Wu, H., Zhang, Y., Wei, Z., 2014. A flexible electrode based on a three-dimensional graphene network-supported polyimide for lithium-ion batteries. J. Mater. Chem. A 2, 10842–10846.

Mini, P.A., Balakrishnan, A., Nair, S.V., Subramanian, K.R., 2011. Highly super capacitive electrodes made of graphene/poly(pyrrole). Chem. Commun. 47, 5753–5755.

Nakahara, K., Iriyama, J., Iwasa, S., Suguro, M., Satoh, M., Cairns, E.J., 2007a. Cell properties for modified PTMA cathodes of organic radical batteries. J. Power Sources 165, 398–402.

Nakahara, K., Iriyama, J., Iwasa, S., Suguro, M., Satoh, M., Cairns, E.J., 2007b. High-rate capable organic radical cathodes for lithium rechargeable batteries. J. Power Sources 165, 870–873.

Nasef, M.M., Hegazy, E.-S.A., 2004. Preparation and applications of ion exchange membranes by radiation-induced graft copolymerization of polar monomers onto non-polar films. Prog. Polym. Sci. 29, 499–561.

Novoselov, K.S., Geim, A.K., Morozov, S., Jiang, D., Zhang, Y., Dubonos, S.A., Grigorieva, I., Firsov, A., 2004. Electric field effect in atomically thin carbon films. Science 306, 666–669.

Nyholm, L., Nystrom, G., Mihranyan, A., Stromme, M., 2011. Toward flexible polymer and paper-based energy storage devices. Adv. Mater. 23, 3751–3769.

Oyaizu, K., Nishide, H., 2009. Radical polymers for organic electronic devices: a radical departure from conjugated polymers? Adv. Mater. 21, 2339–2344.

Oyama, N., Tatsuma, T., Sato, T., Sotomura, T., 1995. Dimercaptan–polyaniline composite electrodes for lithium batteries with high energy density. Nature 373, 598–600.

Park, J.-H., Park, W., Kim, J.H., Ryoo, D., Kim, H.S., Jeong, Y.U., Kim, D.-W., Lee, S.-Y., 2011. Close-packed poly(methyl methacrylate) nanoparticle arrays-coated polyethylene separators for high-power lithium-ion polymer batteries. J. Power Sources 196, 7035–7038.

Patil, U., Lee, S.C., Kulkarni, S., Sohn, J.S., Nam, M.S., Han, S., Jun, S.C., 2015. Nanostructured pseudocapacitive materials decorated 3D graphene foam electrodes for next generation supercapacitors. Nanoscale 7, 6999–7021.

Paul, D., Robeson, L., 2008. Polymer nanotechnology: nanocomposites. Polymer 49, 3187–3204.

Pech, D., Brunet, M., Durou, H., Huang, P., Mochalin, V., Gogotsi, Y., Taberna, P.-L., Simon, P., 2010. Ultrahigh-power micrometre-sized supercapacitors based on onion-like carbon. Nat. Nanotechnol. 5, 651–654.

Qiao, J., Fu, J., Lin, R., Ma, J., Liu, J., 2010. Alkaline solid polymer electrolyte membranes based on structurally modified PVA/PVP with improved alkali stability. Polymer 51, 4850–4859.

Ren, J., Li, L., Chen, C., Chen, X., Cai, Z., Qiu, L., Wang, Y., Zhu, X., Peng, H., 2013. Twisting carbon nanotube fibers for both wire-shaped micro-supercapacitor and micro-battery. Adv. Mater. 25, 1155–1159.

Ryou, M.H., Lee, Y.M., Park, J.K., Choi, J.W., 2011. Mussel-inspired polydopamine-treated polyethylene separators for high-power Li-ion batteries. Adv. Mater. 23, 3066–3070.

Ryou, M.H., Lee, D.J., Lee, J.N., Lee, Y.M., Park, J.K., Choi, J.W., 2012. Excellent cycle life of lithium-metal anodes in lithium-ion batteries with mussel-inspired polydopamine-coated separators. Adv. Energy Mater. 2, 645–650.

Saunier, J., Alloin, F., Sanchez, J.-Y., Maniguet, L., 2004. Plasticized microporous PVdF separators for lithium ions batteries. Part III: gel properties and irreversible modifications of Poly(vinylidene fluoride) membranes under swelling in liquid electrolyte. J. Polym. Sci. Part B 42, 2308–2317.

Sengodu, P., Deshmukh, A.D., 2015. Conducting polymers and their inorganic composites for advanced Li-ion batteries: a review. RSC Adv. 5, 42109–42130.

Senyarich, S., Viaud, P., 2000. Method of forming a separator for alkaline electrolyte secondary electric cell. US Patent 6042970.

Shi, J.-L., Fang, L.-F., Li, H., Zhang, H., Zhu, B.-K., Zhu, L.-P., 2013. Improved thermal and electrochemical performances of PMMA modified PE separator skeleton prepared via dopamine-initiated ATRP for lithium ion batteries. J. Membr. Sci. 437, 160–168.

Sil, A., Sharma, R., Ray, S., 2015. Mechanical and thermal characteristics of PMMA-based nanocomposite gel polymer electrolytes with CNFs dispersion. Surf. Coat. Technol. 271, 201–206.

Sivakkumar, S., Kim, D.-W., 2007. Polyaniline/carbon nanotube composite cathode for rechargeable lithium polymer batteries assembled with gel polymer electrolyte. J. Electrochem. Soc. 154, A134–A139.

Song, Z., Zhou, H., 2013. Towards sustainable and versatile energy storage devices: an overview of organic electrode materials. Energy Environ. Sci. 6, 2280–2301.

Song, Z., Zhan, H., Zhou, Y., 2010. Polyimides: promising energy-storage materials. Angew. Chem. 122, 8622–8626.

Song, Z., Xu, T., Gordin, M.L., Jiang, Y.-B., Bae, I.-T., Xiao, Q., Zhan, H., Liu, J., Wang, D., 2012. Polymer–graphene nanocomposites as ultrafast-charge and -discharge cathodes for rechargeable lithium batteries. Nano Lett. 12, 2205–2211.

Stephan, A.M., Kumar, S.G., Renganathan, N., Kulandainathan, M.A., 2005. Characterization of poly(vinylidene fluoride–hexafluoropropylene) (PVdF–HFP) electrolytes complexed with different lithium salts. Eur. Polym. J. 41, 15–21.

Tang, C., Hackenberg, K., Fu, Q., Ajayan, P.M., Ardebili, H., 2012. High ion conducting polymer nanocomposite electrolytes using hybrid nanofillers. Nano lett. 12, 1152–1156.

Tang, P., Han, L., Zhang, L., 2014. Facile synthesis of graphite/PEDOT/MnO2 composites on commercial supercapacitor separator membranes as flexible and high-performance supercapacitor electrodes. ACS Appl. Mater. Interfaces 6, 10506–10515.

Tarascon, J.M., Armand, M., 2001. Issues and challenges facing rechargeable lithium batteries. Nature 414, 359–367.

Wang, K., Huang, J., Wei, Z., 2010. Conducting polyaniline nanowire arrays for high performance supercapacitors. J. Phys. Chem. C 114, 8062–8067.

Wang, G., Lu, X., Ling, Y., Zhai, T., Wang, H., Tong, Y., Li, Y., 2012a. LiCl/PVA gel electrolyte stabilizes vanadium oxide nanowire electrodes for pseudocapacitors. ACS Nano 6, 10296–10302.

Wang, J.-G., Yang, Y., Huang, Z.-H., Kang, F., 2012b. Rational synthesis of MnO_2/conducting polypyrrole@carbon nanofiber triaxial nano-cables for high-performance supercapacitors. J. Mater. Chem. 22, 16943–16949.

Wang, K., Meng, Q., Zhang, Y., Wei, Z., Miao, M., 2013a. High-performance two-ply yarn supercapacitors based on carbon nanotubes and polyaniline nanowire arrays. Adv. Mater. 25, 1494–1498.

Wang, L., Ye, Y., Lu, X., Wen, Z., Li, Z., Hou, H., Song, Y., 2013b. Hierarchical nanocomposites of polyaniline nanowire arrays on reduced graphene oxide sheets for supercapacitors. Sci. Rep. 3, 3568.

Wang, Y., Tao, S., An, Y., Wu, S., Meng, C., 2013c. Bio-inspired high performance electrochemical supercapacitors based on conducting polymer modified coral-like monolithic carbon. J. Mater. Chem. A 1, 8876.

Wang, K., Wu, H., Meng, Y., Wei, Z., 2014a. Conducting polymer nanowire arrays for high performance supercapacitors. Small 10, 14–31.

Wang, Q., Yan, J., Fan, Z., Wei, T., Zhang, M., Jing, X., 2014b. Mesoporous polyaniline film on ultra-thin graphene sheets for high performance supercapacitors. J. Power Sources 247, 197–203.

Wang, S.-H., Kuo, P.-L., Hsieh, C.-T., Teng, H., 2014c. Design of poly(acrylonitrile)-based gel electrolytes for high-performance lithium ion batteries. ACS Appl. Mater. Interfaces 6, 19360–19370.

Wang, X., Lu, X., Liu, B., Chen, D., Tong, Y., Shen, G., 2014d. Flexible energy-storage devices: design consideration and recent progress. Adv. Mater. 26, 4763–4782.

Wang, L., Feng, X., Ren, L., Piao, Q., Zhong, J., Wang, Y., Li, H., Chen, Y., Wang, B., 2015a. Flexible solid-state supercapacitor based on a metal–organic framework interwoven by electrochemically-deposited PANI. J. Am. Chem. Soc. 137, 4920–4923.

Wang, Y., Zeng, J., Li, J., Cui, X., Al-Enizi, A.M., Zhang, L., Zheng, G., 2015b. One-dimensional nanostructures for flexible supercapacitors. J. Mater. Chem. A 3, 16382–16392.

Wei, W., Cui, X., Chen, W., Ivey, D.G., 2011. Manganese oxide-based materials as electrochemical supercapacitor electrodes. Chem. Soc. Rev. 40, 1697–1721.

Wu, H., Shevlin, S.A., Meng, Q., Guo, W., Meng, Y., Lu, K., Wei, Z., Guo, Z., 2014. Flexible and binder-free organic cathode for high-performance lithium-ion batteries. Adv. Mater. 26, 3338–3343.

Xia, X., Hao, Q., Lei, W., Wang, W., Sun, D., Wang, X., 2012. Nanostructured ternary composites of graphene/Fe_2O_3/polyaniline for high-performance supercapacitors. J. Mater. Chem. 22, 16844.

Xia, C., Chen, W., Wang, X., Hedhili, M.N., Wei, N., Alshareef, H.N., 2015. Highly stable supercapacitors with conducting polymer core-shell electrodes for energy storage applications. Adv. Energy Mater. 5, 1401805.

Xie, K., Li, J., Lai, Y., Zhang, Z., Liu, Y., Zhang, G., Huang, H., 2011. Polyaniline nanowire array encapsulated in titania nanotubes as a superior electrode for supercapacitors. Nanoscale 3, 2202–2207.

Xu, J., Wang, K., Zu, S.-Z., Han, B.-H., Wei, Z., 2010. Hierarchical nanocomposites of polyaniline nanowire arrays on graphene oxide sheets with synergistic effect for energy storage. ACS Nano 4, 5019–5026.

Xu, C., Sun, J., Gao, L., 2011. Synthesis of novel hierarchical graphene/polypyrrole nanosheet composites and their superior electrochemical performance. J. Mater. Chem. 21, 11253.

Yan, J., Wei, T., Shao, B., Fan, Z., Qian, W., Zhang, M., Wei, F., 2010. Preparation of a graphene nanosheet/polyaniline composite with high specific capacitance. Carbon 48, 487–493.

Yan, Y., Cheng, Q., Wang, G., Li, C., 2011. Growth of polyaniline nanowhiskers on mesoporous carbon for supercapacitor application. J. Power Sources 196, 7835–7840.

Yan, Y., Cheng, Q., Zhu, Z., Pavlinek, V., Saha, P., Li, C., 2013. Controlled synthesis of hierarchical polyaniline nanowires/ordered bimodal mesoporous carbon nanocomposites with high surface area for supercapacitor electrodes. J. Power Sources 240, 544–550.

Yan, J., Wang, Q., Wei, T., Fan, Z., 2014. Recent advances in design and fabrication of electrochemical supercapacitors with high energy densities. Adv. Energy Mater. 4, 1300816.

Yang, C.-C., Hsu, S.-T., Chien, W.-C., 2005. All solid-state electric double-layer capacitors based on alkaline polyvinyl alcohol polymer electrolytes. J. Power Sources 152, 303–310.

Yang, Y., Wang, C., Yue, B., Gambhir, S., Too, C.O., Wallace, G.G., 2012. Electrochemically synthesized polypyrrole/graphene composite film for lithium batteries. Adv. Energy Mater. 2, 266–272.

Yang, C.-C., Lian, Z.-Y., Lin, S., Shih, J.-Y., Chen, W.-H., 2014a. Preparation and application of PVDF-HFP composite polymer electrolytes in LiNi 0.5 Co 0.2 Mn 0.3 O_2 lithium-polymer batteries. Electrochim. Acta 134, 258–265.

Yang, J., Xiong, P., Zheng, C., Qiu, H., Wei, M., 2014b. Metal–organic frameworks: a new promising class of materials for a high performance supercapacitor electrode. J. Mater. Chem. A 2, 16640–16644.

Yao, W., Zhou, H., Lu, Y., 2013. Synthesis and property of novel MnO2@polypyrrole coaxial nanotubes as electrode material for supercapacitors. J. Power Sources 241, 359–366.

Ye, Y.-S., Rick, J., Hwang, B.-J., 2013. Ionic liquid polymer electrolytes. J. Mater. Chem. A 1, 2719–2743.

Yoon, S.-B., Kim, K.-B., 2013. Effect of poly(3,4-ethylenedioxythiophene) (PEDOT) on the pseudocapacitive properties of manganese oxide (MnO_2) in the Pedot/MnO2/multiwall carbon nanotube (MWNT) composite. Electrochim. Acta 106, 135–142.

Young, W.-S., Epps, III, T.H., 2012. Ionic conductivities of block copolymer electrolytes with various conducting pathways: Sample preparation and processing considerations. Macromolecules 45, 4689–4697.

Yu, G., Hu, L., Liu, N., Wang, H., Vosgueritchian, M., Yang, Y., Cui, Y., Bao, Z., 2011a. Enhancing the supercapacitor performance of graphene/MnO_2 nanostructured electrodes by conductive wrapping. Nano Lett. 11, 4438–4442.

Yu, H., Wu, J., Fan, L., Xu, K., Zhong, X., Lin, Y., Lin, J., 2011b. Improvement of the performance for quasi-solid-state supercapacitor by using PVA–KOH–KI polymer gel electrolyte. Electrochim. Acta 56, 6881–6886.

Yu, G., Xie, X., Pan, L., Bao, Z., Cui, Y., 2013. Hybrid nanostructured materials for high-performance electrochemical capacitors. Nano Energy 2, 213–234.

Yu, D., Goh, K., Zhang, Q., Wei, L., Wang, H., Jiang, W., Chen, Y., 2014. Controlled functionalization of carbonaceous fibers for asymmetric solid-state micro-supercapacitors with high volumetric energy density. Adv. Mater. 26, 6790–6797.

Yu, Z., Tetard, L., Zhai, L., Thomas, J., 2015. Supercapacitor electrode materials: nanostructures from 0 to 3 dimensions. Energy Environ. Sci. 8, 702–730.

Yuan, C., Su, L., Gao, B., Zhang, X., 2008. Enhanced electrochemical stability and charge storage of MnO_2/carbon nanotubes composite modified by polyaniline coating layer in acidic electrolytes. Electrochim. Acta 53, 7039–7047.

Yuan, L., Yao, B., Hu, B., Huo, K., Chen, W., Zhou, J., 2013. Polypyrrole-coated paper for flexible solid-state energy storage. Energy Environ. Sci. 6, 470–476.

Zhai, Y., Dou, Y., Zhao, D., Fulvio, P.F., Mayes, R.T., Dai, S., 2011. Carbon materials for chemical capacitive energy storage. Adv. Mater. 23, 4828–4850.

Zhang, S.S., 2007. A review on the separators of liquid electrolyte Li-ion batteries. J. Power Sources 164, 351–364.

Zhang, L.L., Zhao, X.S., 2009. Carbon-based materials as supercapacitor electrodes. Chem. Soc. Rev. 38, 2520–2531.

Zhang, J., Kong, L.-B., Cai, J.-J., Luo, Y.-C., Kang, L., 2010a. Nano-composite of polypyrrole/modified mesoporous carbon for electrochemical capacitor application. Electrochim. Acta 55, 8067–8073.

Zhang, L.L., Li, S., Zhang, J., Guo, P., Zheng, J., Zhao, X.S., 2010b. Enhancement of electrochemical performance of macroporous carbon by surface coating of polyaniline†. Chem. Mater. 22, 1195–1202.

Zhang, Q., Li, Y., Feng, Y., Feng, W., 2013a. Electropolymerization of graphene oxide/polyaniline composite for high-performance supercapacitor. Electrochim. Acta 90, 95–100.

Zhang, Q., Uchaker, E., Candelaria, S.L., Cao, G., 2013b. Nanomaterials for energy conversion and storage. Chem. Soc. Rev. 42, 3127–3171.

Zhang, D., Miao, M., Niu, H., Wei, Z., 2014. Core-spun carbon nanotube yarn supercapacitors for wearable electronic textiles. ACS Nano 8, 4571–4579.

Zhang, K., Hu, H., Yao, W., Ye, C., 2015. Flexible and all-solid-state supercapacitors with long-time stability constructed on PET/Au/polyaniline hybrid electrodes. J. Mater. Chem. A 3, 617–623.

Zhong, J., Yang, Z., Mukherjee, R., Varghese Thomas, A., Zhu, K., Sun, P., Lian, J., Zhu, H., Koratkar, N., 2013. Carbon nanotube sponges as conductive networks for supercapacitor devices. Nano Energy 2, 1025–1030.

Zhong, C., Deng, Y., Hu, W., Qiao, J., Zhang, L., Zhang, J., 2015a. A review of electrolyte materials and compositions for electrochemical supercapacitors. Chem. Soc. Rev. 44, 7484–7539.

Zhong, J., Fan, L.-Q., Wu, X., Wu, J.-H., Liu, G.-J., Lin, J.-M., Huang, M.-L., Wei, Y.-L., 2015b. Improved energy density of quasi-solid-state supercapacitors using sandwich-type redox-active gel polymer electrolytes. Electrochim. Acta 166, 150–156.

Zhou, C., Zhang, Y., Li, Y., Liu, J., 2013. Construction of high-capacitance 3D CoO@polypyrrole nanowire array electrode for aqueous asymmetric supercapacitor. Nano Lett. 13, 2078–2085.

Zhu, Y., Murali, S., Stoller, M.D., Ganesh, K., Cai, W., Ferreira, P.J., Pirkle, A., Wallace, R.M., Cychosz, K.A., Thommes, M., 2011. Carbon-based supercapacitors produced by activation of graphene. Science 332, 1537–1541.

Zhu, L., Lei, A., Cao, Y., Ai, X., Yang, H., 2013. An all-organic rechargeable battery using bipolar polyparaphenylene as a redox-active cathode and anode. Chem. Commun. 49, 567–569.

7

Light Emitting Based on Polymer

As an important branch of electronic devices, light-emitting devices have witnessed a rapid development recently. This chapter will focus on various light-emitting devices based on polymers and summarize their working mechanism, structure, fabrication process, and performance. The new advancements and challenges will also be discussed.

7.1 Light-Emitting Conjugated Polymers

7.1.1 Introduction

The phenomenon of electroluminescence was first observed by Destriau in a research on ZnS powder in 1936 (Destriau, 1936). Then the research of electroluminescence was focused on inorganic materials until 1963, when Pope and coworkers fabricated the first organic light-emitting diode (OLED) with single crystal anthracene (Pope et al., 1963). However, due to the high driving voltage and low quantum efficiency, organic electroluminescent materials did not arouse wide interest then. In 1987, Tang and Van Slyke demonstrated a double-layered thin-film OLED with high external quantum efficiency (1%), power efficiency (1.5 lmW^{-1}) and brightness (>1000 cdm^{-2}) (Tang and Vanslyke, 1987). It is acknowledged as a breakthrough in this field, which raises attention toward organic electroluminescent materials. However, the OLEDs based on organic small molecules exhibit some disadvantages. (1) Organic small light-emitting molecules are prone to recrystallization, which is harmful to the stability of the material. (2) The processing condition for fabricating organic small molecule film is rigorous since thermal evaporation is always needed. Additionally, this fabrication is incompatible for constructing OLEDs with three-dimensionally irregular surface. (3) The flexibility is poor. Hence, shortly after that, Friend and Burroughes made use of spin coating technique to realize polymer light-emitting diodes (PLEDs) based on poly(*p*-phenylene vinylene) in 1990 (Burroughes et al., 1990). In 1995, Pei and coworkers further developed polymer light-emitting electrochemical cells (PLECs) based on electroluminescent polymers and ionically conductive species (Pei et al., 1995). Compared to organic small molecule materials, light-emitting conjugated polymer materials are able to form films through solution methods like spin coating and printing. As expected, the light-emitting conjugated polymers have attracted great interests in the field of low-cost and large-area manufacturing light-emitting devices.

7.1.2 Photophysics of Conjugated Polymer

The semiconducting properties of conjugated polymers are derived from alternating single and double bonds. The π-electrons are delocalized across the adjacent p-orbitals.

The delocalized π-electrons can lower the energy of the molecule and enhance the stability. The bonding (π) and antibonding (π*) orbitals create the delocalized valence and conduction band which support the mobile charge carriers. When conjugated polymer is sandwiched between two electrodes in an external circuit, the charge carriers (electrons and holes) are injected. The electrons and holes recombine when they become close, which produces excitons. If the excitons undergo a radiative transition from excited to ground states, photons will be emitted. The light emission induced by electricity is called electroluminescence. The color of the light is determined by the energy gap between the highest occupied molecular orbital (HOMO) and the lowest unoccupied molecular orbital (LUMO) of the material.

According to quantum mechanics, electron is the fermion with 1/2 spin quantum number, which is the eigenvalue of spin angular momentum. In a system comprised of two electrons, the total spin quantum number S is 0 or 1. The spin wave function of the eigenstates S = 0 and S = 1 can be demonstrated by:

$$\sigma_- = \frac{1}{\sqrt{2}}\{\uparrow_1\downarrow_2 - \downarrow_1\uparrow_2\}$$

$$\sigma_+ = \frac{1}{\sqrt{2}}\{\uparrow_1\downarrow_2 + \downarrow_1\uparrow_2\}$$

$$\sigma_+ = \uparrow_1\uparrow_2$$

$$\sigma_+ = \downarrow_1\downarrow_2$$

The arrows represent the spin state, and the index 1 and 2 on the arrows represent electron 1 and 2, respectively. The subscript "–" shows the wave function is symmetric and this excited state is called singlet state (S = 0), while the subscript "+" shows the wave function is antisymmetric and this excited state is called triplet state (S = 1).

According to photophysics, ground state is the stable state of the molecule. As long as all electrons in a molecule arrange according to Pauli exclusion principle, lowest energy principle and Hund's rule, the molecule is at the ground state. The ground state of organic molecule is singlet state, which is showed as S_0. After the electron transition to the excited state, if the spin state of the electron is the same as that of the ground state electron, the excited state of the molecule is showed as singlet state S_1. If the spin state is reverse, the excited state of the molecule is showed as triplet state T_1. In the process of exciton production induced by electricity, the ratio of excitons in excited singlet state to excited triplet state is 1/3.

7.1.2.1 Fluorescence and Phosphorescence

The fluorescence and phosphorescence phenomena are both the result of radiative transition of which the final state is ground state. The difference is that the initial state of fluorescence transition is excited singlet state, while the initial state of phosphorescence is excited triplet state. During the electrical excitation process, an electron jumps from ground state to the second excited singlet state S_2. This electron may come back to the first excited single state S_1 through vibrational relaxation or internal conversion. Since the vibrational relaxation process is very fast, the initial states of single transitions are almost all S_1, thus resulting in $S_1 \rightarrow S_0$ process. The emission of photon will occur during the transition, and this is electrofluorescence process. Different from electrofluorescence, the initial state of the transition in electrophosphorescence is excited triplet state T_1. Phosphorescence is intrinsically a slow (>6 s) and less efficient process. Excited triplet state cannot be easily produced by the transition from ground state, and only when there is good coupling between S_1 and T_1, the intersystem crossing from S_1 to T_1 is able to realize. In addition, the radiative transition from T_1 to ground state is spin forbidden, and the intensity of phosphorescence is quite low, thus having nearly no contribution to light emission.

As described earlier, the statistical ratio of singlet and triplet states generated through electron-hole recombination in double charge injection process is 1/3. As a result, the theoretical highest internal quantum efficiency of polymer light-emitting device is 25%. However, when phosphorescent material, such as heavy metal-organic complex is introduced, the spin-orbit coupling will lead to a mixture of singlet and triplet states which can remove the spin forbidden nature of radiative transition process of the triplet state (Hung and Chen, 2002). The radiative decay speed of triplet state exciton is decreased and the radiative decay time of single state exciton is extended. In addition, the intersystem crossing efficiency from single to triplet states is elevated and phosphorescence with high efficiency can be generated. Then the electroluminescence process takes use of single and triplet state excitons at the same time, and the theoretical highest internal quantum efficiency is improved to 100%.

During the transition of excitons from excited to ground states, the energy releasing process includes internal conversion, intersystem crossing, photochemical reaction, and other nonradiative transition patterns, except for the radiative decay which generates fluorescence or phosphorescence. These different energy decay patterns occur at the same time and compete with each other, in which the radiative decay only occupies a small portion. As a result, among the total energy that the molecule absorbs, the part which ends up with luminescence is limited. Furthermore, during the process that the light exits from the emitting layer, there will be reflection and refraction effect which lead to energy loss from organic layer, metal layer, and glass. As a result, the external quantum efficiency of the device is greatly limited.

7.1.3 Categories of Light-Emitting Conjugated Polymers

Light-emitting polymers generally share a conjugated structure along the polymer backbone. The π-electrons are delocalized, which is the precondition for the conductivity of the polymer. The structure of conjugated polymer is easily tuned by different methods. Modification of monomeric units can improve the solubility of the polymer and regulate the mobility of charge carriers in the polymer materials. Designing of molecular structure is able to control the size of HOMO-LUMO energy gap, resulting in full color display. Doping with organic small molecules and metal-organic complex is also a common method to regulate the luminescence property of the polymer.

7.1.3.1 Poly(p-phenylene vinylene)

Poly(*p*-phenylene vinylene) (PPV) is the first polymer material that act as emission layer in PLEDs. At present PPV and its derivatives are the most profoundly researched with good properties and are commercially applicable. PPV is insoluble in common organic solvents, but by modifying pendant groups it can be dissolved in many organic solvents. PPV has a high molecule weight, which helps it to generate high quality film with a high luminous efficiency. Introducing suitable substitute groups to the backbone of the molecule and adjusting the conjugation length are possible ways to obtain luminescence with different colors. For example, poly[2-methoxy,5-(2′-ethlhexyloxy)-*p*-phenylenevinylene] (MEH-PPV) is an orange-red emitter, alkoxy group of which brings good solubility in common organic solvents. Moreover, the energy band of MEH-PPV matches well with metal electrodes of PLEDs (Braun et al., 1991).

However, the low electron mobility of MEH-PPV results in imbalance injection of electrons and holes and a shift of charge recombination region from center of the emission layer to the cathode, which is prone to reduce exciton quenching. Introducing electron-withdrawing substitutes as pendant groups or making *p*-phenylene vinylene copolymerize with electron transporting monomers are able to enhance the electron transporting capability of the material, thus improving the balance of injection of charge carriers and consequently increasing the quantum efficiency of the device. Fluorine is thus introduced to the monomer units of PPV and its derivatives to obtain novel light-emitting polymer materials poly(*p*-phenylenedifluorovinylene) (PPDFV) and poly(2-dimethyloctylsilyl-*p*-phenylenedifluorovinylene) (DMOS-PPDFV). Their emission wavelengths locate near 565 and 540 nm (green), respectively. Introducing fluorine is to increase the electron affinity and the electron injection capability of the polymer. The highest luminous efficiency of PLEDs which used DMOS-PPDFV as emitting layer reached 2.7 cdA^{-1} at 6.5 V, 7 times higher than that of the devices using DMOS-PPV without fluorine as emitting layer (Jin et al., 2004). A new PPV-based polymer OX2-PPV with oxadiazole and pyridine units was synthesized, which could be dissolved well in organic solvent and functioned as yellow emitter (Mikroyannidis et al., 2004).

7.1.3.2 Polyfluorene

Polyfluorene (PF) exhibits bright blue fluorescence with an emission maximum at about 2.9 eV (428 nm). It is a promising candidate for blue emitter (Klaerner and Miller, 1998). The structural unit of PF has pairs of phenylene rings locked in a coplanar format, which is beneficial for thermal and light stability. Since PF has a large energy gap, it is possible to synthesize copolymer containing chain segments with narrow energy gap and realize emission in the full visible spectrum through energy transfer (Bernius et al., 2000). For fluorene homopolymers, intermolecular interactions, defects in the chemical structure and polarons can lead to fluorescence quenching. Moreover, the rigid structure of PF may generate excimer, resulting in long wavelength emitting aggregates and thus lowering the color purity. The C-9 of fluorene provides an active site for modification. PF-12 with polyphenylene dendrimer substituents on C-9 was developed to suppress quenching and long wavelength emitting (Setayesh et al., 2001).

7.1.3.3 Poly(p-phenylene)

Poly(*p*-phenylene) (PPP) is a typical blue emitter with an emission maximum at about 2.7 eV (460 nm) (Grem et al., 1992). Like PPV, PPP is insoluble in common organic solvents, and films of PPV should be prepared by precursor route. Introducing alkyl or alkoxy groups can obtain soluble PPP derivatives. However the substitutes tend to bring steric hindrance and increase twisting between benzene rings in the backbone, thus changing the conjugation length and affect device property. One solution is to synthesize ladder-type PPP (LPPP), which brings better coplanarity and higher solubility as well as stability. Introducing methyl substitute at bridge carbon (m-LPPP) leads to a suppression of aggregation in LPPP (Tasch et al., 1996).

7.1.3.4 Polycarbazole

Polycarbazole is a blue emission material but carbazole homopolymer suffers low luminous efficiency. Carbazole is usually copolymerized with fluorene and in the copolymer carbazole units are used to destroy the linear structure of PF and decrease intermolecular interactions, thus consequently suppressing long wavelength emitting. Carbazole-fluorene copolymer 2,7-fluorene-*co*-3,9-carbazole copolymer (3,9-PFCz) was developed and the PLEDs constructed with this kind of material demonstrated high efficiency and stable blue light emission (Liu et al., 2007). Besides electroluminescence property, polycarbazole also possesses good hole transporting capability. *N*-hexylcarbazole-*co*-3,4-ethylenedioxythiophene copolymer poly(HCz-*co*-EDOT) was found to have better hole transporting capability than carbazole homopolymer (Beouch et al., 2001). Poly(vinyl carbazole), in which the carbazole groups are in the side chain, is usually used as hole transporting layer in PLEDs. Chemical structures of the relative polymers are shown in Fig. 7.1.

7.2 Polymer Light-Emitting Diodes

7.2.1 Introduction

Electroluminescence is the phenomenon that generates light through electrical excitation rather than black body radiation. In 1990, Friend and Burroughes developed PLEDs based on PPV through spin coating technique (Burroughes et al., 1990). Considering that PLEDs are compatible for solution methods, they have great potential to develop large-area light-emitting displays. In 1992, Heeger and coworkers successfully prepared PLEDs on a flexible substrate (polyethylene terephthalate, PET), making it possible to create

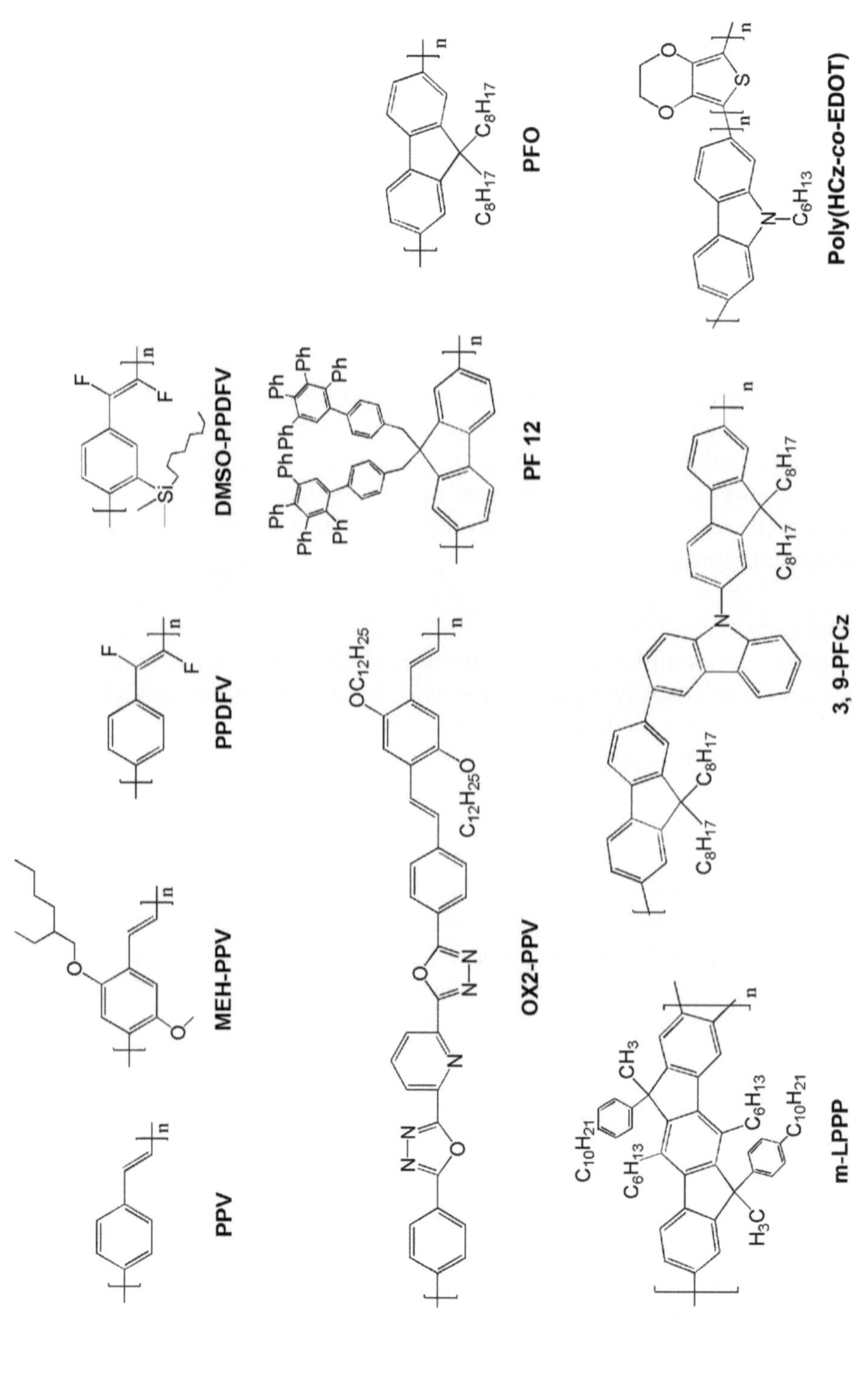

FIGURE 7.1 Light-emitting conjugated polymers.

flexible display devices (Gustafsson et al., 1992). In 1994, Kido and Matsumoto realized white light-emitting polymer electroluminescent devices with power efficiency of 0.83 lmW^{-1} (Kido et al., 1994).

In the early days, PLEDs had encountered many challenges, such as low efficiency and instability. But with the rapid development of electroluminescent materials and device structures, PLEDs have achieved great breakthroughs and demonstrated broad application prospects in the fields of display and lighting source. Compared to inorganic light-emitting diodes, PLEDs have many unique advantages including numerous applicable materials, low cost, compatibility to solution process, adaptability for large-area fabrication, and high flexibility.

7.2.2 Structure and Mechanism

The structure and mechanism of PLED are illustrated in Fig. 7.2A. When bias voltage is applied across the device, holes and electrons overcome the barrier of the interface and are injected from the anode and cathode to the organic layer. They transfer into the HOMO of the hole transporting layer (HTL) and the LUMO of the electron transporting layer (ETL), respectively. Then the charge carries are motivated by the external electric field and transport to the emission layer, forming excitons by recombination. These excitons are in the excited state and are unstable. Finally the excitons jump from the excited state to the ground state, giving rise to light emission. The color of the light basically depends on the energy difference between LUMO and HOMO of the emitting organic material. Thus the emission color can be tuned across the whole visible spectrum by changing the active organic materials. And in the following part several vital issues about this light-emitting process will be discussed.

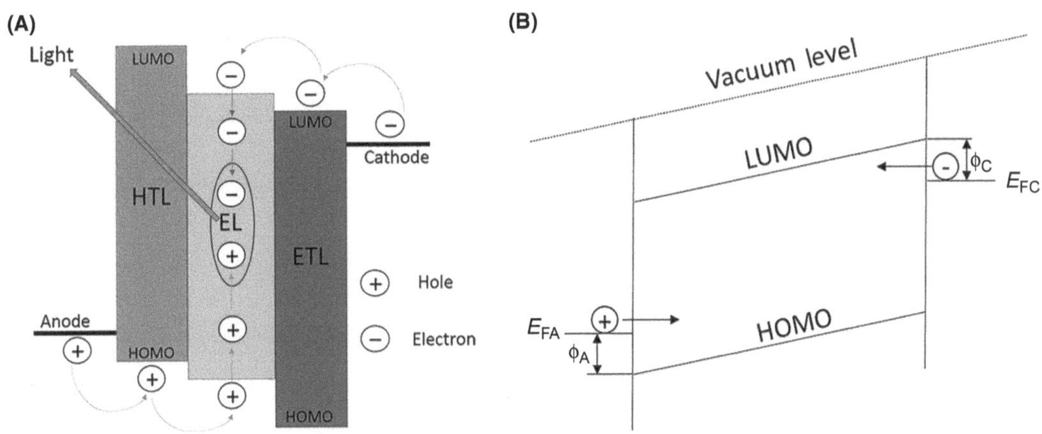

FIGURE 7.2 (A) Schematic illustration to the working mechanism of a typical PLED. (B) Energy level of a PLED.

7.2.2.1 Charge Carrier Injection and Transport

The effective charge carrier injection and transport are crucial to achieve the low driving voltage and high luminous efficiency. First, to have Ohmic contact between the electrodes and the organic layers during the charge carrier injection process and to maximize the mobility of carriers during the charge carrier transport process are essential for lowering the driving voltage. Second, if the charge carrier injection is unbalanced, one type of charge carrier (holes or electrons) will be excess which will fail to realize light emission. Moreover, the interactions between charge carriers and excitons during charge carrier transport process will lead to nonradiative recombination (Hung and Chen, 2002).

The charge carrier injection is the process that the charge transfers from high conductive material to low conductive material. When the work function of the electrode and the energy level of the organic materials are well matched to each other, the interface contact will be Ohmic contact. If the charge mobility of the organic material is relatively low at the same time, the charge carrier injection will be easy and the current will be limited by the intrinsic properties of the organic materials. This is called space charge limited current. In fact, the work function of the metal electrode and the energy level of the organic semiconductor usually do not match very well, so it is hard to generate good Ohmic contact between the electrode and the organic semiconductor. That means there exists large interface barrier and the current injected from the interface is much smaller than the space charge limited current. It is then called injection limited current. In this situation, Fowler–Nordheim tunneling theory is widely accepted to describe the charge carrier injection process (Parker, 1994).

As shown in Fig. 7.2B, E_{FA} (E_{FC}) is the Fermi level of the metal anode (cathode). There are no electrons occupied over the Fermi level at room temperature, and the work function (W) of the metal electrode can be defined as the energy difference between vacuum level and Fermi level. While the electron affinity (E_A) and the ionization energy (I_E) are defined as the energy difference between LUMO or HOMO of organic semiconductor and vacuum level, respectively. Usually the Fermi level of the metal electrode is higher than HOMO and lower than LUMO, which leading to interface barrier (Φ) of anode $\Phi_A = I_E - W_A$ and cathode $\Phi_C = W_C - E_A$, here W_A and W_C represent the work function of anode and cathode, respectively. As a consequence, holes (electrons) cannot hop to HOMO (LUMO). However, when bias voltage is applied across the device, HOMO (LUMO) of organic semiconductor will slant, and then the holes (electrons) distributed around Fermi level will have considerable possibility to cross a triangular barrier and get injected into HOMO (LUMO). When the external electric field becomes higher, the energy level will slant badly and the barrier will become lower. The relationship between the current of the device I and the electric field strength E can be demonstrated in Fowler–Nordheim formula:

$$I \propto E^2 \exp\left(-\frac{\kappa}{E}\right), \quad \kappa = \frac{8\pi\sqrt{2m^*}\varphi^{2/3}}{3qh}$$

Here κ is a parameter that depends on the barrier shape, m^* is the effective mass of the charge carriers, φ is the barrier height, q is charge quantity, and h is Planck constant.

In summary, injection of the charge carrier requires bias voltage to overcome the interface barrier between electrode and organic material. Methods, such as photoelectron spectroscopy have been widely employed to measure the electronic structure and chemical property at the metal/organic interface. It is found that there are chemical reaction, dipoles and atomic diffusion around the interface region. So to determine the charge carrier injection barrier does not just mean calculating the energy difference between the metal work function and the energy level of the organic materials due to the nonnegligible chemical reactions and dipoles. The interface barrier has great influence on the performances of the device. The lowest voltage that is applied to overcome the barrier is called turn-on voltage. Decreasing the interface barrier can surely lower the turn-on voltage and increase the luminous efficiency, and the strategies include n-type or p-type doping of organic semiconductors, designing the chemical structure of the organic material to tune the HOMO (LUMO), and introducing buffer layers (ETL and HTL) with suitable energy levels to match the work function of the metal electrode.

The charge carrier transport is the process that the injected holes and electrons transfer to the emission layer, driven by the external electric field. The charge carrier mobility of organic molecular materials is relatively low (10^{-8}–10^{-2} $cm^2 V^{-1} s^{-1}$), and electron mobility is generally orders of magnitude lower than hole mobility (Kepler et al., 1995). The reason is that in organic semiconductors, the interactions among organic molecules are weak noncovalent interactions, such as Van der Waals force, and the overlap of π-orbitals is not good. The electrons are restricted in single molecule and it is difficult for electrons to hop from one molecule to another. Moreover, organic material layers are generally amorphous, which also results in low mobility. As a consequence, charge carrier injection and transport are both limited factors of PLED to have low turn-on voltage and high luminous efficiency. In addition, injection rate at a contact-limited electrode is proved to be proportional to the charge carrier mobility in organic materials (Scott and Malliaras, 1999). Different from inorganic semiconductors, Hall effect cannot be used to measure the mobility in organic semiconductor. Generally time of flight technique is used to accomplish this task, but the results are not accurate due to the structure defect in organic materials (Chen et al., 2000). Similar to charge carrier injection, the strategy to improve charge carrier transport is introducing appropriate HTL and ETL to realize effective transfer of holes and electrons at the same time.

7.2.3 Performance

In single-layered PLEDs, conjugated polymer is sandwiched between cathode and anode. This single layer structure is favored for its simplicity and easy fabrication. However, it requires the light-emitting material to possess both high quantum efficiency and relatively equal electron and hole mobility, otherwise it will induce imbalance injection of electrons and holes and reduce device efficiency. In a multilayered structure, electron blocking layer, hole injection layer, hole transporting layer, electron transporting layer, electron injection layer, and hole blocking layer are more or less introduced to realize effective and balanced charge carrier injection. Among these layers, charge injection layer is used to decrease the barrier between electrode and charge carrier transporting layer, thus enhancing injected charge carrier numbers; while charge blocking layer is used to confine most of the charge carriers to the emission layer, thus increasing quantum efficiency.

7.2.3.1 Emission Layer

As discussed in Section 7.1, properties of conjugated polymers can be tuned by monomeric unit modification, molecular structure design and doping. These methods can be categorized as molecular engineering strategies which are specialized as repeated unit choice and planarization, resonance structure stabilization, cross conjugation, and making use of donor–acceptor alternating sequence structure. In addition to molecular engineering, Lewis acid can be added to conjugated polymer if it possesses basic functional groups. The reaction of Lewis acid-base is applied to adjust optoelectronic properties of polymer F8Py (Fig. 7.3A) (Zalar et al., 2012). For polymer F8Py, the comonomer pyridine provides a pair of electrons that Lewis acids can bind to. Lewis acid $B(C_6F_5)_3$ was chosen because it had strong Lewis acidity and it was stable in air. With the increasing content of $B(C_6F_5)_3$, the electroluminescence peak underwent a red shift. When 0.02 mol equivalent $B(C_6F_5)_3$ was added, the F8Py-$B(C_6F_5)_3$ peak grew obviously while electroluminescence from F8Py only contributed less than 7% to the total electroluminescence. As a result the emission color shifted from blue to green. As we can see from current density–voltage (*J–V*) and luminance–voltage (*L–V*) curves (Fig. 7.3B), addition of $B(C_6F_5)_3$ from 0.01 to 0.02 mol equivalents resulted in obvious shift to higher biases and thus higher turn-on voltages. At the same time, reduction of external quantum efficiency was not observed when the addition of $B(C_6F_5)_3$ was less than 0.02 mol equivalent. Ultraviolet photoelectron spectroscopy test verified the trend of increasing turn-on voltages. The reason was that once $B(C_6F_5)_3$ was added, HOMO energy level was lowered for F8Py-$B(C_6F_5)_3$, thus increasing the hole injection barrier from 0.3 to 0.6 eV (interfacial effects were not considered). In summary, the electroluminescence properties of PLEDs constructed by F8Py-$B(C_6F_5)_3$ can be straightforwardly tuned through Lewis acid–base interactions without adversely affecting the external quantum efficiency.

7.2.3.2 Hole Injection/Transporting Layer

Hole injection/transporting layers are usually electron blocking layers. They are important parts in PLEDs since they can facilitate holes injection from the anode and confine the electrons in the emission layer (avoid exciton quenching) at the same time. Since PLEDs are constructed in multilayered structure, emission layer should be spin coated on the top of the HTL which must prevents itself from the erosion by solvents. The most commonly used HTL in PLEDs is poly(3,4-ethylenedioxythiophene):poly(styrene sulfonate) (PEDOT:PSS). This material is beneficial because of its high electrical conductivity and hole mobility, and it has been proved to reduce the roughness of indium tin oxide (ITO) after coating. But PEDOT:PSS also has some obvious drawbacks. First, PEDOT:PSS is aqueous dispersed solution with strong acidity which will corrode ITO gradually, and it is

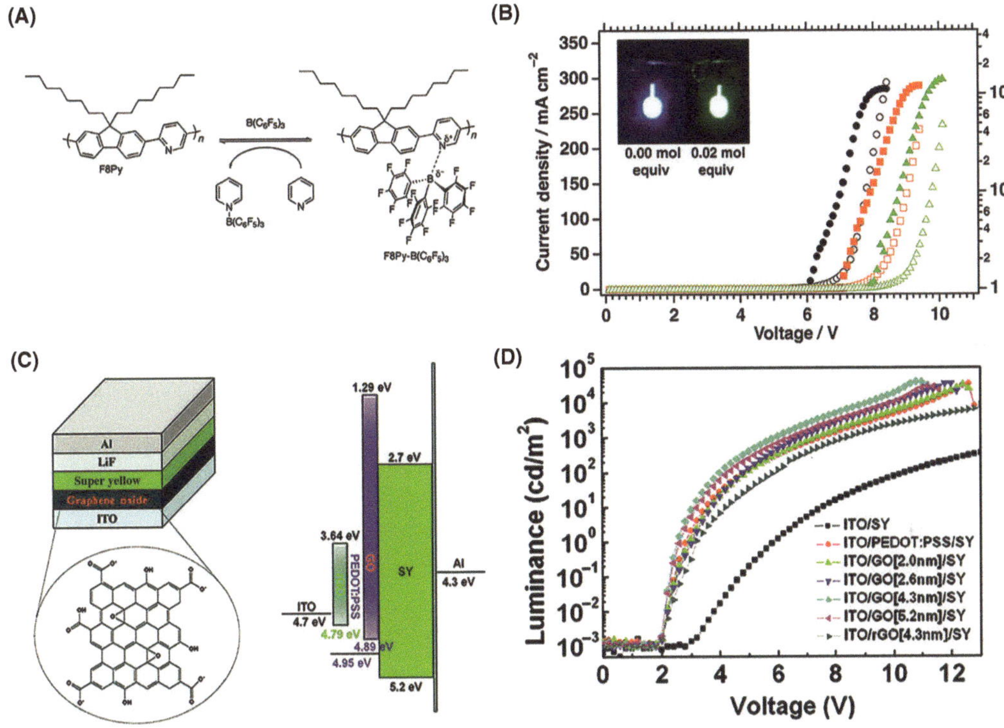

FIGURE 7.3 PLEDs with novel emission layer and HTL. (A) Reversible formation of an adduct between Lewis acidic B-$(C_6F_5)_3$ and Lewis basic F8Py in solution. (B) *J–V* (open symbols) and *L–V* (closed symbols) plots for PLEDs without (black circles) and with 0.01 (red squares) and 0.02 mol equivalents (green triangles) $B(C_6F_5)_3$. (C) Schematic illustration to the structure of PLED with a graphene oxide layer and its energy diagrams. (D) *L–V* plots for PLEDs with various charge transport layers. *(A and B) Reproduced with permission from reference Zalar, P., Henson, Z.B., Welch, G.C., Bazan, G.C., Nguyen, T.Q., 2012. Color tuning in polymer light-emitting diodes with lewis acids. Angew. Chem. 124, 7613–7616. Copyright 2012, Wiley-VCH. (C and D) Reproduced with permission from reference Lee, B. R., Kim, J.-W., Kang, D., Lee, D.W., Ko, S.-J., Lee, H.J., Lee, C.-L., Kim, J.Y., Shin, H.S., Song, M.H., 2012. Highly efficient polymer light-emitting diodes using graphene oxide as a hole transport layer. ACS Nano, 6, 2984–2991. Copyright 2012, American Chemical Society.*

detrimental for the long-term stability of the device. Second, its LUMO level is rather low, which results in insufficient electron blocking capability. In an experiment when a PPV electron blocking layer was introduced between emissive PF layer and PEDOT:PSS, the maximum luminous efficiency of the PLED was almost doubled (Morgado et al., 2002).

To overcome the challenges emerged in the HTL, a series of polymerizable perfluorocyclobutane (PFCB)-based copolymers were developed. They exhibited electron-rich triaryl amine side chains that were favorable for hole transporting and electron blocking (Liu et al., 2006). PFCB film was prepared through a precursor method in which di-trifluorovinylphenylether (DTFV) solution was first coated on ITO and then thermally converted into solid film. Among the many kinds of copolymers, bis(*N*,*N*′-diphenyl-*N*,*N*′-bis(3-butylphenyl)-(1,1′-biphenyl)-4,4′amine)-PFCB (BTPD-PFCB) had the highest triaryl amine group density and the best hole transporting capability. The transmittance of BTPD-PFCB was over 93% in the entire visible spectrum, which was higher than PEDOT:PSS. The work function of PEDOT:PSS was approximately 5.2 eV, while the work function of BTPD-PFCB was around 5.3 eV. Thus for emitting material with a relatively low HOMO level, the hole injection from HTL to emission layer could be enhanced. In the PLED constructed with a green emitting fluorene copolymer as emission layer, the maximum luminance and luminous efficiency were 52,000 cdm^{-2} and 4.7 cdA^{-1} when PEDOT:PSS was used as the HTL, respectively. The maximum luminescence and luminous efficiency were improved to 59,400 cdm^{-2} and 18 cdA^{-1}, respectively, when BTPD-PFCB was used as HTL (Jiang et al., 2002). When the thermally cross-linked hole transporting material polystyrene-bis(diphenylamino)biphenyl-PFCB (PS-TPD-PFCB) was used as HTL in the PLED comprised of green emitter, the maximum luminance reached 20,600 cdm^{-2}, 60% higher than that of the PLED constructed with PEDOT:PSS. At the same time the maximum luminous efficiency was increased from 2.13 to 3.82 cdA^{-1}.

There was no doubt that PFCB-based copolymers demonstrated higher hole transporting and electron blocking properties when they were used as HTL in PLEDs compared with PEDOT:PSS. However, PFCB films needed to be prepared through a complicated precursor method. Metal oxide thin films, such as WO_3, MoO_3, and V_2O_5 are also candidate replacements for PEDOT:PSS because of their high electrical conductivity, optical transmittance, air stability, and mechanical stability. Unfortunately, these metal oxide films must be prepared through vacuum evaporation technique which also limits their applications in PLEDs. Thus to find out a kind of solution-processable materials is of great importance. Carbon materials are nowadays widely used in organic electronic devices by virtue of their high electrical conductivity, mechanical flexibility, and optical transmittance. Solution-processable graphene oxide (GO) was found to be a potential material to replace PEDOT:PSS as HTL since the GO interlayer improved the charge carrier injection balance by preventing electrons transfer from emission layer to ITO, and prevented radiative exciton quenching at the same time (Lee et al., 2012). A GO thin film was prepared by spin coating method. The structure and energy diagram of the PLED are shown in Fig. 7.3C. Additional repetitive spin coating processes resulted in full coverage of GO films, and the GO film thickness was determined by the spin coating times. Here GO contained some functional groups, such as epoxy and hydroxyl groups which would destroy the conjugation structure of the hexagonal graphene lattice. As a result, GO demonstrated insulative properties with a large band gap (3.6 eV). After reduction process which partially removed the functional groups and made GO become reduced GO (rGO), the band gap was reduced to 1.15 eV. The work function of GO and rGO were about 4.89 eV and 4.79 eV (according to UPS measurements), respectively.

The performances of PLEDs with different kinds of HTL layers were measured in terms of J–V, L–V (Fig. 7.3D), luminous efficiency–voltage (E–V) and power efficiency–voltage (P–V) curves. For PLEDs with PEDOT:PSS as HTL, the maximum luminance was 33,800 cdm^{-2} (at 12.6 V) and the maximum luminous efficiency was 8.7 cdA^{-1} (at 9.6 V). For PLEDs with rGO as HTL, the maximum luminance was 8300 cdm^{-2} (at 13.0 V) and the maximum luminous efficiency was 5.0 cdA^{-1} (at 8.6 V), which were both much lower than those with

PEDOT:PSS. The reason was ascribed to that rGO had lower electrical conductivity and hole injection ability because of higher contact barrier between emission layer and HTL. The performance of PLEDs with GO as HTL was closely relative to the thickness of GO film. For the optimized PLEDs, the thickness of GO film was 4.3 nm. This device demonstrated maximum luminance of 39,000 cdm^{-2} (at 10.8 V), maximum luminous efficiency of 19.1 cdA^{-1} (at 6.8 V), and maximum power efficiency of 11.0 lmW^{-1} (at 4.4 V), which were improved by 120, 220, and 280% compared to reference devices with PEDOT:PSS as HTL, respectively. Actually the large band gap of GO endowed good electron blocking ability to itself, which could improve the electron-hole recombination probability. As a result, a higher performance was obtained. When GO films were thinner, it was difficult for the material to realize a full coverage, which was negative for electron blocking behavior. When GO films were thicker, the electrical conductivity was increased and it had bad effect on hole transporting process.

7.2.3.3 Electron Injection/Transporting Layer

Electron injection/transporting layers are introduced in PLEDs to reduce the electron injection barrier between the cathode and the emission layer. Thus ETLs should possess a good electron affinity. At the same time, ETLs should have large ionization potential to block holes. Apart from inorganic materials and small organic molecule materials, polymers, such as conjugated polyelectrolytes (CPEs) are also found to effectively reduce the electron injection barrier. The application of CPEs is favorable since it can remove low work function metals, such as Ca and Ba which are sensitive to the ambient environment and thus lowering the lifetime of the devices. A possible mechanism to explain the reduction of electrons injection barrier is that a permanent dipole is created between the cathode and the ETL (Crispin et al., 2002). Poly(9,9-bis(2-(2-ethoxyethoxy)ethyl)fluorene) (PF_{EO}) is a kind of neutral CPEs which is able to form a uniform film on the surface of emission layer with the help of polyfluorene backbone. However, this neutral CPE material showed much lower current density than the devices with Ca/Al as the cathode, which was caused by weak dipole between polyethylene oxide (PEO) and the cathode.

Besides neutral CPEs, cationic and anionic CPEs are also reported to function as cathode modifiers. When cationic CPEs are used as ETL in PLEDs, the counter ions will transfer to the interface between emission layer and CPEs driven by the applied voltage. The quenching effect of the mobile counter ions will decrease the luminous efficiency of the PLEDs. When anionic CPEs are used as ETL in PLEDs, the counter ions will move toward the cathode to obtain strong oriented dipole, thus creating a double layer between the mobile counter ions and the cathode to facilitate the electron injection process. It is obvious that anionic CPEs are more suitable to act as cathode modifier compared to cationic CPEs. Novel anionic CPEs with both PEO derivative side chains and sulfonate sodium side chains ($PF_{EO}SO_3Na$) were developed (Fig. 7.4A) (Zhu et al., 2012). As $PF_{EO}SO_3Na$/Al was applied as cathode in PLEDs, the current density of the devices was obviously elevated in comparison with Al, PEO/Al, and PF_{EO}/Al related devices at the same applied voltage which proved the effect of strong dipole brought by anionic CPEs. Although the current

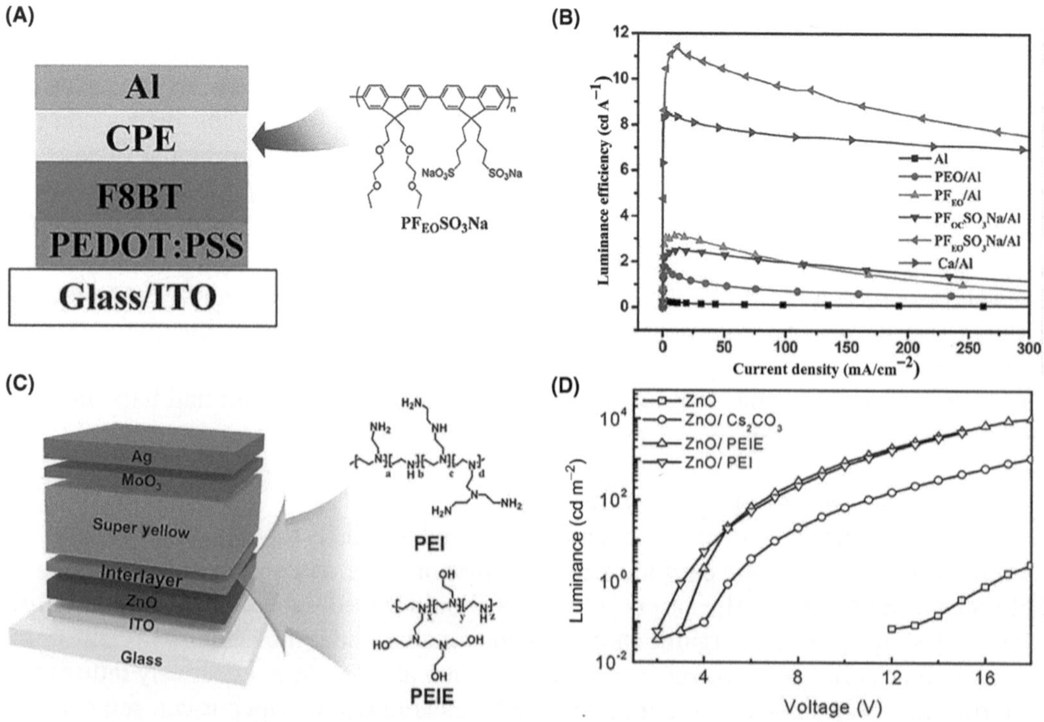

FIGURE 7.4 PLEDs with novel ETLs. (A) PLED architecture and molecular structure of the $PF_{EO}SO_3Na$. (B) *E–J* plots of the devices with different cathodes. (C) The PLED structure using PEI and PEIE on ZnO layer. (D) *L–V* plots of inverted PLEDs without interlayer, Cs_2CO_3, PEIE, and PEI. *(A and B) Reproduced with permission from reference Zhu, X., Xie, Y., Li, X., Qiao, X., Wang, L., Tu, G., 2012. Anionic conjugated polyelectrolyte–wetting properties with an emission layer and free ion migration when serving as a cathode interface layer in polymer light emitting diodes (PLEDs). J. Mater. Chem. 22, 15490–15494. Copyright 2012, Royal Society of Chemistry. (C and D) Reproduced with permission from reference Kim, Y.H., Han, T.H., Cho, H., Min, S.Y., Lee, C.L., Lee, T.W., 2014. Polyethylene imine as an ideal interlayer for highly efficient inverted polymer light-emitting diodes. Adv. Funct. Mater. 24, 3808–3814. Copyright 2014, Wiley-VCH.*

density of $PF_{EO}SO_3Na$/Al devices was lower than that of the devices with Ca/Al as the cathode, the turn-on voltage of which was 2.4 V, as low as that of Ca/Al devices. This low turn-on voltage revealed that $PF_{EO}SO_3Na$ was able to lower the work function of the cathode to the level of Ca. Additionally, the maximum luminous efficiency of $PF_{EO}SO_3Na$/Al based PLEDs reached 11.4 cdA^{-1}, nearly 35% higher than the reference devices based on Ca/Al cathode (Fig. 7.4B). This was because $PF_{EO}SO_3Na$ modification layer also attributed to the hole blocking effect when improving electron injection process. In order to research the effect of PEO derivative groups, devices based on $PF_{OC}SO_3Na$ without PEO derivative side chains were fabricated and measured. The result showed that the performance of devices with $PF_{OC}SO_3Na$ as modification layer was not good, thus proving the importance of PEO derivative groups which facilitated the ions migration under the driving voltage. Note that the performance of PLEDs was highly related to the thickness of $PF_{EO}SO_3Na$ layer because of its insulated properties. For a thicker $PF_{EO}SO_3Na$ film, the luminous efficiency of the devices dropped greatly due to the decreased electrical conductivity. While for a thinner $PF_{EO}SO_3Na$ film, the performance of the devices was extremely poor since PF_{EO}-SO_3Na film could not fully cover the surface of the emission layer.

Compared with the conventional structure, the design of inverted structure removes the air-sensitive electron injection materials (such as Ca and LiF) to solve the device degradation problem in ambient environment. In inverted PLEDs, air-stable metal oxide is usually used as electron injection layer. Among this kind of materials, ZnO is the most widely investigated because of its favorable n-type properties arising from interstitial Zn and oxygen vacancies (Bolink et al., 2007). Unfortunately the LUMO energy level of ZnO is much deeper than that of emissive polymer materials, thus resulting in large electron injection barrier into the emission layer. Cs_2CO_3 interlayer is able to decrease the work function of ZnO by virtue of interfacial dipoles (Park et al., 2011). However, Cs_2CO_3 is air-sensitive which is unfavorable for the device lifetime and roll-to-roll process.

Apart from inorganic materials, branched polyethyleneimine (PEI), and polyethyleneimine ethoxylated (PEIE) were reported to function as electron injection layer for inverted PLEDs (Fig. 7.4C) (Kim et al., 2014). These two air-stable polymers were comprised of amine groups which were able to create strong dipoles between PEI and the ZnO surface. The strong molecule dipoles were favorable to decrease the electron injection barrier and thus improve the electron injection process. Additionally, this polymer interlayer contributed to hole blocking and exciton quenching blocking between emission layer and ZnO. UPS measurements revealed that PEI (4 nm) and PEIE (4 nm) shifted the work function of ZnO from 4.4 to 2.47 eV and 3.29 eV, respectively. The decrease of electron injection barrier helped to facilitate electron injection. At the same time, fluorescence lifetime tended to be relative to the energy level difference between the LUMO of the emission layer and the work function of electron injection layer. Here the fluorescence lifetime was 0.34 ns in devices based on ZnO with the highest energy level difference (1.65 eV), 0.42 ns in devices based on ZnO/PEIE (8 nm) (0.61 eV), and 0.45 ns in devices based on ZnO/PEI (8 nm) (0.31 eV). This phenomenon could be explained by the fact that large energy level difference tented to induce exciton quenching. According to the fluorescence lifetime measurements, PEI was more effective in exciton quenching blocking compared to PEIE.

It must be considered the work function reduction was closely relative to the thickness of the polymer interlayer. For example, the work function of ZnO increased from 2.47 eV (4 nm PEI) to 3.39 eV (16 nm PEI). This was because with the increasing polymer interlayer thickness, the random molecule dipoles countervailed with each other and thus weakened the interfacial dipole effect between ZnO and the interlayer. When polymer interlayer was used in the inverted structure, the device performance was obviously enhanced. Fig. 7.4D demonstrates the *L–V* curves of PLEDs. The turn-on voltage (voltage at luminance of 1 cdm^{-2}) and operating voltage (voltage at luminance of 1000 cdm^{-2}) of PLEDs based on PEI (3 V/10 V) and PEIE (3.8 V/10 V) interlayers were all much lower than PLEDs with Cs_2CO_3 (5.14 V/18 V) and without interlayer (16.3 V turn-on voltage), indicating improvement of electron injection and exciton quenching blocking caused by PEI and PEIE. While at the same applied voltage, the luminance and luminous efficiency of PLEDs based on PEI and PEIE were also obviously higher than the PLEDs with Cs_2CO_3 or without electron injection interlayer.

7.2.3.4 Polymer White Light-Emitting Devices

In recent years, polymer white light-emitting devices (PWLEDs) have attracted great research interests because of their possible application in large-area full-color displays, back-lighting sources for liquid crystal display and solid-state lighting sources. In order to realize full-color light emission, three primary colors (i.e., red, blue, and green) or arbitrary two complementary colors (e.g., yellow and blue) must be mixed together. Therefore multilayered PLEDs with several different emission layers are needed. Polymer materials are suitable to be deposited by solution processes which are low-cost and can be applied to fabricate PLEDs with existing large-area coating or printing techniques. However there remain significant challenges to prepare multilayered polymer films. Considering that many kinds of organic semiconducting molecules can be solved in the same solvent, depositing the second organic film on the established first one will probably lead to morphology damage or intermixing effect. There are some strategies to deal with this problem. One of the approaches is to take use of cross-linkable polymers that can transfer to insoluble state after thermal curing or photocuring. But the addition of special curing agents in this method must be considered since they may have adverse effects on the stability of the device. The other similar approach is introducing high temperature annealing treatment to make the polymer thin film insoluble. However it is really hard to control the morphology and thickness of the polymer film during annealing. One more reasonable approach is making use of orthogonal materials, that is, hydrophobic and hydrophilic materials which are deposited from nonpolar and polar solvents, respectively (Gong et al., 2005).

Based on orthogonal materials, PWLEDs comprising of conjugated blue light-emitting polymer and ionic transition-metal complexes (iTMCs) were developed (Sessolo et al., 2013). The iTMCs were explored to possess high photoluminescence quantum yields and the emission colors of iTMCs were tunable through the spectral range (Bolink et al., 2008). The high solubility of iTMCs in polar solvents helped them to be incorporated into the multilayered PLEDs. For this design, 6-phenyl-2,2′-bipyridine-bis[2-(phenyl)pyridinato]iridium(III) hexafluorophosphate [Ir(ppy)$_2$(Hpbpy)](PF$_6$) and conjugated polymer CB02 were combined together to construct PWLEDs. Bright white emission with tunable color temperatures was achieved through changing the thickness of iTMCs and polymer films. In the PWLEDs, conjugated polymer CB02 acted as blue emitter, while iTMCs acted as orange emitter. Atomic force microscope (AFM) test showed that the film of iTMCs remained smooth after rinsing with mesitylene, proving that iTMCs film was not damaged after depositing a second film whose solvent was mesitylene. In order to research the influence of the thickness of emission layer on the electroluminescence property, three devices were constructed. The thickness of iTMCs film was ranged from 20 nm (device *X*) to 25 nm (device *Y*) and 30 nm (device *Z*) while the thickness of blue emitter CB02 was kept at 70 nm. Actually this tiny difference of film thickness demonstrated intense influence on the emission spectra of the PWLEDs. The luminous efficiency (at voltage of 10 V) was increased from 0.41 cdA^{-1} (device *X*) to 0.46 cdA^{-1} (device *Y*) and 0.90 cdA^{-1} (device *Z*). The phenomenon was expected as iTMCs could generate more efficient electroluminescence compared to conjugated polymer CB02. However, the luminous efficiency of the devices was relatively low and it was probably because of exciton quenching process. Additionally, the triplet energy of iTMCs (phosphorescent material) and polymer (fluorescent material) were similar. As a result, the triplet excitons created in the iTMCs or migrated from polymer to the iTMCs tented to transfer to the nonemission triplet states in the polymer. Finally, a large amount of excitons would undergo nonradiative transition which decreased the luminous efficiency.

The electroluminescence spectra and emission colors demonstrated in Commission Internationale de L'Eclairage (CIE) coordinates were studied based on different thickness of $[Ir(ppy)_2(Hpbpy)](PF_6)$ layers and the applied voltage. For device *X*, a strong peak at 475 nm and a spectra shoulder at 580 nm corresponded to the emission of CB02 (blue) and $[Ir(ppy)_2(Hpbpy)](PF_6)$ (orange), respectively. As the applied voltage increased to 7 V, the orange emission from iTMCs was enhanced and the color coordinates transferred to get nearer to the white point (0.324, 0.337; white point: 1/3, 1/3). Higher voltages continued to increase the orange emission, but there was just a small shift in the corresponding color coordinates. Note that the blue emission from polymer was the main part at low voltage, showing that the recombination region extended to the polymer rather than was located at the heterojunction, indicating the thickness of iTMCs (20 nm) in device *X* was relatively thinner than the ideal value. For device *Y* (iTMCs 25 nm), the orange and blue emission had nearly the same intensity when applied voltage was 7 V, while the color coordinates were (0.353, 0.350). With the increasing applied voltage, the blue emission gradually decreased. At the same time, the variations of color coordinates were limited till the voltage was 10 V, indicating a more stable white light electroluminescence than that of device *X*. When the thickness of iTMCs layer was increased to 30 nm (device *Z*), the blue emission was quite weak and the device indicated obvious voltage-independent emission spectra for which the color coordinates were located at the edge of white and orange zones. In summary, device *Y* was the best choice for constructing PWLEDs and the emission color could be easily tuned by adjusting the thickness of iTMCs layer.

Apart from PWLEDs with double emission layer structure, single emission layer structure is also used to construct PWLEDs. Single layer structure is usually realized by mixing different emission materials together, and a host–guest system is commonly adopted. Host–guest system is favored because the theoretical internal quantum efficiency of electrophosphorescent PLEDs is 100%, much higher than that of electrofluorescent PLEDs. Although PWLEDs are favored for their highly cost-effective characteristic and ease to fabricate large-area devices through solution processes, their applications are limited by the relatively low device performance when acting as solid-state lighting sources. For organic white light-emitting devices (OWLEDs) which were fabricated through vacuum deposition techniques, their power efficiency reached that of fluorescent lamps (40–70 lmW^{-1}). But for most of the PWLEDs, their power efficiency is lower than incandescent light bulbs.

In order to enhance the device performance of PWLEDs, novel PWLEDs were developed through double doping of blue emitter (FIrpic) and yellow emitter (iridium complexes) (Fig. 7.5A) into PVK (host material) with the presence of electron transporting material 1,3-bis[(4-tertbutylphenyl)-1,3,4-oxadiazolyl]phenylene (OXD-7) (Wu et al., 2009). The

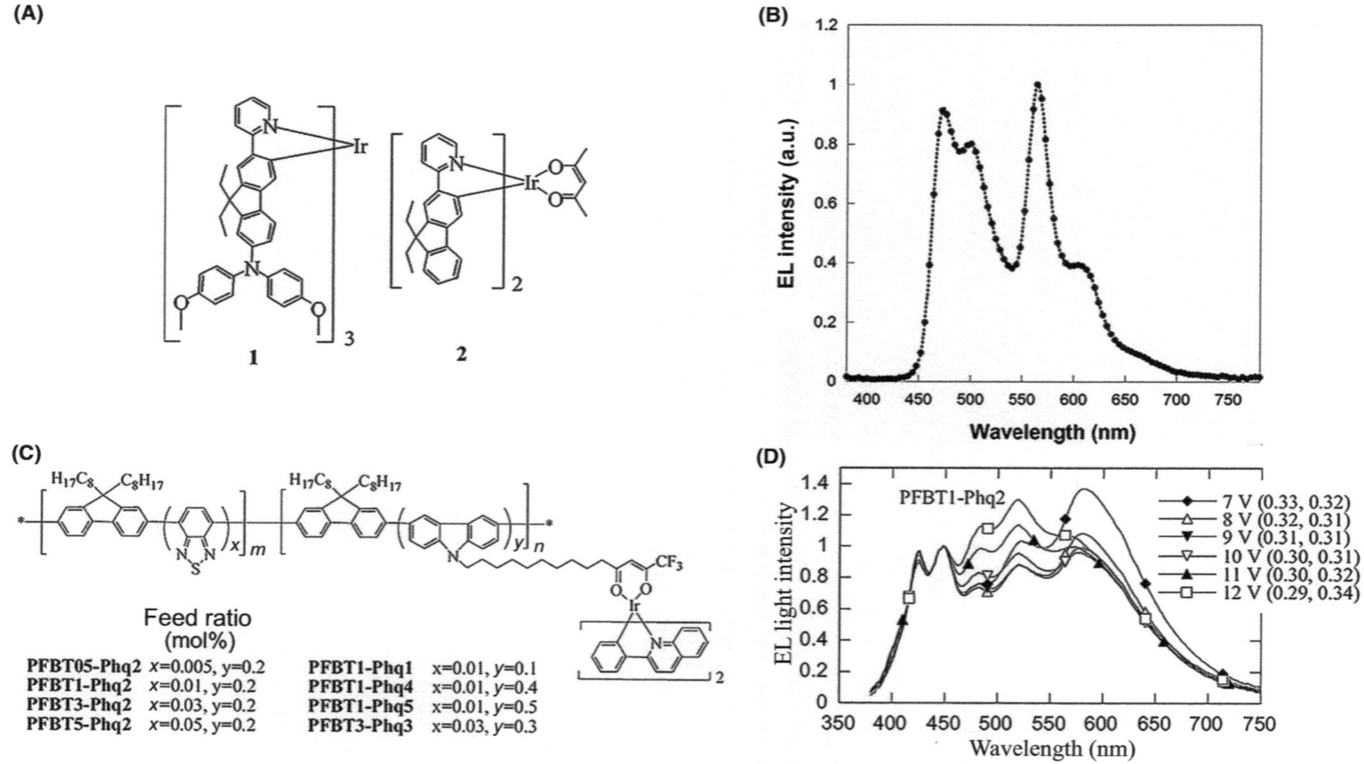

FIGURE 7.5 PWLEDs with single emission layer. (A) Chemical structures of yellow phosphors 1 and 2. (B) The EL spectrum of the PWLED (FIrpic/1 = 30/1) at 12 mA cm^{-2}, with CIE coordinates of (0.320, 0.428), in which PVK (63 wt%)/OXD-7 (27 wt%)/FIrpic (10 wt%) was fixed for the matrix. (C) Chemical structure of PFBT-Phq, the copolymer from fluorene (F), BT, and the 2-phenylquinoline iridium complex (Phq). (D) EL spectra based on the copolymer PFBT1-Phq2 at various applied voltages with corresponding CIE coordinates. *(A and B) Reproduced with permission from reference Wu, H., Zhou, G., Zou, J., Ho, C.L., Wong, W.Y., Yang, W., Peng, J., Cao, Y., 2009. Efficient polymer white-light-emitting devices for solid-state lighting. Adv. Mater. 21, 4181–4184. Copyright 2009, Wiley-VCH. (C and D) Reproduced with permission from reference Jiang, J., Xu, Y., Yang, W., Guan, R., Liu, Z., Zhen, H., Cao, Y., 2006. High-efficiency white-light-emitting devices from a single polymer by mixing singlet and triplet emission. Adv. Mater. 18, 1769–1773. Copyright 2006, Wiley-VCH.*

electroluminescence properties of the single emission layer PWLEDs could be tuned by controlling the blending ratio between FIrpic and iridium complexes. Fig. 7.5B demonstrates electroluminescence spectra of the device at 12 $mAcm^{-2}$ when blending ratio between FIrpic and yellow phosphors 1 is 30/1. The two spectrum peaks at 470 nm and 566 nm correspond to blue and yellow emissions from FIrpic and yellow phosphors 1, respectively, with CIE coordinates at (0.320, 0.428). When blending ratio between FIrpic and yellow phosphors 1 was adjusted to 20/1, the best electroluminescence performance was achieved. The maximum luminous efficiency reached 42.9 cdA^{-1} at 7.2 V, and the maximum power efficiency reached 20.3 lmW^{-1} at 6.0 V. At the operating state where the forward viewing luminance was 1000 cdm^{-2}, the luminous efficiency and power efficiency were 41.7 cdA^{-1} and 16.8 lmW^{-1}. Actually for solid-state lighting applications, all photons should be considered since the lighting fixtures could be incorporated to redirect the photons to forward viewing direction. Hence a factor of 1.7–2.3 should be introduced to the pervious forward viewing efficiencies to achieve even higher total luminous and power efficiencies. These values surely exceeded that of incandescent light bulbs, indicating that PWLEDs were possible to realize low-cost and large-area solid-state lighting.

In the single emission layer, phosphorescent blue-emitting FIrpic and yellow-emitting iridium complexes were simultaneously mixed in the PVK/OXD-7 host matrix. PVK was favored because of its high triplet state (3.0 eV), ease to generate thin film and hole transporting ability, while OXD-7 was added to facilitate electron transport. Furthermore, in this host–guest system carrier trapping excitation mechanism was dominant since the LUMO energy level of FIrpic and yellow phosphors 1 were lower than that of PVK while the HOMO energy level of FIrpic and yellow phosphors 1 were higher than that of PVK. This single emission layer possessed obvious advantages, for example, the device structure was simpler than double emission layer, and only solution processes were needed with low cost. However, it should be noted that this kind of PWLEDs cannot demonstrate purely white light emitting and the short lifetime limited their applications since PVK host and FIrpic are unstable during operating. Hence more stable polymer host and blue emitters remained to be developed.

For most PWLEDs, the realization of white light emitting is by virtue of blend of different polymers or small molecule doping. For example, by blending 0.5% of MEH-PPV with PFO to function as the emission layer of PWLEDs, the power efficiency of the devices reached 16 lmW^{-1} with CIE coordinates of (0.36, 0.40) (Huang et al., 2006). However, this blend or doping method is not favored since the emission color is highly dependent on the applied voltage and it is not easy to control the doping level. Hence it is of great importance to develop single component emission material to construct PWLEDs. Simultaneous emission of red, green, and blue in a single polymer was realized and the CIE coordinates of the PWLEDs were (0.31, 0.34) with luminous efficiency of 1.59 cdA^{-1} (Liu et al., 2005). Doping 1,8-naphthalimide on the polyfluorene backbone to construct PWLEDs with CIE coordinates of (0.32, 0.36) was reported and the luminous efficiency of which reached 3.8 cdA^{-1} (Tu et al., 2006). However, the performances of PWLEDs based on single component emission layer were obviously below the blend or doping system.

In order to enhance the device performance based on single component emission material, a new approach was found to realize PWLEDs through combination of fluorescence and phosphorescence emitting species in the copolymer (Jiang et al., 2006). The copolymer was comprised of fluorene (F) backbone and a small number of benzothiadiazole (BT) and 2-phenylquinoline iridium complexes (Phq) (PFBT-Phq) (Fig. 7.5C). PF with high band gap acted as host matrix and blue emitter while BT in the backbone and Phq on the side chain acted as singlet green emitter and triplet red emitter, respectively. When the content of BT and Phq were in the range of 0.01–0.03 mol% and around 0.2 mol%, respectively, white light emitting was achieved. The CIE coordinates of PFBT1-Phq2 and PFBT3-Phq2 were (0.34, 0.33) and (0.32, 0.33), which were really close to pure white light emitting. Fig. 7.5D depicts the electroluminescence spectra shift of PFBT1-Phq2. As the applied voltage was increased from 7 to 12 V, the CIE coordinates shifted from (0.33, 0.32) to (0.29, 0.34), indicating that the spectra was relatively stable as the operating states changed. The maximum luminous efficiency of 6.1 cdA^{-1} was obtained for PFBT5-Phq2 with CIE coordinates of (0.32, 0.44) and the maximum luminance was 10,110 cdm^{-2} at current density of 345 $mAcm^{-2}$. However, this emission could only be considered as greenish-white since more charges were trapped in BT units, indicating that the content of BT units was too high. It was similar to PFBT1-Phq5 which demonstrated reddish-white emission with CIE coordinates of (0.44, 0.38). In order to increase the purity of white emission, the contents of BT and Phq must be limited. For the PWLEDs based on PFBT3-Phq3, when the contents of BT and Phq were appropriate, the CIE coordinates were (0.31, 0.34) with luminous efficiency of 4.6 cdA^{-1}, which significantly balanced the white emission purity and device efficiency.

7.2.4 Functionality

The properties of PLEDs are closely related to the substrates. Conventional PLEDs are constructed on ITO glass which is intrinsically brittle and unable to bear any bending or stretching deformations. Since polymer materials possess some advantages, such as light weight, flexibility and robustness, it is possible to realize flexible or stretchable PLEDs based on flexible or stretchable substrates. This is quite important because flexible displays and solid-state lighting sources are of increasing interest nowadays for many potential applications. The PLEDs must be highly flexible to endure the arbitrary mechanical deformations induced by the host materials when they are integrated into malleable materials. Besides flexibility and stretchability, PLEDs are also able to be endowed with other functionalities, such as shape memory properties to meet the demands in different situations.

7.2.4.1 Stretchability

Early in 1992, Heeger and coworkers fabricated PLEDs on flexible PET substrates, making it possible to create flexible display devices (Gustafsson et al., 1992). Later on different flexible PLEDs were constructed on polymer anode/PET or other flexible substrates (Huang et al., 2007). However the highly stretchable PLEDs were not realized until fabrication

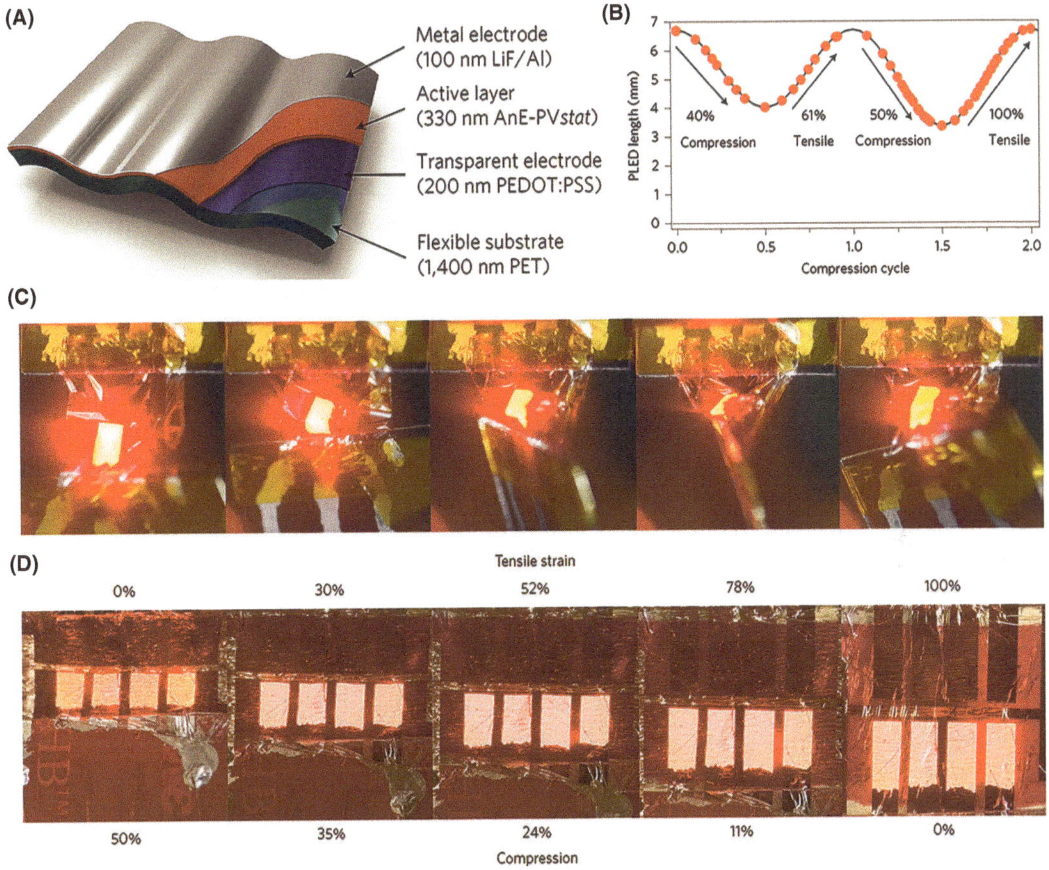

FIGURE 7.6 Schematic structure and characteristic of stretchable PLEDs. (A) Compositional schematic of ultrathin PLEDs with each layer thickness drawn to scale. The schematic shows periodic bending with a radius of curvature of ~5 mm for reference. (B) Length of the PLED during two compression cycles. (C) Photographs of a free-standing ultrathin PLED operating during crumpling. The device is suspended between two pieces of glasses that are moved closer together and simultaneously twisted by 90 degree. (D) Photographs of an ultrathin PLED adhered to a prestrained elastomer tape. The photographs from left to right represent the wrinkled PLED (unstrained elastomer) state being stretched to the flat PLED (stretched elastomer) state. *Reproduced with permission from reference White, M.S., Kaltenbrunner, M., Głowacki, E.D., Gutnichenko, K., Kettlgruber, G., Graz, I., Aazou, S., Ulbricht, C., Egbe, D.A., Miron, M.C., 2013. Ultrathin, highly flexible and stretchable PLEDs. Nat. Photonics 7, 811–816. Copyright 2013, Nature Publishing Group.*

of PLEDs on ultrathin PET foil (1.4 μm) substrates was realized (White et al., 2013). The schematic structure of the PLED is displayed in Fig. 7.6A. Poly(*p*-phenylene-ethynylene)-*alt*-poly(*p*-phenylene-vinylene) derivative with linear octyloxy and 2-ethylhexyloxy side groups (AnE-PV*stat*) functioned as red emitter for red light-emitting PLEDs while poly[2-methoxy-5(3,7-dimethyloctyloxy)]-1,4-phenylenevinylene (MDMO-PPV) functioned as orange emitter for orange light-emitting PLEDs. The highest luminance and luminous efficiency for AnE-PV*stat*-based devices were 113 cdm^{-2} and 0.043 cdA^{-1}, respectively. AnE-PV*stat*-based devices demonstrated CIE coordinates of (0.698, 0.302), which corresponded to red emission. For MDMO-PPV-based devices the highest luminance and luminous efficiency were 122 cdm^{-2} and 0.23 cdA^{-1}, respectively, and they demonstrated CIE coordinates of (0.631, 0.368), which corresponded to orange emission. Both of the two kinds of devices reached the luminance over 100 cdm^{-2}, exceeding the requirement for indoor displays. Although the fabrication of this ultrathin PLEDs was accomplished when the PET foil was directly adhered to the PDMS coated glass, the devices could be easily peeled off and apply to other kinds of surfaces. Both the free standing and surface supported devices could realize light emission.

According to the first order approximation, the surface bending strain of electronics could be denoted as $S = h/2R$ (h: the thickness of the film; R: bending radius of curvature). When the electronic was enduring the same bending strain, the reduction of h would lead to the same reduction of R. Since the electronics based on polymer with a thickness of 125 μm had critical bending radius of about 1 mm,(Rowell et al., 2006) the ultrathin PLEDs with the 1.4 μm substrate could have critical bending radius of about 10 μm. This small bending radius was favored when the PLEDs were incorporated in textiles since they must undergo extreme and complicated deformations including twisting, wrinkling, and crumpling. To test the device performance in different kinds of deformations, the PLED foil was located between two glass slides. One of the slide was gradually moved to the other and twisted by 90 degree at the same time, leading to the buckling, crumpling, and twisting deformations (Fig. 7.6C). It was obvious that the PLEDs were operated normally when subjected to these deformations. In order to realize stretchable devices, the freestanding PLED foils were paved on the 100% prestrained elastomer. The PLED foils would undergo compression process as the folds were generated, without losing the luminance. Then the PLEDs could be stretched to 100% tensile strain, as shown in Fig. 7.6D. Note that the tensile strain of 100% was limited by the elastomer, rather than the PLED foils. The compression and stretching processes were able to be repeated for many times. Fig. 7.6B demonstrates two compressing/stretching processes when the compressive strains are 40 and 50%, respectively. Earlier in 2011, stretchable polymer light-emitting electrochemical cells (PLECs) based on the electrodes comprised of polymer and carbon nanotubes (CNTs) were reported. But the tensile strain of the devices was limited to 45% and the stretching process was only realized over 70°C. The other problem was that the slow response of PLECs was not favorable for their applications in displaying. Compared to the stretchable PLECs, the stretching of ultrathin PLED foils could be realized at room temperature which was of great importance for practical uses.

It must be noted that the stretching ability of PLED foils was realized by virtue of the prestrain of the foils on the elastomer rather than the intrinsic stretching ability of the material itself. In 2014, the first fully stretchable PWLEDs emerged without prestrain based on the novel stretchable transparent conductive electrodes which were comprised of GO-modified silver nanowires (AgNW) imbedded in polyurethane acrylate (PUA) (Liang et al., 2014). The GO-modified AgNW network demonstrated sheet resistance of 14 Ωsq^{-1} with 88% transmittance (at 550 nm). The devices emitted white light from both sides and the luminance reached 1100 cdm^{-2} while the maximum luminous efficiency was 2.0 cdA^{-1}. After 100 stretching cycles from 0 to 40% tensile strain, the PWLEDs could still work with moderate performance, and the maximum tensile strain the devices could endure was up to 130%. This kind of stretchable PLEDs had obvious advantages toward PLED foils since the folds caused by prestrain did not exist here to have any influence on the performance of the devices.

7.2.4.2 Shape-Memory Property

Deformable electronic devices are needed since they can accommodate to irregular surfaces with various shapes, and this advantage is of great value in some novel applications, such as wearable displays and biomedical devices. Shape-memory property of electronics is mostly based on the intrinsic characteristic of the device substrate. In 2011, shape-memory PLEDs were realized with AgNW/cross-linked polyacrylate electrodes (Yu et al., 2011c). For the AgNW film directly coated on the glass, the surface roughness was high (height variation larger than 500 nm) because of the protruding AgNW. However, in the process to prepare AgNW/polymer film, the acrylate monomers penetrated into the AgNW networks, filling the space among AgNW networks and the blank region between AgNW and the substrate. Finally, a cross-linked polymer network was generated through polymerization to imbed most of the nanowires. As a result, the surface of the AgNW/polymer was quite smooth with height variation less than 5 nm. This smooth surface was desired to construct PLEDs. The transmittance of the electrodes changed with the sheet resistance. When the sheet resistance was 100 Ωsq^{-1}, the transmittance was 91% (at 550 nm), 30 Ωsq^{-1} at 86%, and 12 Ωsq^{-1} at 82%.

This AgNW/polymer electrode demonstrated shape-memory property with the help of polyacrylate with a glass transition temperature (T_g) of 110°C. When the electrode was heated to above T_g, it could endure bending or stretching and the deformation would be frozen when cooling below T_g. After the electrode was heated to above T_g again, it could relax back to its initial shape. With the increase of compressive strain, the sheet resistance of the electrode decreased slightly. While with the increase of tensile strain, the sheet resistance increased gradually to nearly 3 times higher than that of the initial value at tensile strain of 16%. Shape-memory PLEDs were fabricated with AgNW/polymer electrodes according to the structure design of polyacrylate/AgNW/PEDOT:PSS/SY-PPV/CsF/Al. The AgNW/polymer PLEDs reached high luminance of 8470 cdm^{-2} at 8 V with maximum luminous efficiency of 14.0 cdA^{-1}. Bending/recovery tests were applied to measure the shape-memory property of the devices (Fig. 7.7A). After 10 convex and concave bending/recovery cycles with 8% strain, the current density and luminance remained unchanged. And the PLEDs could successfully deform and recover to the initial shape with the change of temperature while maintaining the luminance of 300 cdm^{-2} (Fig. 7.7B).

7.3 Polymer Light-Emitting Electrochemical Cells

7.3.1 Introduction

PLECs represent a dynamic research subject toward their promising applications in future lighting and display devices. Unlike PLEDs, the active materials of PLECs are composed of ionically conductive species and the electroluminescent conjugated polymer.

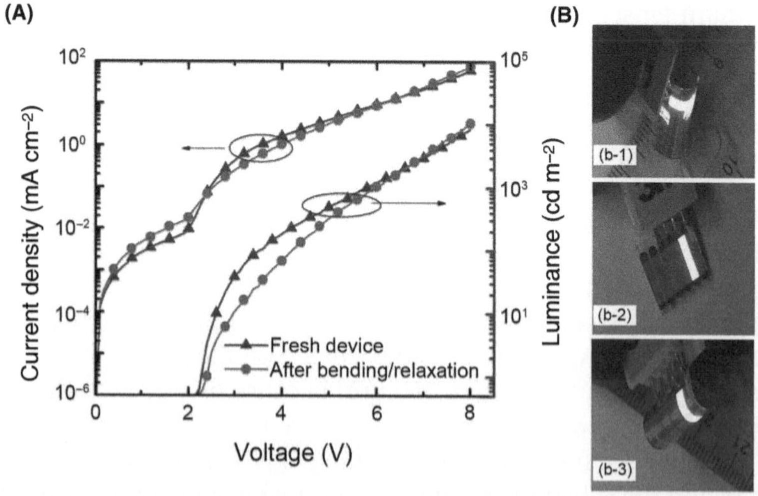

FIGURE 7.7 Characteristics of shape-memory PLEDs. (A) *J–V–L* plots of the shape-memory PLEDs before and after 10 convex bending-recovery cycles and 10 concave bending–recovery cycles up to 8% compressive/tensile strain in the PLEDs. (B) Photographs of shape-memory PLEDs emitting at around 300 cd m^{-2}, bent to a 2.5 mm radius convex (b-1), recovered to flat shape after heating at 120°C for 1 min (b-2), and bent in the opposite direction as in (b-1) to a 2.5 mm radius concave (b-3). *Reproduced with permission from reference Yu, Z., Zhang, Q., Li, L., Chen, Q., Niu, X., Liu, J., Pei, Q., 2011c. Highly flexible silver nanowire electrodes for shape-memory polymer light-emitting diodes. Adv. Mater. 23, 664–668. Copyright 2011, Wiley-VCH.*

The working mechanism of PLECs involves in situ electrochemical doping and the formation of a light-emitting PIN junction in the emission layer. Therefore, the contact resistance and bulk resistance of PLECs are significantly lower than that in PLEDs, resulting in promising advantages, such as low operating voltage and high electroluminescent power efficiency. The formation of PIN junction effectively removes the energy barriers and facilitates the injection of both electrons and holes without the use of low work function cathode. More importantly, PLECs are relatively insensitive to the variations of the active layer thickness, which can be effectively constructed in relatively rougher surfaces. Since the initial discovery in 1995 by Pei and coworkers (Pei et al., 1995), a variety of PLECs based on different structures and materials with improved performance and processability have been developed. Here, we will briefly introduce the working mechanism, device structures and performance of PLECs, as well as the novel device applications including stretchable, flexible, and weavable PLECs.

7.3.2 Mechanism

Although the appearance of a working PLEC is similar to a PLED based on the same electroluminescent polymer, the working mechanism of PLECs is fundamentally different from that of PLEDs. The emissive material in PLECs consists of electroluminescent polymer and electrolyte. The electrolyte is a solid-state ionic conductor consisting of lithium salt and ion-solvation polymer. Before the application of a voltage bias, the dissolved ions

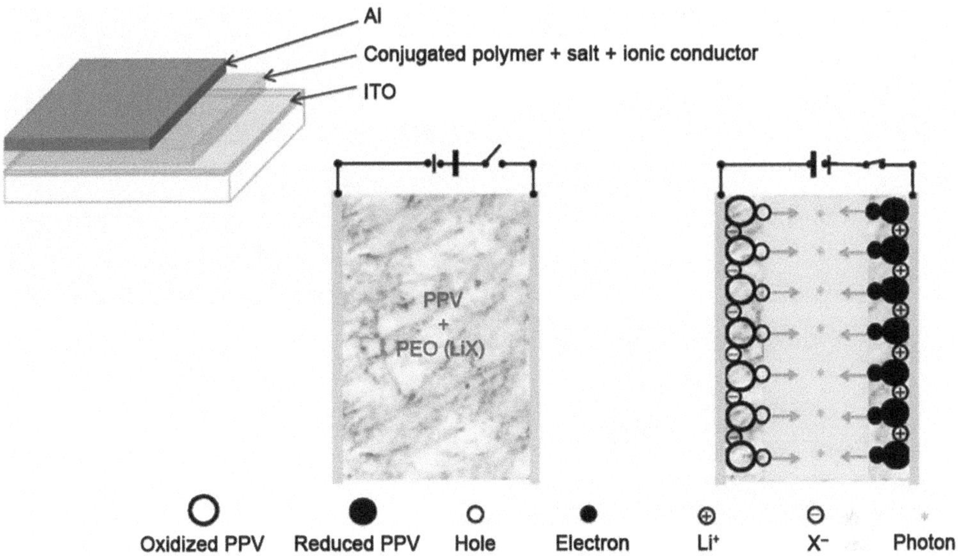

FIGURE 7.8 Schematic illustration to the structure and working mechanism of PLEC. The emission layer comprises of a blend of a light-emitting conjugated polymer (PPV), a salt (LiX), and an ionic conductor (PEO). *Reproduced with permission from reference Yu, Z., Li, L., Gao, H., Pei, Q., 2013. Polymer light-emitting electrochemical cells: Recent developments to stabilize the pin junction and explore novel device applications. Sci. China Chem. 56, 1075–1086. Copyright 2013 Springer.*

are randomly distributed in the active layer. After applying a voltage bias larger than E_g/e, where E_g is the energy gap of the electroluminescent polymer and e is the elementary charge, electrochemical reactions occur in the active layer. The electroluminescent polymer is oxidized near the anode and reduced near the cathode. At the same time, the anions and cations of the dissolved salt are transported by the ionic conductor to the electrode interfaces respectively, resulting in the formation of n-doped and p-doped electroluminescent polymer. The doped polymer is highly conductive, thus allowing the continuous injection of the charge carriers. Therefore, under a constant bias, the in situ electrochemical doping of the electroluminescent polymer can be continued and results in the steady expansion of the doped regions. Eventually, the p-doping and n-doping fronts meet with the region of undoped polymer to form a PIN junction (Fig. 7.8) (Yu et al., 2013).

The processes of dynamic doping and PIN junction formation have been proved by using time-lapse fluorescence imaging technique (Hu and Gao, 2011). The electroluminescent polymer in PLECs is doped upon applying a voltage and therefore dramatic increase in conductivity is obtained. This allows PLECs to operate in planar structure with a large interelectrode spacing, which offers an advantage for illuminating the operating mechanism by direct observation. As shown in Fig. 7.9, time-lapse fluorescence imaging technique was applied to observe the dynamic junctions of a 10.4 mm MEH-PPV:PEO:$CsClO_4$ planar PLEC during turn on and cooling process. The images in Fig. 7.9 were taken after the device had been operated continuously by applying a 400 V bias at 335 K for about

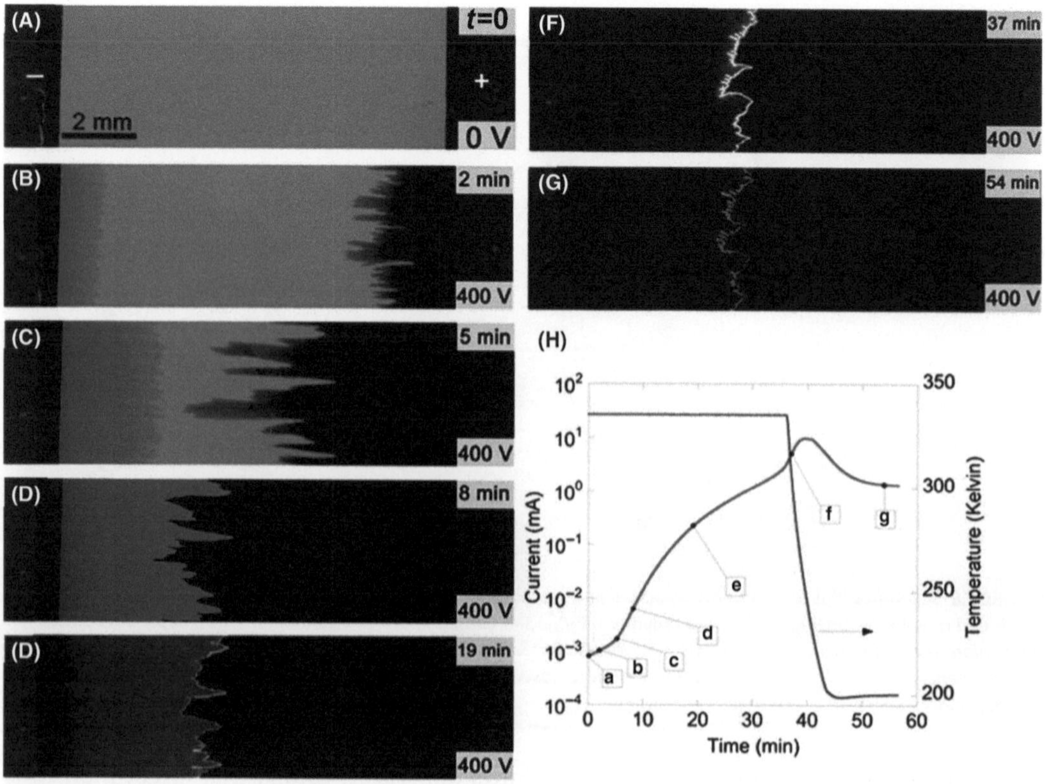

FIGURE 7.9 Time-lapse fluorescence imaging of a 10.4 mm MEH-PPV:PEO:$CsClO_4$ planar PLEC during turn on and cooling. A fixed DC bias of 400 V was applied to turn on the cell, which was at 335 K and under UV illumination. Time since the DC bias was applied to the cell: (A) No bias; (B) 2 min; (C) 5 min; (D) 8 min; (E) 19 min; (F) 37 min; (G) 54 min. Panel H shows cell current (red) and temperature (blue) as a function of time since the DC bias was applied. Uniform enhancement (level adjustment in Photoshop) has been applied to images A–G. *Reproduced with permission from reference Hu, Y., Gao, J., 2011. Direct imaging and probing of the p–n junction in a planar polymer light-emitting electrochemical cell. J. Am. Chem. Soc. 133, 2227–2231. Copyright 2011, American Chemical Society.*

an hour. The active layer displayed orange-red photoluminescence under UV illumination, and there was no doping before applying a bias (Fig. 7.9A). When a bias voltage was applied, the electrochemical doping of MEH-PPV was initiated from the interface of the electrode (Fig. 7.9B). With the increase of time, the doping was propagated toward the opposite electrodes and the doping level increased continually (Fig. 7.9C–F), resulting in the enhanced intensity of electroluminescence (Fig. 7.9H). The formation of PIN junction was accompanied by an increased current, which was due to the decrease of the contact resistance and the bulk resistance in the device. When the current was increased into a milliampere scale, the device was cooled to 200 K and the bias was maintained at 400 V (Fig. 7.9H). As shown in Fig. 7.9G, the electroluminescence remained highly visible after the greatly decreased temperature, suggesting the formation of stable PIN junction in the device.

The working mechanism of PLECs was also proved by using different methods including optical microscope for direct imaging and optical beam induced current measurement (Pei et al., 1995; Dane et al., 2005; Li et al., 1998; Gao and Dane, 2005). It was demonstrated that the migration of the doping fronts were closely related to the applied bias voltage, the mobility of electrons and ions, and also the thickness of active layer. Furthermore, the response time of PLECs was related to the binding energy between cations and anions. The formation of PIN junction effectively removed the energy barriers thereby electrons and holes could be efficiently injected without the use of low work function cathode. The efficient light emission could be generated in the undoped part of the active layer by the radiative recombination of the injected electrons and holes. It should be noted that the thickness of the emission layer was decreased remarkably due to the formation of the PIN junction. Therefore, PLECs could be constructed in relatively rougher surfaces, and millimeter-thick active materials could also be applicable in PLECs.

7.3.3 Structure

7.3.3.1 Sandwiched Structure

PLEDs are typically made of multiple layers of organic materials, sandwiched between a transparent anode and an air-unstable cathode which is often coupled with an electron-injection layer. Such complex architectures are needed for efficient injection of charge carrier across the device. In contrast, PLECs have a much simpler architecture. A typical PLEC consists of an active layer of a conjugated polymer and electrolyte blend sandwiched between two opposite electrodes, one of which is usually ITO. The migration of the ions in the active layer under an applied bias allows for efficient and balanced injection of electrons and holes from air-stable electrodes into the emission layer. Due to the simple single-layer architecture, PLECs can be easily fabricated via solution-based technologies, such as spin coating, spray coating, dip coating, printing, and slot-die coating, which are promising for large-area illumination and displays.

7.3.3.2 Planar Structure

The doped regions in the PLECs have high conductivity and good electrical contact with the electrodes, which allows PLECs to be constructed in a planar structure. In this structure, the electrodes can be prepared onto a substrate with a gap between them before the active layer is cast. Therefore, the electrodes can be generated through pattern-based techniques, such as photolithography and microcontact printing. The electrodes can also be fabricated in an interdigitated pattern and thereby achieve the required luminance over a large area. Compared to the sandwiched structure, the performance of the planar PLECs is relatively insensitive to pinholes or other defects in the emission layer. The planar PLECs offer a new type of light-emitting devices with unique advantages for displays. Such displays can be hybridized with integrated circuits and thereby enable to fabricate integrated devices with new approaches.

The planar PLECs were first demonstrated with an average interelectrode spacing of 15 μm (Yu and Heeger, 1997; Yu et al., 1997). The devices were turned on at a bias voltage greater than 2 V, and showed a luminance of 100 cdm^{-2} at voltage well below 5 V. This performance was comparable to that observed in devices of the sandwiched structure. After this, the interelectrode spacing of planar PLECs was further expanded to be up to 1–2 cm (Gao and Dane, 2004; Hu et al., 2006). Furthermore, a variety of colors could be achieved by simply placing two different light-emitting planar PLECs in series onto the same substrate. Just as we discussed earlier, planer PLECs were also usually applied to research on the working mechanism of the devices. It should be noted that, due to the large polarization potentials at the interfaces between electrode and polymer, the turn-on voltage of large spacing planar PLECs is typically over 100 V at room temperature.

7.3.3.3 Bilayer Structure

The bilayer PLECs can be divided into two categories. The first category is constructed with a polymer bilayer and both layers can emit light. Another one is constructed with a conjugated polymer layer and an electrolyte layer and only the former one can emit light.

For the first category of the bilayer PLECs, the PIN junction can be placed in either of the two layers, depending on the bias. Therefore, the devices can emit two different colors. A two color light-emitting PLEC was developed which consisted of a polymer bilayer structure sandwiched between two electrodes (Yang and Pei, 1997). The structure of this device was ITO/PPV/MEH-PPV/Al, in which both polymer layers were blended with an electrolyte consisting PEO and $LiCF_3SO_3$. When biased at one polarity (ITO as the anode), the voltage induced PIN junction was completely inside the MEH-PPV layer, and the device emitted orange light of MEH-PPV. When biased at the opposite polarity (Al as the anode), the PIN junction was completely inside the PPV layer, and green light from the PPV layer was observed.

The second category of the bilayer PLECs is fundamentally different from the first one. For this kind of bilayer PLECs, the ion- and electron-transport channels are separated, therefore solving the problem of phase separation between the conjugated polymer and the electrolyte. A bilayer PLEC was fabricated on glass substrates on which two Au electrodes with an interelectrode gap of 300 μm had been preprepared (Sandstrom et al., 2010). The bilayer film structure was formed by coating a conjugated polymer (Super Yellow) onto the Au electrodes and then an electrolyte (PEO + $LiCF_3SO_3$) on top of the conjugated polymer layer. In the devices, phase separation occurred in the highest degree, which was stable during operating. Therefore, the operation performance of the bilayer PLECs could be efficiently improved compared with the PLECs with a sandwiched structure.

7.3.4 Performance

The working mechanism of PLECs discussed earlier is responsible for their unique structures and characteristics, which have both advantages and drawbacks in performance compared to PLEDs. The PLECs exhibit a "bipolar" behavior due to the insensitivity to the work function of electrode. Under both forward and reverse bias, PLECs show strong electroluminescence, which has been observed in masses of PLECs constructed with various materials and structures (Pei et al., 1996; Holzer et al., 1999a; Huang et al., 2001; Wenzl et al., 2004; Tang and Edman, 2010). The randomly distributed ions in electroluminescent polymer will redistribute as responded to the electrical field when the bias voltage is applied. As a result, the mobile ions will be accumulated at the interface of electrode and facilitate the injection of charge carriers via tunneling. Moreover, the electrochemical doping of the active layer results in the increased conduction of electrode/polymer contact. Therefore, the current flow and the turn-on voltage in PLECs are independent of the work function of the electrodes.

Furthermore, the electroluminescence turn-on voltage of PLECs is lower than that of PLEDs. The turn-on voltage of a PLEC is found to be approximately the same as the energy gap of the electroluminescent polymer (turn-on voltage $\approx E_g/e$). For instance, the turn-on voltage of PPV and MEH-PPV based devices are 2.4 and 2.1 V, respectively, which are consistent with the energy gap of the materials (Yu et al., 1996). The above result is expected because turning on a PLEC is essentially based on the formation of a light-emitting PIN junction. In order to achieve and maintain the PIN junction by electrochemical doping, the turn-on voltage is required to be equal to E_g/e. Due to the low driving voltage, the luminance of a PLEC can easily reach hundreds of cdm^{-2} below 4 V. For example, a PLEC based on a polyfluorene derivative represents a maximum luminance of 190 cdm^{-2} with a turn-on voltage at 3.1 V (Pei and Yang, 1996). Furthermore, PLECs exhibit a higher quantum efficiency than that of PLEDs based on the same electroluminescent polymer.

PLECs also have significant drawbacks. Since the formation and decay of the PIN junction are relative to the external bias, PLECs generally display slower response than that of PLEDs. PLECs often take seconds or even minutes to reach maximum device current and electroluminescent intensity when driven at fixed voltage bias. Moreover, the PIN junction relaxes once the bias voltage is removed as a result of the internal discharging and ionic redistribution. Furthermore, overdoping occurs when PLECs are driven at a voltage beyond the electrochemical stability window of the active materials, which leads to a decrease of the exciton-to-photon conversion. Therefore, it is essential to stabilize the dynamic PIN junction to achieve instantaneous light emission of PLECs and operate it at the peak performance level. To this end, different strategies have been developed to fix the PIN junction. An efficient and easy method by controlling the polymerization of the ionic dopants was realized (Yu et al., 2009, 2011b). The active materials of the devices comprised a curable ionic conductor, such as methacrylate oligomers. When operated at a high current density, this molecule could be polymerized and cross-linked readily. Therefore, those ionic conduction channels would be cut off and the fixation of the PIN junction was accomplished.

As shown in Fig. 7.10A, the PLEC was fabricated with a simple sandwiched structure of ITO/polymer/Al. The active layer contained an alkoxyphenyl substituted poly(1,4-phenylene vinylene) (AP-PPV), trimethylolpropane trimethacrylate (TMPTMA), and lithium trifluoromethane sulfonate (LiTf). When a bias voltage was initially applied, the PIN junction was formed by electrochemical doping (Fig. 7.10B). At the same time, the ionic conduction channels would be cut off due to the polymerizing of small molecules

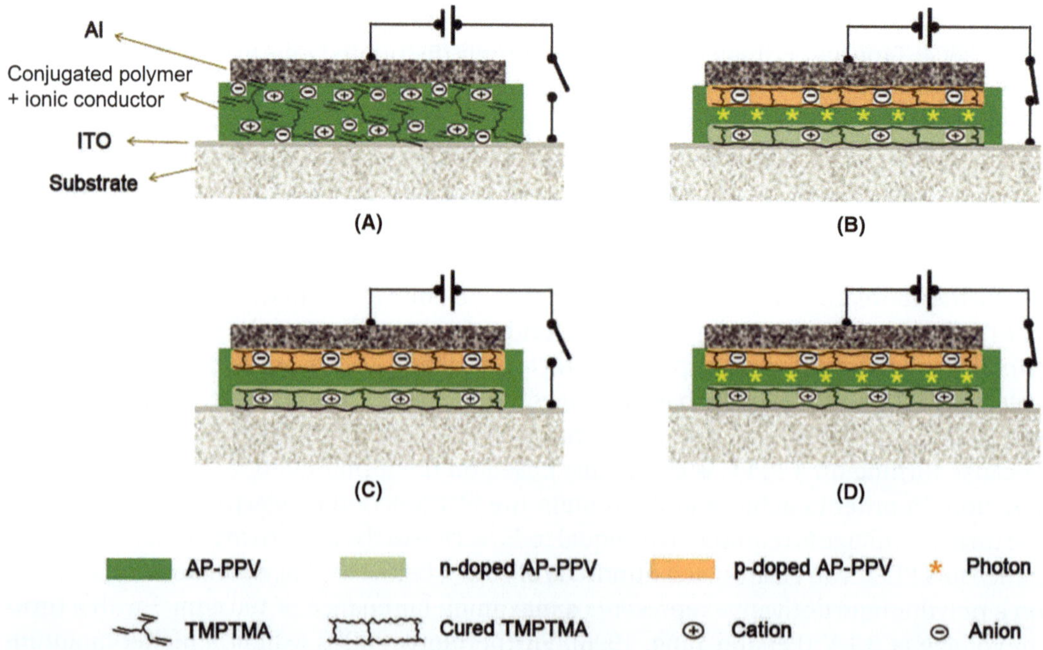

FIGURE 7.10 Schematic illustration to the formation of a stable PIN junction in a polymer thin film containing a conjugated polymer, an electrolyte and curable ion-conducting small molecules. (A) Fresh device, (B) initial charging and light emission, (C) voltage OFF after initial charging, (D) voltage ON and light emission. *Reproduced with permission from reference Yu, Z., Sun, M., Pei, Q., 2009. Electrochemical formation of stable pin junction in conjugated polymer thin films. J. Phys. Chem. B 113, 8481–8486. Copyright 2009, American Chemical Society.*

and become ionically nonconductive, thus the PIN junction was stabilized even after the external bias was withdrawn (Fig. 7.10C). The resulting devices had shown an improved performance. After idling for 72 h without biasing, the diode properties and emission intensity did not change, and the current and light emission turned on immediately (Fig. 7.10D). The stability of the junction was also confirmed by photovoltaic and impedance measurements. All the procedures used in this process were carried out at room temperature, which was beneficial for easy fabrication.

Another major drawback of PLECs is the short operating lifetime compared with that of PLEDs. In fact, the half-life of PLECs was typically on the order of 100 h or less (Cao et al., 1996, 1997; Yang et al., 1996), while the half-life of PLEDs exceeded 10,000 h (Parker et al., 1999). Considering that both PLECs and PLEDs are constructed with the same electroluminescent polymer and structure, the dramatic difference in stability can only be attributed to the electrochemical doping. When the bias voltage is higher than 1 V, overoxidation would occur in the process of electrochemical doping and result in irreversible degradation in the polymer. The stability of PLECs is also adversely affected by the phase separation of the active layer which contains incompatible nonpolar conjugated polymer and polar polymer electrolyte. The existence of phase separation will lead to a highly localized current that can accelerate the degradation of polymer (Habrard et al., 2004). It has

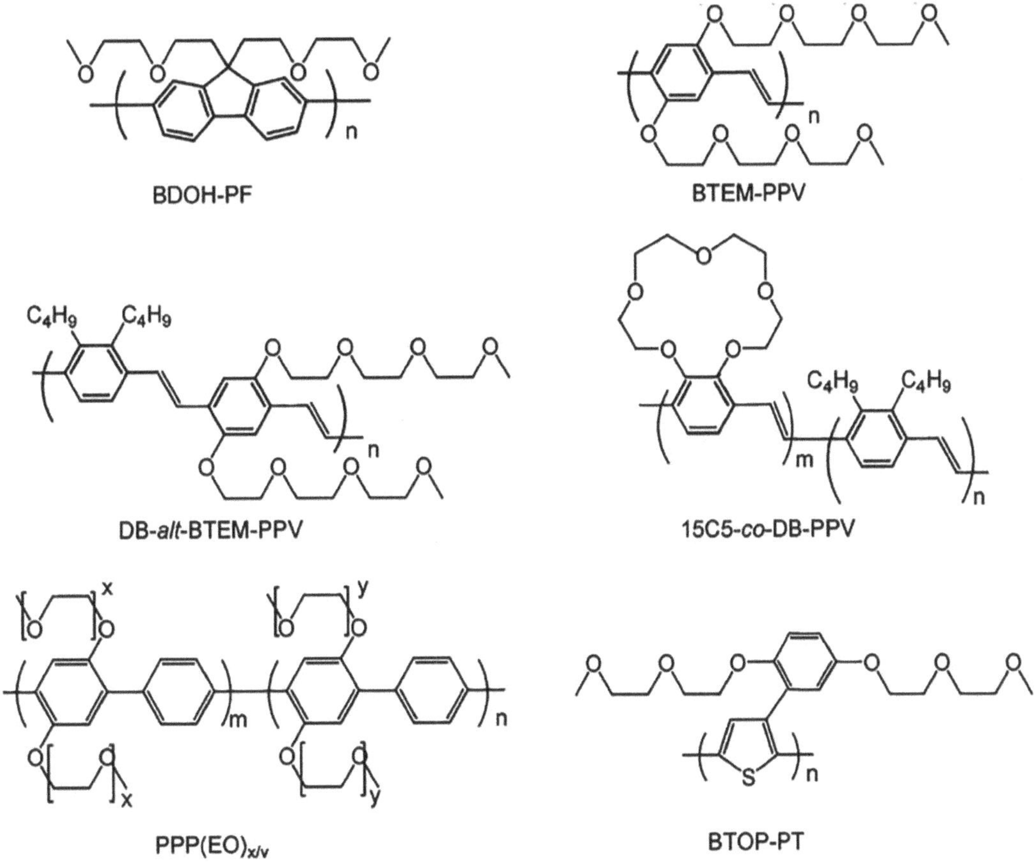

FIGURE 7.11 Chemical structures of electroluminescent polymers (copolymers) with ion-transporting side groups.

been shown that the addition of a bifunctional liquid compound or using a low molecular weight electrolyte could achieve finer phase morphology and therefore significantly improve the performance and operating lifetime of PLECs (Cao et al., 1997; Yang et al., 1996). The operating lifetime of PLECs could also be significantly improved by cooling a fully formed PIN junction to freeze the counter ions, which was first realized in 1997 and they were called frozen-junction PLECs (Gao et al., 1997).

Some researchers (Yang and Pei, 1997; Pei and Yang, 1996) introduced ethylene oxide oligomers as side groups of light-emitting conjugated polymers which proposed an alternative way to improve the blend morphology in PLECs, for example, poly[9,9-bis(3,6-dioxaheptyl)-fluorene-2,7-dily] (BDOH-PF) (Fig. 7.11). As the side groups were PEO-like units, the use of PEO in the PLECs could be drastically reduced or completely eliminated. The side groups also increased the solubility of BDOH-PF in common organic solvents. The conjugated poly(1,4-phentlene) main chain was capable of transporting electrons and holes, while the bis(3,6-dioxaheptyl) side groups functioned to dissolve ions and promote ion transporting which was necessary for the operation of PLECs. The problem of phase separation could be efficiently solved, thereby significantly improved the performance of the devices. They further demonstrated an efficient blue-green PLECs based on this novel polymer and a dissolved lithium salt, and achieved an external quantum efficiency of 4%, as well as a power efficiency of 12 lmW^{-1} at 3.1 V. The brightness of these devices reached 200 $cdcm^{-2}$ at 3.1 V and exceeded 1000 cdm^{-2} at 3.5 V.

Following BDOH-PF, the strategy of introducing polyether-type side groups has been expanded extensively to PPV, PPP and PT based light-emitting conjugated polymer so far (Holzer et al., 1999b; Tasch et al., 1999; Morgado et al., 1999, 2001a,b; Lauter et al., 1997; Johansson et al., 1999). Similarly, the crown ether was also introduced onto PPV backbone as side groups (Morgado et al., 2000). The chemical structures of some polymers with polyether-type side groups are shown in Fig. 7.11. All devices based on these conjugated polymers exhibited typical PLEC behaviors. Importantly, these devices consistently showed fast response, low turn-on voltage, and high power efficiency due to the improved morphology. Namely the performance of those PLECs was usually better than that of PLEDs fabricated from the similar light-emitting conjugated polymers.

Another effective approach of controlling the blend morphology and improving the performance of PLECs is developing a new electrolyte that can be dissolved in nonpolar solvents. Crown ethers are known to form stable complexes with alkali metal cations, which are soluble in nonpolar solvents. Researchers successfully demonstrated high performance PLECs with complexes of several lithium salts and substituted crown ethers as the solid electrolyte (Cao et al., 1997). Ionic liquids are also useful as electrolyte in PLECs due to their unique advantages: they are soluble in common nonpolar solvents, chemically and electrochemically stable, almost nonvolatile, and easily delocalized to supply cations and anions. A highly lipophilic ionic liquid tetrahexylammonium bis(trifluoromethyl-sulfonyl) imide was used as the electrolyte to construct PLECs (Panozzo et al., 2002; Ouisse et al., 2002), and they exhibited fast response and high power efficiency.

7.3.5 Functionality

Making polymer light-emitting devices in soft, light weight, and stretchable formats will enable many new applications pertaining to portable and wearable electronics. PLECs have an intrinsically simple device structure and do not necessitate low-work function cathodes due to the unique device characteristics. Given their advantages, flexible, lightweight, and stretchable PLECs are developed, which represents a major step toward the realization of portable and wearable light-emitting electronics.

7.3.5.1 Stretchability

Stretchability has been perceived as a useful characteristic of next-generation electronics. Unlike conventional devices, stretchable electronics can adapt arbitrary and movable surfaces, thus extending their applications in smart-phones and wearable electronic textiles (Kim et al., 2010; Yu et al., 2011a). The largest challenge of stretchable PLECs resides in transparent electrode that requires to be highly conductive and stretchable. ITO is a widely used transparent conductive electrode material but its intrinsic brittleness limits the flexibility. Various types of conductive materials, such as CNT, graphene, and conductive polymers have been investigated to replace ITO (Yu et al., 2011a; Hecht et al., 2011; Vosgueritchian et al., 2012). It has been shown that single-walled carbon nanotube (SWCNT)-polymer composite electrodes can sustain large strain deformation without much decrease of conductivity (Yu et al., 2011a). The interpenetrating network of SWCNT and the polymer matrix lead to low sheet resistance and high optical transmittance. The sheet resistance of the composites was slightly changed when stretched by up to 50%. An intrinsically stretchable PLEC was fabricated based on a pair of transparent composite electrodes sandwiching an electroluminescent polymer active layer. This device could tolerate linear strains up to 45% without damaging the electroluminescent properties. In addition, a light-emitting device using an elastic electroluminescent blend sandwiched between two metallic electrodes was also reported (Filiatrault et al., 2012). The elastic electroluminescent blend was composed of a ionic transition metal complex Ru(dtb-bpy)$_3$(PF$_6$)$_2$ and ~25 vol% elastomeric matrix of PDMS. The anode was an ultrathin gold layer coated on PDMS substrate, and the opposite electrode was gallium-indium eutectic alloy encased in an epoxy resin. The device can be deformed by up to 27% and endure dynamic deformations of 15% stretching strain with uniform light emission. However, there were still significant obstacles for obtaining a stretchable display based on these light-emitting architectures due to the limited stretchability and conductivity of electrodes, complicated fabrication process and low performance.

Given the low sheet resistance, high transmittance, and solution processability, percolation networks of AgNW are considered as highly promising candidates to rival brittle and expensive sputtered ITO. In recent years, a number of works have employed AgNW percolation networks to fabricate stretchable transparent conductive electrodes, which could be potentially useful for a wide range of applications including stretchable displays and flexible photovoltaic devices (Liang et al., 2014; Kim et al., 2015). An elastomeric polymer light-emitting device comprising an electroluminescent polymer layer sandwiched between a pair of AgNW-polymer composite electrodes was fabricated by a simple and all-solution-based process (Liang et al., 2013). The device exhibited rubbery elasticity and could emit light when exposed to strains as large as 120%. In the PLECs, the composite electrodes were prepared by inlaying AgNW percolation network in the rubbery PUA matrix. The polymer matrix of composite electrode was siliconized urethane acrylate oligomer and ethoxylated bisphenol A dimethacrylate copolymer, which had high transmittance, high stretchability and good bonding force with AgNW. The inset of Fig. 7.12A presents a top-view SEM image of the conductive surface of a 15 Ωsq^{-1} AgNW-PUA composite electrode. It is clear to see that the nanowires are inlaid into the surface layer of the PUA matrix. The resulting AgNW-PUA composite electrode had high transmittance, high surface smoothness, good electrical conductivity, and high stretchability. When the sheet resistance of the composite electrodes increased from 10 to 15 and 25 Ωsq^{-1}, the transmittance accordingly increased from 77% to 83% and 91% at a wavelength of 550 nm. This performance could be comparable to commercial ITO/glass. The AgNW-PUA composite electrodes also

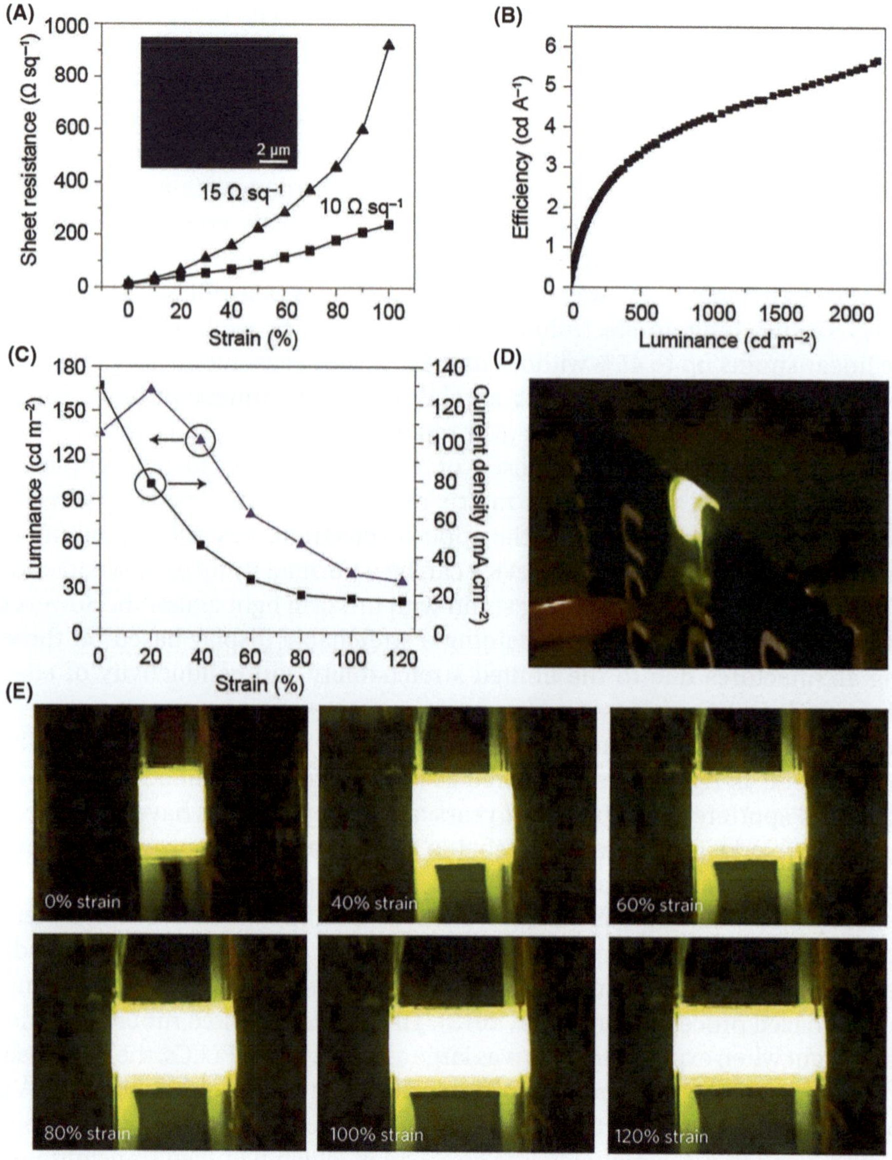

FIGURE 7.12 Characterization of composite electrode and stretchable PLEC. (A) Sheet resistance for AgNW–PUA composite electrodes with increasing strains (sample area, 7 × 10 mm^2; stretch speed, 1 mm s^{-1}). The inset shows a SEM image of the conductive surface of a 15 Ωsq^{-1} AgNW–PUA composite electrode. (B) *E–L* plot of the PLEC. (C) Current density and luminance characteristics of a PLEC at 12 V with increasing strains. (D) Photograph of a PLEC (original emission area of 3.0 × 7.0 mm^2, biased at 12 V) wrapped around the edge of 0.4-mm-thick cardboard. (E) Photographs of a PLEC (original emission area, 5.0 × 4.5 mm^2) biased at 14 V at specified strains. *Reproduced with permission from reference Liang, J., Li, L., Niu, X., Yu, Z., Pei, Q., 2013. Elastomeric polymer light-emitting devices and displays. Nat. Photonics 7, 817–824. Copyright 2013, Nature Publishing Group.*

had high stretchability. In order to test the stretchability of the composite electrodes, 10 and 15 Ωsq^{-1} samples were subjected to repeated stretching and relaxing, and the sheet resistances with increasing strains are shown in Fig. 7.12A. The 10 Ωsq^{-1} sample showed a steady increase in resistance to 235 Ωsq^{-1} at 100% strain. For the 15 Ωsq^{-1} sample, the sheet resistance exhibited a steady increase to about 500 Ωsq^{-1} at 80% strain, after which the sheet resistance increased steeply, but still remained below 1 $K\Omega sq^{-1}$ at 100% strain. In addition, the sheet resistance could still be restored to 45 Ωsq^{-1} after 1500 cycles of continuous stretching–relaxing cycles at 30% strain. Therefore, the composite electrodes with high transmittance, good electrical conductivity, and high stretchability were promising to be applied to the fabrication of the polymer light-emitting devices.

Based on this composite electrode, PLECs were fabricated by an all-solution process. The PLEC devices were made in four steps. (1) A thin PEDOT:PSS layer was spin coated onto AgNW-PUA composite electrode with 15 Ωsq^{-1} sheet resistance as the anode. (2) A solution of a conjugated polymer (Super Yellow) and electrolyte blend was then spin-coated onto the anode. (3) The resulting film was laminated with a second 15 Ωsq^{-1} AgNW-PUA composite electrode (as cathode) to complete the device fabrication. (4) Finally, the device was laminated between a pair of precross-linked urethane liquid rubber compound layers to accomplish the encapsulation.

The *E-L* characteristic curve of the resulting PLEC is shown in Fig. 7.12B. In a typical operation, the devices were precharged under a constant 9 mA cm^{-2} for 10 min to allow the complete establishment of a PIN junction in the emission layer. The luminous efficiency of the PLEC increased rapidly with brightness, and reached 5.7 cdA^{-1} at 2200 cdm^{-2}. It should be noted that the PLEC exhibited fairly uniform light emission, even at very low brightness. When the turn-on bias voltage was applied, the in situ electrochemical doping took place on either side of the electroluminescent polymer and a light-emitting PIN junction was formed. The PIN junction formation efficiently improved the injection of charge carriers from the AgNW network, thereby resulting in uniform light emission over the entire active area.

This kind of PLECs was composed of stretchable components, and they were therefore stretchable. Along the length direction, the device could be stretched to 120%. Fig. 7.12C presents the current density-luminance-strain characteristic curves of the PLECs at a bias voltage of 12 V. It shows that the current density of the device decreased with increasing strain, which could be accounted for the increase in sheet resistance of the composite electrode. On the other hand, the brightness exhibited a gradual increase at the initial 20% strain. The increase in brightness of the device with small strain might be a result of a more balanced injection of electrons and holes. Fig. 7.12E demonstrates photographs of the PLECs at various elongations during operation at 14 V. The deformed devices emitted uniformly and brightly throughout the entire emissive luminous areas, even when being stretched to 120%. The PLEC could also be collapsible due to the highly flexible AgNW-PUA composite electrodes and no mechanical or electrical damage to the device was observed under bending or folding. The bent PLEC displayed uniform and bright emission even when it was wrapped around the edge of 0.4-mm-thick cardboard (Fig. 7.12D). Therefore, the resulting PLEC exhibited rubbery elasticity and was able to emit light at both stretching and bending states.

7.3.5.2 Weavability

In contrast to the conventional planar electronic devices, fiber-shaped electronic devices have recently attracted increasing interests due to their unique advantages. For instance, they are lightweight, deformable, and can be woven into flexible clothes or integrated into other flexible structures, which enable applications in electronic textiles and various complex devices (Chen et al., 2013). Although fiber-shaped electrochemical transistors (Hamedi et al., 2007), organic field-effect transistors (Wang et al., 2012), multifunctional sensors (Peng et al., 2009), energy harvesting, and storage devices (Chen et al., 2013) have been extensively studied, it still remains challenging to make a fiber-shaped light-emitting device.

In recent years, fiber-shaped OLEDs have been demonstrated (O'Connor et al., 2007). However, the electrical characteristics and luminous efficiency of the fiber-shaped OLEDs are much lower than those of planar analogs. Meanwhile, the fabrication method of rotational vacuum deposition has limited suitability for large-scale production, and it is difficult for accurately controlling of the thickness of each layer in the fiber-shaped OLED through a continuous layer-by-layer depositing process. In contrast to OLEDs, the PLECs have a simplest structure with the electroluminescent active layer sandwiched between two electrodes to form a light-emitting PIN junction in situ for the injection of both electrons and holes. Therefore, PLECs can be constructed on relatively rougher surfaces. Moreover, PLECs do not require a low work function cathode which is sensitive to air and have relatively low turn-on voltages because the formation of PIN junction effectively removes the energy barriers of the electroluminescent polymer.

A weavable fiber-shaped PLEC was developed by sandwiching an emission layer between a modified metal wire cathode and an aligned CNT sheet anode (Zhang et al., 2015). As expected, the fiber-shaped PLEC was thin, light weight, flexible, and deformable, and it was able to be woven into flexible clothes, which predicts a promising application in electronic textiles. The schematic structure of a fiber-shaped PLEC is shown in Fig. 7.13A. The fiber-shaped PLEC was made in three steps. First, a stainless steel wire was dip-coated with a thin layer of ZnO nanoparticles. Second, the emissive layer of a conjugated polymer and electrolyte blend was deposited on the modified steel wire, also using a dip-coating method. Finally, an optically transparent aligned CNT sheet was uniformly wrapped onto the outer surface of the electroluminescent polymer layer to produce a fiber-shaped PLEC. A photograph of a fiber-shaped PELC with curved shape is shown as the inset in Fig. 7.13A. The aligned CNT sheet played a critical role in the successful fabrication of the fiber-shaped PLEC. The aligned CNT sheet was flexible and could be closely wrapped on the fiber-based substrate predominantly by van der Waals forces. Due to the highly aligned structure, the CNT sheet presented a high electrical conductivity on the order of 10^2–10^3 S cm^{-1}. The aligned CNT sheet was optically transparent with a transmittance over 87% at wavelength above 550 nm with thickness of 18 nm.

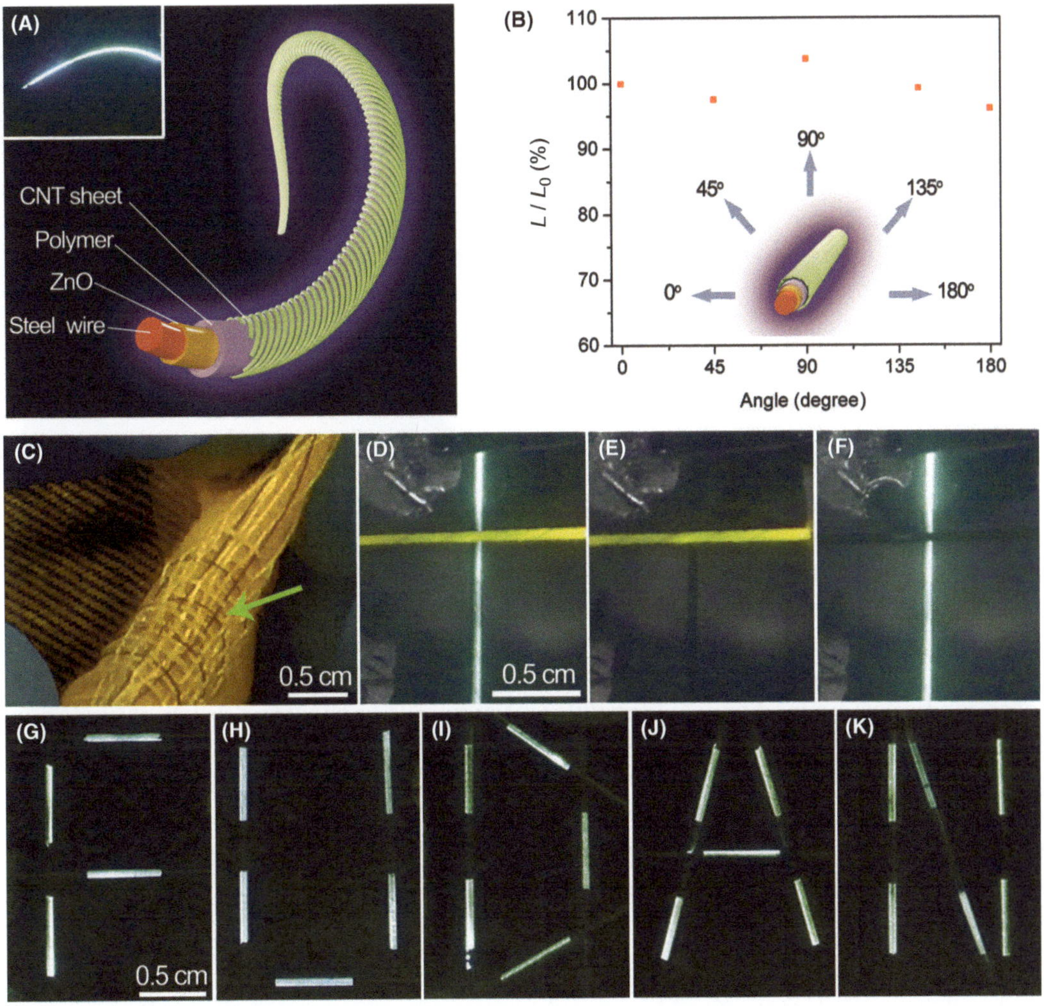

FIGURE 7.13 Schematic structure and characteristic of fiber-shaped PLECs. (A) Schematic illustration to a fiber-shaped PLEC (inset, photograph of a fiber-shaped PLEC biased at 10 V). (B) The dependence of luminance on the observation angle for a fiber-shaped PLEC. L_0 and L correspond to luminance measured at 0 degree and other angles, respectively. (C) Textile under bending and twisting. (D–F) Two fiber-shaped PLECs with different colors being selectively lit. (G–K) Fiber-shaped PLECs being woven into a "FUDAN" pattern (biased at 9 V). *Reproduced with permission from reference Zhang, Z., Guo, K., Li, Y., Li, X., Guan, G., Li, H., Luo, Y., Zhao, F., Zhang, Q., Wei, B., 2015. A colour-tunable, weavable fibre-shaped polymer light-emitting electrochemical cell. Nat. Photonics, 9, 233–238. Copyright 2015, Nature Publishing Group.*

The fiber-shaped PLEC exhibited a high electroluminescent performance after charging under a constant current of 15 mA cm^{-2} for 10 min. For instance, the blue light-emitting fiber-shaped PLEC showed a luminance of 1 cdm^{-2} at 5.6 V and reached a peak value of 609 cdm^{-2} at 13 V. The luminous efficiency increased rapidly with brightness, and reached 0.83 cdA^{-1} at the high end of the brightness range, with an external quantum efficiency of 0.35%. The precharged device displayed a rapid turn-on respond that was similar to that of conventional PLED.

This fiber-shaped PLEC had some unique and promising advantages due to the one-dimensional structure. For instance, it emitted light in all directions, and the brightness was almost independent on observing angle (Fig. 7.13B). The fiber-shaped PLEC also exhibited high flexibility. After bending with a radius of curvature of 6 mm for 100 cycles, the brightness of the device was maintained by above 90%. No obvious damage of the surface structure of the active layer and aligned CNT layer was observed after deformation. Due to its flexibility and weavability, the fiber-shaped PLECs could be easily woven into various flexible structures, such as textile (Fig. 7.13C). The fiber-shaped PLECs could also be selectively lit (Fig. 7.13D–F). As shown in Fig. 7.13G–K, a "FUDAN" pattern is achieved by weaving the fiber-shaped PLECs together.

7.4 Summary

Recently, light-emitting devices based on polymers, such as PLEDs and PLECs have aroused great interests regarding their many promising advantages, such as wide viewing angle, self illumination, high color saturation, lightweight and thin. Furthermore, since polymer materials are able to generate films through solution methods, such as coating and printing, there is great potential to realize low-cost and large-area polymer displays and lighting sources with the help of roll-to-roll techniques. By virtue of the various categories of polymer materials, the luminescence properties of polymer light-emitting devices can be easily tuned by selecting different materials to construct different functional layers.

Due to the intrinsic stretchability of polymers, flexible and stretchable polymer light-emitting devices are realized on flexible substrates. These kinds of devices have attracted significant attentions at both academy and industry since they are suggested to be applied into portable, foldable, and wearable electronics. Wearable electronics are considered to be the next-generation electronics in the future. Among the polymer light-emitting devices with different structures, fiber-shaped PLECs which possess both flexibility and weavability demonstrate high application prospect in the field of wearable light-emitting devices for display, lighting source, and phototherapy. With the rapid progress of relevant theoretical researches and techniques, we firmly believe that wearable light-emitting devices, such as light-emitting clothes, are bound to be achieved in the future.

References

Beouch, L., Van, F.T., Stephan, O., Vial, J.C., Chevrot, C., 2001. Optical and electrochemical properties of soluble N-hexylcarbazole-*co*-3,4-ethylenedioxythiophene copolymers. Synth. Met. 122, 351–358.

Bernius, M., Inbasekaran, M., Woo, E., Wu, W., Wujkowski, L., 2000. Light-emitting diodes based on fluorene polymers. Thin Solid Films 363, 55–57.

Bolink, H.J., Coronado, E., Repetto, D., Sessolo, M., 2007. Air stable hybrid organic-inorganic light emitting diodes using ZnO as the cathode. Appl. Phys. Lett. 91, 223501–1223501.

Bolink, H.J., Coronado, E., Costa, R.D., Lardi S, N., Ortí, E., 2008. Near-quantitative internal quantum efficiency in a light-emitting electrochemical cell. Inorg. Chem. 47, 9149–9151.

Braun, D., Heeger, A.J., Kroemer, H., 1991. Improved efficiency in semiconducting polymer light-emitting diodes. J. Electron. Mater. 20, 945–948.

Burroughes, J.H., Bradley, D., Brown, A.R., Marks, R.N., Friend, R.H., 1990. Light-emitting diodes based on conjugated polymers. Nature 347, 539–541.

Cao, Y., Yu, G., Heeger, A.J., Yang, C., 1996. Efficient, fast response light-emitting electrochemical cells: Electroluminescent and solid electrolyte polymers with interpenetrating network morphology. Appl. Phys. Lett. 68, 3218–3220.

Cao, Y., Pei, Q., Andersson, M.R., Yu, G., Heeger, A.J., 1997. Light-emitting electrochemical cells with crown ether as solid electrolyte. J. Electrochem. Soc. 144, L317–L320.

Chen, B., Lee, C.-S., Lee, S.-T., Webb, P., Chan, Y.-C., Gambling, W., Tian, H., Zhu, W., 2000. Improved time-of-flight technique for measuring carrier mobility in thin films of organic electroluminescent materials. Jpn. J. Appl. Phys. 39, 1190–1192.

Chen, T., Qiu, L., Yang, Z., Peng, H., 2013. Novel solar cells in a wire format. Chem. Soc. Rev. 42, 5031–5041.

Crispin, X., Geskin, V., Crispin, A., Cornil, J., Lazzaroni, R., Salaneck, W.R., Bredas, J.-L., 2002. Characterization of the interface dipole at organic/metal interfaces. J. Am. Chem. Soc. 124, 8131–8141.

Dane, J., Tracy, C., Gao, J., 2005. Direct observation of a frozen junction in polymer light-emitting electrochemical cells. Appl. Phys. Lett. 86, 153509–1153509.

Destriau, G., 1936. Scintillation of zinc sulphide with ü-rays. J. Chem. Phys. 33, 587–625.

Filiatrault, H.L., Porteous, G.C., Carmichael, R.S., Davidson, G.J., Carmichael, T.B., 2012. Stretchable light-emitting electrochemical cells using an elastomeric emissive material. Adv. Mater. 24, 2673–2678.

Gao, J., Dane, J., 2004. Visualization of electrochemical doping and light-emitting junction formation in conjugated polymer films. Appl. Phys. Lett. 84, 2778–2780.

Gao, J., Dane, J., 2005. Imaging the doping and electroluminescence in extremely large planar polymer light-emitting electrochemical cells. J. Appl. Phys. 98, 063513–163513.

Gao, J., Yu, G., Heeger, A.J., 1997. Polymer light-emitting electrochemical cells with frozen pin junction. Appl. Phys. Lett. 71, 1293–1295.

Gong, X., Wang, S., Moses, D., Bazan, G.C., Heeger, A.J., 2005. Multilayer polymer light-emitting diodes: white-light emission with high efficiency. Adv. Mater. 17, 2053–2058.

Grem, G., Leditzky, G., Ullrich, B., Leising, G., 1992. Realization of a blue-light-emitting device using poly(*p*-phenylene). Adv. Mater. 4, 36–37.

Gustafsson, G., Cao, Y., Treacy, G., Klavetter, F., Colaneri, N., Heeger, A.J., 1992. Flexible light-emitting diodes made from soluble conducting polymers. Nature 357, 477–479.

Habrard, F., Ouisse, T., Stephan, O., Armand, M., Stark, M., Huant, S., Dubard, E., Chevrier, J., 2004. Conjugated polymer/molten salt blends: The relationship between morphology and electrical aging. J. Appl. Phys. 96, 7219–7224.

Hamedi, M., Forchheimer, R., Inganas, O., 2007. Towards woven logic from organic electronic fibres. Nat. Mater. 6, 357–362.

Hecht, D.S., Hu, L., Irvin, G., 2011. Emerging transparent electrodes based on thin films of carbon nanotubes, graphene, and metallic nanostructures. Adv. Mater. 23, 1482–1513.

Holzer, L., Wenzl, F., Tasch, S., Leising, G., Winkler, B., Dai, L., Mau, A., 1999a. Ionochromism in a light-emitting electrochemical cell with low response time based on an ionic conductive poly-phenylene vinylene. Appl. Phys. Lett. 75, 2014–2016.

Holzer, L., Winkler, B., Wenzl, F.P., Tasch, S., Dai, L., Mau, A.W.H., Leising, G., 1999b. Light-emitting electrochemical cells and light-emitting diodes based on ionic conductive poly(phenylene vinylene): a new chemical sensor system. Synth. Met. 100, 71–77.

Hu, Y., Gao, J., 2011. Direct imaging and probing of the p − n junction in a planar polymer light-emitting electrochemical cell. J. Am. Chem. Soc. 133, 2227–2231.

Hu, Y., Tracy, C., Gao, J., 2006. High-resolution imaging of electrochemical doping and dedoping processes in luminescent conjugated polymers. Appl. Phys. Lett. 88, 123507–1123507.

Huang, C., Huang, W., Guo, J., Yang, C.-Z., Kang, E.-T., 2001. A novel rigid-rod alternating poly(*p*-phenylenevinylene) derivative with oligo(ethylene oxide) side chains. Polymer 42, 3929–3938.

Huang, J., Li, G., Wu, E., Xu, Q., Yang, Y., 2006. Achieving high-efficiency polymer white-light-emitting devices. Adv. Mater. 18, 114–117.

Huang, J., Wang, X., Bradley, D., 2007. Efficient flexible polymer light emitting diodes with conducting polymer anodes. J. Mater. Chem. 17, 3551–3554.

Hung, L., Chen, C., 2002. Recent progress of molecular organic electroluminescent materials and devices. Mater. Sci. Eng. R Rep. 39, 143–222.

Jiang, X., Liu, S., Liu, M., Herguth, P., Jen, A.Y., Fong, H., Sarikaya, M., 2002. Perfluorocyclobutane-based arylamine hole-transporting materials for organic and polymer light-emitting diodes. Adv. Funct. Mater. 12, 745–751.

Jiang, J., Xu, Y., Yang, W., Guan, R., Liu, Z., Zhen, H., Cao, Y., 2006. High-efficiency white-light-emitting devices from a single polymer by mixing singlet and triplet emission. Adv. Mater. 18, 1769–1773.

Jin, Y., Kim, J., Lee, S., Kim, J.Y., Park, S.H., Lee, K., Suh, H., 2004. Novel electroluminescent polymers with fluoro groups in vinylene units. Macromolecules 37, 6711–6715.

Johansson, T., Mammo, W., Andersson, M., Ingan S, O., 1999. Light-emitting electrochemical cells from oligo (ethylene oxide)-substituted polythiophenes: evidence for in situ doping. Chem. Mater. 11, 3133–3139.

Kepler, R.G., Beeson, P.M., Jacobs, S.J., Anderson, R.A., Sinclair, M.B., Valencia, V.S., Cahill, P.A., 1995. Electron and hole mobility in tris(8-hydroxyquinolinolato-N1,O8) aluminum. Appl. Phys. Lett. 66, 3618–3620.

Kido, J., Hongawa, K., Okuyama, K., Nagai, K., 1994. White light-emitting organic electroluminescent devices using the poly(N-vinylcarbazole) emitter layer doped with three fluorescent dyes. Appl. Phys. Lett. 64, 815–817.

Kim, R.-H., Kim, D.-H., Xiao, J., Kim, B.H., Park, S.-I., Panilaitis, B., Ghaffari, R., Yao, J., Li, M., Liu, Z., 2010. Waterproof AlInGaP optoelectronics on stretchable substrates with applications in biomedicine and robotics. Nat. Mater. 9, 929–937.

Kim, Y.H., Han, T.H., Cho, H., Min, S.Y., Lee, C.L., Lee, T.W., 2014. Polyethylene imine as an ideal interlayer for highly efficient inverted polymer light-emitting diodes. Adv. Funct. Mater. 24, 3808–3814.

Kim, Y.H., M Ller-Meskamp, L., Leo, K., 2015. Ultratransparent polymer/semitransparent silver grid hybrid electrodes for small-molecule organic solar cells. Adv. Energy Mater. 5, 1401822–11401822.

Klaerner, G., Miller, R., 1998. Polyfluorene derivatives: effective conjugation lengths from well-defined oligomers. Macromolecules 31, 2007–2009.

Lauter, U., Meyer, W.H., Wegner, G., 1997. Molecular composites from rigid-rod poly(*p*-phenylene) s with oligo (oxyethylene) side chains as novel polymer electrolytes. Macromolecules 30, 2092–2101.

Lee, B.R., Kim, J.-W., Kang, D., Lee, D.W., Ko, S.-J., Lee, H.J., Lee, C.-L., Kim, J.Y., Shin, H.S., Song, M.H., 2012. Highly efficient polymer light-emitting diodes using graphene oxide as a hole transport layer. ACS Nano 6, 2984–2991.

Li, Y., Gao, J., Yu, G., Cao, Y., Heeger, A.J., 1998. AC impedance of polymer light-emitting electrochemical cells and light-emitting diodes: a comparative study. Chem. Phys. Lett. 287, 83–88.

Liang, J., Li, L., Niu, X., Yu, Z., Pei, Q., 2013. Elastomeric polymer light-emitting devices and displays. Nat. Photonics 7, 817–824.

Liang, J., Lu, L., Tong, K., Zhi, R., Wei, H., Niu, X., Chen, Y., Pei, Q., 2014. Silver nanowire percolation network soldered with graphene oxide at room temperature and its application for fully stretchable polymer light-emitting diodes. ACS Nano 8, 1590–1600.

Liu, J., Zhou, Q., Cheng, Y., Geng, Y., Wang, L., Ma, D., Jing, X., Wang, F., 2005. The first single polymer with simultaneous blue, green, and red emission for white electroluminescence. Adv. Mater. 17, 2974–2978.

Liu, M.S., Niu, Y.H., Luo, J., Chen, B., Kim, T.D., Bardecker, J., Jen, A.K.Y., 2006. Material and interface engineering for highly efficient polymer light emitting diodes. J. Macromol. Sci. Part C 46, 7–26.

Liu, R., Xiong, Y., Zeng, W., Wu, Z., Du, B., Yang, W., Sun, M., Cao, Y., 2007. Extremely color-stable blue light-emitting polymers based on alternating 2,7-fluorene-co-3,9-carbazole copolymer. Macromol. Chem. Phys. 208, 1503–1509.

Mikroyannidis, J.A., Spiliopoulos, I.K., Kasimis, T.S., Kulkarni, A.P., Jenekhe, S.A., 2004. New poly(*p*-phenylene vinylene) derivatives with two oxadiazole rings per repeat unit: synthesis, photophysical properties, electroluminescence, and metal ion recognition. J. Polymer Sci. Part A 42, 2112–2123.

Morgado, J., Friend, R., Cacialli, F., Chuah, B., Moratti, S., Holmes, A., 1999. Light-emitting devices based on a poly(*p*-phenylene vinylene) derivative with ion-coordinating side groups. J. Appl. Phys. 86, 6392–6395.

Morgado, J., Cacialli, F., Friend, R.H., Chuah, B.S., Moratti, S.C., Holmes, A.B., 2000. Luminescence properties of PPV-based copolymers with crown ether substituents. Synth. Met. 111, 449–452.

Morgado, J., Cacialli, F., Friend, R., Chuah, B., Rost, H., Holmes, A., 2001a. Light-emitting devices based on a poly(*p*-phenylenevinylene) statistical copolymer with oligo(ethylene oxide) side groups. Macromolecules 34, 3094–3099.

Morgado, J., Friend, R.H., Cacialli, F., Chuah, B.S., Rost, H., Moratti, S.C., Holmes, A.B., 2001b. Light-emitting electrochemical cells based on poly(*p*-phenylene vinylene) copolymers with ion-transporting side groups. Synth. Met. 122, 111–113.

Morgado, J., Friend, R., Cacialli, F., 2002. Improved efficiency of light-emitting diodes based on polyfluorene blends upon insertion of a poly(*p*-phenylene vinylene) electron-confinement layer. Appl. Phys. Lett. 80, 2436–2438.

O'connor, B., An, K.H., Zhao, Y., Pipe, K.P., Shtein, M., 2007. Fiber shaped light emitting device. Adv. Mater. 19, 3897–3900.

Ouisse, T., Armand, M., Kervella, Y., Stephan, O., 2002. Fully transparent, organic light-emitting electrochemical cells. Appl. Phys. Lett. 81, 3131–3133.

Panozzo, S., Armand, M., Stephan, O., 2002. Light-emitting electrochemical cells using a molten delocalized salt. Appl. Phys. Lett. 80, 679–681.

Park, J.S., Lee, B.R., Jeong, E., Lee, H.-J., Lee, J.M., Kim, J.-S., Kim, J.Y., Woo, H.Y., Kim, S.O., Song, M.H., 2011. High performance polymer light-emitting diodes with N-type metal oxide/conjugated polyelectrolyte hybrid charge transport layers. Appl. Phys. Lett. 99, 163305–1163305.

Parker, I.D., 1994. Carrier tunneling and device characteristics in polymer light-emitting diodes. J. Appl. Phys. 75, 1656–1666.

Parker, I., Cao, Y., Yang, C., 1999. Lifetime and degradation effects in polymer light-emitting diodes. J. Appl. Phys. 85, 2441–2447.

Pei, Q., Yang, Y., 1996. Efficient photoluminescence and electroluminescence from a soluble polyfluorene. J. Am. Chem. Soc. 118, 7416–7417.

Pei, Q., Yu, G., Zhang, C., Yang, Y., Heeger, A.J., 1995. Polymer light-emitting electrochemical cells. Science 269, 1086–1088.

Pei, Q., Yang, Y., Yu, G., Zhang, C., Heeger, A.J., 1996. Polymer light-emitting electrochemical cells: In situ formation of a light-emitting pn junction. J. Am. Chem. Soc. 118, 3922–3929.

Peng, H., Sun, X., Cai, F., Chen, X., Zhu, Y., Liao, G., Chen, D., Li, Q., Lu, Y., Zhu, Y., 2009. Electrochromatic carbon nanotube/polydiacetylene nanocomposite fibres. Nat. Nanotechnol. 4, 738–741.

Pope, M., Kallmann, H.P., Magnante, P., 1963. Electroluminescence in organic crystals. J. Chem. Phys. 38, 2042–2043.

Rowell, M.W., Topinka, M.A., Mcgehee, M.D., Prall, H.-J., Dennler, G., Sariciftci, N.S., Hu, L., Gruner, G., 2006. Organic solar cells with carbon nanotube network electrodes. Appl. Phys. Lett. 88, 233506–1233506.

Sandstrom, A., Matyba, P., Inganäs, O., Edman, L., 2010. Separating ion and electron transport: the bilayer light-emitting electrochemical cell. J. Am. Chem. Soc. 132, 6646–6647.

Scott, J.C., Malliaras, G.G., 1999. Charge injection and recombination at the metal–organic interface. Chem. Phys. Lett. 299, 115–119.

Sessolo, M., Tordera, D., Bolink, H.J., 2013. Ionic iridium complex and conjugated polymer used to solution-process a bilayer white light-emitting diode. ACS Appl. Mater. Interfaces 5, 630–634.

Setayesh, S., Grimsdale, A.C., Weil, T., Enkelmann, V., M Llen, K., Meghdadi, F., List, E.J., Leising, G., 2001. Polyfluorenes with polyphenylene dendron side chains: toward non-aggregating, light-emitting polymers. J. Am. Chem. Soc. 123, 946–953.

Tang, S., Edman, L., 2010. Quest for an appropriate electrolyte for high-performance light-emitting electrochemical cells. J. Phys. Chem. Lett. 1, 2727–2732.

Tang, C.W., Vanslyke, S.A., 1987. Organic electroluminescent diodes. Appl. Phys. Lett. 51, 913–915.

Tasch, S., Niko, A., Leising, G., Scherf, U., 1996. Highly efficient electroluminescence of new wide band gap ladder-type poly(para-phenylenes). Appl. Phys. Lett. 68, 1090–1092.

Tasch, S., Holzer, L., Wenzl, F.P., Gao, J., Winkler, B., Dai, L., Mau, A.W.H., Sotgiu, R., Sampietro, M., Scherf, U., 1999. Light-emitting electrochemical cells with microsecond response times based on PPPs and novel PPVs. Synth. Met. 102, 1046–1049.

Tu, G., Mei, C., Zhou, Q., Cheng, Y., Geng, Y., Wang, L., Ma, D., Jing, X., Wang, F., 2006. Highly efficient pure-white-light-emitting diodes from a single polymer: polyfluorene with naphthalimide moieties. Adv. Funct. Mater. 16, 101–106.

Vosgueritchian, M., Lipomi, D.J., Bao, Z., 2012. Highly conductive and transparent PEDOT:PSS films with a fluorosurfactant for stretchable and flexible transparent electrodes. Adv. Funct. Mater. 22, 421–428.

Wang, S., Kappl, M., Liebewirth, I., M Ller, M., Kirchhoff, K., Pisula, W., M Llen, K., 2012. Organic field-effect transistors based on highly ordered single polymer fibers. Adv. Mater. 24, 417–420.

Wenzl, F.P., Pachler, P., Suess, C., Haase, A., List, E.J., Poelt, P., Somitsch, D., Knoll, P., Scherf, U., Leising, G., 2004. The influence of the phase morphology on the optoelectronic properties of light-emitting electrochemical cells. Adv. Funct. Mater. 14, 441–450.

White, M.S., Kaltenbrunner, M., Głowacki, E.D., Gutnichenko, K., Kettlgruber, G., Graz, I., Aazou, S., Ulbricht, C., Egbe, D.A., Miron, M.C., 2013. Ultrathin, highly flexible and stretchable PLEDs. Nat. Photonics 7, 811–816.

Wu, H., Zhou, G., Zou, J., Ho, C.L., Wong, W.Y., Yang, W., Peng, J., Cao, Y., 2009. Efficient polymer white-light-emitting devices for solid-state lighting. Adv. Mater. 21, 4181–4184.

Yang, Y., Pei, Q., 1997. Efficient blue-green and white light-emitting electrochemical cells based on poly[9, 9-bis (3, 6-dioxaheptyl)-fluorene-2, 7-diyl]. J. Appl. Phys. 81, 3294–3298.

Yang, Y., Pei, Q., Heeger, A.J., 1996. Efficient blue polymer light-emitting diodes from a series of soluble poly(paraphenylene)s. J. Appl. Phys. 79, 934–939.

Yu, G., Heeger, A.J., 1997. High efficiency photonic devices made with semiconducting polymers. Synth. Met. 85, 1183–1186.

Yu, G., Yang, Y., Cao, Y., Pei, Q., Zhang, C., Heeger, A.J., 1996. Measurement of the energy gap in semiconducting polymers using the light-emitting electrochemical cell. Chem. Phys. Lett. 259, 465–468.

Yu, G., Pei, Q., Heeger, A.J., 1997. Planar light-emitting devices fabricated with luminescent electrochemical polyblends. Appl. Phys. Lett. 70, 934–936.

Yu, Z., Sun, M., Pei, Q., 2009. Electrochemical formation of stable pin junction in conjugated polymer thin films. J. Phys. Chem. B 113, 8481–8486.

Yu, Z., Niu, X., Liu, Z., Pei, Q., 2011a. Intrinsically stretchable polymer light-emitting devices using carbon nanotube-polymer composite electrodes. Adv. Mater. 23, 3989–3994.

Yu, Z., Wang, M., Lei, G., Liu, J., Li, L., Pei, Q., 2011b. Stabilizing the dynamic p-i-n junction in polymer light-emitting electrochemical cells. J. Phys. Chem. Lett. 2, 367–372.

Yu, Z., Zhang, Q., Li, L., Chen, Q., Niu, X., Liu, J., Pei, Q., 2011c. Highly flexible silver nanowire electrodes for shape-memory polymer light-emitting diodes. Adv. Mater. 23, 664–668.

Yu, Z., Li, L., Gao, H., Pei, Q., 2013. Polymer light-emitting electrochemical cells: recent developments to stabilize the pin junction and explore novel device applications. Sci. China Chem. 56, 1075–1086.

Zalar, P., Henson, Z.B., Welch, G.C., Bazan, G.C., Nguyen, T.Q., 2012. Color tuning in polymer light-emitting diodes with lewis acids. Angew. Chem. 124, 7613–7616.

Zhang, Z., Guo, K., Li, Y., Li, X., Guan, G., Li, H., Luo, Y., Zhao, F., Zhang, Q., Wei, B., 2015. A colour-tunable, weavable fibre-shaped polymer light-emitting electrochemical cell. Nat. Photonics 9, 233–238.

Zhu, X., Xie, Y., Li, X., Qiao, X., Wang, L., Tu, G., 2012. Anionic conjugated polyelectrolyte–wetting properties with an emission layer and free ion migration when serving as a cathode interface layer in polymer light emitting diodes (PLEDs). J. Mater. Chem. 22, 15490–15494.

8

Electrically Driving Sensors Based on Polymer

8.1 Electromechanical Actuators

Electromechanical actuators enabling the direct conversion of electrical energy to mechanical energy are promising candidates for the practical applications including robotics, artificial muscles, biomimetic devices, and sensors (Ma et al., 2013; Madden, 2007; Bauer et al., 2014). The electrochemical actuators based on the inorganic materials, such as ferroelectric ceramics and shape-memory alloys have been widely explored due to their powerful stress output and rapid responsiveness (Zhang and Li, 2012; Madden et al., 2004). However, their practical applications are greatly hindered by the small actuation strain (<10%), heavy weight and high cost. To this end, electromechanical actuators based on polymeric materials and their composites have attracted increasing attentions due to the advantages of polymeric materials, including synthesis controllability, high flexibility, large strain, and low cost. According to the actuating mechanism, the electromechanical actuators based on polymeric materials are mainly divided into three categories, that is, electrochemically driving, electrostatically driving, and electrothermally driving actuators.

8.1.1 Electrostatically Driving Actuators

The electrostatically driving actuator is one of the most explored polymer actuators driven by electrical field due to the large strain, rapid responsiveness, and good controllability. The electrostatic actuator based on polymeric materials typically consists of a dielectric elastomer film laminated by two electrode layers (Fig. 8.1A), that is similar to the configuration of a planer capacitor (Madden, 2007; Brochu and Pei, 2010). It has been verified that the electromechanical actuation of the dielectric film is driven by the electrostatic effects of charges on the film surface. More specifically, upon application of an electrical field on the electrodes, the charges are redistributed on the two sides of the film and an electrostatic attraction is thus generated between the opposite electrodes, while electrostatic repulsion is generated in each electrode. Consequently, the thickness of the film is decreased and the planer area is expanded. Particularly, due to the fact that most dielectric films are essentially incompressible, the decrease of thickness could additionally increase the expansion of planar area (Brochu and Pei, 2010). Based on the electrostatic effect, it has been shown that the generated stress is proportional to the dielectric constant of the elastomeric film and the square of the electrical field applied upon (Zhao and Wang, 2014; Brochu and Pei, 2010). To achieve a large and rapid actuation strain, the dielectric film should be thin and elastomeric besides the excellent dielectric performances. Generally,

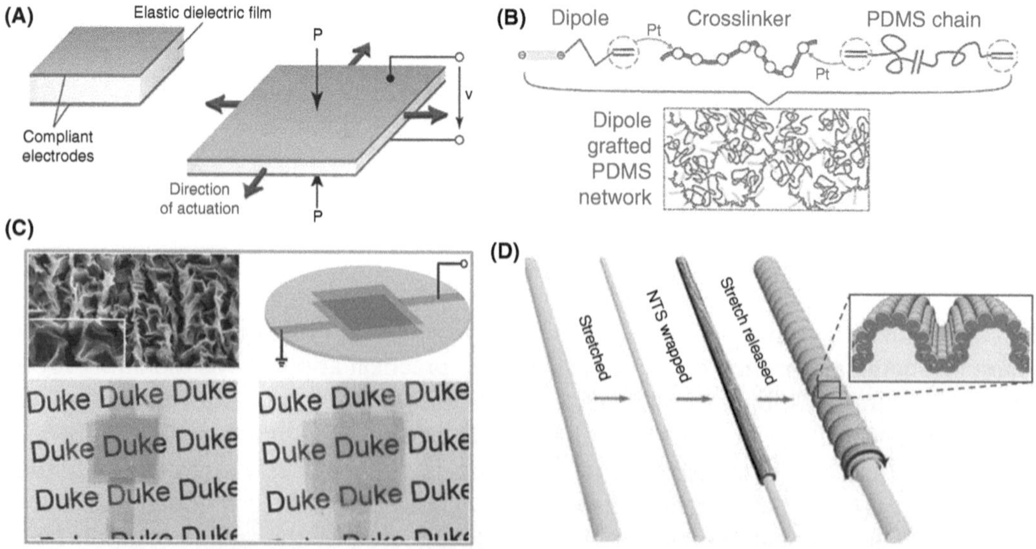

FIGURE 8.1 Electrostatically driving actuators. (A) Schematic illustration to the actuation mechanism of typical electrostatic actuators. (B) The dielectric elastomer (PDMS) grafted with dipole groups through cross-linking process. (C) A transparent electrostatically driving actuators made from crumpled graphene films as electrodes. (D) Electrostatically driving fiber actuator by wrapping aligned CNT sheets around a prestretched dielectric polymer fiber. *(A) Reproduced with permission from reference Madden, J. D., 2007. Mobile robots: motor challenges and materials solutions. Science 318, 1094–1097. Copyright 2007, AAAS. (B) Reproduced with permission from reference Kussmaul, B., Risse, S., Kofod, G., Wach , R., Wegener, M., Mccarthy, D. N., Kr Ger, H., Gerhard, R., 2011. Enhancement of dielectric permittivity and electromechanical response in silicone elastomers: molecular grafting of organic dipoles to the macromolecular network. Adv. Funct. Mater. 21, 4589–4594. Copyright 2011, Wiley-VCH. (C) Reproduced with permission from reference Zang, J. F., Ryu, S., Pugno, N., Wang, Q. M., Tu, Q., Buehler, M. J., Zhao, X. H., 2013. Multifunctionality and control of the crumpling and unfolding of large-area graphene. Nat. Mater. 12, 321–325. Copyright 2013, Nature Publishing Group. (D) Reproduced with permission from reference Liu, Z. F., Fang, S., Moura, F. A., Ding, J. N., Jiang, N., Di, J., Zhang, M., Lepro, X., Galvao, D. S., Haines, C. S., Yuan, N. Y., Yin, S. G., Lee, D. W., Wang, R., Wang, H. Y., Lv, W., Dong, C., Zhang, R. C., Chen, M. J., Yin, Q., Chong, Y. T., Zhang, R., Wang, X., Lima, M. D., Ovalle-Robles, R., Qian, D., Lu, H., Baughman, R. H., 2015. Hierarchically buckled sheath-core fibers for superelastic electronics, sensors, and muscles. Science 349, 400–404. Copyright 2015, AAAS.*

a large strain (10–100%) with a rapid strain rate could be generated by the electrostatic actuators under a driving voltage ranged from kilovolts to megavolts.

8.1.1.1 Dielectric Elastomers

A variety of polymers with a high dielectric constant, such as acrylic acrylate adhesive films (e.g., VHB series from 3 M), polydimethylsiloxane (PDMS), polyurethanes (PUs), isoprene, and fluoroelastomers have been explored to serve as dielectric layer in electrostatically driving actuators (Mirfakhrai et al., 2007). For instance, the actuator made of prestretched VHB 4910 elastomer from 3 M could generate a large actuation strain up to 158% under an electric field of 412 kV mm^{-1} (Pelrine et al., 2000). Although the electrostatically driving actuators exhibit large strain and rapid responsiveness, they still suffer from the limitation of ultrahigh operation voltage up to 100 V μm^{-1}.

The relationship between the driving voltage and actuation strain (s) could be expressed by Eq. (8.1):

$$s = -\frac{p}{Y} = -\frac{\varepsilon' \varepsilon_0}{Y} E^2 = -\frac{\varepsilon' \varepsilon_0}{Y} \left(\frac{U}{d} \right)^2 \tag{8.1}$$

where s, p, Y, ε', ε_0, E, U, and d represent actuation strain, electrostatic pressure, elastic modulus, dielectric permittivity, vacuum permittivity (8.854×10^{-12} F m^{-1}), electric-field strength, driving voltage, and the thickness of dielectric film, respectively. It indicates that two general strategies could be used to reduce the operation voltage, that is, decreasing the thickness to enhance the electric-field strength (E) and directly increasing the dielectric permittivity (ε'). It has been found that the dielectric loss and dielectric breakdown strength of the elastomer film could be well maintained by reducing the thickness, but the stress or strain output of the actuator would be weakened (Brochu and Pei, 2010). Besides, the breakdown of the actuator is likely to occur because of the local high electric field of the thinner films. Therefore, it is still unsatisfactory to decrease the operation voltage by reducing the thickness of the dielectric films.

Alternatively, great efforts have been made to enhance the dielectric constant of the films to decrease the operation voltage. It has been reported that the dielectric permittivity of the dielectric film could be efficiently enhanced through incorporation of ceramic nanoparticles with a high dielectric permittivity, conductive nanofillers, or organic dipole groups into the polymer matrix (Brochu and Pei, 2010). For instance, the TiO_2 nanoparticles coated with silicone oil could be composited into the commercially SEBS (polystyrene-*co*-ethylene-*co*-butylene-*co*-styrene) through a solution mixing method (Stoyanov et al., 2011). Based on the dielectric mixing rules, the resulting composite film exhibited an enhanced dielectric permittivity due to the higher permittivity of TiO_2 than that of the matrix. Compared with the matrix film, a decrease of 27% in electric-field strength was achieved for the composite film with an optimum nanoparticle volume percentage of 15%. Note that a large amount of nanoparticles are required to achieve a high dielectric constant; the aggregation of the nanoparticles in the matrix is inevitable although a variety of modified methods have been explored, thus causing a degraded mechanical properties of the films. The incorporation of the conductive fillers (e.g., metallic and polymeric nanoparticles) encapsulated with an insulating layer also suffered from the similar drawbacks of aggregation in the matrix (Opris et al., 2011; Molberg et al., 2010; Hu et al., 2014). As an alternative, the dipole at molecular level could be introduced into the elastomer matrix through chemical grafting rather than traditional blending and compositing (Kussmaul et al., 2011; Risse et al., 2012). This method could yield homogeneous elastomer film at molecular level and efficiently prevent the aggregation problem. For instance, a push–pull dipole molecule, *N*-allyl-*N*-methyl-*p*-nitroaniline, was grafted to the cross-linkers of silicone through a one-step film formation process (Fig. 8.1B) (Kussmaul et al., 2011). The dielectric permittivity of the resulting silicone with a dipole content of 13.4% could be greatly enhanced by a factor up to 5.9, due to the alignment of dipoles upon application of a voltage. Therefore, a smaller electric-field strength was needed to achieve a given actuation strain. Besides introduction of dipole molecule, the polar functional groups could also be incorporated into the elastomer matrix to enhance the dielectric permittivity while the glass-transition temperature was maintained to be relatively low. The high-molecular-weight polymethylvinylsiloxane functionalized with nitrile groups exhibited a high permittivity of up to 10.1 at 10^4 Hz and low modulus of 154 kPa with the strain of 10%. The actuators prepared from them could generate a strain of 20.5% under an electric field as low as 10.8 V μm^{-1}. In addition, the elastomer film showed a self-repair function after breakdown, which is very important for the durability and reliability of the actuators in practical applications.

The prestretching treatment of the dielectric film is also very important for the actuation performances (Brochu and Pei, 2010; Zhao and Wang, 2014). The prestretching could be applied on the dielectric films in uniaxial/biaxial directions, circular or spherical direction. The thickness of a dielectric film undergoes a thinning process when a given voltage is applied upon with the increasing area. In other words, the thickness of the film is decreased while the area is increased. As expected, a higher electric field is generated under the same given voltage, resulting in a higher Maxwell force to make the film further thinner until the electric field exceeds the dielectric strength. At a critical point, the continuous decrease in thickness may make the film thin down abruptly and cause a pull-in effect (Zhao and Wang, 2014). The pull-in effect could be efficiently relieved after a prestretching process because the film may become stiffer and a more moderate decrease in thickness is thus generated during actuation. In addition, the prestretched elastomer film also exhibited rapid actuation responsiveness due to the reduced viscoelastic nature after stretching treatment. Besides, it is also reported that the actuation direction could be tuned by changing the prestretching directions (Pelrine et al., 2000). It is because the fact that the film could be stiffened in the prestretching direction, causing the film to primarily actuate in the nonstretching direction. However, the actuators made from prestretching film generally suffered the disadvantage where a rigid frame is necessary to maintain the stretching state, which significantly increases the mass of the equipment and thus reduces the work/power density.

8.1.1.2 Electrodes

The role of the electrodes in electrostatic actuators is of critical importance for the actuation performances. The ideal electrode materials should process both high electrical conductivity and mechanical stretchability in compliance with the large deformation of the laminated dielectric elastomer during actuation. In addition, the electrodes should also provide a uniform and stable charge distribution on both of the surfaces.

Recently, a large number of stretchable electrode materials for electrostatic actuation have been explored. The thin metal films are pioneered as the earliest candidates due to their high electrical conductivity, but they greatly suffered from the disadvantage of small strain less than 1%. Alternatively, the metallic nanoparticles including titanium, gold, and palladium were coated on the elastomer surfaces through metal ion implantation method.

For instance, the dielectric elastomer coated with gold nanoparticles showed a low area resistance less than 1 kΩ per square and could be stretched up to a strain of 175% (Rosset et al., 2009). Besides, a constant conductivity over 100,000 cyclic strains up to 20% could be well maintained. The metallic nanowires could also be composited into the elastomer matrix to prepare stretchable electrodes with a high electrical conductivity, for example, the compliant silver nanowire-polymer composite electrodes exhibited an area resistance as low as 10 Ω per square and maintained a relatively stable conductivity with a cyclic strain higher than 40% (Yun et al., 2012). Compared with the metal-based electrodes, the carbon-based materials including carbon grease, graphite, and carbon powder are more competitive because they are highly conductive, unexpansive and well compliant with the dielectric elastomers. Recently, carbon nanotubes (CNTs) and graphene have also been considered as promising electrode candidates due to their excellent mechanical and electrical properties. By virtue of the flexibility and transparency of the electrode made from thin film of single-walled CNTs, the resulting actuator could generate a large area strain of 200% with a self-clearing function (Yuan et al., 2008). In addition, the probability of terminal failure caused by defects was efficiently reduced. Similarly, the actuator by using crumpled graphene film as electrode could also achieve a large strain and high transmittance up to 60%, accompanied with a tunable superhydrophobicity property (Fig. 8.1C) (Zang et al., 2013). It has been reported that the aligned CNT sheets could also directly serve as the electrodes due to the superior electrical and mechanical properties to random CNTs (Cakmak et al., 2015). Besides the large actuation strain, the actuators based on aligned CNT sheets exhibited directional motions, caused by the mechanically anisotropic constraint of the CNT sheets. They could also be wrapped around the prestretched elastomer fibers to form fiber-shaped actuators with simultaneous contractive and rotary actuations (Fig. 8.1D) (Liu et al., 2015).

In general, the electrostatically driving actuators could generate a large actuation strain with a rapid responsiveness at milliseconds, accompanied with low power consumption because of the very small current. They have exhibited a promising potential in the fields of artificial muscles, robotics, and mechanical machines (Anderson et al., 2012). Unfortunately, the disadvantage of high operation voltage ranged from kilovolts to megavolts has greatly hindered the practical applications. Therefore, continuous efforts are urgent to efficiently reduce the driving voltage by creating novel dielectric elastomers combined with design optimization of devices.

8.1.2 Electrochemically Driving Actuators

Compared with the electrostatically driving actuators, the electrochemically driving actuators are widely explored due to the low voltage at several volts (Kong and Chen, 2014). The electrochemically driving actuators typically consist of an electrolyte layer laminated by two electrode layers. Typically, they generate a bending motion due to the asymmetrically volume shrinkage and expansion of electrode layers, caused by the ion migrations under an electric field. The ionic electroactive polymers and composites are mainly used to construct electrochemically driving actuators. Among them, conducting polymer and ionic polymer metal composite (IPMC) are mostly explored.

8.1.2.1 Conducting Polymers

It has been found that conducting polymers are semiconducting as they are undoped while conductive as doped (Mirfakhrai et al., 2007; Baughman, 1996). The doping process can be typically achieved through a redox process during the electrochemical process. The volume change of the conducting polymers is generally caused by the ion insertion/deinsertion during the electrochemical doping process. The contraction and expansion of the conducting polymer depends on various parameters including the choice of conducting polymers, electrolytes, and the initial state prior to application of voltage (i.e., oxidation, neutral, or reduction states). For polypyrrole in a neutral state (Carpi et al., 2011) oxidation and reduction simultaneously occur at the anode and the cathode, respectively. For an anion-driven actuation, the volume expansion is generated when the anions are inserted into the oxidized polypyrrole at cathode to neutralize the electron deinsertion. In contrast, an opposite cation-driven actuation is possible if the anion is too large to migrate in electrolyte, and a volume contraction is thus generated at the cathode. It is also reported that the change in bond length, caused by the variation of charge density among the backbone of conducting polymer, could also give rise to the volume changes (Bay et al., 2003). However, this contribution was expected to be smaller than the ionic migration.

At present, the electrochemical actuators based on conducting polymers have been generally constructed with bilayer or multilayer configurations to achieve electromechanically bending motion. For conducting-polymer-based actuator with a bilayer configuration, the expansion/contraction of the conducting polymer layer during the electrochemical process could cause a bending motion away from or toward the side of conducting polymer layer. Polypyrrole, polyaniline (PANI), and polythiophene (PTh) derivatives are mostly explored conducting polymers for electrochemical actuators. The films to be laminated with the conducting polymers are required to be inert, flexible, and insensitive to electrochemical stimuli in volume changes. Adhesive tapes (Otero et al., 1992; Pei and Inganas, 1992) deposited metal films, (Smela et al., 1993; Jager et al., 2000a,b) commercial polymer films (Higgins et al., 2003), and graphene (Liu et al., 2012) have been widely used as the laminated layers. The actuator with a multilayered structure is obtained by laminating two conducting-polymer films with an electrolyte layer (Torop et al., 2014). Upon application of a voltage, contraction, and expansion occur on the two electrodes to cause bending motion.

The actuation speed linearly depends on the migration rate in electrolyte system, that is, the amount of migration ions per unit of time. The strain rate of typical conducting polymers is small ($<1\%\ s^{-1}$), due to the slow migration speed of ions during the electrochemical process (Warren and Madden, 2006). In addition, the conducting-polymer-based actuators suffered from the drawback of short lifetime due to the structural instability of the conducting polymers during the cyclic electrochemical process. To this end, functional or conductive additives have been incorporated into the conducting polymers to enhance the actuation performances. For instance, the introduction of CNTs could efficiently improve the electrical conductivity and mechanical strength of the PANI. The actuator prepared from the reinforced PANI could generate a large actuation stress more than 30 MPa, which is 300 times higher than that of natural skeletal muscle (Spinks et al., 2006). Besides, the use of CNTs could efficiently improve the responsiveness and extend the actuation durability of PANI, (Blighe et al., 2012) probably due to the enhancement of ionic migration and structural stability after introduction of CNTs. However, the CNT-reinforced conducting polymers showed higher modulus and a small actuation strain was thus generated. Alternatively, the carbide-derived carbon could be introduced into the conducting polymers to improve the electrical conductivity and enhance the actuation performances without any decrease of actuation strain. In particular, the energy efficiency could be efficiently improved because of the formation of double-layer capacitance and porous structure in the composite electrode (Torop et al., 2014). In addition to the aforementioned methods, the actuation speed could also be improved by decreasing the thickness of conducting layer and enhancing the interfacial contact between conducting polymer and electrolyte (Madden et al., 2000; Hara et al., 2005; Ding et al., 2003; Smela et al., 1995).

However, the electrochemical actuators made from conducting polymers still suffered from the short cyclic lifetime and slow responsiveness. In particular, compared with the dry-state counterparts, the mechanical performances of the conducting polymers in electrolyte media are significantly degraded (Spinks et al., 2006) which significantly limits the practical applications.

8.1.2.2 Ionic Polymer Metal Composite

The electrochemically driving IPMC actuator generally consists of an ion-exchange polymer film laminated by two electrodes. The bending motions could be generated due to the expansion and contraction of the electrodes, caused by ion migration between electrodes under an electric field. The model of electrochemical IPMC actuator was pioneered in 1999, where the single-walled carbon nanotubes (SWCNTs) served as electrodes while aqueous NaCl solution provided ions for insertion/desertion (Fig. 8.2A) (Baughman et al., 1999). Upon application of voltage, the charges were injected into an SWNT electrode, followed by ion insertion from electrolyte to neutralize the interface between electrode and electrolyte. The dimensional change in bond length was generated by the charge injection caused by quantum chemical and double-layer electrostatic effects.

At the early stage, liquid electrolyte was mainly used in the operation system of IPMC actuator. However, the use of liquid electrolyte greatly hinders the practical applications of the electrochemical actuators due to the heavy weight and complexity of electrochemical system. In addition, the actuation responsiveness is slow in liquid electrolyte because of the slow migration of ions and liquid resistance. In particular, the actuation stability and cyclic lifetime are also greatly affected by the used electrolyte. The aqueous-electrolyte-based actuators generally exhibit a limited cyclic lifetime less than 100 cycles (Kong and Chen, 2014). The use of electrolyte containing volatile solvents has also caused an instable actuation with a short cyclic lifetime due to the leakage and evaporation of the solvents

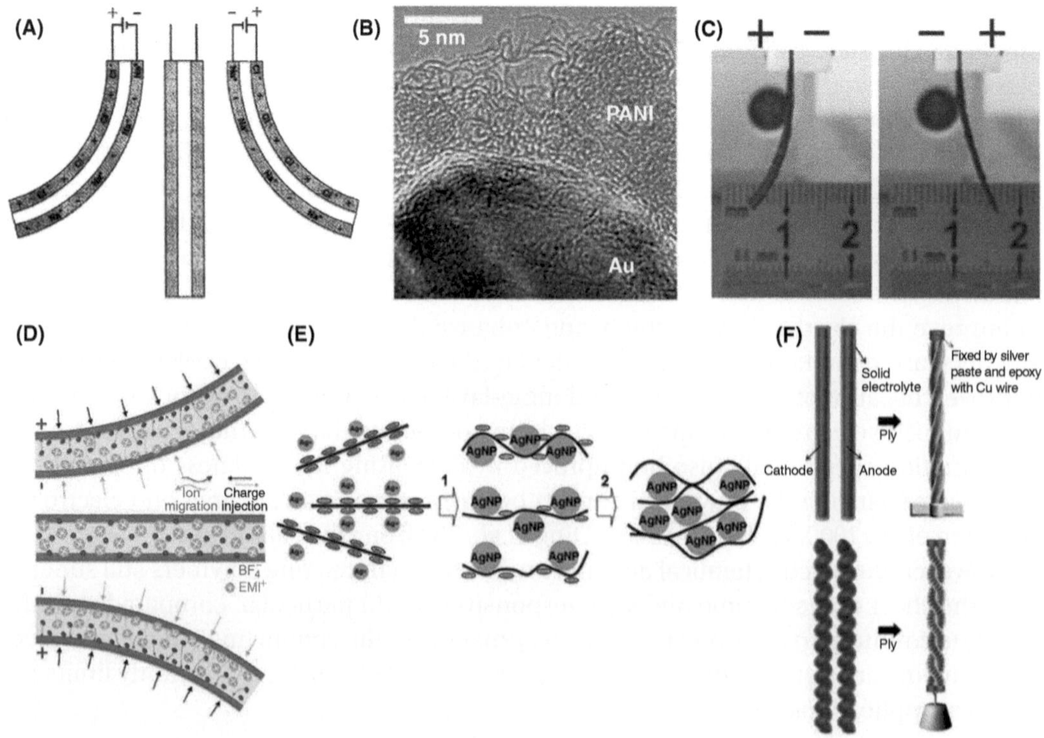

FIGURE 8.2 Electrochemically driving actuators. (A) Schematic view of an electrochemically driving bimorph actuator operated in aqueous NaCl. The actuator is made from two strips of SWNTs as electrodes. (B) Transmission electron microscope (TEM) of a PANI film covered with the ligaments of nanoporous gold. (C) Photographs of bending motion generated by a bimorph actuator upon application of a voltage. (D) Schematic illustration to the bending motion generated by a bimorph actuator upon application of positive and negative electric-field. (E) Schematic illustration to the preparation of RGO/Ag hybrid electrodes. (F) Preparation and configuration of the fiber-shaped actuators. *(A) Reproduced with permission from reference Baughman, R.H., Cui, C.X., Zakhidov, A.A., Iqbal, Z., Barisci, J.N., Spinks, G.M., Wallace, G.G., Mazzoldi, A., De Rossi, D., Rinzler, A.G., Jaschinski, O., Roth, S., Kertesz, M., 1999. Carbon nanotube actuators. Science 284, 1340–1344. Copyright 1999, AAAS. (B) Reproduced with permission from reference Detsi, E., Onck, P., De Hosson, J.T.M., 2013. Metallic muscles at work: high rate actuation in nanoporous gold/polyaniline composites. ACS Nano 7, 4299–4306. Copyright 2013, American Chemical Society. (C) Reproduced with permission from reference Lu, L.H., Chen, W., 2010. Biocompatible composite actuator: a supramolecular structure consisting of the biopolymer chitosan, carbon nanotubes, and an ionic liquid. Adv. Mater. 22, 3745–3748. Copyright 2010, Wiley-VCH. (D) Reproduced with permission from reference Li, J.Z., Ma, W.J., Song, L., Niu, Z.G., Cai, L., Zeng, Q.S., Zhang, X.X., Dong, H.B., Zhao, D., Zhou, W.Y., Xie, S.S., 2011. Superfast-response and ultrahigh-power-density electromechanical actuators based on hierarchal carbon nanotube electrodes and chitosan. Nano Lett. 11, 4636–4641. Copyright 2011, American Chemical Society. (E) Reproduced with permission from reference Lu, L.H., Liu, J.H., Hu, Y., Zhang, Y.W., Chen, W., 2013. Graphene-stabilized silver nanoparticle electrochemical electrode for actuator design. Adv. Mater. 25, 1270–1274. Copyright 2013, Wiley-VCH. (F) Reproduced with permission from reference Lee, J.A., Kim, Y.T., Spinks, G.M., Suh, D., Lepro, X., Lima, M.D., Baughman, R.H., Kim, S.J., 2014. All-solid-state carbon nanotube torsional and tensile artificial muscles. Nano Lett. 14, 2664–2669. Copyright 2014, American Chemical Society.*

during actuation (Shahinpoor and Kim, 2002). It is reported that the introduction of inert and nonvolatile ionic liquid could efficiently enhance the actuation stability and durability (Bennett and Leo, 2004). The resulting actuators exhibited stable and rapid actuations under a high frequency of 100 Hz (Nemat-Nasser, 2002). Recently, the solid-state electrolytes have been further developed to replace the liquid ones. The actuators based on solid-state electrolyte could be operated in air with a stable actuation (Lee et al., 2014). Unfortunately, the actuation responsiveness of the resulting actuators is relatively slow due to the confined ionic conductivity in solid-state electrolytes (Baughman, 2003). In addition, the redox reactions during the electrochemical actuation severely affect the interfacial contact between solid-state electrolyte and electrodes, and a degradation in actuating performances is thus caused. Therefore, an electrolyte-free strategy has further been proposed to fabricate IPMC actuators through introduction of metal/polymer interface (Fig. 8.2B) (Detsi et al., 2013). For instance, the PANI doped with sulfuric acid could be incorporated onto the porous gold electrode. The resulting electrochemical actuator exhibited a large strain rate which was 3 orders of magnitude higher than that of the traditional three-component actuator.

The conductive metal films, such as platinum and gold are firstly used as electrodes in the electrochemical actuators (Tiwari and Garcia, 2011). Although the metal-based electrodes show a good electrical conductivity, they suffered from various drawbacks including expensiveness, heavy weight, structural instability, and limited resources, which significantly limit their practical applications. Alternatively, the electrically conductive carbon nanomaterials have been explored to serve as electrodes. Among them, CNTs has been widely used due to their electrical conductivity, high specific surface area, structural stability, mechanical strength, and flexibility. For instance, the electrodes could be prepared from multiwalled carbon nanotubes (MWCNTs) composited with chitosan. In addition to the good electrical conductivity, the resulting electrodes were highly flexible and could closely contact with the electrolyte layer without obvious gaps between them (Fig. 8.2C) (Lu and Chen, 2010). The resulting actuator with a thickness of 170 μm generated a high bending actuation of 2 mm upon application of a low voltage of 3 V. This composite membrane was flexible without any cracking problem during actuation. Compared with the random CNTs, the CNT film where the CNTs are highly aligned exhibited higher electrical conductivity (2000 S cm^{-1}), high modulus (5 GPa), and high tensile strength (360 MPa). Therefore, the CNT film is promising to serve as electrodes in the IPMC actuator. For instance, the as-prepared CNT films could be separated by the chitosan-based electrolyte (Fig. 8.2D) (Li et al., 2011). Porous structure of the SWCNT films and good compatibility among all components enabled a strong interfacial adhesion between the electrode and the electrolyte layer. The high conductivity of the SWCNT films guaranteed a small surface resistance of the electrode layer 10 Ω/sq., so the voltage drop along the actuator surface was small. Benefited from the high specific surface area of SWCNTs, the charge injection and ion migration between the two electrodes could be efficiently enhanced to cause a rapid responsiveness (19 ms), large stress rate (1080 MPa s^{-1}), and high power density (244 W kg^{-1}).

However, the electrical conductivity of the carbon nanomaterials is vulnerable to degradation due to the inefficient dispersion in a large-scale actuator. To this end, some metallic additives have been incorporated into the carbon-based electrode to enhance the electrical conductivity and actuation stability. For instance, for the IPMC actuator where the reduced graphene oxide was used as electrode, the electrical conductivity of the electrodes could be efficiently improved after introduction of Ag nanoparticles (Fig. 8.2E) (Lu et al., 2013). As a result, both the actuation frequency and stability could be improved. Upon application of a low voltage of 1 V, the actuator could be driven at a wide frequency range (0.01–10 Hz), and no obvious decrease in displacement was observed over 500 cycles of actuation.

IPMC actuators in fiber configuration have been proposed to meet the requirements of wearable electronics, such as flexibility and wearability (Yang et al., 2015; Kim et al., 2013; Jeong et al., 2012; Ahn et al., 2009; Yamada et al., 2011; Choi et al., 2014). In comparison with the simple bending motion for planer actuator, contraction, and rotation could be simultaneously generated by the fiber-shaped actuator. Due to the high flexibility, strong strength, and excellent electrical conductivity (Miao, 2011; De Volder et al., 2013) the CNT fiber is a competitive candidate to serve as the electrodes in fiber-shaped actuators. The CNT fiber dry-spun from CNT array was first used as electrochemical actuators in a three-electrode system (Foroughi et al., 2011). Upon application of a low voltage, the resulting fiber actuator could generate a reversible 15,000 degree rotation with a speed of 590 revolutions per minute. However, the use of liquid electrolyte in the electrochemical system went against their practical applications. Alternatively, solid-state electrolyte was used as ion conductive layer between the CNT fiber electrodes.(Lee et al., 2014) The resulting actuator exhibited large rotation of 53 degree mm^{-1} and contraction of 1.3% with high stability and reversibility. The solid-state fiber actuator is promising in the applications, such as microrobotics, artificial muscles, and smart textiles.

In general, the electrochemically driving actuators could generate powerful and sophisticated actuation under very low driving voltage at several volts. Unfortunately, they suffered from some limitations including slow responsiveness, short cyclic lifetime, and limited electrolyte media. The disadvantages of electrochemical actuators significantly limit their practical applications. Continuous efforts should be focused on the enhancement of actuation stability and responsiveness through creating efficient solid-state electrolytes and optimization of the electrochemical system.

8.1.3 Electrothermally Driving Actuators

The electrothermally driving actuator with a layered structure has been also widely explored for the bending actuation (Chen et al., 2008, 2011; Zhang et al., 2014). Compared with electrothermally driving actuator made from inorganic materials, such as shape-memory alloys, the polymer-based counterpart exhibits competitive superiority including flexibility, lightweight, and large deformation. Generally, the electrothermal actuators based on polymers can be classified into two types. One is single-layered polymer film composited with conductive additives that can expand or contract due to the electrothermal heating (Chen et al., 2008; Sellinger et al., 2010; Hu et al., 2010). The other one is laminated film with a double-layered structure composed of electrically conductive layer and polymer layer (Chen et al., 2011; Seo et al., 2012; Liang et al., 2012; Zhang et al., 2011, 2014). These two layers have a great difference in coefficient of thermal expansion (CTE). Upon application of an electric current, the conductive layer with a low CTE could generate a joule heat, and an obvious expansion of the polymeric layer is caused. Consequently, a bending deformation is thus generated due to the mismatched volume changes of the electrode and polymer layer. The expansion layers are generally made from some polymers with a high CTE, such as PDMS, chitosan, PU, and epoxy. The conductive metallic materials are firstly used as the heating electrodes through various methods, such as spraying, spin coating, and heat pressing. However, they display some limitations, such as heavy weight, rigidity, and finite resource.

Recently, due to their high flexibility, electrical conductivity, and electroheating properties, CNTs have been used as the heating electrodes. For instance, the CNTs could be readily composited into the polymer films, such as chitosan and silicone elastomer to form a highly conductive network (Hu et al., 2010; Chen et al., 2008). Upon application of a pulse voltage, the resulting composite film actuators could generate a tunable vibration due to the thermal expansion and contraction of the polymer matrix, caused by the periodically heating of the CNT conductive network. Compared with the random CNTs, CNT sheets, and fibers where the CNTs are highly aligned exhibited a higher electrical conductivity and mechanical performances. Therefore, they are more competitive to serve as the heating electrodes in electrothermally driving actuators. Typically, the film actuator with a double-layered structure could be obtained by incorporating CNT sheets onto the surface of PDMS film through a curing process (Fig. 8.3A) (Chen et al., 2011). The resulting film actuator could generate a large and tunable bending motion under a low driving voltage less than 700 V m^{-1}. The double-layered film could also be fabricated into an electromechanical gripper to manipulate objects. However, these film actuators generated monotonous electromechanical motion, such as bending or vibration. To solve the problem, some efforts have been made to change the anisotropy of the CNT sheets against the longitudinal axis direction of the PDMS strip. The large-area double-layered film, laminated with highly aligned CNTs and PDMS, could be designed into various shapes, such as T-shaped, U-shaped, and Z-shaped actuators through a laser cutting method. The anisotropy of the CNTs could be readily tuned by varying the cutting direction. Upon application of low voltages, the resulting actuators with different shapes could generate complex motions including bending, twisting, and bionic motions in imitating human hands.

Graphene and its derivatives have also been explored as electrode materials because of their remarkable thermal, electronic and mechanical performances. In particular, the graphene exhibits a negative CTE, which is in contrast to most of conversional materials. By virtue of this, a microactuator with a double-layered structure could be obtained by transferring the as-grown graphene film to the epoxy substrate film through a bisacrificial-layer wet release process (Zhu et al., 2011). The resulting graphene-on-organic microactuator

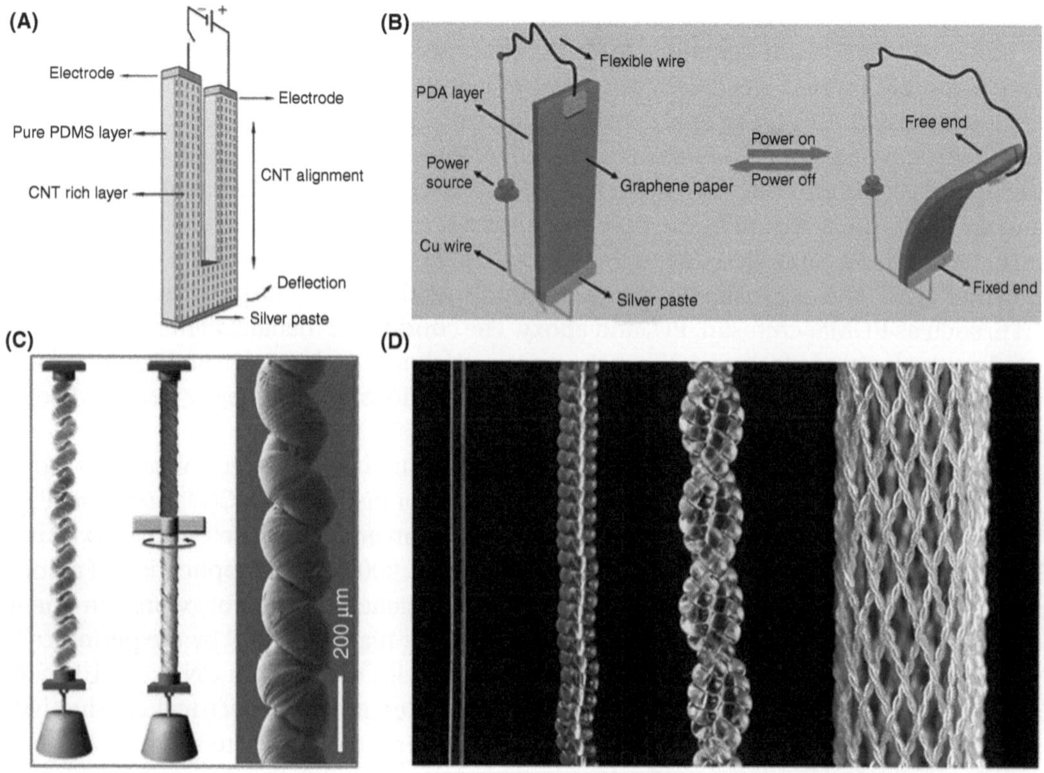

FIGURE 8.3 Electrothermally driving actuators. (A) Schematic illustration to the actuation mechanism and configuration for a typical electrothermally driving actuator. The aligned CNT film serves as the electrode. (B) Schematic illustration to the bending motion combined with color changes generated by an electrothermally driving actuator made from graphene and polydiacetylene. (C) Electrothermally driving actuator from helical CNT fibers filled with paraffin wax. (D) Photographs of electrothermally driving fiber actuator. They are prepared by wrapping CNT sheets around nylon thread. The CNT-sheet-coated fiber could be further woven into smart textiles. *(A) Reproduced with permission from reference Chen, L.Z., Liu, C.H., Liu, K., Meng, C.Z., Hu, C.H., Wang, J.P., Fan, S. S., 2011. High-performance, low-voltage, and easy-operable bending actuator based on aligned carbon nanotube/polymer composites. ACS Nano 5, 1588–1593. Copyright 2011, American Chemical Society. (B) Reproduced with permission from reference Liang, J.J., Huang, L., Li, N., Huang, Y., Wu, Y.P., Fang, S.L., Oh, J., Kozlov, M., Ma, Y.F., Li, F.F., Baughman, R., Chen, Y.S., 2012. Electromechanical actuator with controllable motion, fast response rate, and high-frequency resonance based on graphene and polydiacetylene. ACS Nano 6, 4508–4519. Copyright 2012, American Chemical Society. (C) Reproduced with permission from reference Lima, M.D., Li, N., De Andrade, M.J., Fang, S.L., Oh, J., Spinks, G.M., Kozlov, M.E., Haines, C.S., Suh, D., Foroughi, J., Kim, S.J., Chen, Y.S., Ware, T., Shin, M.K., Machado, L.D., Fonseca, A.F., Madden, J.D.W., Voit, W.E., Galvao, D.S., Baughman, R.H., 2012. Electrically, chemically, and photonically powered torsional and tensile actuation of hybrid carbon nanotube yarn muscles. Science 338, 928–932. Copyright 2012, AAAS. (D) Reproduced with permission from reference Haines, C.S., Lima, M.D., Li, N., Spinks, G.M., Foroughi, J., Madden, J.D.W., Kim, S.H., Fang, S.L., De Andrade, M.J., Goktepe, F., Goktepe, O., Mirvakili, S.M., Naficy, S., Lepro, X., Oh, J., Kozlov, M.E., Kim, S.J., Xu, X.R., Swedlove, B.J., Wallace, G.G., Baughman, R.H., 2014. Artificial muscles from fishing line and sewing thread. Science 343, 868–872. Copyright 2014, AAAS.*

generated large bending amplitude under a low driving voltage of 4 V. In particular, the bending motion could be completed within a short time of 0.02 s, which is 10–100 times rapider than the conventional double-layered electrothermal actuators. The graphene film prepared by the chemical vapor deposition method is difficult to be scaled up. Alternatively, the graphene film could be prepared from graphite at a large scale by using chemical exfoliation method coupled with a reduction process in solution (Liang et al., 2012). The resulting graphene film is highly flexible and electrically conductive, showing promising potential to serve as heating electrodes. For instance, the polydiacetylene crystal with a thermal expansion property could be in situ synthesized on the surface of graphene film to prepare electrothermal bimorph actuator (Fig. 8.3B). The resulting bimorph actuator exhibited a high flexibility and mechanical strength. It could generate a large bending motion with a curvature of 0.37 cm^{-1} when a low current density of 0.74 A mm^{-2} was used. The actuator also exhibited highly reversible and durable swing motions with a rapid responsiveness under a high frequency of 200 Hz. Additionally, the bimorph film could generate electrochromatic phenomenon during actuation, due to the electrically induced conformation changes in polydiacetylene. The actuation of the graphene-based bimorph actuator could be further enhanced by increasing the negative CTE of graphene layer in a spongy structure. The electrothermal actuator based on the spongy graphene exhibited higher bending amplitude with a curvature of about 1.2 cm^{-1} under a low driving voltage.

Electrothermally driving actuators with a unique fiber shape have also attracted increasing attentions. Among them, the fiber-shaped electrothermally driving actuators with helical structures are most widely explored for the simultaneously rotary and contractive actuations. The CNT fiber, where the CNTs are helically aligned have showed promising potentials to serve as the electrothermal actuators, due to its high flexibility and conductivity and specific surface area. The CTE of the CNT fiber was very small (at a level of -10^{-6} °C^{-1}) (Chen et al., 2015b). Therefore, some functional guest materials with a high CTE have been incorporated into the CNT fiber to achieve large electrothermally driving actuation output (Fig. 8.3C). Among them, paraffin is one of the most promising candidates as the guest additive because of its linearly thermal expansion behavior, high thermal stability, and the ability to wet CNT yarns (Lima et al., 2012). In addition, the CNT fiber with a high specific surface area could provide considerable capacity for the infiltration of paraffin. Upon application of a voltage of 40 V cm^{-1}, a 6.9-cm long and half-paraffin-infiltrated CNT fiber could rotate a 16.5-times-heavier paddle with a reversible rotation of ~1,000 degree for over 2 million cycles, accompanied with a rapid speed of ~11,500 revolutions per minute. Besides, a contractive actuation of about 10% could also be achieved. The fiber actuators could be widely used in the fields of motors, contractile muscles, and sensors. Similarly, the paraffin could also be incorporated into other helical fibers to prepare electrothermally rotary and contractive actuators. For instance, the niobium nanowire yarns filled with paraffin could also generate rapid rotation and contraction when a low voltage was used (Mirvakili et al., 2013). Although the electrothermal fiber actuators made from CNTs and inorganic nanowires have shown powerful actuation performances, the high-cost problem has restricted their applications. To this end, the inexpensive polymer fibers including Nylon, polyester, and polyethylene have been transformed into helically coiled structure through a twisting process (Fig. 8.3D) (Haines et al., 2014). The resultant coiled fiber precoated with conductive materials could generate large contractive strain up to 49% when it is subjected to heating or electrothermal treatment. The generated mechanical work (5.3 kW kg^{-1}) was comparable to that of a jet engine. The large-scale polymer fiber could be readily woven into smart textiles to mechanically respond to the change of temperature.

In summary, the electrothermally driving actuators exhibit large actuation under a relative low driving voltage. Compared with the electrochemically driving actuators, they show better actuation stability because it is free of electrolyte. Unfortunately, they exhibit a slow responsiveness due to the unsatisfactory heat conduction of the polymer. Therefore, some efforts should be made to improve the heat conduction by decreasing the thickness of the polymeric layer or introducing functional additives.

8.1.4 Conclusions and Perspectives

Electromechanical actuators attract increasing attentions because of their promising potentials in the fields of artificial muscles, robotics, and biomimetic devices. Among them, the electrostatically, electrochemically, and electrothermally driving actuators are mostly explored. For the typical electromechanical actuators with a layered structure, the electrostatically driving actuator generates mechanical motions by the laminated dielectric polymer caused by Coulomb force; the electrochemical trilayer actuator, composed of an electrolyte layer sandwiched by two electrodes, produces a bending motion due to the volume expansion or shrinkage of electrodes as a result of ion migrations under electric field; the electrothermally driving actuator bends/unbends due to the mismatched thermal expansion coefficients of the two contacted layers. Although these actuators have exhibited some advantages, for example, large actuation strains for electrothermally and electrostatically driving actuators and low driving voltages for electrochemical actuators, there remain some disadvantages for each one of them. For instance, the electrothermally and electrochemically driving actuators suffer from slow responsiveness and weak cyclic life; electrochemically driving actuators are limited by the electrolyte media and electrostatically driving actuators are driven by ultrahigh voltages. To this end, it is critical to develop electromechanical actuators based on a novel mechanism and realize programmable and sophisticated motions.

Recently, the CNTs have been used as building blocks to prepare helical fiber actuators without any guest materials through a dry-spinning method (Fig. 8.4A). Upon application of low driving voltages less than 5 V cm^{-1}, the CNT fibers could simultaneously generate contractive and rotary actuations based on a novel electromagnetic mechanism (Guo et al., 2012; Yuan and Poulin, 2014; Chen et al., 2015a,b). They generated a large stress of more than 200 times that of typical skeletal muscle, accompanied with a rapid responsiveness (<200 ms) (Guo et al., 2012; Chen et al., 2015b). Compared with the electrothermally and electrochemically driving actuators, the electromagnetically driving actuators

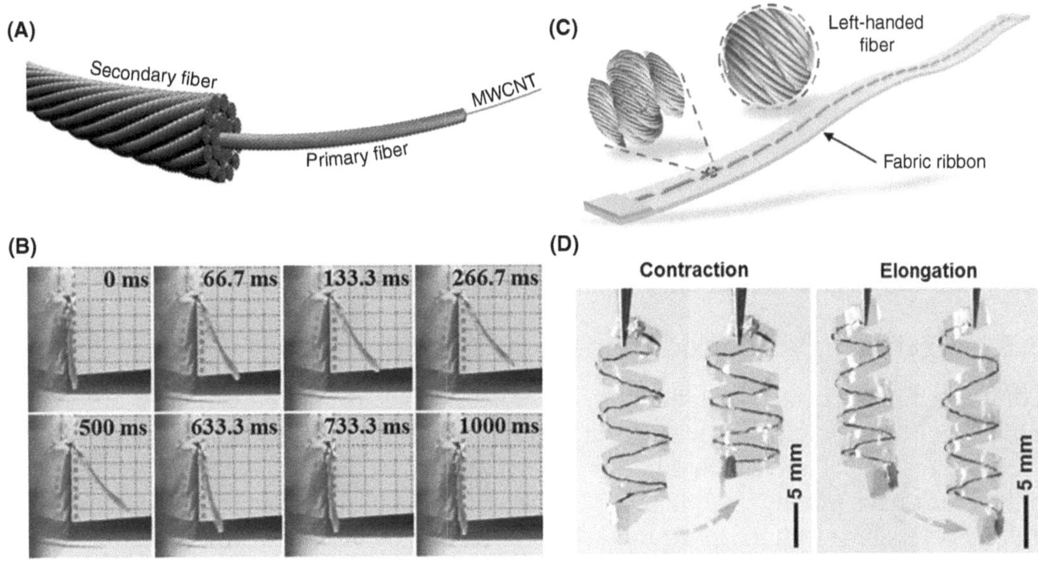

FIGURE 8.4 Electromechanically driving actuators based on a novel electromagnetic mechanism. (A) Schematic illustration to the helical assembly of CNTs into helically primary and secondary fibers without any guest materials. (B) The flapping motion driven by the secondary CNT fiber at a low voltage. (C) The spring-like fiber that was over-twisted from multiply secondary CNT fibers being woven into polymer film through a sewing method. (D) Electromechanical contraction and elongation generated by helical ribbon actuators with left-handedness (left image) and right-handedness (right image), respectively. *(A–B) Reproduced with permission from reference Chen, P.N., He, S.S., Xu, Y.F., Sun, X.M., Peng, H.S., 2015a. Electromechanical actuator ribbons driven by electrically conducting spring-like fibers. Adv. Mater. 27, 4982–4988. Copyright 2015, Wiley-VCH. (C–D) Reproduced with permission from reference Chen, P.N., Xu, Y.F., He, S.S., Sun, X.M., Guo, W.H., Zhang, Z.T., Qiu, L.B., Li, J.F., Chen, D.Y., Peng, H.S., 2015b. Biologically inspired, sophisticated motions from helically assembled, conducting fibers. Adv. Mater. 27, 1042–1047. Copyright 2015, Wiley-VCH.*

exhibited a higher reversibility and durability in actuation because they are composed of aligned CNTs without any guest materials. Besides, they could be operated in various media, such as air, water, and organic solution. A biomimetic flapping motion with a high frequency could be achieved by an artificial wing made from the secondary fibers (Fig. 8.4B). Both contractive strain and rotation output could be greatly enhanced by forming coiled structures (Chen et al., 2015a). The CNT fibers could be woven into polymer films to prepare fabric ribbon actuators (Fig. 8.4C). The resulting ribbon generated electromechanical bending/unbending with rapid responsiveness and high reversibility. The ribbons could be further shaped into helically three-dimensional structures through a heat-setting method to generate programmable elongation and contraction (Fig. 8.4C). These helical assembled fibers are lightweight, flexible, mechanically robust, and electrically conducting, and thus could be employed to design into unique actuating, biomimetic, and sensing devices with programmable and sophisticated motions. Along this way, the electromagnetic mechanism could also be generalized to develop electromechanical actuators through the helical and aligned assembly of other conductive one-dimensional nanostructures, such as nanowires and nanotubes.

8.2 Electrochromic Materials and Devices

Electrochromism is a phenomenon displayed by some materials reversibly changing colors. Various materials can be used to construct electrochromic devices, such as transition metal oxides, liquid crystals, photonic crystals, and polymers (Booth and Casey, 2009; Nicoletta et al., 2005; Arsenault et al., 2007; Garnier et al., 1983). Here, we will focus on the electrochromic materials based on polymers. There are several mechanisms to explain the color changes of polymer electrochromic materials like electro-induced oxidation–reduction and electrothermal chromatic transition and so on.

8.2.1 Electro-Induced Oxidation–Reduction Mechanism

Conjugated polymers, discovered in the late 1970s, have attracted a variety of attentions because of their unique properties, such as electrical conductivity and color versatility. The conjugated polymers with different colors can be used as ideal electrochromic materials, which have potential applications in sensors, mirrors, displayers, and textiles. Most of their reversible electrochromic behaviors are caused by the electro-induced oxidation–reduction, that is, the reversible change of a chemical species between two redox states under a certain voltage (Niklasson and Granqvist, 2007; Beaujuge et al., 2010).

Based on the electro-induced oxidation–reduction mechanism, an electrochromic device usually consists of multilayered structure: substrate, transparent conductor, electrochromic, electrolyte, ion storage/electrochromic layer, transparent conductor, and substrate (Fig. 8.5A) (Thakur et al., 2012). In a typical device, the transparent conductor serves as the working electrode which needs to adhere to the electrochromic layer; the electrolyte plays an important role in conducting ions and supporting redox reactions between the electrodes; the electrochromic layer changes color under voltages; the ion storage layer works as a buffer, that is, when ions in the electrolyte are intercalated into the electrochromic layer, the ion storage layer provides ions to the electrolyte layer and captures ions. With the development of the electrochromic devices, the ion storage layer is gradually replaced by another electrochromic layer. Obviously, the most important part of the device is the electrochromic layer. The colorful responses of conjugated polymers to voltage are derived from the changes of electron states in molecules.

8.2.1.1 Electrochromic Layer

There are various conjugated polymers as electrochromic materials, such as PTh, PANI, polypyrrole (PPy), and their derivatives. The PTh is the first conjugated polymer investigated as an electrochromic material in 1983 (Garnier et al., 1983; Gazard et al., 1983). The unsubstituted PTh switches from red to blue by p-doping and red to dark-green by n-doping. However, the insolubility of PTh is unbeneficial to synthesizing its derivatives, thus severely limiting its further applications. One of the most explored PTh derivatives is poly(3,4-ethylenedioxythiophene) (PEDOT), which can reversibly change colors between dark blue and sky blue under reduced and oxidized states, respectively (Garnier et al., 1983). Moreover, the PEDOT shows various advantages including good conductivity, high stability, and unique electrooptical properties.

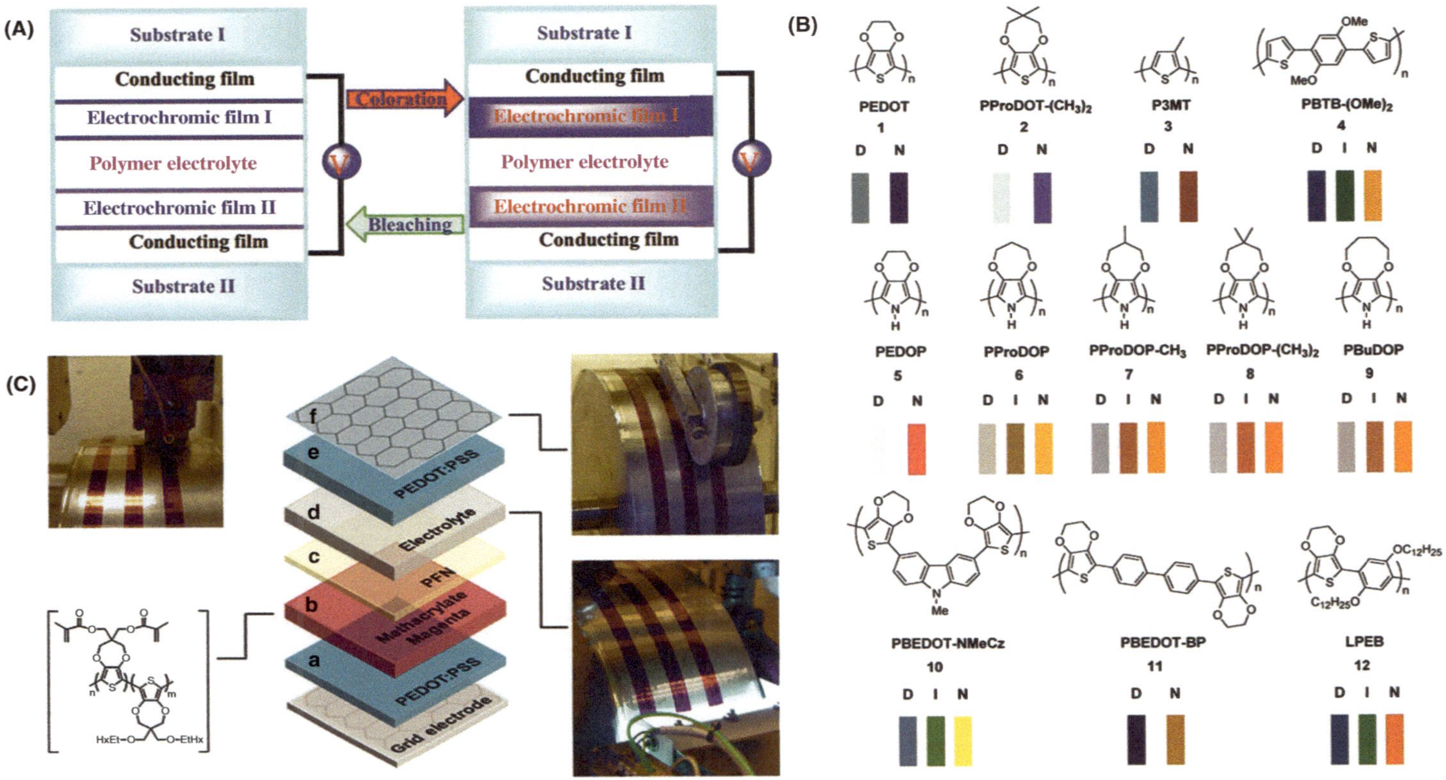

FIGURE 8.5 The electrochromic materials and electrochromic devices based on the electro-induced oxidation–reduction mechanism. (A) Typical configuration and electrochromic transition of an electrochromic device. (B) Chemical structures of some conjugated polymers with colors corresponding to the doped state (D), neutral state (N), and intermediate state (I). (C) Continuous fabrication of a solid-state electrochromic device. *(A) Reproduced with permission from reference Thakur, V.K., Ding, G., Ma, J., Lee, P.S., Lu, X., 2012. Hybrid materials and polymer electrolytes for electrochromic device applications. Adv. Mater. 24, 4071–4096. Copyright 2012, Wiley-VCH. (B) Reproduced with permission from reference Thompson, B.C., Schottland, P., Zong, K., Reynolds, J.R., 2000. In situ colorimetric analysis of electrochromic polymers and devices. Chem. Mater. 12, 1563–1571. Copyright 2000, American Chemical Society. (C) Reproduced with permission from reference Jensen, J., Krebs, F.C., 2014. From the bottom up—flexible solid state electrochromic devices. Adv. Mater. 26, 7231–7234. Copyright 2014, Wiley-VCH.*

PANI is another promising conjugated polymer. Due to its low cost, environmental stability and rapid color switching, PANI has exhibited a wide range of potential applications in many areas. By increasing the applied voltages, PANI generates a gradual color changes among yellow, green, blue, and black, accompanied with a rapid response and high reversibility. In addition, many efforts have been devoted to investigating its derivatives because PANI has an ill-defined structure. However, they show a limited solubility in many organic solvents (Wang et al., 1996). To this end, a series of electrochromic polyamides with well-defined structures are designed and synthesized (Gao et al., 2004). It was found that both the orthogonal tert-butoxycarbonyl group on the oligoaniline segment and the nonconjugated aromatic or aliphatic segment on the polymer backbone played a crucial role in determining the solubility of the polyamides. Compared with PANI which only has two distinct redox states, these copolymers had three distinct redox states.

The traditional electrochromic materials generate limited colors. To enrich the color changes, some other electrochromic polymers have been synthesized to display red (Dyer et al., 2010), orange (Dyer et al., 2010), yellow (Kerszulis et al., 2014; Lin et al., 2015), green (Gunbas et al., 2008; Durmus et al., 2007), blue (Wu et al., 2007; Invernale et al., 2009; Balan et al., 2008), and purple (Reeves et al., 2004). For instance, a series of conjugated polymers which generated different colors at doped, intermediate, and neutral states have been successfully synthesized (Thompson et al., 2000). Their chemical structures and corresponding colors are listed in Fig. 8.5B. The colors of copolymers could be readily tuned by varying the components of monomers (Beaujuge et al., 2008a). The balance in both short-wavelength and long-wavelength absorption bands could be achieved by using the donor-acceptor approach. In addition, several kinds of conjugated polymers could switch from colored neutral states to highly transparent oxidized states (Beaujuge et al., 2008b; Shi et al., 2010; Amb et al., 2010). The π-conjugated polymers including PEDOT, poly(2,5-dimethoxyaniline) (PDMA), and poly(3-methylthiophene) (P3MT) could be incorporated to the stainless steel wires to prepare electrochromic fibers through an electrochemical polymerization (Li et al., 2014). A polymer gel electrolyte was then coated onto the electrochromic layer and another stainless steel wire was twisted afterward. Under a low voltage, these electrochromic fibers exhibited fast (millisecond level) and reversible color switching between dark blue and light blue (PEDOT), yellow and green (PDMA), red and blue (P3MT), as well as green and yellow (PDMA).

Apart from developing electrochromic materials with various colors, improving the electrochromic properties is also important. Preparing inorganic/polymer hybrid materials is one of the promising strategies. There are many reports about the polymer/inorganic hybrid materials with outstanding electrochromic properties, such as NiO/PEDOT (Xia et al., 2009), NiO/PANI (Sonavane et al., 2010b), NiO/PPy (Sonavane et al., 2010a), PEDOT/Au/CdSe (Bhandari et al., 2010), WO_3/PANI (Ma et al., 2014), Au or Ag/PEDOT:PSS (Namboothiry et al., 2007), WO_3/Poly(*p*-phenylenebenzobisthiazole) (Zhu et al., 2010), PEDOT/CNT (Bhandari et al., 2008), and PANI/Graphene (Sheng et al., 2011). These hybrid materials showed rapid responsiveness and high coloration efficiency due to the synergetic properties provided by the components. For instance, it was found that WO_3/PANI arrays with hierarchical nanoplate structures exhibited excellent electrochromic properties (Ma et al., 2014). They were prepared by using an electrochemical polymerization method combined with a hydrothermal process. They exhibited high surface area and short diffusion distances, which was very important to the electrochromic performances of the devices. Furthermore, the WO_3/PANI arrays exhibited reversible multicolor changes from light yellow to green and green to blue from −0.6 to 1.2 V. Beside, the transparency of hybrid materials could also be tuned by changing the voltages from −0.6 to 0.8 V. However, the responsiveness was relatively slow, ranged from 500 to 800 ms. Therefore, the metal oxides/polymer hybrid (NiO/PANI) with a hierarchical structure was prepared to improve the responsiveness.

8.2.1.2 Electrolytes

Another way to improve the performance of the electrochromic devices is enhancing the property of electrolyte layer. Generally, the ionic conductivity of electrolyte is ranged from 10^{-4} to 10^{-7} S cm^{-1} (Thakur et al., 2012). At present, there are various types of electrolytes, such as liquid electrolyte, solid inorganic electrolyte, ceramic electrolyte, polymer electrolyte, and composite electrolyte (Agrawal and Pandey, 2008; Manuel Stephan and Nahm, 2006). Each of them has their own advantages and drawbacks. For instance, polymer electrolyte shows flexibility, high mechanical strength, good processability, and interfacial compatibility. Poly(2-acrylamido-2-methyl-L-propanesulfonic acid) was the first polymer electrolyte used in electrochromic device (Randin, 1982). During the past decades, great efforts have been made to develop and improve the performances of electrolytes. The most commonly used polymer electrolytes are as follows: polymethylmethacrylate (PMMA), polyvinylidenefluoride (PVDF), polyethyleneoxide (PEO), polyethyleneglycol (PEG), polyvinylalcohol (PVA), polyacrylonitrile (PAN), polyvinylchloride (PVC), polyvinylsulfones, and so on.

PEO is one of the most explored electrolytes. The polar ether group in segment makes it form coordinate bonds with many salts. Moreover, PEO is easy to be synthesized and modified. But there are still some disadvantages including low melting point, limited operating temperature range, and high crystallinity. To solve these problems, a novel solid polymer electrolyte was prepared by using layer-by-layer assembly method (Nguyen et al., 2011). The electrolyte consisted of four layers, that is, linear polyethyleneimine (LPEI), polyacrylicacid (PAA), PEO, with an interlapped structure formed through electrostatic and hydrogen bonding forces. PAA served as a bridging molecule to connect LPEI and PEO. PEO has a good solubility of salts, while LPEI exhibits a high ionic transport due to its heteroatomic hydrophilic backbone. In addition, thermal cross-linking of PAA and LPEI could form amide product and break the hydrogen bonds to cause a separated crystalline. The ionic conductivity of the electrolyte in dried state was above 10^{-5} S cm^{-1}, similar to that under a moderate humidity. Then, two types of devices in transmission and reflective modes were designed to investigate the electrochromic performances. The highest modulation for the transmission mode devices were about 38% at a wavelength of 670 nm.

Compared with PEO electrolytes, PDVF, and PMMA electrolytes exhibited higher ionic conductivities. In particular, PMMA has attracted increasing attentions due to its low cost, high solvent retention ability, high transparency, and processibility. The first all-polymer electrochromic device was obtained based on a gel electrolyte and PEDOT-PSS [poly(styrene sulfonate)] electrochromic material (Argun et al., 2003). The fabricated device exhibited a maximum transmittance change of 51% at 540 nm. In addition, this device was fairly stable and only 5% contrast loss was observed after 32,000 cycles.

The fabrication of electrochromic devices at large scale is necessary for the practical applications. Recently, a continuous fabrication process of electrochromic devices has been reported (Jensen and Krebs, 2014). A flexible solid-state electrochromic device was designed and built by sequentially stacking individual layers using slot-die coating and flexographic printing (Fig. 8.5C). First, the conductive PEDOT-PSS layer was coated on the silver grid foil by using slot-die coating method. Then, the electrochromic (methacrylate magenta) and PFN layers were coated to guarantee a sufficient wetting of the liquid electrolyte with electrochromic polymers. The electrolyte layer and another conductive PEDOT-PSS layer were then coated onto the above layers by using slot-die coating. Finally, the silver-grid electrode was deposited at the top layer through flexographic printing. The resulting device showed an optical contrast of 35% and a response time of 25 s under a low voltage from −0.4 to 1.0 V.

8.2.1.3 Applications

The electrochromic devices can be used in various fields including smart windows, energy devices, and displayers. The energy storage smart window could be fabricated by integrating energy-storage and electrochromic functions (Wang et al., 2012). By using aligned PANI nanowire arrays as electrodes, the flexible device generated color change from light yellow green (0 V) to deep blue (1 V). In addition, this smart device could effectively harvest and store renewable energy. A smart electrochemical capacitor with electrochromic function could be prepared by using PANI as electrodes (Chen et al., 2014a). The PANI was electrodeposited onto the aligned CNT sheet. The supercapacitor exhibited different colors of blue, yellow, and green, under a range of voltages from −1 to 1 V during the charging and discharging process (Fig. 8.6A). The color changes were rapid and reversible. In addition, these supercapacitors also exhibited high specific capacitances of 308.4 $F g^{-1}$. The electrochromic fiber-shaped supercapacitor with good flexibility and stretchability was further developed to meet the requirements of wearable application (Fig. 8.6B) (Chen et al., 2014b). These fiber-shaped supercapacitors exhibited rapid and reversible electrochromic transitions between two working stages, thus providing dynamic information on their charging states. Besides, they showed high electrochemical performances under deformations, such as bending and stretching. As expected, the electrochromic materials could also be used as displayers (Fig. 8.6C). PProDOT-$(CH_2OEtHx)_2$ (PProDOT): poly(3,4-propylenedioxythiophene) could serve as an electrochromic layer due to the high contrast, good stability, and rapid responsiveness from magenta neutral state to colorless oxidized state (Vasilyeva et al., 2009). The resulting electrochromic device was fabricated through a solution-based process. They

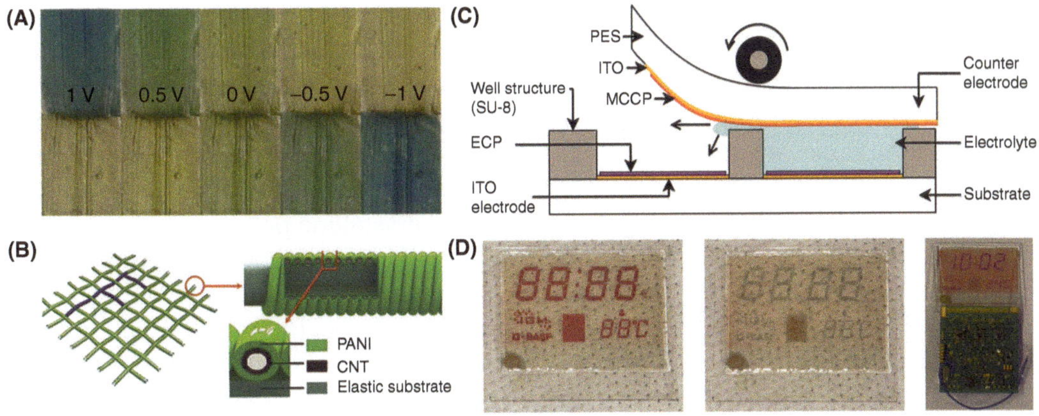

FIGURE 8.6 Applications of electrochromic devices. (A) Color transitions of electrochemical capacitors during a charging-discharging process. (B) Electrochromic fiber-shaped super capacitors. (C) Structure of a segmented solid-state polymer electrochromic displayer. (D) Magenta and colorless states of an electrochromic displayer. C and D. *(A–B) Reproduced with permission from reference Chen, X., Lin, H., Deng, J., Zhang, Y., Sun, X., Chen, P., Fang, X., Zhang, Z., Guan, G., Peng, H., 2014b. Electrochromic fiber-shaped supercapacitors. Adv. Mater. 26, 8126–8132. Copyright 2014, Wiley-VCH. (C–D) Reproduced with permission from reference Remmele, J., Shen, D.E., Mustonen, T., Fruehauf, N., 2015. High performance and long-term stability in ambiently fabricated segmented solid-state polymer electrochromic displays. ACS Appl. Mater. Inter. 7, 12001–12008. Copyright 2015, American Chemical Society.*

could afford a rapid segment switching within a narrow voltage window, accompanied with a long-term stability over 200,000 cycles. In addition, the devices combined with a printed circuit board could be used as a clock (Fig. 8.6D) (Remmele et al., 2015).

8.2.2 Electrothermal Chromatic Mechanism

Another mechanism to realize electrochromism is electrothermal chromatic transition. Electricity is much easier to control compared to heat and most of the conductive materials, including many metals and conjugated polymers, can generate heat upon pass of electric current. Electrochromic device based on the electrothermal mechanism is generally composed of chromic and electrically conducting layers.

8.2.2.1 Electrochromic Materials

The electrochromic materials based on the electrothermal chromatic mechanism include inorganic compounds, organic small molecule, conjugated polymers, and liquid crystalline polymers. The inorganic compounds and organic small molecule include CuI, $PbCrO_4$, VO_2 (Booth and Casey, 2009), $[(C_2H_5)_2NH_2]_2CuCl_4$ (Bloomquist et al., 1988), and triarylmethane derivatives (Nambo and Crudden, 2014; Ren et al., 1996). Although some of them have been commercialized, they share some disadvantages including slow ion diffusion, low coloration efficiencies, high electrical power, and undesirable coloration-bleaching kinetics.

Some conjugated polymers and liquid crystalline polymers which possess thermochromic properties are promising electrochromic materials. In comparison with the transition metal oxides, conjugated polymers have merits of lower redox switching potentials, faster response, higher coloration efficiency and more diversified colors (Thakur et al., 2012). Conjugated polymers consist of a π-electron system in their backbones, leading to excellent electronic and optical properties. They exhibit colors because these conjugated polymers can absorb visible light. The color transition results from the changes of the conjugation length of the π-electron system, in response to numerous external stimuli, such as temperature, light, and electricity. For liquid crystalline polymers, they can form helical superstructures. This periodic structure can reflect the incident light according to the Bragg's law and thus display structural colors. As thermochromic materials, the liquid crystalline polymers mostly have cholesteric phase. The color of the reflected light relies on the pitch length of the helical superstructure and thus changes readily with temperature.

Conjugated polymers, including polydiacetylene (PDA), PU, PTh, and their derivatives, have been widely studied as thermochromic materials. PDA was first synthesized by Wegner (1969), which has intriguing properties associated with their interfacial polymerization, diverse morphologies, and distinct color transition in response to environmental stimuli (Charych et al., 1993; Lu et al., 2001; Peng et al., 2009; Chen et al., 2010). It is easy to synthesize PDA by polymerizing diacetylenic monomers under ultraviolet at ambient temperature. In addition, diverse morphologies (molecular film, nanohelix, nanovesicle, liposome, and nanotube) can be achieved through a self-assembling process. The color change of PDA typically from blue to red is readily identified by naked eyes.

However, the color transition of PDA in response to some stimuli (temperature, pH, and chemical) is irreversible, which hampered their applications in many areas. To this end, many efforts have been made to improve their sensitivity and reversibility. Ways to improve the thermochromatic reversibility include reinforcing hydrogen bonding and modifying functional groups, such as azobenzene, amide moiety, or carboxylic groups (Ahn et al., 2003; Gu et al., 2008). The 10,12-pentacosadiynoic acid (PCDA) does not show reversible thermochromic transitions in response to temperature because of the weak hydrogen bonding of carboxylic groups. Therefore, a series of PCDA monomers were synthesized by introducing amide and carboxylic groups, which could effectively form the intermolecular double hydrogen bonds (Fig. 8.7A–B) (Ahn et al., 2003). Furthermore, some novel azobenzene-substituted PDA derivatives have also been designed and synthesized (Wu et al., 2009). The stability of PDA micelles was efficiently improved due to the enhanced intermolecular interactions among side chains. They were prone to self-assemble into stacks of bilayered two-dimensional structures, which therefore exhibited reversible thermochromic transitions. For instance, PDA films showed chromatic transitions from blue to purple and red at a wide transition temperature range (75–105°C). When the temperature was higher than 105°C, the chromatic transitions became partly reversible. This chromatic transition would be irreversible if the temperature was over 150°C. Based on the fact that the intermolecular forces in peptides are strong because of many carbonyl and imino groups, two kinds of polydiacetylene monomers (i.e., C23-GAGAGAGY and C25-GAGAGAGY) were synthesized by introducing a peptide to the side chain (Guo et al., 2014). The peptide-based amphiphilic polydiacetylene (PA-PDA) materials could be processed into various forms like fiber,

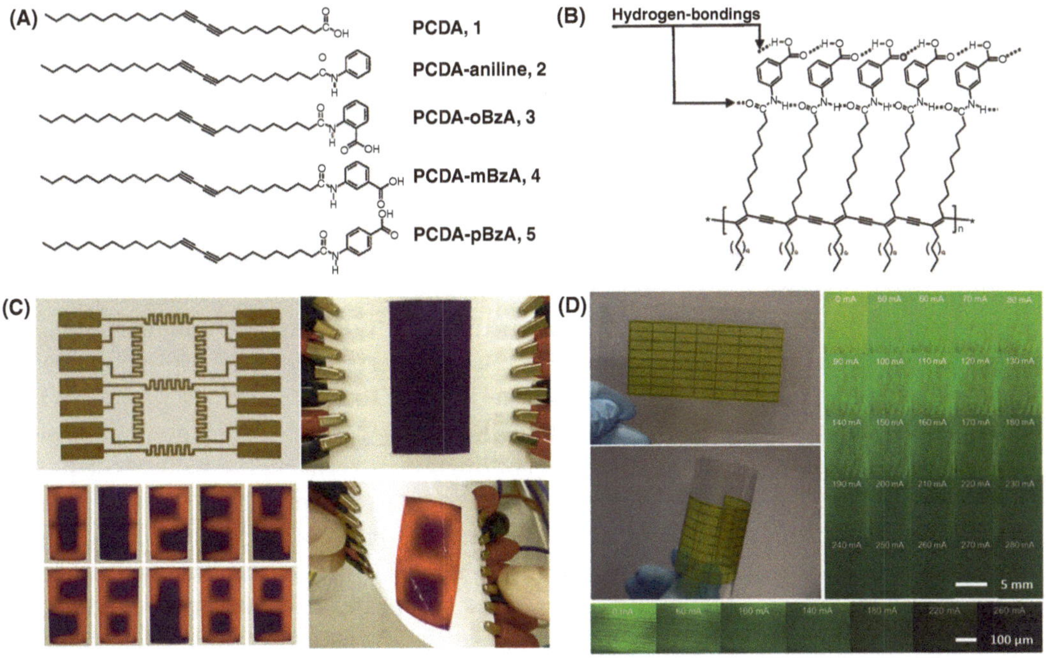

FIGURE 8.7 Chromatic polymers and applications of electrochromic devices based on electrothermal chromatic mechanism. (A) Chemical structures of substituted PDA monomers. (B) Hydrogen-bonds between amide—carbonyl groups and terminal carboxyls. (C) The color transitions of textile prototype at different voltages. (D) The color and fluorescent intensity of CNT/PTP composite films at different currents. *(A–B) Reproduced with permission from reference Ahn, D.J., Chae, E.H., Lee, G.S., Shim, H.Y., Chang, T.E., Ahn, K.D., Kim, J.M., 2003. Colorimetric reversibility of polydiacetylene supramolecules having enhanced hydrogen-bonding under thermal and pH stimuli. J. Am. Chem. Soc. 125, 8976–8977. Copyright 2003, American Chemical Society. (C) Reproduced with permission from reference Laforgue, A., Rouget, G., Dubost, S., Champagne, M.F., Robitaille, L., 2012. Multifunctional resistive-heating and color-changing monofilaments produced by a single-step coaxial melt-spinning process. ACS Appl. Mater. Inter. 4, 3163–3168. Copyright 2012, American Chemical Society. (D) Reproduced with permission from reference Sun, X., Zhang, Z., Lu, X., Guan, G., Li, H., Peng, H., 2013. Electric current test paper based on conjugated polymers and aligned carbon nanotubes. Angew. Chem. Int. Edit. 52, 7776–7780. Copyright 2013, Wiley-VCH.*

membrane, and gel which displayed ultrafast and reversible thermochromatic transitions even at an extremely high temperature (200°C).

Although most of thermochromic polymer materials are conjugated polymers, there still exist some heterochain polymers which also show thermochromic behavior. For example, PU, composed of soft and hard segments, is opaque at ambient temperature due to the scattering of crystal grains and boundaries. After a heating treatment, the soft segments melt into amorphous phase and PU thus became transparent. A novel method was proposed by using PU to fabricate electrothermochromic devices based on CNT sheet (Meng et al., 2011). The CNT sheet layer was transparent and conductive and could be sandwiched between two PU films on glass substrates. These CNT/PU composite films were flexible and switchable from opaque to transparent under 30 V and could be reversibly turned back to the opaque state once turning off the voltage. In addition, the composite also showed lower melting temperature (~47°C) than pure PU film (crystal melting point of 52°C) under a low content (0.2 wt.%) of CNT sheets.

8.2.2.2 Conductor

To trigger the thermochromism using electricity, a conductor is necessary to conduct current. Since high electrical conductivity, metal-based material has been considered as an ideal conductor. For instance, Ag paste film composited with TiO_2 nanoparticles exhibited high conductivity and mechanical flexibility, which could serve as conductor layer in electrochromic devices (Yoon et al., 2011). PDA precursors containing supramolecular diacetylene and surfactant nanocomposites could be readily printed on the conductive paper. After UV-induced photopolymerization, the blue PDA was yielded. The resulted device exhibited reversible and stable color transition between red and blue due to the electroheating under a bias potential of 4–5 V. They were further explored for extending the electrothermochromic property by introducing conductive patterns on the back of the paper. Therefore, the PDA regions contacted with the conductive layer showed a color change from blue to red. Furthermore, some flexible electrothermochromic paper using conducting wires as conductors was also fabricated (Shin et al., 2014). The conductive composite was prepared by mixing TiO_2 nanoparticles and silver paste. Then, it was printed on the back of the photo paper through a pattern mask. An electrothermochromic paper could be finally fabricated after coating with the mixture of PDA monomer and transparent screen printing ink on the front of photo paper. When a voltage of 1.8 V was applied, the regions connected with conducting wires on the back changed from blue to red due to the heat generated from the wires. Upon the application and removal voltages, digits from 0 to 9 could be easily displayed (Fig. 8.7C).

Apart from metals, conducting polymers can also act as conductors due to its electrical conductivity. For example, the conductive PEDOT nanofiber mats could be prepared by using a two-step procedure (Laforgue, 2010). A mixture solution, containing iron (III) tosylate, pyridine, and polyvinylpyrrolidone, was first electrospun into nanofibers. Then, oxidant fibers were converted to PEDOT by base-inhibited vapor-phase polymerization. Due to the conductive and doped PEDOT, the free-standing nanofibers showed a light blue color. The conductivity of PEDOT nanofibers was 60 ± 10 S cm^{-1}. The PEDOT nanofibers effectively generated heat when current was over 100 mA. For instance, PEDOT mats reached a temperature of 105°C under a current of 250 mA. The resistive heating properties of PEDOT nanofiber mats made them suitable in the applications, such as electrochromic fabrics. Therefore, an electrochromic device was fabricated by depositing thermochromic ink onto the surface of PEDOT nanofiber mats. The ink showed a color transition temperature of 37°C and changed from blue to white under a current of 100 mA. However, it was difficult to pass the current through the whole textile. To solve the problem, the multilayer monofilaments with a coaxial configuration were further investigated (Laforgue et al., 2012). The multilayer monofilaments were extruded by a Randcastle coextrusion line, equipped with a unique single-hole die at different temperatures and screw rotation speeds. The multilayer monofilament consisted of three layers: core layer

[PP-MWNT, polypropylene (PP) with MWCNTs], intermediate layer (PP-TiO_2), and sheath layer [PP-TCM-TiO_2, thermochromic microcapsules (TCM)]. The conductivity of the core layer was ~1 S cm^{-1}. Thus, this monofilament was easily to heat up at different voltages. The temperature gradually rose up to over 100°C at voltages above 100 V. The fabric showed a color change from green to beige.

Carbon materials with excellent electrical and mechanical properties are also suitable for conductors. Among them, aligned CNTs have explored as the promising candidate in the electrochromic devices. For instance, the poly[1-phenyl-2-(ptrimethylsilyl) phenylacetylene] (PTP) could be incorporated into the aligned CNTs through a compositing process (Sun et al., 2013). Due to the excition deconfinement and molecular perturbation within the backbone, PTP demonstrates a reversible change in fluorescent intensity induced by heat and solvent (Kwak et al., 2006, 2008). Besides, PTP with silicon and benzene moieties in side chains has higher thermal stability because of a special "rigid jacket" effect formed through the molecular interactions of silicon and benzene moieties in the side chains. For the aligned CNT composited with PTP, the PTP was directly coated on the surface of aligned CNT sheets. The CNT/PTP composite film exhibited a high conductivity (380 S cm^{-1}), tensile strength (300 MPa), and decomposition temperature (500°C). The CNT/PTP composite showed a rapid fluorescent quenching with increasing current from 0 to 32 mA (Fig. 8.7D). By increasing the current from 0 to 280 mA, the flexible composite films exhibited a color transition from yellowish green to dark green when exposed to UV light (Fig. 8.7D). In addition, the CNT/PTP composite films also showed a strong fluorescence under UV light at increasing currents. The color transitions were rapid and highly reversible even after a thousand cycles with currents less than 130 mA.

8.2.3 Other Electrochromic Mechanisms

Besides electricity-induced oxidation–reduction and heating mechanisms, there are other electrochromic mechanisms that give rise to the electrochromic behavior, such as electro-introduced conformational changes of conjugated polymers, reorientation of liquid crystals and change of lattice constant in opals.

8.2.3.1 Electro-Induced Conformational Changes

In general, the electrochromic behavior of PDA is caused by electro-induced heating. However, we recently discovered that electricity can also directly cause the conformational changes of PDA (Peng et al., 2009). A PDA layer was coated onto the surface of a CNT fiber. Due to the aligned arrangement of CNTs, the CNT/PDA composite fibers exhibited high conductivity of 10^2–10^3 S cm^{-1}. They were able to reversibly change color from blue to red in response to current (Fig. 8.8A). It was found that the color change was attributed to electricity rather than electro-induced heating. To verify the mechanism, we coated pure benzophenone (with a melting point of ~44°C) onto the CNT fiber to indicate the temperature. The benzophenone did not melt upon application of current of 30 mA, indicating that the temperature of the CNT composite fiber was less than 44°C. In other word, the electro-induced heating mechanism could be excluded because thermochromism of PDA

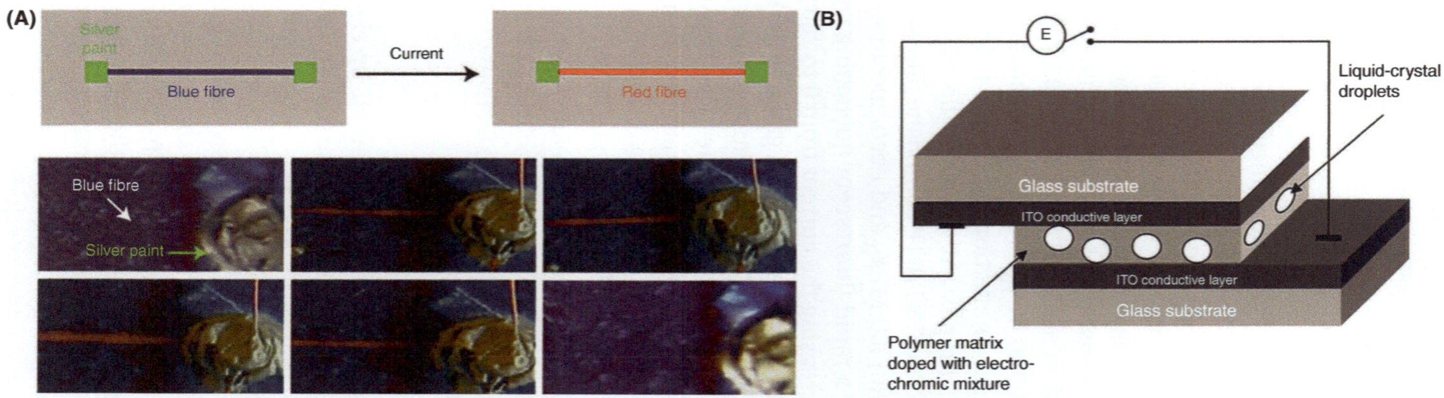

FIGURE 8.8 Electrochromic device based on different mechanisms. (A) Chromatic transitions of a composite CNT/PDA fiber in response to electric current. (B) The multilayer structure of a PDLC devices. *(A) Reproduced with permission from reference Peng, H., Sun, X., Cai, F., Chen, X., Zhu, Y., Liao, G., Chen, D., Li, Q., Lu, Y., Zhu, Y., Jia, Q., 2009. Electrochromatic carbon nanotube/polydiacetylene nanocomposite fibres. Nat. Nanotechnol. 4, 738–741. Copyright 2009, Nature Publishing Group. (B) Reproduced with permission from reference Nicoletta, F.P., Chidichimo, G., Cupelli, D., De Filpo, G., De Benedittis, M., Gabriele, B., Salerno, G., Fazio, A., 2005. Electrochromic polymer-dispersed liquid-crystal film: a new bifunctional device. Adv. Funct. Mater. 15, 995–999. Copyright 2005, Wiley-VCH.*

occurs at 56°C. Therefore, the color transition of CNT/PDA composite fiber was due to the conformational change caused by electricity. When current passed through the composite fiber, the polarization of carboxyl groups resulted in strong interactions between PDA and CNTs, and the π-electron delocalization of PDA backbone was decreased. The conformation of PDA was returned to the original state upon the removal of the current and hence the color of CNT/PDA composite fiber was recovered to the original state.

8.2.3.2 Reorientation of Liquid Crystals

Another type of electrochromic material, polymer-dispersed liquid crystals (PDLCs) that are dispersed micrometre-sized liquid-crystal droplets into the polymer matrix, can switch from translucent to opaque states under electric field. The electric field could reorient the liquid crystal molecules, and a reversible transition in transmittance is generated. As shown in Fig. 8.8B, a functional device that can change its colors by electricity have been fabricated from the PDLC film (Nicoletta et al., 2005). It is composed of nematic liquid crystals and thermoplastic matrices. The as-prepared electrochromic device was opaque due to the light scattering of liquid crystals droplets which were randomly dispersed in the polymer matrix. Upon application of an electric field ($<1\ V\ \mu m^{-1}$), the liquid crystals were reoriented along the direction of electric field at the surface of droplets, which increased transmittance from 2% to 80%. After removal of the electric field, the liquid crystals turned back to the random arrangement. In addition, this electrochromic device also changed its color from light yellow to dark blue due to the redox reaction of the added electrochromic molecules in the liquid crystal. Both color and transmittance changes were fast and exhibited a short relaxation time.

8.2.3.3 Changes in Lattice Constants of Opals

Recently, photonic crystals have attracted increasing attention for their unique electrochromic properties. Photonic crystals possess periodic dielectric, metallodielectric, or even superconductor structures, which affect the propagation of electromagnetic waves by defining allowed and forbidden photonic energy bands. In other words, photonic crystals can affect the motion of photons and reflect narrow bands of wavelengths, and exhibit structural colors according to Bragg's law. Photonic crystals have aroused much attention due to their unique properties, such as multiresponsive colors, highly ordered structures, and nonbleachable structural colors which covered the entire visible range. Photonic crystals can change their colors provided that the lattice constants of crystal structures were varied according to, for example, temperature-induced volume variation and electro-induced redox reactions (Seeboth et al., 2014).

As shown in Fig. 8.9A, a full-color photonic crystal device was designed and fabricated in a form of electrochemical cell (Arsenault et al., 2007). The device was based on a two-component composite, that is, silica microspheres and polyferrocenylsilane (PFS). The silica microspheres with an ordered self-assembly array were deposited on ITO glass substrate and served as the inactive structural scaffold. PFS with pendant unsaturated C═C bonds were incorporated into the lattice spacing of silica microspheres and polymerized under UV irradiation through a thiol-ene reaction. During the electrochemical oxidation reaction in PFS, electrons were drawn out of the PFS backbone, while anions from the organic solvent-based liquid electrolyte were driven into the PFS to neutralize the positive charge. This influx of both solvent and ions into the polymer caused the whole photonic crystal structure to swell and thus red-shifted the reflected peak. The color changes could cover the entire visible range if the sizes of silica microspheres and the cross-linker content were appropriate. By increasing the voltage, the device could change color from blue to green and green to red at 1.5 and 2.0 V, respectively (Fig. 8.9B). The electrically tunable photonic crystal electrochemical cell is promising for the application of full-color displayers.

Although full-color photonic crystal device has been realized, there are still some problems including sluggish switching process and high trigger voltage, which are derived from the fact that the close-packed silica microspheres and polymer matrix impeded the influx of the ions and solvent. To this end, a high-performance electroactive device was further fabricated based on an inverse polymer-gel opal (Puzzo et al., 2009). Two PFS derivatives were infiltrated into the voids among the monodisperse silica microspheres array that were deposited on ITO glass (Fig. 8.9C). After the polymerization, the silica microspheres were etched by HF and then a free-standing inverse polymer-gel opal was obtained. Then, the inverse

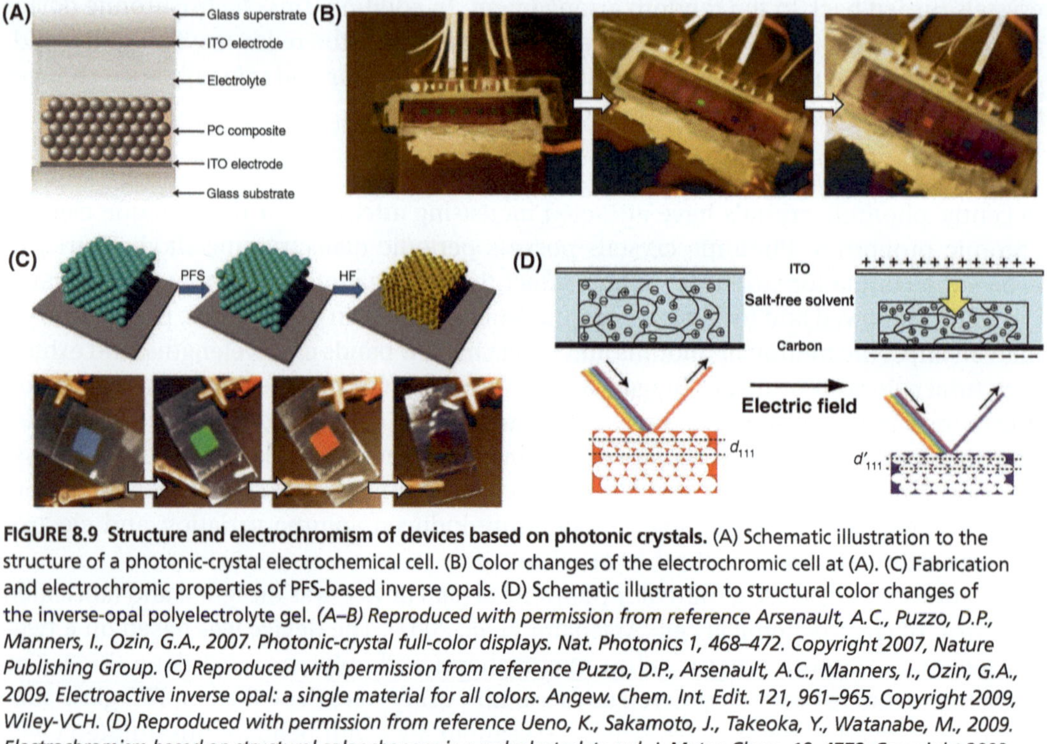

FIGURE 8.9 Structure and electrochromism of devices based on photonic crystals. (A) Schematic illustration to the structure of a photonic-crystal electrochemical cell. (B) Color changes of the electrochromic cell at (A). (C) Fabrication and electrochromic properties of PFS-based inverse opals. (D) Schematic illustration to structural color changes of the inverse-opal polyelectrolyte gel. *(A–B) Reproduced with permission from reference Arsenault, A.C., Puzzo, D.P., Manners, I., Ozin, G.A., 2007. Photonic-crystal full-color displays. Nat. Photonics 1, 468–472. Copyright 2007, Nature Publishing Group. (C) Reproduced with permission from reference Puzzo, D.P., Arsenault, A.C., Manners, I., Ozin, G.A., 2009. Electroactive inverse opal: a single material for all colors. Angew. Chem. Int. Edit. 121, 961–965. Copyright 2009, Wiley-VCH. (D) Reproduced with permission from reference Ueno, K., Sakamoto, J., Takeoka, Y., Watanabe, M., 2009. Electrochromism based on structural color changes in a polyelectrolyte gel. J. Mater. Chem. 19, 4778. Copyright 2009, The Royal Society of Chemistry.*

polymer-gel opal was integrated into a button cell. This inverse polymer-gel opal cell exhibited gradual color changes through blue, green to red under increasing voltages from 1.2 to 2.8 V. In particular, the color changed rapidly due to the decreased diffusion length of ions in the ordered macropores and mesopores of inverse polymer-gel opal structure.

Another problem for photonic crystals is the limited reversibility of stop-band tuning. Although there are several reports to date on stable stop-band tuning, for example, 30 times in nonvolatile display pixels based on the block-copolymer photonic gel, the performance is not high enough (Hwang et al., 2011). The performance of switching reliability could be greatly enhanced by using tuneable colloidal photonic crystals with an ion exchange resin-modified ITO electrode (Han et al., 2014). Colloidal photonic crystal with nonclose-packed array was made from monodisperse polystyrene (PS) microspheres with a diameter of 138 nm which were dispersed in deionized water. The device transited its color from red to green and further to blue under increasing voltages from 0 to 3.2 V. The increased voltage forced colloid nanospheres move toward anode and increased the concentration of nanospheres, resulting in a change of lattice constant of colloid photonic crystal. In addition, this color-tuning switching was much more stable than that of other reports and allowed high cycle performance over 800 times, due to the thin layer of ion exchange resins overcoated on the ITO surface.

There are some other reports about color-tuning photonic crystals or inverse opals, for example, polypyrrole inverse opals with electrically color switching (Xu et al., 2008). Almost all of these reported structures are caused by the volume variation (Puzzo et al., 2009). However, instead of the whole volume, the structural color change could also be caused by the variation of thickness in the polymer-gel layer (Ueno et al., 2009). The inverse polymer-gel opals were fabricated based on the closely packed colloidal PS crystal template. To investigate the electrochromic property of this inverse polymer-gel opal, the inverse opal was placed between two electrodes. When a voltage of 9 V was applied upon, the cell changed color from green to blue due to the reduced lattice spacing (d_{111}) of photonic crystal, caused by the contraction of inverse polymer-gel along the electric-field direction (Fig. 8.9D). Such an interesting anisotropic deformation of the inverse-opal polyelectrolyte gels under a voltage can be applicable to many sensors.

8.2.4 Conclusions and Perspectives

Electrochromic devices that can reversibly change colors in response to electric field are promising in various fields including smart windows, sensors, displays, and military camouflage. The color changes of the electrochromic devices are generally based on the oxidation–reduction or electrothermal chromic mechanism. The conjugated polymers are mainly used as electrochromic materials. The devices based on electrochemical oxidation–reduction mechanism could work under very low voltages. However, these devices are always made of complicated layered structure. In addition, the low response speed, weak cyclic life, and limited colors have seriously hindered their advancements. Although more conjugated polymers and their composites which exhibit richer colors under different states have been investigated, the colors that cover entire visible light are still difficult to achieve. The photonic crystals have attracted increasing attentions due to their unique photonic band gap and exhibit vivid structural colors from infrared to ultraviolet light. As expected, it is readily achieved for any color by using photonic crystals as electrochromic layer. The electrothermal-chromic devices have simple structures with conductor and thermochromic layers. However, these devices always work at high voltage and high temperature and suffered from the disadvantage of short lifetime. To this end, it is crucial to develop novel electrochromic devices with enrich colors, rapid response speed, long cyclic life, as well as low voltage.

References

Agrawal, R.C., Pandey, G.P., 2008. Solid polymer electrolytes: materials designing and all-solid-state battery applications: an overview. J. Phys. D 41, 223001–223018.

Ahn, D.J., Chae, E.H., Lee, G.S., Shim, H.Y., Chang, T.E., Ahn, K.D., Kim, J.M., 2003. Colorimetric reversibility of polydiacetylene supramolecules having enhanced hydrogen-bonding under thermal and pH stimuli. J. Am. Chem. Soc. 125, 8976–8977.

Ahn, B.Y., Duoss, E.B., Motala, M.J., Guo, X.Y., Park, S.I., Xiong, Y.J., Yoon, J., Nuzzo, R.G., Rogers, J.A., Lewis, J.A., 2009. Omnidirectional printing of flexible, stretchable, and spanning silver microelectrodes. Science 323, 1590–1593.

Amb, C.M., Beaujuge, P.M., Reynolds, J.R., 2010. Spray-processable blue-to-highly transmissive switching polymer electrochromes via the donor-acceptor approach. Adv. Mater. 22, 724–728.

Anderson, I.A., Gisby, T.A., Mckay, T.G., O'brien, B.M., Calius, E.P., 2012. Multi-functional dielectric elastomer artificial muscles for soft and smart machines. J. Appl. Phys. 112, 041101.

Argun, A.A., Cirpan, A., Reynolds, J.R., 2003. The first truly all-polymer electrochromic devices. Adv. Mater. 15, 1338–1341.

Arsenault, A.C., Puzzo, D.P., Manners, I., Ozin, G.A., 2007. Photonic-crystal full-color displays. Nat. Photonics 1, 468–472.

Balan, A., Gunbas, G., Durmus, A., Toppare, L., 2008. Donor—acceptor polymer with benzotriazole moiety: enhancing the electrochromic properties of the "donor unit". Chem. Mater. 20, 7510–7513.

Bauer, S., Bauer-Gogonea, S., Graz, I., Kaltenbrunner, M., Keplinger, C., Schwodiauer, R., 2014. 25th anniversary article: a soft future: from robots and sensor skin to energy harvesters. Adv. Mater. 26, 149–161.

Baughman, R.H., 1996. Conducting polymer artificial muscles. Synth. Met. 78, 339–353.

Baughman, R.H., 2003. Materials science—muscles made from metal. Science 300, 268–269.

Baughman, R.H., Cui, C.X., Zakhidov, A.A., Iqbal, Z., Barisci, J.N., Spinks, G.M., Wallace, G.G., Mazzoldi, A., De Rossi, D., Rinzler, A.G., Jaschinski, O., Roth, S., Kertesz, M., 1999. Carbon nanotube actuators. Science 284, 1340–1344.

Bay, L., West, K., Sommer-Larsen, P., Skaarup, S., Benslimane, M., 2003. A conducting polymer artificial muscle with 12% linear strain. Adv. Mater. 15, 310–313.

Beaujuge, P.M., Ellinger, S., Reynolds, J.R., 2008a. The donor-acceptor approach allows a black-to-transmissive switching polymeric electrochrome. Nat. Mater. 7, 795–799.

Beaujuge, P.M., Ellinger, S., Reynolds, J.R., 2008b. Spray processable green to highly transmissive electrochromics via chemically polymerizable donor-acceptor heterocyclic pentamers. Adv. Mater. 20, 2772–2776.

Beaujuge, P.M., Amb, C.M., Reynolds, J.R., 2010. Spectral engineering in π-conjugated polymers with intramolecular donor—acceptor interactions. Acc. Chem. Res. 43, 1396–1407.

Bennett, M.D., Leo, D.J., 2004. Ionic liquids as novel solvents for ionic polymer transducers. Smart Structures And Materials: Electroactive Polymer Actuators And Devices (Eapad), vol. 5385. pp. 210–220.

Bhandari, S., Deepa, M., Srivastava, A.K., Lal, C., Kant, R., 2008. Poly (3, 4-ethylenedioxythiophene) (PEDOT)-Coated MWCNTs tethered to conducting substrates: facile electrochemistry and enhanced coloring efficiency. Macromol. Rapid Commun. 29, 1959–1964.

Bhandari, S., Deepa, M., Sharma, S.N., Joshi, A.G., Srivastava, A.K., Kant, R., 2010. Charge transport and electrochromism in novel nanocomposite films of poly (3, 4-ethylenedioxythiophene)-Au nanoparticles—CdSe quantum dots. J. Phys. Chem. C 114, 14606–14613.

Blighe, F.M., Diamond, D., Coleman, J.N., Lahiff, E., 2012. Increased response/recovery lifetimes and reinforcement of polyaniline nanofiber films using carbon nanotubes. Carbon 50, 1447–1454.

Bloomquist, D.R., Pressprich, M.R., Willett, R.D., 1988. Thermochromism in copper(II) halide salts. 4. Bis(diethylammonium)tetrachlorocuprate(II), structure of the high-temperature phase and physical characterization of its two phases. J. Am. Chem. Soc. 110, 7391–7398.

Booth, J.M., Casey, P.S., 2009. Production of VO2 M1 and M2 nanoparticles and composites and the influence of the substrate on the structural phase transition. ACS Appl. Mater. Interfaces 1, 1899–1905.

Brochu, P., Pei, Q., 2010. Advances in dielectric elastomers for actuators and artificial muscles. Macromol. Rapid Comm. 31, 10–36.

Cakmak, E., Fang, X., Yildiz, O., Bradford, P.D., Ghosh, T.K., 2015. Carbon nanotube sheet electrodes for anisotropic actuation of dielectric elastomers. Carbon 89, 113–120.

Carpi, F., Kornbluh, R., Sommer-Larsen, P., Alici, G., 2011. Electroactive polymer actuators as artificial muscles: are they ready for bioinspired applications? Bioinspir. Biomim. 6, 045006.

Charych, D.H., Nagy, J.O., Spevak, W., Bednarski, M.D., 1993. Direct colorimetric detection of a receptor-ligand interaction by a polymerized bilayer assembly. Science 261, 585–588.

Chen, L.Z., Liu, C.H., Hu, C.H., Fan, S.S., 2008. Electrothermal actuation based on carbon nanotube network in silicone elastomer. Appl. Phys. Lett. 92, 263104.

Chen, X., Kang, S., Kim, M.J., Kim, J., Kim, Y.S., Kim, H., Chi, B., Kim, S.J., Lee, J.Y., Yoon, J., 2010. Thin-film formation of imidazolium-based conjugated polydiacetylenes and their application for sensing anionic surfactants. Angew. Chem. Int. Edit. 49, 1422–1425.

Chen, L.Z., Liu, C.H., Liu, K., Meng, C.Z., Hu, C.H., Wang, J.P., Fan, S.S., 2011. High-performance, low-voltage, and easy-operable bending actuator based on aligned carbon nanotube/polymer composites. ACS Nano 5, 1588–1593.

Chen, X., Lin, H., Chen, P., Guan, G., Deng, J., Peng, H., 2014a. Smart, stretchable supercapacitors. Adv. Mater. 26, 4444–4449.

Chen, X., Lin, H., Deng, J., Zhang, Y., Sun, X., Chen, P., Fang, X., Zhang, Z., Guan, G., Peng, H., 2014b. Electrochromic fiber-shaped supercapacitors. Adv. Mater. 26, 8126–8132.

Chen, P.N., He, S.S., Xu, Y.F., Sun, X.M., Peng, H.S., 2015a. Electromechanical actuator ribbons driven by electrically conducting spring-like fibers. Adv. Mater. 27, 4982–4988.

Chen, P.N., Xu, Y.F., He, S.S., Sun, X.M., Guo, W.H., Zhang, Z.T., Qiu, L.B., Li, J.F., Chen, D.Y., Peng, H.S., 2015b. Biologically inspired, sophisticated motions from helically assembled, conducting fibers. Adv. Mater. 27, 1042–1047.

Choi, C., Lee, J.A., Choi, A.Y., Kim, Y.T., Lepro, X., Lima, M.D., Baughman, R.H., Kim, S.J., 2014. Flexible supercapacitor made of carbon nanotube yarn with internal pores. Adv. Mater. 26, 2059–2065.

De Volder, M.F.L., Tawfick, S.H., Baughman, R.H., Hart, A.J., 2013. Carbon nanotubes: present and future commercial applications. Science 339, 535–539.

Detsi, E., Onck, P., de Hosson, J.T.M., 2013. Metallic muscles at work: high rate actuation in nanoporous gold/polyaniline composites. ACS Nano 7, 4299–4306.

Ding, J., Liu, L., Spinks, G.M., Zhou, D.Z., Wallace, G.G., Gillespie, J., 2003. High performance conducting polymer actuators utilising a tubular geometry and helical wire interconnects. Synth. Met. 138, 391–398.

Durmus, A., Gunbas, G.E., Camurlu, P., Toppare, L., 2007. A neutral state green polymer with a superior transmissive light blue oxidized state. Chem. Commun., 3246–3248.

Dyer, A.L., Craig, M.R., Babiarz, J.E., Kiyak, K., Reynolds, J.R., 2010. Orange and red to transmissive electrochromic polymers based on electron-rich dioxythiophenes. Macromolecules 43, 4460–4467.

Foroughi, J., Spinks, G.M., Wallace, G.G., Oh, J., Kozlov, M.E., Fang, S.L., Mirfakhrai, T., Madden, J.D.W., Shin, M.K., Kim, S.J., Baughman, R.H., 2011. Torsional carbon nanotube artificial muscles. Science 334, 494–497.

Gao, J., Liu, D.G., Sansi Ena, J.M., Wang, H.L., 2004. Synthesis and characterization of electrochromic polyamides with well-defined molecular structures and redox properties. Adv. Funct. Mater. 14, 537–543.

Garnier, F., Tourillon, G., Gazard, M., Dubois, J., 1983. Organic conducting polymers derived from substituted thiophenes as electrochromic material. J. Electroanal. Chem. Interfacial Electrochem. 148, 299–303.

Gazard, M., Dubois, J., Champagne, M., Garnier, F., Tourillon, G., 1983. Electrooptical properties of thin film of polyheterocycles. J. Phys. Colloq. 44, C3-537–C3-542.

Gu, Y., Cao, W., Zhu, L., Chen, D., Jiang, M., 2008. Polymer mortar assisted self-assembly of nanocrystalline polydiacetylene bricks showing reversible thermochromism. Macromolecules 41, 2299–2303.

Gunbas, G.E., Durmus, A., Toppare, L., 2008. Could green be greener? Novel donor–acceptor-type electrochromic polymers: towards excellent neutral green materials with exceptional transmissive oxidized states for completion of RGB color space. Adv. Mater. 20, 691–695.

Guo, W.H., Liu, C., Zhao, F.Y., Sun, X.M., Yang, Z.B., Chen, T., Chen, X.L., Qiu, L.B., Hu, X.H., Peng, H.S., 2012. A novel electromechanical actuation mechanism of a carbon nanotube fiber. Adv. Mater. 24, 5379–5384.

Guo, H., Zhang, J., Porter, D., Peng, H., L Wik, D.W.P.M., Wang, Y., Zhang, Z., Chen, X., Shao, Z., 2014. Ultrafast and reversible thermochromism of a conjugated polymer material based on the assembly of peptide amphiphiles. Chem. Sci. 5, 4189–4195.

Haines, C.S., Lima, M.D., Li, N., Spinks, G.M., Foroughi, J., Madden, J.D.W., Kim, S.H., Fang, S.L., De Andrade, M.J., Goktepe, F., Goktepe, O., Mirvakili, S.M., Naficy, S., Lepro, X., Oh, J., Kozlov, M.E., Kim, S.J., Xu, X.R., Swedlove, B.J., Wallace, G.G., Baughman, R.H., 2014. Artificial muscles from fishing line and sewing thread. Science 343, 868–872.

Han, M.G., Heo, C.J., Shim, H., Shin, C.G., Lim, S.J., Kim, J.W., Jin, Y.W., Lee, S., 2014. Structural color manipulation using tunable photonic crystals with enhanced switching reliability. Adv. Opt. Mater. 2, 535–541.

Hara, S., Zama, T., Takashima, W., Kaneto, K., 2005. Free-standing polypyrrole actuators with response rate of 10.8% s(−1). Synth. Met. 149, 199–201.

Higgins, S.J., Lovell, K.V., Rajapakse, R.M.G., Walsby, N.M., 2003. Grafting and electrochemical characterisation of poly-(3,4-ethylenedioxythiophene) films, on Nafion and on radiation-grafted polystyrenesulfonate-polyvinylidene fluoride composite surfaces. J. Mater. Chem. 13, 2485–2489.

Hu, Y., Chen, W., Lu, L.H., Liu, J.H., Chang, C.R., 2010. Electromechanical actuation with controllable motion based on a single-walled carbon nanotube and natural biopolymer composite. ACS Nano 4, 3498–3502.

Hu, W., Zhang, S.N., Niu, X., Liu, C., Pei, Q., 2014. An aluminum nanoparticle–acrylate copolymer nanocomposite as a dielectric elastomer with a high dielectric constant. J. Mater. Chem. C 2, 1658.

Hwang, K., Kwak, D., Kang, C., Kim, D., Ahn, Y., Kang, Y., 2011. Electrically tunable hysteretic photonic gels for nonvolatile display pixels. Angew. Chem. Int. Edit. 50, 6311–6314.

Invernale, M.A., Seshadri, V., Mamangun, D.M.D., Ding, Y., Filloramo, J., Sotzing, G.A., 2009. Polythieno[3,4-b]thiophene as an optically transparent ion-storage layer. Chem. Mater. 21, 3332–3336.

Jager, E.W.H., Inganas, O., Lundstrom, I., 2000a. Microrobots for micrometer-size objects in aqueous media: potential tools for single-cell manipulation. Science 288, 2335–2338.

Jager, E.W.H., Smela, E., Inganas, O., 2000b. Microfabricating conjugated polymer actuators. Science 290, 1540–1545.

Jensen, J., Krebs, F.C., 2014. From the bottom up—flexible solid state electrochromic devices. Adv. Mater. 26, 7231–7234.

Jeong, G.S., Baek, D.H., Jung, H.C., Song, J.H., Moon, J.H., Hong, S.W., Kim, I.Y., Lee, S.H., 2012. Solderable and electroplatable flexible electronic circuit on a porous stretchable elastomer. Nat. Commun. 3, 10.

Kerszulis, J.A., Amb, C.M., Dyer, A.L., Reynolds, J.R., 2014. Follow the yellow brick road: structural optimization of vibrant yellow-to-transmissive electrochromic conjugated polymers. Macromolecules 47, 5462–5469.

Kim, Y., Zhu, J., Yeom, B., Di Prima, M., Su, X.L., Kim, J.G., Yoo, S.J., Uher, C., Kotov, N.A., 2013. Stretchable nanoparticle conductors with self-organized conductive pathways. Nature 500, 59–63.

Kong, L.R., Chen, W., 2014. Carbon nanotube and graphene-based bioinspired electrochemical actuators. Adv. Mater. 26, 1025–1043.

Kussmaul, B., Risse, S., Kofod, G., Wach, R., Wegener, M., Mccarthy, D.N., Kr Ger, H., Gerhard, R., 2011. Enhancement of dielectric permittivity and electromechanical response in silicone elastomers: molecular grafting of organic dipoles to the macromolecular network. Adv. Funct. Mater. 21, 4589–4594.

Kwak, G., Fukao, S., Fujiki, M., Sakaguchi, T., Masuda, T., 2006. Temperature-dependent, static, and dynamic fluorescence properties of disubstituted acetylene polymer films. Chem. Mater. 18, 2081–2085.

Kwak, G., Lee, W.-E., Jeong, H., Sakaguchi, T., Fujiki, M., 2008. Swelling-induced emission enhancement in substituted acetylene polymer film with large fractional free volume: fluorescence response to organic solvent stimuli. Macromolecules 42, 20–24.

Laforgue, A., 2010. Electrically controlled color-changing textiles using the resistive heating properties of PEDOT nanofibers. J. Mater. Chem. 20, 8233.

Laforgue, A., Rouget, G., Dubost, S., Champagne, M.F., Robitaille, L., 2012. Multifunctional resistive-heating and color-changing monofilaments produced by a single-step coaxial melt-spinning process. ACS Appl. Mater. Interfaces 4, 3163–3168.

Lee, J.A., Kim, Y.T., Spinks, G.M., Suh, D., Lepro, X., Lima, M.D., Baughman, R.H., Kim, S.J., 2014. All-solid-state carbon nanotube torsional and tensile artificial muscles. Nano Lett 14, 2664–2669.

Li, J.Z., Ma, W.J., Song, L., Niu, Z.G., Cai, L., Zeng, Q.S., Zhang, X.X., Dong, H.B., Zhao, D., Zhou, W.Y., Xie, S.S., 2011. Superfast-response and ultrahigh-power-density electromechanical actuators based on hierarchal carbon nanotube electrodes and chitosan. Nano Lett 11, 4636–4641.

Li, K., Zhang, Q., Wang, H., Li, Y., 2014. Red, green, blue (RGB) electrochromic fibers for the new smart color change fabrics. ACS Appl. Mater. Interfaces 6, 13043–13050.

Liang, J.J., Huang, L., Li, N., Huang, Y., Wu, Y.P., Fang, S.L., Oh, J., Kozlov, M., Ma, Y.F., Li, F.F., Baughman, R., Chen, Y.S., 2012. Electromechanical actuator with controllable motion, fast response rate, and high-frequency resonance based on graphene and polydiacetylene. ACS Nano 6, 4508–4519.

Lima, M.D., Li, N., De Andrade, M.J., Fang, S.L., Oh, J., Spinks, G.M., Kozlov, M.E., Haines, C.S., Suh, D., Foroughi, J., Kim, S.J., Chen, Y.S., Ware, T., Shin, M.K., Machado, L.D., Fonseca, A.F., Madden, J.D.W., Voit, W.E., Galvao, D.S., Baughman, R.H., 2012. Electrically, chemically, and photonically powered torsional and tensile actuation of hybrid carbon nanotube yarn muscles. Science 338, 928–932.

Lin, K.W., Ming, S.L., Zhen, S.J., Zhao, Y., Lu, B.Y., Xu, J.K., 2015. Molecular design of DBT/DBF hybrid thiophenes pi-conjugated systems and comparative study of their electropolymerization and optoelectronic properties: from comonomers to electrochromic polymers. Polym. Chem. 6, 4575–4587.

Liu, J., Wang, Z., Xie, X.J., Cheng, H.H., Zhao, Y., Qu, L.T., 2012. A rationally-designed synergetic polypyrrole/graphene bilayer actuator. J. Mater. Chem. 22, 4015–4020.

Liu, Z.F., Fang, S., Moura, F.A., Ding, J.N., Jiang, N., Di, J., Zhang, M., Lepro, X., Galvao, D.S., Haines, C.S., Yuan, N.Y., Yin, S.G., Lee, D.W., Wang, R., Wang, H.Y., Lv, W., Dong, C., Zhang, R.C., Chen, M.J., Yin, Q., Chong, Y.T., Zhang, R., Wang, X., Lima, M.D., Ovalle-Robles, R., Qian, D., Lu, H., Baughman, R.H., 2015. Hierarchically buckled sheath-core fibers for superelastic electronics, sensors, and muscles. Science 349, 400–404.

Lu, L.H., Chen, W., 2010. Biocompatible composite actuator: a supramolecular structure consisting of the biopolymer chitosan, carbon nanotubes, and an ionic liquid. Adv. Mater. 22, 3745–3748.

Lu, Y., Yang, Y., Sellinger, A., Lu, M., Huang, J., Fan, H., Haddad, R., Lopez, G., Burns, A.R., Sasaki, D.Y., 2001. Self-assembly of mesoscopically ordered chromatic polydiacetylene/silica nanocomposites. Nature 410, 913–917.

Lu, L.H., Liu, J.H., Hu, Y., Zhang, Y.W., Chen, W., 2013. Graphene-stabilized silver nanoparticle electrochemical electrode for actuator design. Adv. Mater. 25, 1270–1274.

Ma, K.Y., Chirarattananon, P., Fuller, S.B., Wood, R.J., 2013. Controlled flight of a biologically inspired, insect-scale robot. Science 340, 603–607.

Ma, D., Shi, G., Wang, H., Zhang, Q., Li, Y., 2014. Controllable growth of high-quality metal oxide/conducting polymer hierarchical nanoarrays with outstanding electrochromic properties and solar-heat shielding ability. J. Mater. Chem. A 2, 13541.

Madden, J.D., 2007. Mobile robots: motor challenges and materials solutions. Science 318, 1094–1097.

Madden, J.D., Cush, R.A., Kanigan, T.S., Hunter, I.W., 2000. Fast contracting polypyrrole actuators. Synth. Met. 113, 185–192.

Madden, J.D.W., Vandesteeg, N.A., Anquetil, P.A., Madden, P.G.A., Takshi, A., Pytel, R.Z., Lafontaine, S.R., Wieringa, P.A., Hunter, I.W., 2004. Artificial muscle technology: physical principles and naval prospects. IEEE J. Ocean. Eng. 29, 706–728.

Manuel Stephan, A., Nahm, K.S., 2006. Review on composite polymer electrolytes for lithium batteries. Polymer 47, 5952–5964.

Meng, F., Zhang, X., Xu, G., Yong, Z., Chen, H., Chen, M., Li, Q., Zhu, Y., 2011. Carbon nanotube composite films with switchable transparency. ACS Appl. Mater. Interfaces 3, 658–661.

Miao, M.H., 2011. Electrical conductivity of pure carbon nanotube yarns. Carbon 49, 3755–3761.

Mirfakhrai, T., Madden, J.D.W., Baughman, R.H., 2007. Polymer artificial muscles. Mater. Today 10, 30–38.

Mirvakili, S.M., Pazukha, A., Sikkema, W., Sinclair, C.W., Spinks, G.M., Baughman, R.H., Madden, J.D.W., 2013. Niobium nanowire yarns and their application as artificial muscles. Adv. Funct. Mater. 23, 4311–4316.

Molberg, M., Crespy, D., Rupper, P., N Esch, F., M Nson, J.-A.E., L We, C., Opris, D.M., 2010. High breakdown field dielectric elastomer actuators using encapsulated polyaniline as high dielectric constant filler. Adv. Funct. Mater. 20, 3280–3291.

Nambo, M., Crudden, C.M., 2014. Modular synthesis of triarylmethanes through palladium-catalyzed sequential arylation of methyl phenyl sulfone. Angew. Chem. Int. Edit. 53, 742–746.

Namboothiry, M.A., Zimmerman, T., Coldren, F.M., Liu, J., Kim, K., Carroll, D.L., 2007. Electrochromic properties of conducting polymer metal nanoparticles composites. Synth. Met. 157, 580–584.

Nemat-Nasser, S., 2002. Micromechanics of actuation of ionic polymer-metal composites. J. Appl. Phys. 92, 2899–2915.

Nguyen, C.A., Argun, A.A., Hammond, P.T., Lu, X., Lee, P.S., 2011. Layer-by-layer assembled solid polymer electrolyte for electrochromic devices. Chem. Mater. 23, 2142–2149.

Nicoletta, F.P., Chidichimo, G., Cupelli, D., De Filpo, G., De Benedittis, M., Gabriele, B., Salerno, G., Fazio, A., 2005. Electrochromic polymer-dispersed liquid-crystal film: a new bifunctional device. Adv. Funct. Mater. 15, 995–999.

Niklasson, G.A., Granqvist, C.G., 2007. Electrochromics for smart windows: thin films of tungsten oxide and nickel oxide, and devices based on these. J. Mater. Chem. 17, 127–156.

Opris, D.M., Molberg, M., Walder, C., Ko, Y.S., Fischer, B., N Esch, F.A., 2011. New silicone composites for dielectric elastomer actuator applications in competition with acrylic foil. Adv. Funct. Mater. 21, 3531–3539.

Otero, T.F., Angulo, E., Rodriguez, J., Santamaria, C., 1992. Electrochemomechanical properties from a bilayer—polypyrrole nonconducting and flexible material artificial muscle. J. Electroanal. Chem. 341, 369–375.

Pei, Q.B., Inganas, O., 1992. Conjugated polymers and the bending cantilever method—electrical muscles and smart devices. Adv. Mater. 4, 277–278.

Pelrine, R., Kornbluh, R., Pei, Q.B., Joseph, J., 2000. High-speed electrically actuated elastomers with strain greater than 100%. Science 287, 836–839.

Peng, H., Sun, X., Cai, F., Chen, X., Zhu, Y., Liao, G., Chen, D., Li, Q., Lu, Y., Zhu, Y., Jia, Q., 2009. Electrochromatic carbon nanotube/polydiacetylene nanocomposite fibres. Nat. Nanotechnol. 4, 738–741.

Puzzo, D.P., Arsenault, A.C., Manners, I., Ozin, G.A., 2009. Electroactive inverse opal: a single material for all colors. Angew. Chem. Int. Edit. 121, 961–965.

Randin, J.P., 1982. Ion-containing polymers as semisolid electrolytes in WO 3-based electrochromic devices. J. Electrochem. Soc. 129, 1215–1220.

Reeves, B.D., Grenier, C.R.G., Argun, A.A., Cirpan, A., Mccarley, T.D., Reynolds, J.R., 2004. Spray coatable electrochromic dioxythiophene polymers with high coloration efficiencies. Macromolecules 37, 7559–7569.

Remmele, J., Shen, D.E., Mustonen, T., Fruehauf, N., 2015. High performance and long-term stability in ambiently fabricated segmented solid-state polymer electrochromic displays. ACS Appl. Mater. Interfaces 7, 12001–12008.

Ren, Y., Jager, W.F., Neckers, D.C., 1996. Simultaneous photoinduced color formation and polymerization: the formation of highly colored thin films from colorless precursors due to formation of triarylmethane cations by acid generating decomposition of iodonium salts1. Macromolecules 29, 3751–3757.

Risse, S., Kussmaul, B., Kr Ger, H., Kofod, G., 2012. Synergistic improvement of actuation properties with compatibilized high permittivity filler. Adv. Funct. Mater. 22, 3958–3962.

Rosset, S., Niklaus, M., Dubois, P., Shea, H.R., 2009. Metal ion implantation for the fabrication of stretchable electrodes on elastomers. Adv. Funct. Mater. 19, 470–478.

Seeboth, A., Lotzsch, D., Ruhmann, R., Muehling, O., 2014. Thermochromic polymers—function by design. Chem. Rev. 114, 3037–3068.

Sellinger, A.T., Wang, D.H., Tan, L.S., Vaia, R.A., 2010. Electrothermal polymer nanocomposite actuators. Adv. Mater. 22, 3430–3435.

Seo, D.K., Kang, T.J., Kim, D.W., Kim, Y.H., 2012. Twistable and bendable actuator: a CNT/polymer sandwich structure driven by thermal gradient. Nanotechnology 23, 75501–75507.

Shahinpoor, M., Kim, K.J., 2002. Mass transfer induced hydraulic actuation in ionic polymer-metal composites. J. Intell. Mater. Syst. Struct. 13, 369–376.

Sheng, K., Bai, H., Sun, Y., Li, C., Shi, G., 2011. Layer-by-layer assembly of graphene/polyaniline multilayer films and their application for electrochromic devices. Polymer 52, 5567–5572.

Shi, P., Amb, C.M., Knott, E.P., Thompson, E.J., Liu, D.Y., Mei, J., Dyer, A.L., Reynolds, J.R., 2010. Broadly absorbing black to transmissive switching electrochromic polymers. Adv. Mater. 22, 4949–4953.

Shin, H., Yoon, B., Park, I.S., Kim, J.M., 2014. An electrothermochromic paper display based on colorimetrically reversible polydiacetylenes. Nanotechnology 25, 094011.

Smela, E., Inganas, O., Pei, Q.B., Lundstrom, I., 1993. Electrochemical muscles—micromachining fingers and corkscrews. Adv. Mater. 5, 630–632.

Smela, E., Inganas, O., Lundstrom, I., 1995. Controlled folding of micrometer-size structures. Science 268, 1735–1738.

Sonavane, A., Inamdar, A., Dalavi, D., Deshmukh, H., Patil, P., 2010a. Simple and rapid synthesis of NiO/PPy thin films with improved electrochromic performance. Electrochim. Acta 55, 2344–2351.

Sonavane, A., Inamdar, A., Deshmukh, H., Patil, P., 2010b. Multicolored electrochromic thin films of NiO/PANI. J. Phys. D. 43, 315102.

Spinks, G.M., Mottaghitalab, V., Bahrami-Samani, M., Whitten, P.G., Wallace, G.G., 2006. Carbon-nanotube-reinforced polyaniline fibers for high-strength artificial muscles. Adv. Mater. 18, 637–640.

Stoyanov, H., Kollosche, M., Risse, S., Mccarthy, D.N., Kofod, G., 2011. Elastic block copolymer nanocomposites with controlled interfacial interactions for artificial muscles with direct voltage control. Soft Matter 7, 194–202.

Sun, X., Zhang, Z., Lu, X., Guan, G., Li, H., Peng, H., 2013. Electric current test paper based on conjugated polymers and aligned carbon nanotubes. Angew. Chem. Int. Edit. 52, 7776–7780.

Thakur, V.K., Ding, G., Ma, J., Lee, P.S., Lu, X., 2012. Hybrid materials and polymer electrolytes for electrochromic device applications. Adv. Mater. 24, 4071–4096.

Thompson, B.C., Schottland, P., Zong, K., Reynolds, J.R., 2000. In situ colorimetric analysis of electrochromic polymers and devices. Chem. Mater. 12, 1563–1571.

Tiwari, R., Garcia, E., 2011. The state of understanding of ionic polymer metal composite architecture: a review. Smart Mater. Struct. 20, 83001–83016.

Torop, J., Aabloo, A., Jager, E.W.H., 2014. Novel actuators based on polypyrrole/carbide-derived carbon hybrid materials. Carbon 80, 387–395.

Ueno, K., Sakamoto, J., Takeoka, Y., Watanabe, M., 2009. Electrochromism based on structural color changes in a polyelectrolyte gel. J. Mater. Chem. 19, 4778.

Vasilyeva, S.V., Unur, E., Walczak, R.M., Donoghue, E.P., Rinzler, A.G., Reynolds, J.R., 2009. Color purity in polymer electrochromic window devices on indium—tin oxide and single-walled carbon nanotube electrodes. ACS Appl. Mater. Interfaces 1, 2288–2297.

Wang, L.X., Soczka-Guth, T., Havinga, E., M Llen, K., 1996. Poly (phenylenesulfidephenylenamine) (PPSA)—the "compound" of polyphenylenesulfide with polyaniline. Angew. Chem. Int. Edit. 35, 1495–1497, (in English).

Wang, K., Wu, H., Meng, Y., Zhang, Y., Wei, Z., 2012. Integrated energy storage and electrochromic function in one flexible device: an energy storage smart window. Energy Environ. Sci. 5, 8384.

Warren, M.R., Madden, J.D., 2006. Electrochemical switching of conducting polymers: a variable resistance transmission line model. J. Electroanal. Chem. 590, 76–81.

Wegner, G., 1969. Topochemische reaktionen von monomeren mit konjugierten dreifachbindungen I. mitt.: polymerisation von derivaten des 2.4-hexadiin-1.6 diols im kristallinen zustand. Z. Naturforsch. 246, 824.

Wu, C.G., Lu, M.I., Chang, S.J., Wei, C.S., 2007. A solution-processable high-coloration-efficiency low-switching-voltage electrochromic polymer based on polycyclopentadithiophene. Adv. Funct. Mater. 17, 1063–1070.

Wu, S., Niu, L., Shen, J., Zhang, Q., Bubeck, C., 2009. Aggregation-induced reversible thermochromism of novel azo chromophore-functionalized polydiacetylene cylindrical micelles. Macromolecules 42, 362–367.

Xia, X., Tu, J., Zhang, J., Huang, X., Wang, X., Zhang, W., Huang, H., 2009. Multicolor and fast electrochromism of nanoporous NiO/poly (3, 4-ethylenedioxythiophene) composite thin film. Electrochem. Commun. 11, 702–705.

Xu, L., Wang, J.X., Song, Y.L., Jiang, L., 2008. Electrically tunable polypyrrole inverse opals with switchable stopband, conductivity, and wettability. Chem. Mater. 20, 3554–3556.

Yamada, T., Hayamizu, Y., Yamamoto, Y., Yomogida, Y., Izadi-Najafabadi, A., Futaba, D.N., Hata, K., 2011. A stretchable carbon nanotube strain sensor for human-motion detection. Nat. Nanotechnol. 6, 296–301.

Yang, Z.B., Ren, J., Zhang, Z.T., Chen, X.L., Guan, G.Z., Qin, L.B., Zhang, Y., Peng, H.S., 2015. Recent advancement of nanostructured carbon for energy applications. Chem. Rev. 115, 5159–5223.

Yoon, B., Ham, D.Y., Yarimaga, O., An, H., Lee, C.W., Kim, J.M., 2011. Inkjet printing of conjugated polymer precursors on paper substrates for colorimetric sensing and flexible electrothermochromic display. Adv. Mater. 23, 5492–5497.

Yuan, J.K., Poulin, P., 2014. Fibers do the twist. Science 343, 845–846.

Yuan, W., Hu, L.B., Yu, Z.B., Lam, T., Biggs, J., Ha, S.M., Xi, D.J., Chen, B., Senesky, M.K., Gr Ner, G., Pei, Q., 2008. Fault-tolerant dielectric elastomer actuators using single-walled carbon nanotube electrodes. Adv. Mater. 20, 621–625.

Yun, S., Niu, X., Yu, Z., Hu, W., Brochu, P., Pei, Q., 2012. Compliant silver nanowire-polymer composite electrodes for bistable large strain actuation. Adv. Mater. 24, 1321–1327.

Zang, J.F., Ryu, S., Pugno, N., Wang, Q.M., Tu, Q., Buehler, M.J., Zhao, X.H., 2013. Multifunctionality and control of the crumpling and unfolding of large-area graphene. Nat. Mater. 12, 321–325.

Zhang, S., Li, F., 2012. High performance ferroelectric relaxor-PbTiO3 single crystals: status and perspective. J. Appl. Phys. 111, 031301.

Zhang, X.B., Pint, C.L., Lee, M.H., Schubert, B.E., Jamshidi, A., Takei, K., Ko, H., Gillies, A., Bardhan, R., Urban, J.J., Wu, M., Fearing, R., Javey, A., 2011. Optically- and thermally-responsive programmable materials based on carbon nanotube-hydrogel polymer composites. Nano Lett. 11, 3239–3244.

Zhang, X., Yu, Z.B., Wang, C., Zarrouk, D., Seo, J.W.T., Cheng, J.C., Buchan, A.D., Takei, K., Zhao, Y., Ager, J.W., Zhang, J.J., Hettick, M., Hersam, M.C., Pisano, A.P., Fearing, R.S., Javey, A., 2014. Photoactuators and motors based on carbon nanotubes with selective chirality distributions. Nat. Commun. 5, 149–168.

Zhao, X., Wang, Q., 2014. Harnessing large deformation and instabilities of soft dielectrics: theory, experiment, and application. Appl. Phys. Rev. 1, 021304.

Zhu, J., Wei, S., Alexander, Jr., M., Dang, T.D., Ho, T.C., Guo, Z., 2010. Enhanced electrical switching and electrochromic properties of poly (*p*-phenylenebenzobisthiazole) thin films embedded with nano-WO_3. Adv. Funct. Mater. 20, 3076–3084.

Zhu, S.E., Shabani, R., Rho, J., Kim, Y., Hong, B.H., Ahn, J.H., Cho, H.J., 2011. Graphene-based bimorph microactuators. Nano Lett. 11, 977–981.

9

Flexible Electronic Devices Based on Polymers

9.1 Introduction

Current electronic devices including energy harvesting and storage devices, light-emitting devices and electrically driving sensors are generally rigid, heavy and bulky, and can hardly meet the requirements of flexible electronics. Therefore, flexible, lightweight and shape-controllable electronic devices are regarded as the future trend in the development of next-generation electronic devices (Wang et al., 2014; Li et al., 2014; Zhou et al., 2014). Furthermore, in some specific applications, such as wearable devices, they need to remain highly stable during bending, folding, compressing, and even stretching. To understand the electrical behavior and mechanism of the flexible electronic devices under the mechanical deformation, a lot of studies have been carried out in the past decade. Among the investigated materials, polymers that are mechanically flexible due to their unique structures have been mostly explored for the fabrication of flexible electronic devices. They are particularly promising for flexible fiber-shaped electronic devices that demonstrate unique and promising advantages compared to the planar structure (Chen et al., 2012b). For instance, a planar device can be generally bent along a specific direction and may break under some complex deformations, such as twisting, while a fiber-shaped device can be also twisted besides bending; a fiber-shaped device is lightweight and may be woven into flexible textile that becomes a mainstream research direction in modern electronics (Dalton et al., 2003). This chapter first describes the use of polymers for the planar flexible electronic devices and then highlights the advancement of fiber-shaped electronic devices based on polymers. The remaining challenges and future developments are summarized in the end.

9.2 Flexible Energy Harvesting Devices Based on Polymers

Energy harvesting devices can harvest the surrounding energies, including sunlight, pressure, and human body movements, and convert them into electricity. They show promising applications in environmentally friendly and self-powering flexible electronics. Compared with the traditional rigid energy harvesting devices, the realization of flexibility can greatly enrich their applications in portable or wearable devices. In this part, we will mainly focus on the applications of polymers and their composites in flexible energy harvesting devices including flexible solar cells and piezoelectricity devices.

9.2.1 Flexible Solar Cells Based on Polymers

Solar cells can directly convert solar energy into electricity, which is clean and inexhaustible. Generally, the solar cells are built with a planar sandwiched structure on the rigid substrate, which restricts their applications in many fields. Flexible polymer substrates, such as poly(ethyleneterephthalate) (PET), poly(ethylene naphthalate) (PEN), and polydimethylsiloxane (PDMS), have been widely used to fabricate flexible solar cells. With high stability and flexibility, the solar cells can easily integrate with other portable or wearable devices to significantly extend their applications.

9.2.1.1 Dye-Sensitized Solar Cells

Typically, a dye-sensitized solar cell (DSSC) is composed of a dye-sensitized mesoporous titania electrode, a platinum counter electrode, and an electrolyte containing iodine/iodide. Indium tin oxide (ITO) glass was the commonly used conductive substrate for a DSSC, but its intrinsic rigidity largely limited the flexibility of the resulted device. Therefore, it is necessary to develop the other flexible and conductive substrates aiming at flexible DSSCs.

A lot of flexible DSSCs have been fabricated using a plastic substrate to replace the conventional ITO. Generally, nanostructured metal oxides, such as TiO_2 or ZnO, are prepared in the form of a mesoporous film on a polymer substrate via a low temperature process. For example, a highly bendable DSSC was fabricated using a ZnO nanowire photoanode that was prepared on the ITO-coated PET substrate by a hydrothermal growth method. The ITO-coated PET film was transparent, conductive, and flexible enough as the substrate of the photoanode. It can be bent with a small radius of 2 mm without obvious damage. The as-fabricated flexible DSSCs were stable in photovoltaic performance even after 2000 cycles of bending (Jiang et al., 2008). Similarly, ITO-coated PEN film also was used as the flexible substrate for the growth of mesoporous TiO_2 by chemical sintering method (Weerasinghe et al., 2010). It can be then developed as a family of flexible conductive substrates.

Some flexible integrated devices have also been demonstrated using polymer substrates. A flexible integrated device had been fabricated in a planar comb-teeth architecture based on the PET substrate, which achieved both high mechanical flexibility and light transparency. Two kinds of electrochemical devices, DSSCs and supercapacitors, were integrated into a flexible device. The integrated device showed high electrochemical performances besides flexibility and good mechanical stability during bending (Fig. 9.1A) (Li et al., 2013).

As a critical component of the DSSC, counter electrode is also a key to the flexibility of the device. Some flexible conductive materials, such as conducting polymers, carbon-based nanomaterials and their composites have been employed as flexible counter electrodes to replace the Pt electrode in the fabrication of flexible DSSCs. For instance, after coating with conducting polymer PEDOT, a flexible counter electrode was developed with low sheet resistance and served as an ideal candidate in replacement of the conventional Pt counter electrode (Fig. 9.1B) (Mozer et al., 2010).

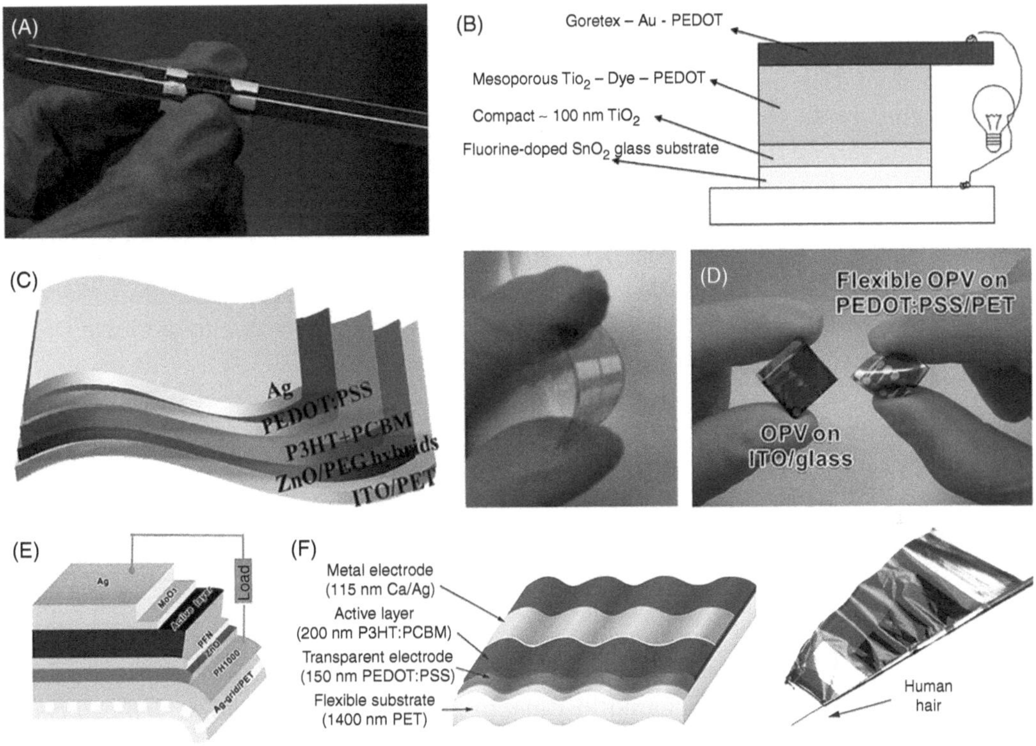

FIGURE 9.1 Flexible solar cells based on different kinds of polymer substrates. (A) Photograph of a flexible DSSC based on ITO-coated PET substrate wrapped on a pen. (B) Schematic illustration of the layer structure of a solid-state DSSC based on PEDOT on Goretex film as a counter electrode. (C) Schematic illustration (left) and photograph (right) of the PSC based on ITO-coated PET substrate. (D) Comparison of PSCs fabricated on conventional ITO/glass and flexible PEDOT:PSS/PET substrate. (E) Schematic illustration of the flexible solar cell based on Ag-grid/PET substrate. (F) Schematics of the ultrathin and flexible organic solar cell (left) based on PET film and the extreme bending demonstration by wrapping it around a 35-μm-radius human hair (right). *(A) Reproduced with permission from reference Li, H., Zhao, Q., Wang, W., Dong, H., Xu, D.S., Zou, G.J., Duan, H.L., Yu, D.P., 2013. Novel planar-structure electrochemical devices for highly flexible semitransparent power generation/storage sources. Nano Lett. 13, 1271–1277. Copyright 2013, American Chemical Society. (B) Reproduced with permission from reference Mozer, A.J., Panda, D.K., Gambhir, S., Romeo, T.C., Winther-Jensen, B., Wallace, G.G., 2010. Flexible and compressible Goretex-PEDOT membrane electrodes for solid-state dye-sensitized solar cells. Langmuir 26, 1452–1455. Copyright 2010, American Chemical Society. (C) Reproduced with permission from reference Hu, T., Li, F., Yuan, K., Chen, Y.W., 2013. Efficiency and air-stability improvement of flexible inverted polymer solar cells using ZnO/Poly(ethylene glycol) hybrids as cathode buffer layers. ACS Appl. Mater. Interfaces 5, 5763–5770. Copyright 2013, American Chemical Society. (D) Reproduced with permission from reference Cho, C.K., Hwang, W.J., Eun, K., Choa, S.H., Na, S.I., Kim, H.K., 2011. Mechanical flexibility of transparent PEDOT:PSS electrodes prepared by gravure printing for flexible organic solar cells. Sol. Energy Mater. Sol. Cells 95, 3269–3275. Copyright 2011, Elsevier. (E) Reproduced with permission from reference Mao, L., Chen, Q., Li, Y.W., Li, Y., Cai, J.H., Su, W.M., Bai, S., Jin, Y.Z., Ma, C.Q., Cui, Z., Chen, L.W., 2014. Flexible silver grid/PEDOT:PSS hybrid electrodes for large area inverted polymer solar cells. Nano Energy 10, 259–267. Copyright 2014, Elsevier. (F) Reproduced with permission from reference Kaltenbrunner, M., White, M.S., Glowacki, E.D., Sekitani, T., Someya, T., Sariciftci, N.S., Bauer, S., 2012. Ultrathin and lightweight organic solar cells with high flexibility. Nat. Commun. 3, 770. Copyright 2012, Nature Publishing Group.*

Electronic devices in the form of textile combine the idiosyncrasies of both electronics and textile, which has essential properties, such as light weight and elasticity. Carbon-based nanomaterials can be coated on the flexible textile substrate to achieve both high flexibility and conductivity. For instance, a composite counter electrode was prepared by incorporating carbon nanotube (CNT) film with PEDOT through electrochemical deposition. The DSSCs with such a counter electrode had an efficiency of 4.0%, which was similar to the efficiency of the conventional sputtered platinum on ITO-coated plastic substrate (Aitola et al., 2012). Furthermore, CNTs were coated onto a polyester fabric using a tape-casting method. Such a conductive fabric demonstrated high flexibility and low surface resistance around 15 Ω sq^{-1}, and was employed as the counter electrode to fabricate DSSCs with power conversion efficiency of 5.69% (Arbab et al., 2015).

9.2.1.2 Polymer Solar Cells

Compared to the commonly used Si-based solar cells, polymer solar cells (PSCs) have promising advantages, such as light weight, flexibility, and easy integration with other electronic devices. Similar to the structure of flexible DSSCs, plastic substrates have been employed to prepare flexible PSCs, such as ITO-coated PET film. For example, a flexible inverted PSC composed of P3HT/PCBM blends on the ITO-coated PET substrate had been developed (Fig. 9.1C). Such a plastic substrate can greatly enhance the flexibility of the PSC compared to the traditional planar counterpart based on ITO glass (Hu et al., 2013). However, the flexibility of the ITO-coated PET electrode was rather limited, which can cause the cracking of ITO coating under large-angle bending.

Conducting polymers can be directly coated onto polymer substrate to prepare flexible electrodes, which can replace ITO layer for the fabrication of flexible PSCs. For instance, a metal free electrode has been developed by depositing PEDOT:PSS (PH500) on PET substrate via a spin-coating method. Such a polymer electrode demonstrated high flexibility and a maximal power conversion efficiency of 2.2% (Zhou et al., 2008). In order to further enhance the stability of the polymer-based electrode, a gravure printing method was developed to print transparent PEDOT:PSS layer onto flexible PET substrate. The gravure-printed PEDOT:PSS/PET electrode had a sheet resistance of 359 Ω sq^{-1} and showed high stability under a bending radius of 10 mm. Moreover, the flexible electrode exhibited a constant sheet resistance over 2000 bending cycles for both outer and inner bending test at a radius of 10 mm. The high twisting and stretching stability also indicated the large potential of these electrodes for the fabrication of flexible PSCs (Fig. 9.1D) (Cho et al., 2011).

Except the conducting polymers, various metal grids were employed to construct composite layers, which were designed to further enhance the electrical conductivity of the flexible electrodes. A flexible ITO-free conductive PET film was prepared by coating modified silver ink on the surface of PET film. Such a flexible electrode demonstrated high flexibility and a maximal power conversion efficiency of 4.8% (Nickel et al., 2014). On the other hand, conducting polymers with high transparency and flexibility were also employed to modify the metal conductive layers. For instance, a roll-to-roll method has been developed to produce large-sized electrodes using composite conducting polymers and metal grids. Metal grids were deposited on the plastic substrate by screen printing with the use of Ag paste to increase the electrical conductivity of the substrate. PEDOT:PSS was then printed on metal grids to produce flexible electrodes. On the basis of this technology, a metal grid/conducting polymer composite electrode was prepared using embedded silver grid with the conducting polymer PEDOT:PSS, demonstrating both high transparency and low sheet resistance (4.5 Ω sq^{-1}). The resulting electrodes can be employed to fabricate flexible PSCs, which showed enhanced flexibility without obvious cracks observed under bending (Fig. 9.1E). Moreover, the power conversion efficiency of this large-area flexible PSC reached 5.85% (Mao et al., 2014).

Apart from serving as substrate and conductive layer, polymers can be also used as functional layers in the fabrication of PSCs. An ultrathin and lightweight polymer-based solar cell with a thickness of less than 2 μm had been developed on a 1.4 μm-thick PET substrate (Fig. 9.1F). The resulting PSC showed a comparable power conversion efficiency of 4.2% with high flexibility. By adhering the flexible PSC to a prestretched elastomer, a stretchable PSC can be also fabricated. Even during repeated compression and stretching, these stretchable PSCs showed stable photovoltaic performances (Kaltenbrunner et al., 2012).

9.2.2 Flexible Piezoelectric Devices Based on Polymers

A piezoelectric device that converts mechanical energy into electricity has attracted intense attentions due to the potential applications in portable and wearable devices. Generally, piezoelectric devices are attached onto human bodies to collect the mechanical energy, such as breathing or walking during daily activities. As a result, flexibility becomes a key factor concerning the practical applications of piezoelectric devices. ITO-coated flexible electrodes based on polymer substrates, such as PET, PDMS or polyimide (PI), have been widely used to produce flexible electronic devices. For instances, based on an aqueous solution method, ZnO nanorods were successfully grown on flexible ITO-coated polyether sulfone substrate. Since the ZnO nanorods and top electrode can be prepared on flexible substrates, a fully flexible device can be fabricated with high promise (Fig. 9.2A). In this system, the top electrodes were prepared by embossing three-dimensional Pd/Au convex arrays on the same flexible PES substrate, which was synthesized by anodization and etching of aluminium (Choi et al., 2009).

Graphene can be also used to prepare flexible electrodes due to its high conductivity and transparency. To achieve high electrical conductivity, a large-sized graphene sheet was synthesized on a nickel-coated SiO_2/Si wafer by chemical vapor deposition and then transferred onto a flexible PEN substrate. It can be employed as the transparent electrode to replace the ITO or ITO-coated substrate. A highly flexible nanogenerator had been fabricated using this graphene-based electrode, and it can effectively work after rolling (Fig. 9.2B) (Choi et al., 2010).

Polymers, such as PET and PI, are also used as flexible substrates to fabricate nanogenerators. For instance, commercial PI films (Kapton) were used as the flexible substrates for

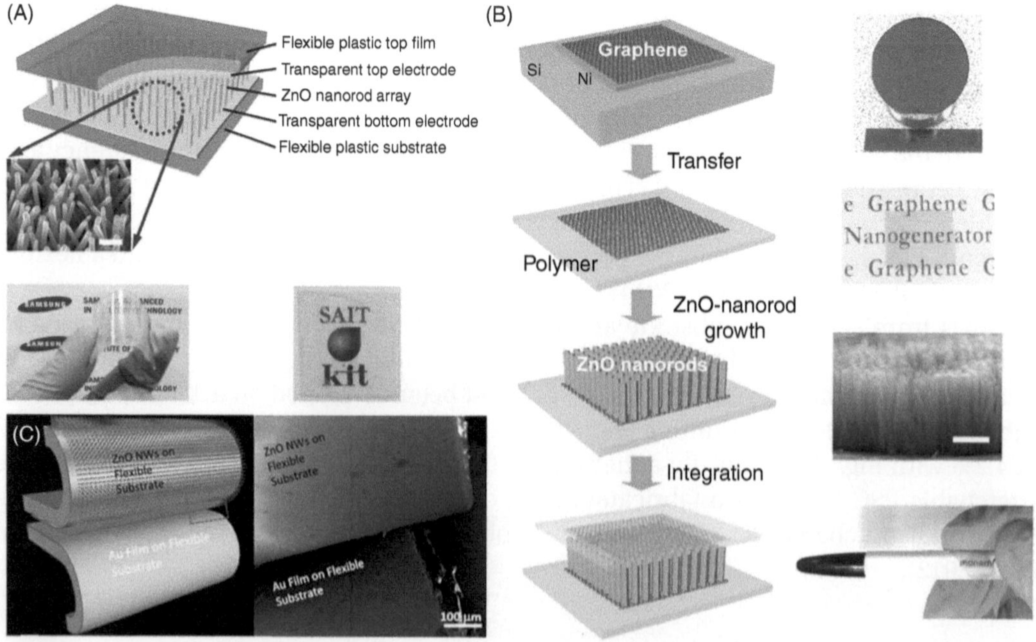

FIGURE 9.2 Flexible piezoelectric devices based on different polymer substrates. (A) Schematic illustration of an integrated generator, scanning electron microscopy image of ZnO nanorod array on a flexible ITO/PES substrate (top) and photographic images of the fully flexible charge-generating nanodevice (bottom). (B) Schematic illustration of the graphene-based fully rollable transparent nanogenerator. (C) Schematic illustration of the flexible nanogenerator with both the ZnO nanowire array and Au film on Kapton (left); scanning electron microscopy image of a junction between the ZnO nanowire array and Au film of the nanogenerator (right). *(A) Reproduced with permission from reference Choi, M.Y., Choi, D., Jin, M.J., Kim, I., Kim, S.H., Choi, J.Y., Lee, S.Y., Kim, J.M., Kim, S.W., 2009. Mechanically Powered transparent flexible charge-generating nanodevices with piezoelectric ZnO nanorods. Adv. Mater. 21, 2185–2189. Copyright 2009, Wiley-VCH. (B) Reproduced with permission from reference choi, D., CHOI, M.Y., Choi, W.M., Shin, H.J., Park, H.K., Seo, J.S., Park, J., Yoon, S.M., Chae, S.J., Lee, Y.H., Kim, S.W., Choi, J.Y., Lee, S.Y., Kim, J.M., 2010. Fully rollable transparent nanogenerators based on graphene electrodes. Adv. Mater. 22, 2187–2192. Copyright 2010, Wiley-VCH. (C) Reproduced with permission from reference Lee, M., Bae, J., Lee, J., Lee, C.S., Hong, S., Wang, Z.L., 2011. Self-powered environmental sensor system driven by nanogenerators. Energy Environ. Sci. 4, 3359–3363. Copyright 2011, The Royal Society of Chemistry.*

both ZnO and Au electrode. The nanogenerator can be stably operated under deforming (Fig. 9.2C) (Lee et al., 2011). Later, a p-type PEDPT:PSS was grown on a flexible ITO-coated PET substrate and served as the top contact to form a p–n junction with the ZnO nanorods to produce flexible nanogenerators (Briscoe et al., 2012).

In summary, a variety of flexible energy harvesting devices including solar cells and piezoelectric devices have been developed to meet the requirement of next-generation portable or wearable devices. However, the power conversion efficiencies of electronic devices were largely limited by the high electrical resistance of flexible electrodes based on polymer substrates. Therefore, more efforts are required to develop new flexible electrodes with higher electrical conductivities.

9.3 Flexible Energy Storage Devices Based on Polymers

Flexible energy storage devices have also attracted increasing attentions, and the key challenges to fabricate them lies in the selection of suitable electrode materials and the design of appropriate structures to achieve stable mechanical and electrochemical performances during bending, twisting, or even stretching. One efficient solution is to introduce polymers or their composites to enhance the flexibility. Supercapacitors and lithium-ion batteries (LIBs) are mostly investigated at both academy and industry. This section focuses on the application of polymers and their composites in the recent advancement of flexible supercapacitors and LIBs.

9.3.1 Flexible Supercapacitors Based on Polymers

Supercapacitors can store energy at a high rate by forming electrochemical double layers of charges or through pseudocapacitive surface redox reactions, which results in a high power density with long cyclic stability, low cost, and environmental friend. Supercapacitors have been widely used for a variety of consuming electronic products because they can store and release electric energy rapidly. With the vigorous development of portable and wearable devices, the demand of compatible supercapacitors triggers a continuously increasing interest in flexible supercapacitors in the past years. Generally, a planar supercapacitor includes a sandwiched structure and is composed of two electrodes and electrolyte. Polymer may be used as electrodes or electrolytes or both in a flexible supercapacitor.

9.3.1.1 Flexible Substrate

Polymers can be directly employed as substrates to fabricate supercapacitors. PDMS and PET are the mostly used substrates in the realization of flexible supercapacitors. In the case of a PDMS film, it may produce flexible and transparent thin-film supercapacitors (Fig. 9.3A) with a high cyclic stability after 10,000 cycles under bending (Jung et al., 2012). Single-walled carbon nanotubes can be further incorporated into the PDMS film to further enhance the mechanical, electronic, and thermal stability (Yuksel et al., 2014). Similarly, conductive materials, such as graphene and CNT, are also coated onto the PET film to achieve both high flexibility and electrical conductivity (Fig. 9.3B). No obvious decreases in electrochemical performance were observed from the as-fabricated supercapacitor when the bending angle increased from 0 to 120 degree (Fei et al., 2014). Based on the same strategy, a variety of metal nanoparticles has been also deposited on the PET substrate. The electrical resistances of composite films remained almost unchanged after bending for 10,000 cycles. These conductive composite films were demonstrated as promising electrodes for flexible supercapacitors with high electrochemical performances (Yeo et al., 2014).

Besides PDMS and PET, the other polymer materials also attract increasing interests and have been synthesized by various methods. For instance, localized pulsed laser irradiation was used to rapidly convert the pristine PI surface into an electrically

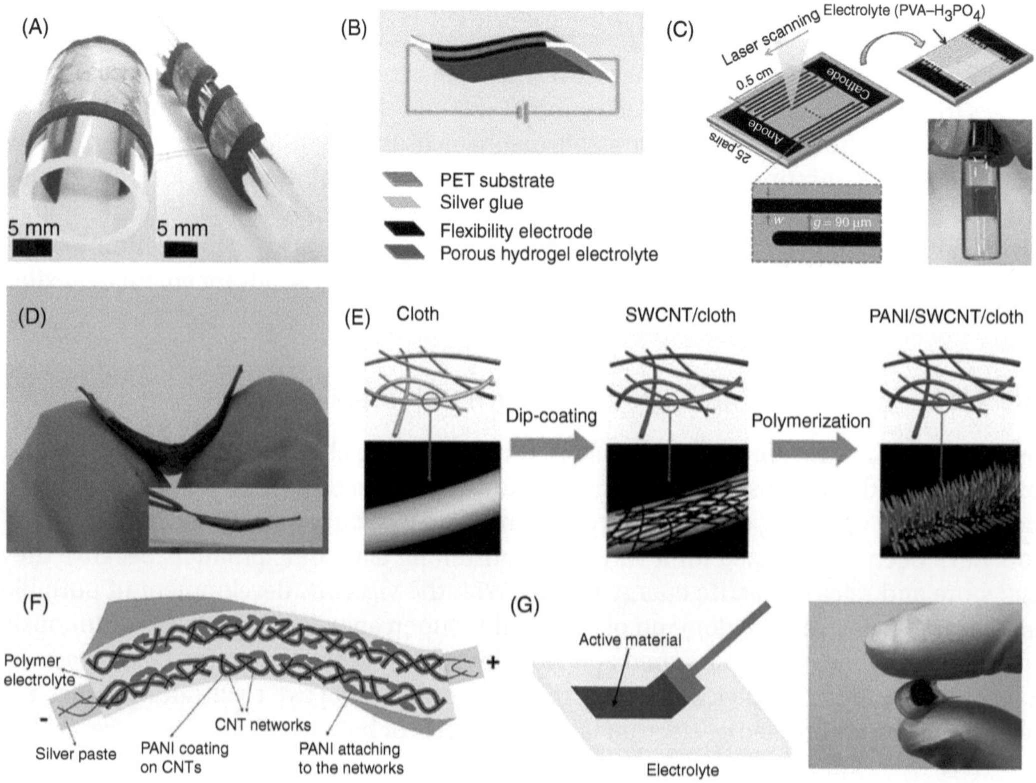

FIGURE 9.3 Flexible supercapacitors based on different polymers or polymer composites. (A) Photograph of the PDMS-based flexible supercapacitors wrapped on the glass tube. (B) Schematic illustration of flexible supercapacitor based on PET film. (C) Schematic illustration of the microsupercapacitor fabricated by laser carbonization of PI film. (D) Photograph of the flexible supercapacitor based on paper electrode. (E) Schematic illustration of the preparation of PANI/CNT/cloth electrode. (F) Schematic illustration of the PANI/CNT composite electrode well solidified in the polymer gel electrolyte. (G) Schematic illustration of the two-step construction of the supercapacitor sheet based on polymer gel electrolyte film. *(A) Reproduced with permission from reference Jung, H.Y., Karimi, M.B., Hahm, M.G., Ajayan, P.M., Jung, Y.J., 2012. Transparent, flexible supercapacitors from nano-engineered carbon films. Sci. Rep. 2, 65. Copyright 2012, Nature Publishing Group. (B) Reproduced with permission from reference Fei, H.J., Yang, C.Y., Bao, H., Wang, G.C., 2014. Flexible all-solid-state supercapacitors based on graphene/carbon black nanoparticle film electrodes and cross-linked poly(vinyl alcohol)-H_2SO_4 porous gel electrolytes. J. Power Sources 266, 488–495. Copyright 2014, Elsevier. (C) Reproduced with permission from reference In, J.B., Hsia, B., Yoo, J.-H., Hyun, S., Carraro, C., Maboudian, R., Grigoropoulos, C.P., 2015. Facile fabrication of flexible all solid-state micro-supercapacitor by direct laser writing of porous carbon in polyimide. Carbon 83, 144–151. Copyright 2015, Elsevier. (D) Reproduced with permission from reference Kang, Y.J., Chung, H., Han, C.H., Kim, W., 2012b. All-solid-state flexible supercapacitors based on papers coated with carbon nanotubes and ionic-liquid-based gel electrolytes. Nanotechnology 23, 065401. Copyright 2012, IOP Publishing. (E) Reproduced with permission from reference Wang, K., Zhao, P., Zhou, X.M., Wu, H.P., Wei, Z.X., 2011a. Flexible supercapacitors based on cloth-supported electrodes of conducting polymer nanowire array/SWCNT composites. J. Mater. Chem. 21, 16373–16378. Copyright 2011, The Royal Society of Chemistry. (F) Reproduced with permission from reference Meng, C., Liu, C., Chen, L., Hu, C., Fan, S., 2010. Highly flexible and all-solid-state paperlike polymer supercapacitors. Nano Lett. 10, 4025–4031. Copyright 2010, American Chemical Society. (G) Reproduced with permission from reference Liu, Q.A., Nayfeh, M.H., Yau, S.T., 2010. Brushed-on flexible supercapacitor sheets using a nanocomposite of polyaniline and carbon nanotubes. J. Power Sources 195, 7480–7483. Copyright 2010, Elsevier.*

conductive porous carbon structure with a preset pattern. The interdigitated electrode patterns were produced directly on the PI sheet by programmed laser scanning. Based on this method, the PI sheet can act as both a flexible substrate and a precursor for the carbonization to produce the flexible conductive electrode for the fabrication of flexible supercapacitors (Fig. 9.3C). Compared to the electrodes with conductive materials simply coated on polymer substrates, this integrated electrode showed enhanced stability and flexibility. It can stably work under decreasing bending radii from 25 to 3.5 mm (In et al., 2015).

Natural polymers, such as nonwoven clothes made from wood fibers and papers, also can be used as substrates that are flexible and lightweight. CNTs are coated onto a paper to obtain flexible CNT/paper electrode. Due to the high flexibility of the paper substrate and good distribution of CNTs, the specific capacitance and power and energy densities were varied in less than 5% after bending for 100 cycles at a bending radius of 4.5 mm (Fig. 9.3D) (Kang et al., 2012b). Furthermore, due to the higher mechanical strength and better chemical stability, bacterial nanocellulose papers were also employed to make flexible supercapacitors. An all-solid-state flexible supercapacitor had been fabricated from bacterial nanocellulose, CNT, and ionic liquid-based gel electrolytes with a good flexibility and mechanical stability (Kang et al., 2012a). Similarly, nonwoven clothes with super rough surfaces have been dipped into a CNT ink to produce the CNT-coated clothes. Conducting polymers, such as polyaniline (PANI) nanowires were further deposited onto the surface of the CNT/cloth to obtain the PANI/CNT/cloth composite electrode (Fig. 9.3E). Such a flexible composite electrode can be directly used to fabricate flexible supercapacitors without using any insulating binders (Wang et al., 2011a).

9.3.1.2 Flexible Electrode

Conducting polymers have been demonstrated with high specific capacitances and good electrical conductivities, so they can be directly used as flexible electrodes to fabricate supercapacitors (Wang et al., 2011b). A variety of conducting polymers, such as polypyrrole (PPy), PANI and polythiophene (PTP), have been explored. However, bare conducting polymers typically suffer from poor stability after several charge and discharge processes. A commonly used strategy to solve this problem is to incorporate more conductive materials, such as CNT and graphene with better mechanical and thermal stability. These composite electrodes not only display high electrical conductivity but also produce high specific capacitance, good rate capability, and long cyclic stability. For instance, CNT networks served as templates grow PANI on their outer surfaces (Fig. 9.3F). After the in situ chemical polymerization of aniline monomers, PANI was produced and formed a uniform coating layer around the CNTs. The resulting supercapacitor was flexible and showed high specific capacitance of 350 F g^{-1} (Meng et al., 2010). Additionally, a "skeleton/skin" strategy has been developed for the preparation of freestanding, thin and flexible CNT/PANI hybrid films. These CNT/PANI hybrid films can be used as both electrodes and charge collectors for a flexible supercapacitor, without the use of additional metallic current collectors or polymer substrates (Niu et al., 2012).

9.3.1.3 Polymer Gel Electrolyte

The use of solid-state electrolyte instead of the liquid electrolyte is another key strategy to produce flexible supercapacitors that can stably work under and after bending. Polymer gel electrolytes were the mostly used solid-state electrolytes due to their high ionic conductivity, good mechanical strength, high flexibility, low cost, and nontoxicity. In addition, the use of a polymer gel electrolyte can avoid the leakage problem and reduce the packaging complexity. A polymer gel electrolyte is typically composed of three main components, that is, a polymer matrix as the host, an organic or aqueous solvent as the plasticizer, and a supporting electrolytic salt. Poly(ethylene oxide) (PEO), poly(vinyl alcohol) (PVA), and poly(vinylidene fluoride) (PVDF) are the popular host polymers for preparing gel electrolytes.

Typically, the solid-state electrolyte can function as both a substrate and separator in the fabrication of a flexible supercapacitor by two steps. For instance, a freestanding PVA/H_3PO_4 film was obtained by first casting the gel electrolyte on a glass slide and then peeled off from the glass. It had been further sandwiched between two composite electrodes that were prepared from conducting polymers and nanoparticles to produce the flexible supercapacitor (Fig. 9.3G) (Liu et al., 2010, 2013b).

9.3.2 Flexible Lithium-Ion Batteries Based on Polymer

LIBs are widely studied for high energy density, long-term stability, low memory effect, and slow loss of charge when not in use. The realization of flexibility may greatly promote their applications in many emerging fields, such as electronic skins and sensors. A number of designs for flexible LIBs have been reported in recent years. Similar to the strategy in fabricating flexible supercapacitors, the flexibility in LIBs can be achieved also by using flexible substrate, electrode, electrolyte, and separator. Polymers may meet the requirements in each of the above parts for the LIBs.

9.3.2.1 Flexible Substrate

Various polymers were suitable to employ as flexible substrates to make the LIBs flexible. For example, PDMS was used to prepare flexible LIB via thin film fabrication and layer transfer techniques. This LIB was first constructed on a brittle mica substrate and then transferred onto a flexible PDMS substrate to realize the high flexibility (Fig. 9.4A). This flexible LIB exhibited stable electrochemical properties with the operated voltage maintained by 99.8% after 20,000 bending cycles at a radius of 3.1 mm and could power a flexible organic light-emitting diode (LED) system under repeatedly bending (Fig. 9.4B) (Koo et al., 2012). Without scarifying the flexibility, the storage capacity of LIB can be enhanced by 670% through building porous structures in the PDMS film (Lee et al., 2012). Apart from the flexibility, the LIB can also possess the stretchability by replacing the electrodes and active components on the PDMS substrate with stretchable ones (Fig. 9.4C) (Weng et al., 2015).

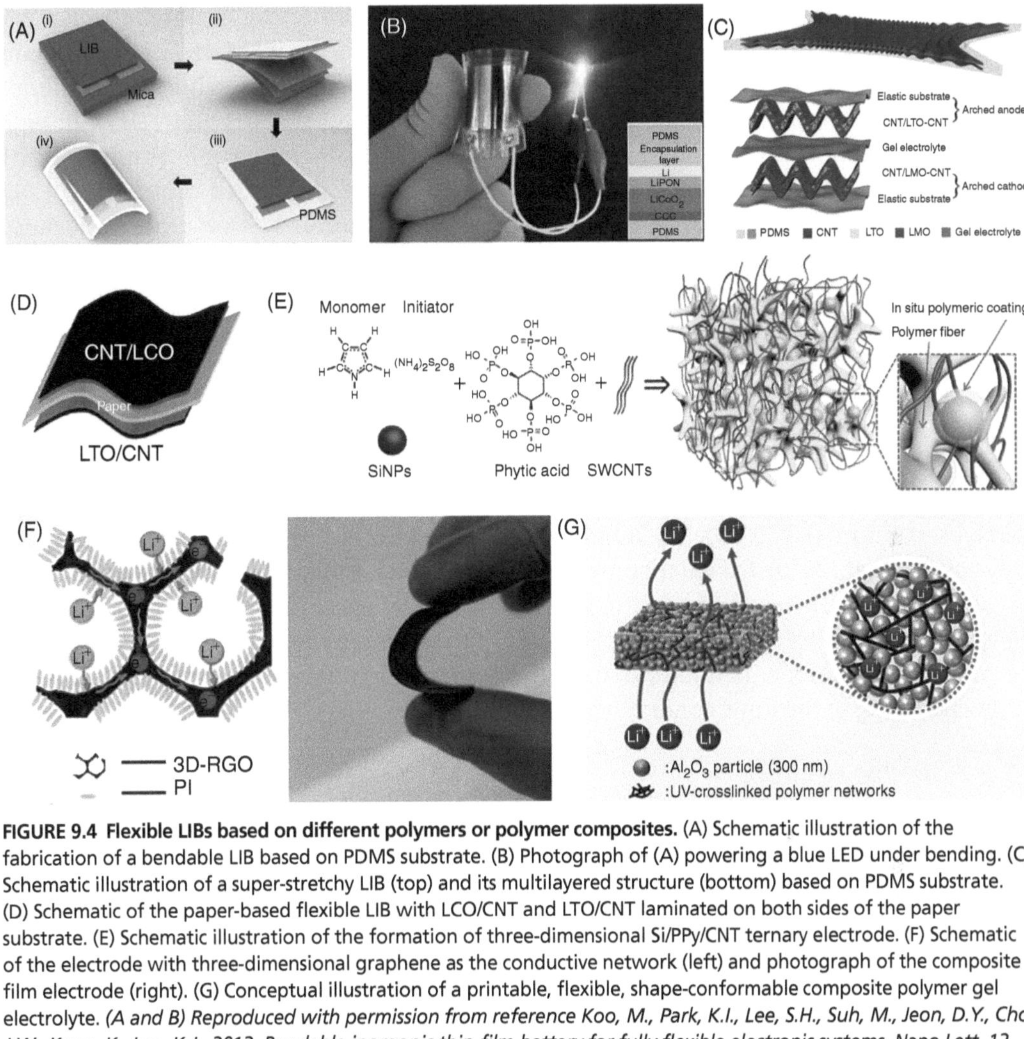

FIGURE 9.4 Flexible LIBs based on different polymers or polymer composites. (A) Schematic illustration of the fabrication of a bendable LIB based on PDMS substrate. (B) Photograph of (A) powering a blue LED under bending. (C) Schematic illustration of a super-stretchy LIB (top) and its multilayered structure (bottom) based on PDMS substrate. (D) Schematic of the paper-based flexible LIB with LCO/CNT and LTO/CNT laminated on both sides of the paper substrate. (E) Schematic illustration of the formation of three-dimensional Si/PPy/CNT ternary electrode. (F) Schematic of the electrode with three-dimensional graphene as the conductive network (left) and photograph of the composite film electrode (right). (G) Conceptual illustration of a printable, flexible, shape-conformable composite polymer gel electrolyte. *(A and B) Reproduced with permission from reference Koo, M., Park, K.I., Lee, S.H., Suh, M., Jeon, D.Y., Choi, J.W., Kang, K., Lee, K.J., 2012. Bendable inorganic thin-film battery for fully flexible electronic systems. Nano Lett. 12, 4810–4816. Copyright 2012, American Chemical Society. (C) Reproduced with permission from reference Weng, W., Sun, Q., Zhang, Y., He, S.S., Wu, Q.Q., Deng, J., Fang, X., Guan, G.Z., Ren, J., Peng, H.S., 2015. A gum-like lithium-ion battery based on a novel arched structure. Adv. Mater. 27, 1363–1369. Copyright 2015, Wiley-VCH. (D) Reproduced with permission from reference Hu, L.B., Wu, H., La Mantia, F., Yang, Y.A., Cui, Y., 2010. Thin, flexible secondary Li-ion paper batteries. ACS Nano 4, 5843–5848. Copyright 2010, American Chemical Society. (E) Reproduced with permission from reference Liu, B.R., Soares, P., Checkles, C., Zhao, Y., Yu, G.H., 2013a. Three-dimensional hierarchical ternary nanostructures for high-performance Li-ion battery anodes. Nano Lett. 13, 3414–3419. Copyright 2013, American Chemical Society. (F) Reproduced with permission from reference Meng, Y.N., Wu, H.P., Zhang, Y.J., Wei, Z.X., 2014. A flexible electrode based on a three-dimensional graphene network-supported polyimide for lithium-ion batteries. J. Mater. Chem. A 2, 10842–10846. Copyright 2014, The Royal Society of Chemistry. (G) Reproduced with permission from reference Kil, E.H., Choi, K.H., Ha, H.J., Xu, S., Rogers, J.A., Kim, M.R., Lee, Y.G., Kim, K.M., Cho, K.Y., Lee, S.Y., 2013. Imprintable, bendable, and shape-conformable polymer electrolytes for versatile-shaped lithium-ion batteries. Adv. Mater. 25, 1395–1400. Copyright 2013, Wiley-VCH.*

PET is also a commonly used substrate material for constructing flexible LIBs. For example, a porous PET combined with polymer electrolyte was incorporated as a compliant skeleton to enhance flexibility of the LIB. The achieved LIB exhibited stable electrochemical performance even under a severely wrinkled state (Choi et al., 2014b). PVDF with high mechanical robustness can also be employed as a flexible substrate. For example, PEDOT film was used to enhance the electrochemical performance of aligned multiwalled CNT array in LIB. However, this CNT/PEDOT composite film cannot be freestanding because of its poor mechanical robustness. By incorporating the PVDF on the surface of composite film, they can be easily peeled off from the underlying substrate without any visual signs of damage (Chen et al., 2007).

Papers or textiles also play a key role in the flexible LIBs by acting as substrates and/or separators, due to their high flexibility, foldability, good recyclability, and easy fabrication (Zhang et al., 2014). A paper-based thin and flexible LIB was reported by using commercial paper as a separator and single-walled CNT films as a current collector. This commercial paper functions as both a substrate and separator membrane with lower impedance than commercial separators (Fig. 9.4D). No failure was observed for the paper LIB after bending for 50 cycles (Hu et al., 2010). Except for only being used as the flexible substrate, some polymers, for example, PEO, can be employed to link the active materials to prepare flexible electrodes. For example, PEO was used to link the MoS_2-RGO hybrid paper to enhance the flexibility of the composite electrode. In addition, the electrochemical performance can be remarkably enhanced owing to the ionic conductivity contribution by PEO (Liu et al., 2013c).

9.3.2.2 Flexible Electrode

Conducting polymers are promising basic backbones to construct flexible electrodes for LIBs because of their high flexibility, conformability, and versatility. Furthermore, the other active materials can be incorporated into the conducting polymers to form high-performance flexible electrodes. For example, a novel three-dimensional nanoarchitecture composed of PPy-Si core–shell nanofibers was achieved by the deposition of Si onto the electropolymerized PPy nanofibers. This core–shell structure indicated a high cyclic stability after repeated lithium insertion and extraction (Du et al., 2012).

A ternary PPy/Si/CNT hydrogel framework has been developed to further enhance the performance of flexible LIBs. In this framework, CNT could offer a continuous electron transport network and high porosity to accommodate the volume expansion of Si nanoparticles which were encapsulated within the conductive PPy framework, resulting in a great improvement in cyclic performance and mechanical stability (Fig. 9.4E) (Liu et al., 2013a). The Si nanoparticles can also be replaced by metal oxides. For example, a three-dimensionally interconnected hybrid hydrogel system has been developed by using CNT, conducting polymer, and TiO_2 particles. The achieved electrode showed negligible change of electrical resistance after bending for 500 cycles at a bending radius of 3.5 mm (Chen et al., 2014c).

PI is a kind of organic carbonyl compound with high theoretical capacity and good thermal stability, which can also be used as active material to fabricate flexible electrodes.

A flexible film electrode has been designed by using three-dimensional graphene foam as a conductive network and PI as an active material (Fig. 9.4F). Compared with the electrode prepared by a conventional method, the hierarchical structure of graphene composed of PI in the proposed electrode resulted in lower resistance for the reaction with lithium ions. The structure of the three-dimensional composite film electrode showed high flexibility and demonstrated high potential as a cathode for the flexible LIB (Meng et al., 2014). Furthermore, highly conductive single-walled CNT film was introduced as a current collector to prepare a binder-free flexible electrode (Wu et al., 2014).

9.3.2.3 Polymer Gel Electrolyte

For the flexible LIB, polymer gel electrolytes usually consisted of liquid electrolytes embedded in polymer matrices, which were prepared by solvent evaporation of premixed mixtures of liquid electrolytes and polymers dissolved in organic solvents. A liquid electrolyte in LIB consists of lithium salts, such as $LiPF_6$, $LiBF_4$, or $LiClO_4$, in an organic solvent, such as ethylene carbonate, dimethyl carbonate, and diethyl carbonate. A lot of polar polymers including PEO, PVDF, PMMA, and PAN have been employed as the polymer host to prepare polymer gel electrolytes, because of their good affinity with liquid components and high flexibility as the solid component. Among various polymer gel electrolytes, PEO was the mostly used polymer due to a high solubility for lithium salts. A solid PEO electrolyte also acting as a separator layer was sandwiched between the $LiFePO_4$ cathode and $Li_4Ti_5O_{12}$ anode to fabricate the flexible and stretchable LIB (Liu et al., 2012). However, for the solid electrolytes, a high flexibility always comes at a price of sacrificing conductivity. Therefore, improving the flexibility with high ionic conductivity is the most challenging and critical issue for the advancement of polymer gel electrolytes.

A variety of approaches, such as the incorporation of nanoparticles or blending with other polymers, have been employed to prepare high-performance polymer gel electrolytes with high flexibility. For example, one-dimensional CNTs can grow within two-dimensional montmorillonite clay platelets to create hybrid three-dimensional nanofillers, which were added to the PEO electrolyte so as to enhance its ion conductivity and mechanical properties (Tang et al., 2012). While, a facile and scalable approach was further developed to fabricate highly bendable polymer electrolytes, which consisted of a UV-cured polymer matrix, high-boiling-point liquid electrolyte, and Al_2O_3 nanoparticles (Fig. 9.4G). This composite polymer electrolyte was mechanically stable even under complicated deformations. Moreover, the maze patterns do not distort even after being subjected to five cycles of bending stress (Kil et al., 2013).

Recently, a new kind of plastic crystal composite electrolytes with the polymer matrix as a mechanically reinforcing framework was designed to prepare flexible LIBs with high mechanical strength. For example, a UV-curable semiinterpenetrating polymer network matrix was used to form the plastic crystal composite electrolyte, which indicated a remarkable improvement in flexibility (Ha et al., 2012).

To conclude, flexible supercapacitors and LIBs have been vigorously developed in recent years, which promise a great potential for applications in portable, wearable, and flexible energy storage systems. However, the relatively poor performance and complicated fabrication may hinder their further development. Thus designing an effective, low-cost, and scalable strategy to construct the flexible energy storage devices will be one of the most promising breakthrough directions to work on.

9.4 Flexible Light-Emitting Devices Based on Polymer

Compared with conventional rigid LEDs, flexible LEDs not only possess the low power consumption, but possess the compatibility with flexible substrates, which is promising for a wide range of applications in aerospace, transport, biomedicine, and electronics. In the following part, various flexible LEDs are introduced by using polymer or polymer composite as the flexible substrates or electrodes.

Typically, ITO-coated PET is one of the most investigated polymer substrates in flexible LEDs. The achieved flexible LEDs could be bent by a radius of 0.5 cm without obvious degradation as shown in Fig. 9.5A (Gu et al., 1997). However, the deposition temperature is a little high, which may reduce the device performance. Therefore, indium zinc oxide (IZO) was developed which can be deposited on the polycarbonate (PC) substrate at a temperature below 50°C. Such an IZO-coated PC flexible substrate showed very small resistance change during 5000 bending cycles, indicating high mechanical robustness and flexibility (Fig. 9.5B). The LEDs fabricated on this IZO-coated PET substrate exhibited similar performance and efficiency to a device fabricated on conventional ITO-coated glass (Kang et al., 2007).

Although the flexible LEDs based on ITO-coated PET substrates possess high electroluminescent performance, the flexibility of them cannot fully match the requirements for practical applications in highly flexible electronics determined by the inorganic property of ITO material. Therefore, a series of other conductive materials, such as metal nanowires, CNT, graphene, and conducting polymers were also employed to further improve the flexibility of electrodes and devices. For example, a conducting polymer was inkjet-printed onto PET textiles to use as the electrode (Fig. 9.5C) (Hu et al., 2011). Furthermore, the conducting CNT was introduced to form a CNT/PEDOT:PSS composite electrode, which greatly enhances the electrical conductivity. The surface resistivity of the CNT/PEDOT electrode decreased with decreasing bend radius, while the ITO/PET sheet increased dramatically as the bending radius was smaller than 9 mm (Fig. 9.5D) (Hu et al., 2012).

These conductive materials composited with conducting polymers can also be imparted into other elastomeric polymer substrates to make the LEDs more flexible. For example, sliver nanowires were first spray-coated onto the glass, followed by spin-coating PDMS onto it to form a flexible electrode, and then electroluminescent layers were sandwiched between two as-prepared electrodes to achieve the flexible LED. The LEDs could maintain the functionality under large mechanical deformations, such as stretching, flexing, or twisting (Fig. 9.5E). Moreover, this LED could be stretched to as high as 100% (Wang et al., 2015). Replacing the PDMS with PU that is a commonly used shape-memory material, the formed LED could be shape-memorable to meet more requirements.

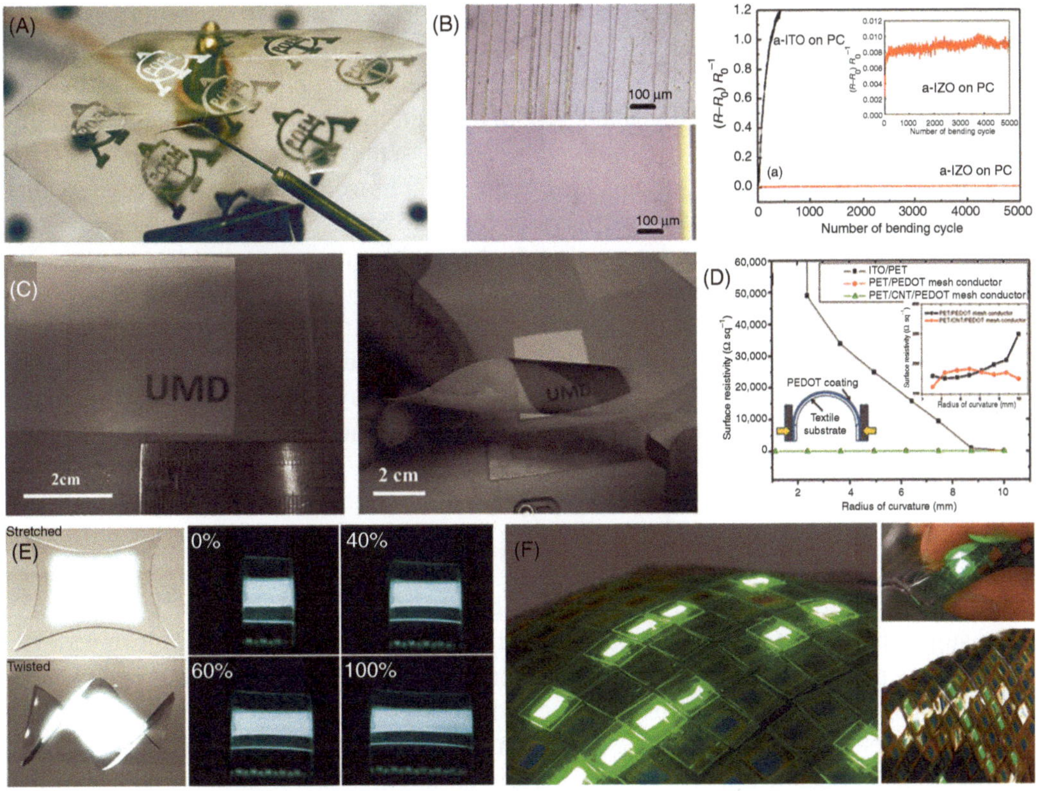

FIGURE 9.5 Flexible LEDs based on different polymer substrates. (A) Photograph of an array of nine unpackaged flexible LEDs based on ITO-coated PET film. (B) Optical micrographs of ITO-coated PC and IZO-coated PC substrates after bending test. (C) Transparent PET mesh fabric before and after printed with six layers of PEDOT:PSS. (D) Surface resistivity of PET mesh conductor as a function of bending radius. (E) Photographs of the luminous LED (based on PDMS substrate) under mechanical manipulations of stretching and twisting. (F) A demonstration of the stretchable display based on PDMS substrate, which can be spread over arbitrary curved surfaces. It worked even under folding and crumpling. *(A) Reproduced with permission from reference Gu, G., Burrows, P.E., Venkatesh, S., Forrest, S.R., Thompson, M.E., 1997. Vacuum-deposited, nonpolymeric flexible organic light-emitting devices. Opt. Lett. 22, 172–174. Copyright 1997, OSA Publishing. (B) Reproduced with permission from reference Kang, J.W., Jeong, W.I., Kim, J.J., Kim, H.K., Kim, D.G., Lee, G.H., 2007. High-performance flexible organic light-emitting diodes using amorphous indium zinc oxide anode. Electrochem. Solid State Lett. 10, J75–J78. Copyright 2007, ECS. (C) Reproduced with permission from reference Hu, B., Li, D.P., Ala, O., Manandhar, P., Fan, Q.G., Kasilingam, D., Calvert, P.D., 2011. Textile-based flexible electroluminescent devices. Adv. Funct. Mater. 21, 305–311. Copyright 2011, Wiley-VCH. (D) Reproduced with permission from reference Hu, B., Li, D.P., Manandharm, P., Fan, Q.G., Kasilingam, D., Calvert, P., 2012. CNT/conducting polymer composite conductors impart high flexibility to textile electroluminescent devices. J. Mater. Chem. 22, 1598–1605. Copyright 2012, The Royal Society of Chemistry. (E) Reproduced with permission from reference Wang J.X., Yan, C.Y., Chee, K.J., Lee, P.S., 2015. Highly stretchable and self-deformable alternating current electroluminescent devices. Adv. Mater. 27, 2876–2882. Copyright 2015, Wiley-VCH. (F) Reproduced with permission from reference Sekitani, T., Nakajima, H., Maeda, H., Fukushima, T., Aida, T., Hata, K., Someya, T., 2009. Stretchable active-matrix organic light-emitting diode display using printable elastic conductors. Nat. Mater. 8, 494–499. Copyright 2009, Nature Publishing Group.*

Flexible LEDs previously mentioned are mainly fabricated by incorporating the flexible active and electrode materials onto the flexible polymer substrates to realize the intrinsic flexibility. However, these fabrications require the constructing materials to be flexible which significantly limits the selectivity of applied materials. Therefore, a novel and scalable strategy was designed to develop flexible LEDs, that is, integrating plenty of small and rigid LEDs onto an elastic and flexible conductor achieved by dispersing single-walled CNTs in a highly elastic fluorinated copolymer rubber. The resulting flexible LED could be stretched by 30~50% and spread over a hemisphere without any mechanical or electrical damage (Fig. 9.5F) (Sekitani et al., 2009).

In conclusion, flexible LEDs based on polymer substrates have been widely studied. They possess high flexibility compared to the conventional rigid LEDs and demonstrate promising applications in the portable and wearable displays. However, the performance of current flexible LEDs should be enhanced so as to expand their applications in flexible light and displays.

9.5 Flexible Electrochromic Devices Based on Polymers

Flexible electrochromic devices (ECDs) are becoming increasing important for their promising applications in many areas, such as the portable and wearable electronic devices, including smart windows, functional supercapacitors, and flexible displays. Typically, an ECD consists of four parts of substrate, conductive electrode, electrochromic material, and electrolyte. Enormous efforts have been made to improve the flexibility of ECDs including utilizing flexible polymer substrates and conductive materials.

Similar to the flexible LEDs, flexible ECDs can also be realized by depositing conductive electrodes and active materials onto the flexible polymer substrates including PET, PDMS, PU as well as papers. For realizing the flexible ECDs, preparing the flexible conductive electrodes is critically important. The commonly used electrode materials contain ITO, CNT, graphene, metal nanowires, and conducting polymers. For instance, a flexible inkjet-printed ECD has been developed by printing PANI/silica or PEDOT/silica colloidal composite ink on the ITO-coated PET film (Fig. 9.6A) (Shim et al., 2008). However, the optical contrast of the ECDs based on ITO-coated PET substrate is poor, which cannot meet the requirements for practical applications. After a lot of investigations, reduced graphene oxide (RGO) film was found to effectively solve aforementioned problems by electrodepositing them on the ITO-coated PET substrates. The resulting optical contrast was remarkably enhanced. Moreover, the switching voltage between colored and bleached states was reduced because of the oxidization of RGO films (Fig. 9.6B) (Palenzuela et al., 2014). A flexible ECD based on aligned CNT sheets and PET substrates was also successfully developed. To fabricate this device, the electrochromic material, poly(1-phenyl-2-(*p*-trimethylsilyl)phenylacetylene) (PTP), was directly coated onto the multiwalled CNT to fabricate CNT/PTP composite film. The resulting composite film showed a rapid reversible color change from yellowish green to dark green in response to applied electric currents. This flexible ECD based on CNT/PET electrode substrate could be bent for over 100 cycles without breaking (Sun et al., 2013).

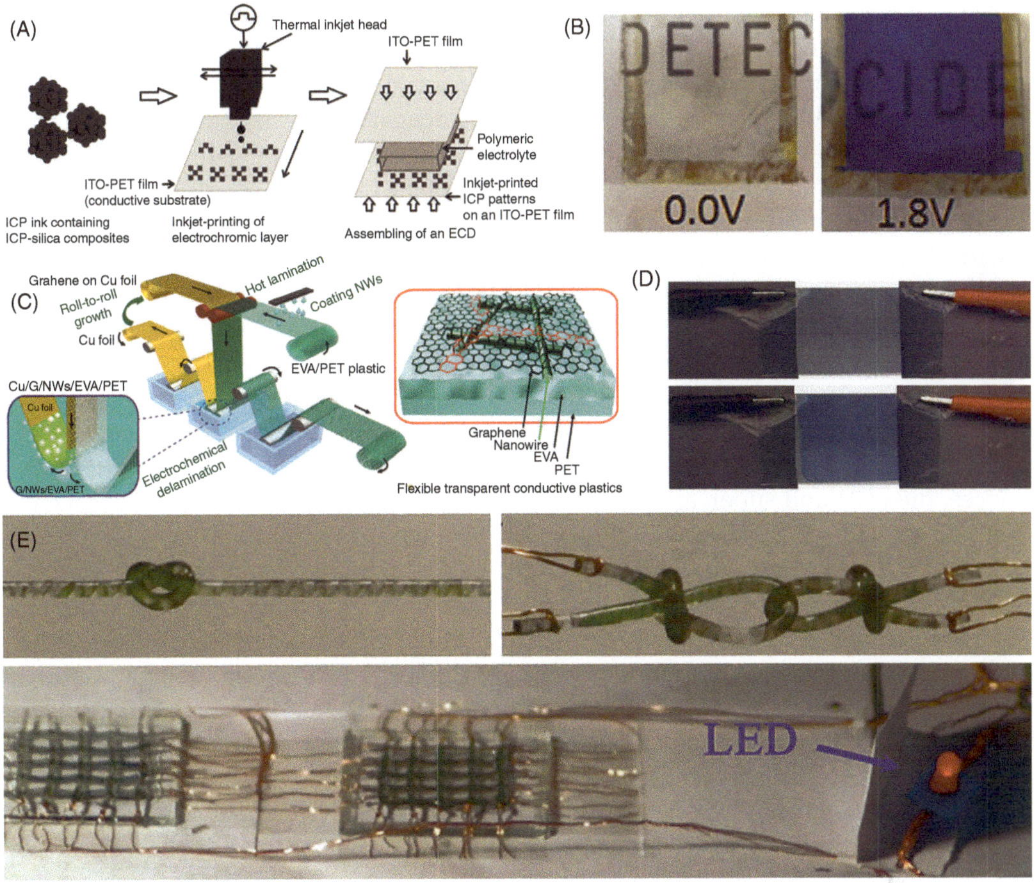

FIGURE 9.6 Flexible electrochromic displays based on different substrates. (A) The schematic procedure for the fabrication of an ECD using an inkjet-printed electrochromic layer. (B) Photographs of the RGO based electrochromic device in the bleached (left) and colored state (right). (C) Schematic diagram of graphene and metal nanowire hybrid films produced by a continuous roll-to-roll process. (D) Typical photograph of an electrochromic device in the transparent state and dark blue state. (E) Multifunctional flexible fiber-shaped supercapacitors based on PDMS. *(A) Reproduced with permission from reference Shim, G.H., Han, M.G., Sharp-Norton, J.C., Creager, S.E., Foulger, S.H., 2008. Inkjet-printed electrochromic devices utilizing polyaniline–silica and poly(3,4-ethylenedioxythiophene)–silica colloidal composite particles. J. Mater. Chem. 18, 594–601. Copyright 2008, The Royal Society of Chemistry. (B) Reproduced with permission from reference Palenzuela, J., Vinuales, A., Odriozola, I., Cabanero, G., Grande, H.J., Ruiz, V., 2014. Flexible viologen electrochromic devices with low operational voltages using reduced graphene oxide electrodes. ACS Appl. Mater. Interfaces 6, 14562–14567. Copyright 2014, American Chemical Society. (C) Reproduced with permission from reference Deng, B., Hsu, P.C., Chen, G., Chandrashekar, B.N., Liao, L., Ayitimuda, Z., Wu, J., Guo, Y., Lin, L., Zhou, Y., Aisijiang, M., Xie, Q., Cui, Y., Liu, Z., Peng, H., 2015. Roll-to-roll encapsulation of metal nanowires between graphene and plastic substrate for high-performance flexible transparent electrodes. Nano Lett. 15, 4206–4213. Copyright 2015, American Chemical Society. (D) Reproduced with permission from reference Mecerreyes, D., Marcilla, R., Ochoteco, E., Grande, H., Pomposo, J.A., Vergaz, R., S Nchez Pena, J.M., 2004. A simplified all-polymer flexible electrochromic device. Electrochim. Acta 49, 3555–3559. Copyright 2004, Elsevier. (E) Reproduced with permission from reference Chen, X.L., Lin, H.J., Deng, J., Zhang, Y., Sun, X.M., Chen, P.N., Fang, X., Zhang, Z. T., Guan, G.Z., Peng, H.S., 2014b. Electrochromic fiber-shaped supercapacitors. Adv. Mater. 26, 8126–8132. Copyright 2014, Wiley-VCH.*

Metal nanowires are also promising alternatives to develop flexible ECDs in light of their high optical transparency and electrical conductivity (Søndergaard et al., 2013). For one instance, a flexible ECD was made to exhibit a rapid switching based on flexible silver grid-printed PET substrates (Jensen et al., 2014). Encapsulating silver nanowires between graphene and PET substrates produced flexible ECDs with high-performances through a continuous and large-scale roll-to-roll process (Fig. 9.6C) (Deng et al., 2015).

PEDOT is also a commonly used conducting polymer for constructing flexible ECDs due to its high electrical conductivity and flexibility. More importantly, PEDOT can also be simultaneously used as an electrochromic layer in flexible ECDs. Therefore, the fabrication of flexible ECDs is greatly simplified. The achieved flexible ECD exhibited a chromatic contrast in the entire visible range between 0 and 3 V (Fig. 9.6D) (Mecerreyes et al., 2004). PANI is another typical material used as electrochromic and conductive materials to develop flexible ECDs. Meanwhile, PANI can also serve as an electrochemical active component in supercapacitors as mentioned previously. Due to its multifunctions in various devices, the ECD and supercapacitor can be integrated together so that one device can combine the color-tunable ability and energy storage for many potential applications, for example, smart windows (Wang et al., 2012). PANI was first deposited onto the CNT/PDMS to prepare a flexible electrode, followed by sandwiching the gel electrolyte between two as-prepared flexible electrodes to achieve a flexible ECD that can store energy. The resulting ECD showed reversible color changes from yellow to green and then to blue under applied voltages (Chen et al., 2014a). Based on the same materials, ECDs can further be designed at a fiber format and then they were woven into textiles with different patterns, as shown in Fig. 9.6E (Chen et al., 2014b).

In summary, a great number of flexible ECDs have been developed and show promising applications in displays and sensors. Meanwhile, different functions, for example, energy storage can be integrated into the ECDs to smartly change colors. Many other electrochemical properties are being attempted to add into the ECDs for multifunctional applications.

9.6 Flexible Fiber-Shaped Electronic Devices

Various electronic devices have been developed in a planer structure for tremendous applications in aerospace, transport, biomedicine, and electronics. However, for many new emerging fields, electronic devices only being bent two-dimensionally are usually not enough, which requires being twistable in three dimensions. Recently, a new family of fiber-shaped electronic devices including solar cells, LEDs, supercapacitors, and LIBs has been developed to meet this requirement for applications in portable and wearable electronic devices. Compared with the conventional two-dimensional or three-dimensional structures, these one-dimensional electronic devices show some unique and promising advantages, such as high flexibility, light weight, wearability, and high-performance. Fiber-shaped electronic devices typically exhibit diameters ranged from micrometers to millimeters, which are generally thin and lightweight. Due of their unique fiber-shaped structure, they are highly flexible and can be woven or knitted into fabrics or textiles. In this section, we mainly focus on the novel flexible fiber-shaped electronic devices constructed by carbon-based materials or polymer composites. The relationship between the fiber structure and the performance, such as flexibility, stability, and electrochemical properties, are summarized; the perspectives on the future development of fiber-shaped electronic devices will be also discussed.

9.6.1 Fiber-Shaped Energy Harvesting Devices

The newly developed fiber-shaped solar cells have attracted much interest at both academy and industry owing to their angle-independent collection for incident light. Recently, we have developed a series of high-performance and flexible fiber-shaped solar cells by using the unique carbon nanostructure fiber electrodes. Typically, these solar cells are classified in three types, that is, DSSC, PSC, and perovskite solar cell.

In a typical DSSC, the platinum wire is generally employed as the counter electrode due to its high electrocatalytic activity. However, the high cost and low flexibility has largely limited its practical application in the fiber-shaped DSSC. Therefore, carbon nanostructure fibers, such as CNT fibers and graphene fibers with high electrocatalytic activity and flexibility have been recognized as potential electrode materials in fiber-shaped DSSCs. For constructing fiber-shaped DSSC, a dry-spinning CNT fiber served as a counter electrode was wrapped around a fiber working electrode that absorbed the dyes. Here the working electrode is formed by growing TiO_2 nanotubes on the outer surface of titanium wire. After injection of the liquid electrolyte, a fiber-shaped DSSC was achieved and shown in Fig. 9.7A(Chen et al., 2012d). Compared with the conventional planar solar cells, such a fiber-shaped DSSC can be easily integrated into various flexible or wearable devices by a convenient weaving technology. Furthermore, various CNT or graphene composite fibers, such as CNT/RGO (Sun et al., 2014a), CNT/graphene (Fang et al., 2014), graphene/Pt (Yang et al., 2013b) have also been developed as the counter electrodes to improve the performance of the fiber-shaped DSSCs. For example, a graphene fiber electrodeposited with platinum nanoparticles was employed as the counter electrode to produce the fiber-shaped DSSCs. Due to the high catalysis of nanostructured graphene and platinum, this flexible DSSC achieved a certified recorded energy conversion efficiency of 8.45% in fiber-shaped photovoltaic devices (Yang et al., 2013b). The titanium wire can be replaced by CNT fiber to enhance the flexibility of fiber-shaped DSSCs (Fig. 9.7B) (Chen et al., 2012c).

Different from the conventional fiber-shaped DSSCs with a twisted structure, a coaxial DSSC was developed by wrapping a thin CNT film around the modified Ti wire as the counter electrode (Fig. 9.7C). Unlike the CNT fiber electrode, this CNT film electrode ensured a full contact with the underlying active layer to form the coaxial structure. Due to the layered structure, the contact between two electrodes can be easily maintained during bending. Such a coaxial DSSC was stable and can be bent to high angles up to 90 degree reversibly without performance degradation (Zhang et al., 2011).

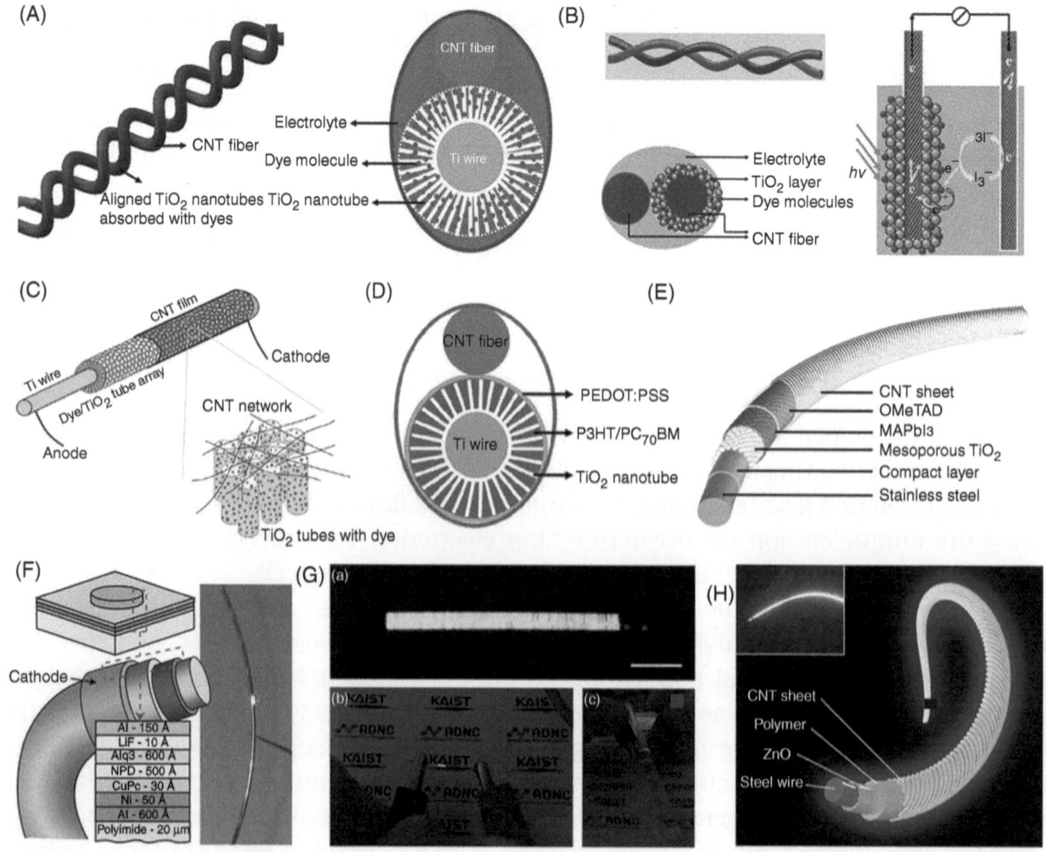

FIGURE 9.7 Flexible fiber-shaped solar cells and LEDs. (A) Schematic illustration of the fiber-shaped DSSC with an aligned TiO_2 nanotube-modified titanium wire as the working electrode. (B) Schematic illustration of the fiber-shaped DSSC from two CNT fibers with a CNT/TiO_2 composite fiber as the working electrode. (C) Illustration of the coaxial DSSC based on a TiO_2 nanotube-modified titanium wire, and a transparent CNT film served as counter electrode. (D) Schematic illustration of the polymer photovoltaic fiber by top view. (E) Schematic illustration of the fiber-shaped perovskite solar cell. (F) Schematic illustration and photograph of the fiber-shaped OLED. (G) Microscopic image of the fiber-based PLED operated at 8 V (top) as well as before and wrapping onto an acrylic cylinder with a radius of 2.5 mm (bottom). (H) Schematics of the structure of a fiber-shaped PLEC. Inset, photograph of a fiber-shaped PLEC biased at 10 V. *(A) Reproduced with permission from reference Chen, T., Qiu, L.B., Kia, H.G., Yang, Z.B., Peng, H.S., 2012d. Designing aligned inorganic nanotubes at the electrode interface: towards highly efficient photovoltaic wires. Adv. Mater. 24, 4623–4628. Copyright 2012, Wiley-VCH. (B) Reproduced with permission from reference Chen, T., Qiu, L.B., Cai, Z.B., Gong, F., Yang, Z.B., Wang, Z.S., Peng, H.S., 2012c. Intertwined aligned carbon nanotube fiber based dye-sensitized solar cells. Nano Lett. 12, 2568–2572. Copyright 2012, American Chemical Society. (C) Reproduced with permission from reference Zhang, S., Ji, C.Y., Bian, Z.Q., Liu, R.H., Xia, X.Y., Yun, D.Q., Zhang, L.H., Huang, C.H., Cao, A.Y., 2011. Single-wire dye-sensitized solar cells wrapped by carbon nanotube film electrodes. Nano Lett. 11, 3383–3387. Copyright 2011, American Chemical Society. (D) Reproduced with permission from reference Chen, T., Qiu, L., Li, H., Peng, H., 2012a. Polymer photovoltaic wires based on aligned carbon nanotube fibers. J. Mater. Chem. 22, 23655–23658. Copyright 2012, The Royal Society of Chemistry. (E) Reproduced with permission from reference Qiu, L.B., Deng, J., Lu, X., Yang, Z.B., Peng, H.S., 2014. Integrating perovskite solar cells into a flexible fiber. Angew. Chem. Int. Ed. 53, 10425–10428. Copyright 2014, Wiley-VCH. (F) Reproduced with permission from reference O'connor, B., An, K.H., Zhao, Y., Pipe, K.P., Shtein, M., 2007. Fiber shaped organic light emitting device. Adv. Mater. 19, 3897–3900. Copyright 2007, Wiley-VCH. (G) Reproduced with permission from reference Kwon, S., Kim, W., Kim, H., Choi, S., Park, B.-C., Kang, S.-H., Choi, K.C., 2015. High luminance fiber-based polymer light-emitting devices by a dip-coating method. Adv. Electron. Mater. 1, 1500103. Copyright 2015, Wiley-VCH. (H) Reproduced with permission from reference Zhang, Z.T., Guo, K.P., Li, Y.M., Li, X.Y., Guan, G.Z., Li, H.P., Luo, Y.F., Zhao, F.Y., Zhang, Q., Wei, B., Pei, Q.B., Peng, H.S., 2015a. A colour-tunable, weavable fibre-shaped polymer light-emitting electrochemical cell. Nat. Photon. 9, 233–238. Copyright 2015, Nature Publishing Group.*

PSCs are also an important branch of organic photovoltaic devices that show great importance in harvesting solar energy. Fig. 9.7D shows the structure of a typical fiber-shaped PSC. A TiO_2 nanotube-modified titanium wire was used as the electron-transporting layer and cathode. Then, this modified titanium wire was dip-coated with P3HT:PCBM and PEDOT/PSS layers which act as the photoactive layer and hole-transporting layer. Finally, a CNT fiber or film was wrapped around the primary fiber electrode to produce the fiber-shaped PSC. Due to the unique fiber structure, the parameters of this PSC (such as fill factor and power conversion efficiency) remained almost unchanged under different incident angles with the same light intensity (Chen et al., 2012a).

Recently, perovskite solar cells have attracted increasing interests for the encouraging high power conversion efficiency, and they are also fabricated into a fiber shape. A fiber-shaped perovskite solar cell was prepared by using steel wire as the anode where the functional layers were coated layer by layer (Fig. 9.7E). A transparent aligned CNT film was used as the transparent conductive electrode, which played a significant role in the fabrication of the fiber-shaped perovskite solar cells. By employing the CNT film-coated PDMS fiber as a flexible substrate, a highly flexible and stretchable fiber-shaped perovskite solar cell could be produced by inserting the conductive PDMS fiber into the spring-like perovskite solar cell (Qiu et al., 2014). In order to further increase the flexibility of fiber-shaped perovskite solar cells, a CNT fiber was employed as a fiber substrate to load perovskite active materials without using the relatively rigid metal wires. Then, a bare CNT fiber was twisted on the perovskite/CNT fiber to form a fiber-shaped perovskite solar cell sealed by a transparent PMMA protecting layer. This fiber-shaped perovskite solar cell can be wrapped onto a capillary tube with curvature radius of 0.3 mm without obvious performance degradation (Li et al., 2015).

9.6.2 Fiber-Shaped Light-Emitting Devices

Fiber-shaped flexible LEDs have been developed for enlarging the kinds of portable and wearable electronics, which shows promising applications in microelectronics and biomedicine fields. Compared with the conventional planar LEDs with rigid or flexible film substrates, fiber-shaped LEDs showed unique applications because of the softness, light weight and wearability. Similar to the layered structure of planar LEDs, a fiber-shaped LED has been prepared by conformably depositing the charge-transporting layer and electroluminescent layers, metallic anode and cathode onto a polyimide-coated silica fiber (Fig. 9.7F). Such a fiber-shaped LED exhibited electrical characteristics and luminescence efficiency comparable to the planar LEDs, and offered a novel and promising device geometry with angularly uniform emission intensity across the entire emission spectrum (O'connor et al., 2007).

According to the similar method, a fiber-shaped polymer light-emitting diode (PLED) was developed on the PET fiber by a dip-coating method except for the thermal-evaporation of the outermost Al semitransparent electrode. This PET-based fiber-shaped PLED showed a high luminance of over 1000 cd m^{-2}, which was high enough for application in wearable displays. Such a fiber-shaped PLED can stably work even under being wrapped on a cylinder with a radius of 2.5 mm (Fig. 9.7G). (Kwon et al., 2015).

However, the fabrication of these described fiber-shaped LEDs is complicated, which cannot meet the requirements for large-scale applications. Therefore, a fiber-shaped polymer light-emitting electrochemical cell (PLEC) by using an all-solution-based method had been developed. This fiber-shaped PLEC showed a coaxial structure which included a modified metal wire cathode and a conducting aligned CNT sheet anode, with an electroluminescent polymer layer sandwiched between them (Fig. 9.7H). These fiber-shaped PLECs were lightweight, flexible and soft, which can be woven into light-emitting textiles for large-scale applications.(Zhang et al., 2015a) Because of the unique property of PLECs, the two electrodes can be made from the same materials, such as CNT fiber. Without using metal wires, this CNT-based PLEC showed enhanced flexibility and wearability (Zhang et al., 2015b).

9.6.3 Fiber-Shaped Energy Storage Devices

Fiber-shaped supercapacitors and LIBs have also been developed as new types of flexible energy storage devices to meet the requirement of portable and wearable energy storage devices. In contrast to the conventional planar energy storage devices, these fiber-shaped energy storage devices can be readily woven into textiles. Recently, carbon-based fiber electrodes have been widely investigated to construct flexible fiber-shaped energy storage devices.

Similar to the conventional supercapacitors, the fiber-shaped supercapacitors are composed of two electrodes and a gel electrolyte. Typically, a fiber-shaped supercapacitor can be easily achieved by twisting two fiber electrodes precoated with a gel electrolyte. Therefore, the fiber electrodes are critically important for enhancing the performances of fiber-shaped supercapacitors. Graphene fibers have been widely explored to make flexible fiber-shaped supercapacitors. For instance, the polyelectrolyte-wrapped graphene fiber or graphene/CNT composite fiber was first prepared by a coaxial wet-spinning assembly strategy. The polyelectrolyte sheath can prevent the short circuit between the two fiber electrodes after twisting and provided access for ion penetration in the supercapacitors. The fiber-shaped supercapacitors were flexible enough to shuttle back and forth without fracture in the cowoven cloth (Kou et al., 2014).

Recently, fiber-shaped supercapacitors are also constructed by disordered CNT fibers. However, the performance of the supercapacitors based on the disordered CNT fiber is poor due to the relatively low electrical conductivity. Fortunately, aligned CNT fibers can be also prepared by a dry-spinning process, and the aligned structure offers high electrical conductivity, high tensile strength and flexibility, which are ideal substitutes of metal fibers for the fabrication of flexible fiber-shaped supercapacitors (Fig. 9.8A). As expected, the electrochemical performance of resulting supercapacitor was enhanced. Meanwhile, this fiber-shaped supercapacitor showed excellent flexibility (Fig. 9.8B) (Ren et al., 2013a).

However, the storage capability from the fiber electrode based on bare carbon nanomaterial is poor owing to the double-layer capacitance mechanism. Therefore, pseudocapacitive materials, such as metallic oxide and conducting polymer can also be incorporated into the CNT fiber to enhance the capacitance. For example, CNT/PANI composite fibers

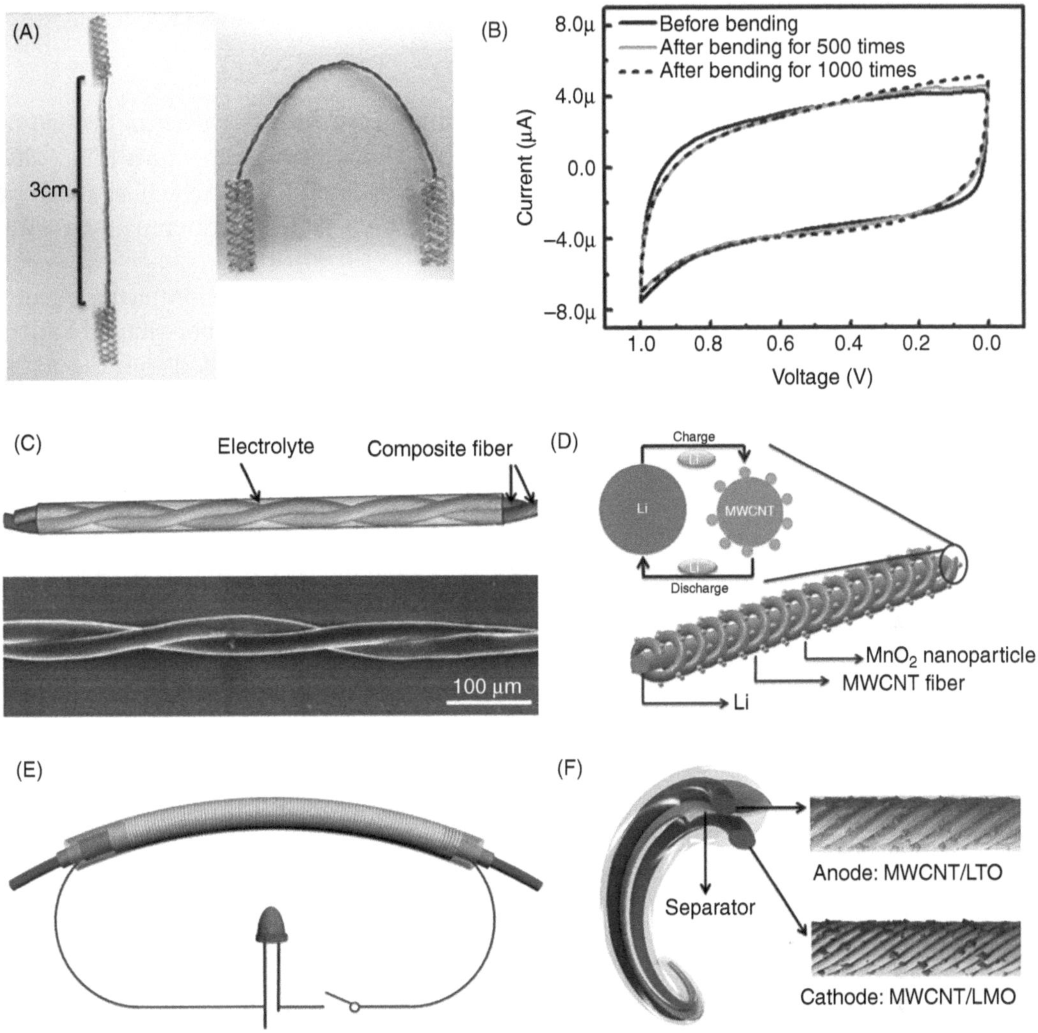

FIGURE 9.8 Fiber-shaped supercapacitors and LIBs based on carbon nanomaterials. (A) Photographs of a fiber-shaped supercapacitor based on CNT fiber before and after bending. (B) Cyclic voltammograms of (A) before and after bending for 500 and1000 cycles. (C) Schematic illustration and scanning electron microscopy image of a fiber-shaped supercapacitor by twisting two aligned CNT/PANI composite fibers. (D) Schematic illustration of a fiber-shaped LIB fabricated by twisting an aligned CNT/MnO_2 composite fiber and Li wire as positive and negative electrodes, respectively. (E) Schematic illustration of a coaxial fiber-shaped LIB based on cotton fiber. (F) Schematic illustration of a fiber-shaped LIB with aligned CNT/LTO and CNT/LMO composite yarns paired as the anode and cathode, respectively. *(A and B) Reproduced with permission from reference Ren, J., Bai, W.Y., Guan, G.Z., Zhang, Y., Peng, H.S. 2013a. Flexible and weaveable capacitor wire based on a carbon nanocomposite fiber. Adv. Mater. 25, 5965–5970. Copyright 2013, Wiley-VCH. (C) Reproduced with permission from reference Cai, Z.B., Li, L., Ren, J., Qiu, L.B., Lin, H.J., Peng, H.S., 2013. Flexible, weavable and efficient microsupercapacitor wires based on polyaniline composite fibers incorporated with aligned carbon nanotubes. J. Mater. Chem. A 1, 258–261. Copyright 2013, The Royal Society of Chemistry. (D) Reproduced with permission from reference Ren, J., Li, L., Chen, C., Chen, X.L., Cai, Z.B., Qiu, L.B., Wang, Y.G., Zhu, X.R., Peng, H.S., 2013b. Twisting carbon nanotube fibers for both wire-shaped micro-supercapacitor and micro-battery. Adv. Mater. 1155–1159. Copyright 2013, Wiley-VCH. (E) Reproduced with permission from reference Weng, W., Sun, Q., zhang, Y., LIN, H.J., Ren, J., Lu, X., Wang, M., Peng, H.S., 2014. Winding aligned carbon nanotube composite yarns into coaxial fiber full batteries with high performances. Nano Lett. 14, 3432–3438. Copyright 2014, American Chemical Society. (F) Reproduced with permission from reference Ren, J., Zhang, Y., Bai, W.Y., Chen, X.L., Zhang, Z.T., Fang, X., Weng, W., Wang, Y.G., Peng, H.S., 2014. Elastic and wearable wire-shaped lithium-ion battery with high electrochemical performance. Angew. Chemie Int. Ed. 53, 7864–7869. Copyright 2014, Wiley-VCH.*

have been prepared through an easy electrodeposition process. After twisting two composite fibers together, we may obtain metal-free fiber-shaped supercapacitors (Fig. 9.8C) (Cai et al., 2013). Other conducting polymers, such as PPy and PEDOT, which also display high specific capacitances, have been deposited onto the CNT sheet or fiber to prepare the composite fiber electrodes (Lee et al., 2013).

Transition metal oxides, such as MnO_2, RuO_2, and $Ni(OH)_2$ have been intensively studied as electrode materials, which can also be employed to prepare fiber-shaped supercapacitors. A CNT/MnO_2 composited fiber electrode has been developed with the novel design of pore-trapped MnO_2. Owing to the three-dimensional porosity inside the CNT fiber, MnO_2 nanoparticles can be easily trapped in the pore during the electrodeposition process. Such a composite fiber was constructed by the high conductive CNT fiber and high capacitive MnO_2 nanoparticles, resulting in high-performance fiber-shaped supercapacitors (Choi et al., 2014a).

Besides improving the capacitance of the fiber-shaped supercapacitors, many efforts have been devoted to integrate encouraging functions to the supercapacitors to extend their applications. For example, the resulting fiber-shaped supercapacitors can be wrapped around an elastic polymer fiber, for example, PDMS, to form the spring-like supercapacitors with the stretchability as high as 100% (Yang et al., 2013a). Through changing the PDMS fiber substrate with a PU fiber, the resulting supercapacitors can possess self-memory ability (Sun et al., 2014b).

Based on the similar strategy, a series of fiber-shaped LIBs also has been developed. With the incorporation of active materials, CNT fibers can be employed as the cathodes or anodes in the fiber-shaped batteries. At the first attempt, a fiber-shaped LIB was fabricated from a CNT/MnO_2 composite fiber as cathode. The high flexibility, mechanical strength and electrical conductivity of the fiber electrode were impervious to the incorporation of MnO_2 nanoparticles. After being twisted with a metallic lithium wire, a flexible fiber-shaped LIB can be fabricated (Fig. 9.8D). Due to the high flexibility of the CNT composite fiber cathode, no obvious damage in fiber structure and electrical conductivity decay can be traced after bending for 100 cycles (Ren et al., 2013b).

However, the lithium fiber electrode significantly limited the practical application of such a fiber-shaped LIB. Therefore, preparing a suitable fiber electrode to replace the lithium fiber as anode in fiber-shaped LIBs is badly needed. To this end, a new CNT/Si composite fiber electrode was synthesized to replace the lithium fiber. The silicon was deposited on an aligned CNT sheet by electron beam evaporation, followed by scrolling into a fiber. The composite CNT/Si fiber exhibited a good flexibility and no obvious damage can be observed after the repeated deformation (Lin et al., 2014). A fiber-shaped full LIB was finally achieved by winding CNT/LMO and CNT/Si hybrid fibers onto a cotton fiber, followed by coating a gel electrolyte (Fig. 9.8E). It was metal-free, lightweight and flexible, and can be woven into traditional textiles (Weng et al., 2014).

Meanwhile, a more stable fiber-shaped LIB was constructed based on CNT/LTO and CNT/LMO fiber electrodes without substrate. Benefited from the flexible fiber electrodes, the resulting fiber-shaped full LIB exhibited even a better flexibility and can be bent into various shapes without fracture in the structure and degradation in the electrochemical performances. Similar to the stretchable fiber-shaped supercapacitors, a flexible and stretchable full LIB was designed by fixing the anode and cathode fibers in parallel and spirally wound in helix onto an elastic fiber, which was coated by a thin layer of gel electrolyte and sealed into a heat-shrinkable tube (Fig. 9.8F). This fiber-shaped LIB can be stretched by 100% because of its spring-like structure, PDMS substrate and polymer gel electrolyte. The specific capacity can be maintained by over 80% after 200 cycles of stretching with the strain of 100% (Ren et al., 2014).

Based on their light weight and flexibility, fiber-shaped electronic devices can be greatly used in the portable and wearable devices. However, these academic achievements have not yet been translated into industrial success, which was limited by the complex fabrication process and low electrochemical performance originated from the high internal resistance of fiber electrode. Therefore, designing a simple and scalable approach to develop fiber electrodes with high-performances is encouraging and meaningful for the future advancement of fiber-shaped devices.

9.7 Perspective

In this chapter, we have reviewed the recent development of the flexible electronic devices, such as solar cells, supercapacitors, LIBs, LEDs, and electrically driving sensors. A new and promising structure by designing them into a fiber shape is also highlighted at the end. A variety of polymers and polymer composites have been employed as substrates, electrodes, or electrolytes to construct flexible electronic devices. The incorporation of polymers proves a key to realize the flexibility. However, the insulating or poorly conductive property at nature has limited the improvement in the other properties, such as energy conversion and storage capability. Therefore, it is important to develop new polymer materials with better combined properties for enhancing the electronic performances of flexible devices.

Moreover, the realization of a large-scale production is recognized possibly as the greatest challenge for the development of flexible electronic devices. For instance, the power conversion efficiency of a fiber-shaped solar cell and specific capacitance of a fiber-shaped supercapacitor are severely decreased with the increasing length from the current centimeters at a lab scale to meters or even kilometers at industry. Furthermore, the available fabrications are relatively too complex and time-consuming from a viewpoint of applications. A lot of efforts have to be made to solve these technical problems besides addressing the above scientific issues.

References

Aitola, K., Borghei, M., Kaskela, A., Kemppainen, E., Nasibulin, A.G., Kauppinen, E.I., Lund, P.D., Ruiz, V., Halme, J., 2012. Flexible metal-free counter electrode for dye solar cells based on conductive polymer and carbon nanotubes. J. Electroanal. Chem. 683, 70–74.

Arbab, A.A., Sun, K.C., Sahito, I.A., Qadir, M.B., Jeong, S.H., 2015. Multiwalled carbon nanotube coated polyester fabric as textile based flexible counter electrode for dye sensitized solar cell. Phys. Chem. Chem. Phys. 17, 12957–12969.

Briscoe, J., Stewart, M., Vopson, M., Cain, M., Weaver, P.M., Dunn, S., 2012. Nanostructured p–n junctions for kinetic-to-electrical energy conversion. Adv. Energy Mater. 2, 1261–1268.

Cai, Z.B., Li, L., Ren, J., Qiu, L.B., Lin, H.J., Peng, H.S., 2013. Flexible, weavable and efficient microsupercapacitor wires based on polyaniline composite fibers incorporated with aligned carbon nanotubes. J. Mater. Chem. A 1, 258–261.

Chen, J., Liu, Y., Minett, A.I., Lynam, C., Wang, J.Z., Wallace, G.G., 2007. Flexible, aligned carbon nanotube/conducting polymer electrodes for a lithium-ion battery. Chem. Mater. 19, 3595–3597.

Chen, T., Qiu, L., Li, H., Peng, H., 2012a. Polymer photovoltaic wires based on aligned carbon nanotube fibers. J. Mater. Chem. 22, 23655–23658.

Chen, T., Qiu, L., Yang, Z., Cai, Z., Ren, J., Li, H., Lin, H., Sun, X., Peng, H., 2012b. An integrated "energy wire" for both photoelectric conversion and energy storage. Angew. Chem. 124, 12143–12146.

Chen, T., Qiu, L.B., Cai, Z.B., Gong, F., Yang, Z.B., Wang, Z.S., Peng, H.S., 2012c. Intertwined aligned carbon nanotube fiber based dye-sensitized solar cells. Nano Lett. 12, 2568–2572.

Chen, T., Qiu, L.B., Kia, H.G., Yang, Z.B., Peng, H.S., 2012d. Designing aligned inorganic nanotubes at the electrode interface: towards highly efficient photovoltaic wires. Adv. Mater. 24, 4623–4628.

Chen, X., Lin, H., Chen, P., Guan, G., Deng, J., Peng, H., 2014a. Smart, stretchable supercapacitors. Adv. Mater 26, 4444–4449.

Chen, X.L., Lin, H.J., Deng, J., Zhang, Y., Sun, X.M., Chen, P.N., Fang, X., Zhang, Z.T., Guan, G.Z., Peng, H.S., 2014b. Electrochromic fiber-shaped supercapacitors. Adv. Mater. 26, 8126–8132.

Chen, Z., To, J.W.F., Wang, C., Lu, Z.D., Liu, N., Chortos, A., Pan, L.J., Wei, F., Cui, Y., Bao, Z.N., 2014c. A Three-dimensionally interconnected carbon nanotube-conducting polymer hydrogel network for high-performance flexible battery electrodes. Adv. Energy Mater. 4, 1400207.

Cho, C.K., Hwang, W.J., Eun, K., Choa, S.H., Na, S.I., Kim, H.K., 2011. Mechanical flexibility of transparent PEDOT:PSS electrodes prepared by gravure printing for flexible organic solar cells. Sol. Energy Mater. Sol. Cells 95, 3269–3275.

Choi, M.Y., Choi, D., Jin, M.J., Kim, I., Kim, S.H., Choi, J.Y., Lee, S.Y., Kim, J.M., Kim, S.W., 2009. Mechanically powered transparent flexible charge-generating nanodevices with piezoelectric ZnO nanorods. Adv. Mater. 21, 2185–2189.

Choi, D., Choi, M.Y., Choi, W.M., Shin, H.J., Park, H.K., Seo, J.S., Park, J., Yoon, S.M., Chae, S.J., Lee, Y.H., Kim, S.W., Choi, J.Y., Lee, S.Y., Kim, J.M., 2010. Fully rollable transparent nanogenerators based on graphene electrodes. Adv. Mater. 22, 2187–2192.

Choi, C., Lee, J.A., Choi, A.Y., Kim, Y.T., Lepro, X., Lima, M.D., Baughman, R.H., Kim, S.J., 2014a. Flexible supercapacitor made of carbon nanotube yarn with internal pores. Adv. Mater. 26, 2059–2065.

Choi, K.H., Cho, S.J., Kim, S.H., Kwon, Y.H., Kim, J.Y., Lee, S.Y., 2014b. Thin, deformable, and safety-reinforced plastic crystal polymer electrolytes for high-performance flexible lithium-ion batteries. Adv. Funct. Mater. 24, 44–52.

Dalton, A.B., Collins, S., Munoz, E., Razal, J.M., Ebron, V.H., Ferraris, J.P., Coleman, J.N., Kim, B.G., Baughman, R.H., 2003. Super-tough carbon-nanotube fibres—these extraordinary composite fibres can be woven into electronic textiles. Nature 423, 703–1703.

Deng, B., Hsu, P.C., Chen, G., Chandrashekar, B.N., Liao, L., Ayitimuda, Z., Wu, J., Guo, Y., Lin, L., Zhou, Y., Aisijiang, M., Xie, Q., Cui, Y., Liu, Z., Peng, H., 2015. Roll-to-roll encapsulation of metal nanowires between graphene and plastic substrate for high-performance flexible transparent electrodes. Nano Lett. 15, 4206–4213.

Du, Z.J., Zhang, S.C., Liu, Y., Zhao, J.F., Lin, R.X., Jiang, T., 2012. Facile fabrication of reticular polypyrrole-silicon core-shell nanofibers for high performance lithium storage. J. Mater. Chem. 22, 11636–11641.

Fang, X., Yang, Z.B., Qiu, L.B., Sun, H., Pan, S.W., Deng, J., Luo, Y.F., Peng, H.S., 2014. Core-sheath carbon nanostructured fibers for efficient wire-shaped dye-sensitized solar cells. Adv. Mater. 26, 1694–1698.

Fei, H.J., Yang, C.Y., Bao, H., Wang, G.C., 2014. Flexible all-solid-state supercapacitors based on graphene/carbon black nanoparticle film electrodes and cross-linked poly(vinyl alcohol)-H_2SO_4 porous gel electrolytes. J. Power Sources 266, 488–495.

Gu, G., Burrows, P.E., Venkatesh, S., Forrest, S.R., Thompson, M.E., 1997. Vacuum-deposited, nonpolymeric flexible organic light-emitting devices. Opt. Lett. 22, 172–174.

Ha, H.J., Kil, E.H., Kwon, Y.H., Kim, J.Y., Lee, C.K., Lee, S.Y., 2012. UV-curable semi-interpenetrating polymer network-integrated, highly bendable plastic crystal composite electrolytes for shape-conformable all-solid-state lithium ion batteries. Energy Environ. Sci. 5, 6491–6499.

Hu, L.B., Wu, H., La Mantia, F., Yang, Y.A., Cui, Y., 2010. Thin, flexible secondary Li-Ion paper batteries. ACS Nano 4, 5843–5848.

Hu, B., Li, D.P., Ala, O., Manandhar, P., Fan, Q.G., Kasilingam, D., Calvert, P.D., 2011. Textile-based flexible electroluminescent devices. Adv. Funct. Mater. 21, 305–311.

Hu, B., Li, D.P., Manandharm, P., Fan, Q.G., Kasilingam, D., Calvert, P., 2012. CNT/conducting polymer composite conductors impart high flexibility to textile electroluminescent devices. J. Mater. Chem. 22, 1598–1605.

Hu, T., Li, F., Yuan, K., Chen, Y.W., 2013. Efficiency and air-stability improvement of flexible inverted polymer solar cells using ZnO/poly(ethylene glycol) hybrids as cathode buffer layers. ACS Appl. Mater. Interfaces 5, 5763–5770.

In, J.B., Hsia, B., Yoo, J.-H., Hyun, S., Carraro, C., Maboudian, R., Grigoropoulos, C.P., 2015. Facile fabrication of flexible all solid-state micro-supercapacitor by direct laser writing of porous carbon in polyimide. Carbon 83, 144–151.

Jensen, J., H Sel, M., Kim, I., Yu, J.-S., Jo, J., Krebs, F.C., 2014. Fast switching ITO free electrochromic devices. Adv. Funct. Mater. 24, 1228–1233.

Jiang, C.Y., Sun, X.W., Tan, K.W., Lo, G.Q., Kyaw, A.K.K., Kwong, D.L., 2008. High-bendability flexible dye-sensitized solar cell with a nanoparticle-modified ZnO-nanowire electrode. Appl. Phys. Lett., 92.

Jung, H.Y., Karimi, M.B., Hahm, M.G., Ajayan, P.M., Jung, Y.J., 2012. Transparent, flexible supercapacitors from nano-engineered carbon films. Sci. Rep. 2, 65.

Kaltenbrunner, M., White, M.S., Glowacki, E.D., Sekitani, T., Someya, T., Sariciftci, N.S., Bauer, S., 2012. Ultrathin and lightweight organic solar cells with high flexibility. Nat. Commun. 3, 770.

Kang, J.W., Jeong, W.I., Kim, J.J., Kim, H.K., Kim, D.G., Lee, G.H., 2007. High-performance flexible organic light-emitting diodes using amorphous indium zinc oxide anode. Electrochem. Solid State Lett. 10, J75–J78.

Kang, Y.J., Chun, S.J., Lee, S., Kim, B.Y., Kim, J.H., Chung, H., Lee, S.Y., Kim, W., 2012a. All-Solid-state flexible supercapacitors fabricated with bacterial nanocellulose papers, carbon nanotubes, and triblock-copolymer ion gels. ACS Nano 6, 6400–6406.

Kang, Y.J., Chung, H., Han, C.H., Kim, W., 2012b. All-solid-state flexible supercapacitors based on papers coated with carbon nanotubes and ionic-liquid-based gel electrolytes. Nanotechnology 23, 065401.

Kil, E.H., Choi, K.H., Ha, H.J., Xu, S., Rogers, J.A., Kim, M.R., Lee, Y.G., Kim, K.M., Cho, K.Y., Lee, S.Y., 2013. Imprintable, bendable, and shape-conformable polymer electrolytes for versatile-shaped lithium-ion batteries. Adv. Mater. 25, 1395–1400.

Koo, M., Park, K.I., Lee, S.H., Suh, M., Jeon, D.Y., Choi, J.W., Kang, K., Lee, K.J., 2012. Bendable inorganic thin-film battery for fully flexible electronic systems. Nano Lett. 12, 4810–4816.

Kou, L., Huang, T.Q., Zheng, B.N., Han, Y., Zhao, X.L., Gopalsamy, K., Sun, H.Y., Gao, C., 2014. Coaxial wet-spun yarn supercapacitors for high-energy density and safe wearable electronics. Nat. Commun. 5, 536–538.

Kwon, S., Kim, W., Kim, H., Choi, S., Park, B.-C., Kang, S.-H., Choi, K.C., 2015. High luminance fiber-based polymer light-emitting devices by a dip-coating method. Adv. Electron. Mater. 1, 1500103.

Lee, M., Bae, J., Lee, J., Lee, C.S., Hong, S., Wang, Z.L., 2011. Self-powered environmental sensor system driven by nanogenerators. Energy Environ. Sci. 4, 3359–3363.

Lee, H., Yoo, J.K., Park, J.H., Kim, J.H., Kang, K., Jung, Y.S., 2012. A Stretchable polymer-carbon nanotube composite electrode for flexible lithium-ion batteries: porosity engineering by controlled phase separation. Adv. Energy Mater. 2, 976–982.

Lee, J.A., Shin, M.K., Kim, S.H., Cho, H.U., Spinks, G.M., Wallace, G.G., Lima, M.D., Lepro, X., Kozlov, M.E., Baughman, R.H., Kim, S.J., 2013. Ultrafast charge and discharge biscrolled yarn supercapacitors for textiles and microdevices. Nat. Commun. 4, 1970.

Li, H., Zhao, Q., Wang, W., Dong, H., Xu, D.S., Zou, G.J., Duan, H.L., Yu, D.P., 2013. Novel planar-structure electrochemical devices for highly flexible semitransparent power generation/storage sources. Nano Lett. 13, 1271–1277.

Li, L., Wu, Z., Yuan, S., Zhang, X.B., 2014. Advances and challenges for flexible energy storage and conversion devices and systems. Energy Environ. Sci. 7, 2101–2122.

Li, R., Xiang, X., Tong, X., Zou, J.Y., Li, Q.W., 2015. Wearable double-twisted fibrous perovskite solar cell. Adv. Mater. 27, 3831–3835.

Lin, H.J., Weng, W., Ren, J., Qiu, L.B., Zhang, Z.T., Chen, P.N., Chen, X.L., Deng, J., Wang, Y.G., Peng, H.S., 2014. Twisted aligned carbon nanotube/silicon composite fiber anode for flexible wire-shaped lithium-ion battery. Adv. Mater. 26, 1217–1222.

Liu, Q.A., Nayfeh, M.H., Yau, S.T., 2010. Brushed-on flexible supercapacitor sheets using a nanocomposite of polyaniline and carbon nanotubes. J. Power Sources 195, 7480–7483.

Liu, Y., Gorgutsa, S., Santato, C., Skorobogatiy, M., 2012. Flexible, solid electrolyte-based lithium battery composed of $LiFePO_4$ cathode and $Li_4Ti_5O_{12}$ anode for applications in smart textiles. J. Electrochem. Soc. 159, A349–A356.

Liu, B.R., Soares, P., Checkles, C., Zhao, Y., Yu, G.H., 2013a. Three-dimensional hierarchical ternary nanostructures for high-performance Li-ion battery anodes. Nano Lett. 13, 3414–3419.

Liu, Q., Nayfeh, O., Nayfeh, M.H., Yau, S.T., 2013b. Flexible supercapacitor sheets based on hybrid nanocomposite materials. Nano Energy 2, 133–137.

Liu, Y.T., Zhu, X.D., Duan, Z.Q., Xie, X.M., 2013c. Flexible and robust MoS_2-graphene hybrid paper cross-linked by a polymer ligand: a high-performance anode material for thin film lithium-ion batteries. Chem. Commun. 49, 10305–10307.

Mao, L., Chen, Q., Li, Y.W., Li, Y., Cai, J.H., Su, W.M., Bai, S., Jin, Y.Z., Ma, C.Q., Cui, Z., Chen, L.W., 2014. Flexible silver grid/PEDOT:PSS hybrid electrodes for large area inverted polymer solar cells. Nano Energy 10, 259–267.

Mecerreyes, D., Marcilla, R., Ochoteco, E., Grande, H., Pomposo, J.A., Vergaz, R., S Nchez Pena, J.M., 2004. A simplified all-polymer flexible electrochromic device. Electrochim. Acta 49, 3555–3559.

Meng, C., Liu, C., Chen, L., Hu, C., Fan, S., 2010. Highly flexible and all-solid-state paperlike polymer supercapacitors. Nano Lett. 10, 4025–4031.

Meng, Y.N., Wu, H.P., Zhang, Y.J., Wei, Z.X., 2014. A flexible electrode based on a three-dimensional graphene network-supported polyimide for lithium-ion batteries. J. Mater. Chem. A 2, 10842–10846.

Mozer, A.J., Panda, D.K., Gambhir, S., Romeo, T.C., Winther-Jensen, B., Wallace, G.G., 2010. Flexible and compressible Goretex-PEDOT membrane electrodes for solid-state dye-sensitized solar cells. Langmuir 26, 1452–1455.

Nickel, F., Haas, T., Wegner, E., Bahro, D., Salehin, S., Kraft, O., Gruber, P.A., Colsmann, A., 2014. Mechanically robust, ITO-free, 4.8% efficient, all-solution processed organic solar cells on flexible PET foil. Sol. Energy Mater. Sol. Cells 130, 317–321.

Niu, Z.Q., Luan, P.S., Shao, Q., Dong, H.B., Li, J.Z., Chen, J., Zhao, D., Cai, L., Zhou, W.Y., Chen, X.D., Xie, S.S., 2012. A "skeleton/skin" strategy for preparing ultrathin free-standing single-walled carbon nanotube/polyaniline films for high performance supercapacitor electrodes. Energy Environ. Sci. 5, 8726–8733.

O'connor, B., An, K.H., Zhao, Y., Pipe, K.P., Shtein, M., 2007. Fiber shaped organic light emitting device. Adv. Mater. 19, 3897–3900.

Palenzuela, J., Vinuales, A., Odriozola, I., Cabanero, G., Grande, H.J., Ruiz, V., 2014. Flexible viologen electrochromic devices with low operational voltages using reduced graphene oxide electrodes. ACS Appl. Mater. Interfaces 6, 14562–14567.

Qiu, L.B., Deng, J., Lu, X., Yang, Z.B., Peng, H.S., 2014. Integrating perovskite solar cells into a flexible fiber. Angew. Chem. Int. Ed. 53, 10425–10428.

Ren, J., Bai, W.Y., Guan, G.Z., Zhang, Y., Peng, H.S., 2013a. Flexible and weaveable capacitor wire based on a carbon nanocomposite fiber. Adv. Mater. 25, 5965–5970.

Ren, J., Li, L., Chen, C., Chen, X.L., Cai, Z.B., Qiu, L.B., Wang, Y.G., Zhu, X.R., Peng, H.S., 2013b. Twisting carbon nanotube fibers for both wire-shaped micro-supercapacitor and micro-battery. Adv. Mater. 25, 1155–1159.

Ren, J., Zhang, Y., Bai, W.Y., Chen, X.L., Zhang, Z.T., Fang, X., Weng, W., Wang, Y.G., Peng, H.S., 2014. Elastic and wearable wire-shaped lithium-ion battery with high electrochemical performance. Angew. Chemie Int. Ed. 53, 7864–7869.

Sekitani, T., Nakajima, H., Maeda, H., Fukushima, T., Aida, T., Hata, K., Someya, T., 2009. Stretchable active-matrix organic light-emitting diode display using printable elastic conductors. Nat. Mater. 8, 494–499.

Shim, G.H., Han, M.G., Sharp-Norton, J.C., Creager, S.E., Foulger, S.H., 2008. Inkjet-printed electrochromic devices utilizing polyaniline–silica and poly(3,4-ethylenedioxythiophene)–silica colloidal composite particles. J. Mater. Chem. 18, 594–601.

Søndergaard, R.R., H Sel, M., J Rgensen, M., Krebs, F.C., 2013. Fast printing of thin, large area, ITO free electrochromics on flexible barrier foil. J. Polym. Sci. Part B: Polym. Phys. 51, 132–136.

Sun, X., Zhang, Z., Lu, X., Guan, G., Li, H., Peng, H., 2013. Electric current test paper based on conjugated polymers and aligned carbon nanotubes. Angew. Chemie Int. Ed. 52, 7776–7780.

Sun, H., You, X., Deng, J.E., Chen, X.L., Yang, Z.B., Ren, J., Peng, H.S., 2014a. Novel graphene/carbon nanotube composite fibers for efficient wire-shaped miniature energy devices. Adv. Mater. 26, 2868–2873.

Sun, H., You, X., Jiang, Y.S., Guan, G.Z., Fang, X., Deng, J., Chen, P.N., Luo, Y.F., Peng, H.S., 2014b. Self-healable electrically conducting wires for wearable microelectronics. Angew. Chem. Int. Ed. 53, 9526–9531.

Tang, C.Y., Hackenberg, K., Fu, Q., Ajayan, P.M., Ardebili, H., 2012. High ion conducting polymer nanocomposite electrolytes using hybrid nanofillers. Nano Lett. 12, 1152–1156.

Wang, K., Zhao, P., Zhou, X.M., Wu, H.P., Wei, Z.X., 2011a. Flexible supercapacitors based on cloth-supported electrodes of conducting polymer nanowire array/SWCNT composites. J. Mater. Chem. 21, 16373–16378.

Wang, K., Zou, W.J., Quan, B.G., Yu, A.F., Wu, H.P., Jiang, P., Wei, Z.X., 2011b. An all-solid-state flexible micro-supercapacitor on a chip. Adv. Energy Mater. 1, 1068–1072.

Wang, K., Wu, H., Meng, Y., Zhang, Y., Wei, Z., 2012. Integrated energy storage and electrochromic function in one flexible device: an energy storage smart window. Energy Environ. Sci. 5, 8384.

Wang, X.F., Lu, X.H., Liu, B., Chen, D., Tong, Y.X., Shen, G.Z., 2014. Flexible energy-storage devices: design consideration and recent progress. Adv. Mater. 24, 4763–4782.

Wang, J.X., Yan, C.Y., Chee, K.J., Lee, P.S., 2015. Highly stretchable and self-deformable alternating current electroluminescent devices. Adv. Mater. 27, 2876–2882.

Weerasinghe, H.C., Sirimanne, P.M., Franks, G.V., Simon, G.P., Cheng, Y.B., 2010. Low temperature chemically sintered nano-crystalline TiO_2 electrodes for flexible dye-sensitized solar cells. J. Photochem. Photobiol. a-Chem. 213, 30–36.

Weng, W., Sun, Q., Zhang, Y., Lin, H.J., Ren, J., Lu, X., Wang, M., Peng, H.S., 2014. Winding aligned carbon nanotube composite yarns into coaxial fiber full batteries with high performances. Nano Lett. 14, 3432–3438.

Weng, W., Sun, Q., Zhang, Y., He, S.S., Wu, Q.Q., Deng, J., Fang, X., Guan, G.Z., Ren, J., Peng, H.S., 2015. A gum-like lithium-ion battery based on a novel arched structure. Adv. Mater. 27, 1363–1369.

Wu, H.P., Shevlin, S.A., Meng, Q.H., Guo, W., Meng, Y.N., Lu, K., Wei, Z.X., Guo, Z.X., 2014. Flexible and binder-free organic cathode for high-performance lithium-ion batteries. Adv. Mater. 26, 3338–3343.

Yang, Z.B., Deng, J., Chen, X.L., Ren, J., Peng, H.S., 2013a. A highly stretchable, fiber-shaped supercapacitor. Angew. Chem. Int. Ed. 52, 13453–13457.

Yang, Z.B., Sun, H., Chen, T., Qiu, L.B., Luo, Y.F., Peng, H.S., 2013b. Photovoltaic wire derived from a graphene composite fiber achieving an 8.45% energy conversion efficiency. Angew. Chem. Int. Ed. 52, 7545–7548.

Yeo, J., Kim, G., Hong, S., Kim, M.S., Kim, D., Lee, J., Lee, H.B., Kwon, J., Suh, Y.D., Kang, H.W., Sung, H.J., Choi, J.H., Hong, W.H., Ko, J.M., Lee, S.H., Choa, S.H., Ko, S.H., 2014. Flexible supercapacitor fabrication by room temperature rapid laser processing of roll-to-roll printed metal nanoparticle ink for wearable electronics application. J. Power Sources 246, 562–568.

Yuksel, R., Sarioba, Z., Cirpan, A., Hiralal, P., Unalan, H.E., 2014. Transparent and flexible supercapacitors with single walled carbon nanotube thin film electrodes. Acs Appl. Mater. Interfaces 6, 15434–15439.

Zhang, S., Ji, C.Y., Bian, Z.Q., Liu, R.H., Xia, X.Y., Yun, D.Q., Zhang, L.H., Huang, C.H., Cao, A.Y., 2011. Single-wire dye-sensitized solar cells wrapped by carbon nanotube film electrodes. Nano Lett. 11, 3383–3387.

Zhang, J.J., Yue, L.P., Hu, P., Liu, Z.H., Qin, B.S., Zhang, B., Wang, Q.F., Ding, G.L., Zhang, C.J., Zhou, X.H., Yao, J.H., Cui, G.L., Chen, L.Q., 2014. Taichi-inspired rigid-flexible coupling cellulose-supported solid polymer electrolyte for high-performance lithium batteries. Sci. Rep. 4, 6727.

Zhang, Z.T., Guo, K.P., Li, Y.M., Li, X.Y., Guan, G.Z., Li, H.P., Luo, Y.F., Zhao, F.Y., Zhang, Q., Wei, B., Pei, Q.B., Peng, H.S., 2015a. A colour-tunable, weavable fibre-shaped polymer light-emitting electrochemical cell. Nat. Photon. 9, 233–238.

Zhang, Z.T., Zhang, Q., Guo, K.P., Li, Y.M., Li, X.Y., Wang, L., Luo, Y.F., Li, H.P., Zhang, Y., Guan, G.Z., Wei, B., Zhu, X.R., Peng, H.S., 2015b. Flexible electroluminescent fiber fabricated from coaxially wound carbon nanotube sheets. J. Mater. Chem. C 3, 5621–5624.

Zhou, Y.H., Zhang, F.L., Tvingstedt, K., Barrau, S., Li, F.H., Tian, W.J., Inganas, O., 2008. Investigation on polymer anode design for flexible polymer solar cells. Appl. Phys. Lett. 92, 233308.

Zhou, G., Li, F., Cheng, H.M., 2014. Progress in flexible lithium batteries and future prospects. Energy Environ. Sci. 7, 1307–1338.

10

Summary and Perspective

Since Hermann Staudinger proposed the modern conception of polymers in the early 1920s, in the following years, polymer evolved from a buzzword to a household name. Accompanied by the wide spread of industrial products—plastics, rubbers, and fibers—academic researches have made solid progress in devising novel synthesis strategies, inventing high-performance polymers and deepening the understanding of the structure–property relationship. In 1970s, the discovery of conductive polymers has broken through the stereotype that polymers are isolating, which enormously extends the application of polymers. Academically, the conception of polymer has been significantly enriched by their successful penetration into biological, optical, material, and electronic applications, which makes it necessary to systematically summarize the established achievements in a specific subject to underpin its development.

In this book, we particularly focus on the polymers that are applied in electronic devices, which are under intensive study in recent years. Polymers prefixed with the word "electronic" do not represent a new class of polymer independent from, for example, conductive polymers or biopolymers. "Electronic polymers" are a combination of polymers—such as conjugated polymers, conductive polymers, and polymer gels—with various properties that are used as different components in electronic devices, such as photoactive sensitizers, pseudocapacitive materials, and structural substrates. As the achievements in electronic devices keep renewing the interest in research, the repertoire of electronic polymer is still expanding, so that we are not able to exhaustively include all the polymers that were applied in electronic devices. But the framework established and paradigms demonstrated in this book may provide a comprehensive picture of the applications of polymers in electronic devices. The research in electronic devices is highly multidisciplinary. We hope this book could benefit readers with different academic backgrounds, not least those who are studying polymers and would like to extend their products to electronic applications or those working at electronics and looking for polymer materials to develop new devices.

In the epilog of this book, we would like first to recapitulate what we have discussed in the chapters as a summary. The organization of this book follows a distinct line starting from the synthesis methodology, structure–property relationship to the diverse applications in various electronic devices, including energy harvesting and storage devices, luminescent devices, sensors as well as flexible devices. Chapter 2 introduces the synthesis strategies for a particular type of polymers—conjugated polymers that are widely used in polymer light-emitting diodes, polymer solar cells (PSCs), and polymer electrochromic displays (PECDs). The purity, molecular weight, and regioregularity of conjugated polymers are mightily dictated by their synthetic routes, which are strongly relevant to their electronic properties. Moreover, from the viewpoint of application, their processability, electronic properties, and stability have be taken into consideration so that the rational designs and modifications on backbones, side chains, and substitutes are discussed accordingly. As the polymers exert a significant impact on performances of the electronic devices, in Chapter 3, we try to elucidate the correlation between the chemical components, molecular structure, and aggregation morphology of polymers and their electronic properties. Emphatically, we have demonstrated the structural and morphological variety of polymers used in electronic devices as well as how different structures and morphologies influence their performances. Chapter 4 focuses on polymer composites with various nanomaterials, such as metals, carbonaceous materials, and inorganic compounds incorporated as second component. The uniform dispersion, proper content, certain alignment, and interfacial interaction are important to maximize the synergetic effect of polymer and guest component which leads to the reinforcement in mechanical strength, electrical conductivity, and electrochemical performances.

In Chapters 5–9, we have introduced the polymers applied in various electronic devices including photovoltaic cells, thermoelectric generator, piezoelectric transducer, triboelectric nanogenerator, lithium ion batteries, supercapacitors, polymer light-emitting diodes, polymer light-emitting electrochemical cells, electromechanical actuators, electrochromic devices, and their flexible counterparts, in which polymers serve as active materials, functional components, or supporting substrates. The variety of electronic devices presented in this book profoundly highlights the merits of polymers—the diversity. The diversity in chemical structure makes their properties tailorable, tuneable, and controllable by modification in backbones, side chains, and substitutes. The diversity in morphology promises various forms of macroassemblies ranging from different dimensions that is beneficial for fabricating flexible devices. In this regard, the research in electronic polymers is dedicated to think about that how the diversity of polymer can benefit the advancement of electronic devices.

While the examples presented in this book may paint an encouraging picture for electronic polymers, it is clear for us working in this area that we have just started and many challenges are awaiting resolutions. As a perspective, we would like to end this book with a couple of general questions regarding electronic polymers.

Can they be synthesized through an economical route? Normally, many state-of-the-art polymers are synthesized through a sophisticated route that entails the use of delicate catalysts like transition-metal catalysts. Indeed, these catalysts are able to promise a high selectivity and high regularity yet their high costs will make the synthesis route less transferable to an industrial scale. Devising a novel synthesis route is definitely an achievement but only scientifically. Considering how these methods can be translated to benefit industrial production will become increasingly important.

Can they be produced in large scale? When we are talking about yield, we might be better bear in mind that the yield is specified to single step, specific configuration, and from lab-scale batch. Things may be entirely different when scaled up to pilot batch and for manufacture, given the multiple procedures and side products. Thus, solely acclaiming a high yield is exhilarated but less meaningful to predict the application. The more important concern is “are they scalable?”

Can they be continuously processable? One of the important motivations for polymer-based electronic devices is that the polymers are compatible for continuous processing technology, such as roll-to-roll and ink-jet printing. A series of time-consuming labor work for polymer preparation will undermine its competitiveness to other rival materials.

Can they be resistant to the environmental variations? Some polymers are extremely sensitive to moisture, light, or oxygen that not only impose strict requirements on their preparations but also make the device vulnerable to surroundings. Additionally, the degradation of polymer will also weaken the endurability and stability of the device.

Apparently, more work need to be done to answer these questions.

Index

FSC
www.fsc.org
MIX
Papier | Fördert
gute Waldnutzung
FSC® C083411

Kolibri 360 GmbH
Ferdinand-Jühlke-Straße 7
99095 Erfurt, Deutschland
produktsicherheit@kolibri360.de